Manuf[...]
Systems Design and Analysis

Manufacturing Systems Design and Analysis

Context and techniques

Second edition

B. Wu

Department of Manufacturing and Engineering Systems,
Brunel University, Middlesex, UK

CHAPMAN & HALL

London · Glasgow · Weinheim · New York · Tokyo · Melbourne · Madras

Published by Chapman & Hall, 2–6 Boundary Row, London SE1 8HN, UK

Chapman & Hall, 2–6 Boundary Row, London SE1 8HN, UK

Blackie Academic & Professional, Wester Cleddens Road, Bishopbriggs, Glasgow G64 2NZ, UK

Chapman & Hall GmbH, Pappelallee 3, 69469 Weinheim, Germany

Chapman & Hall USA, One Penn Plaza, 41st Floor, New York NY 10119, USA

Chapman & Hall Japan, ITP-Japan, Kyowa Building, 3F, 2-2-1 Hirakawacho, Chiyoda-ku, Tokyo 102, Japan

Chapman & Hall Australia, Thomas Nelson Australia, 102 Dodds Street, South Melbourne, Victoria 3205, Australia

Chapman & Hall India, R. Seshadri, 32 Second Main Road, CIT East, Madras 600 035, India

First edition 1992

Second edition 1994

Typeset in 10/11 pt Palatino by Best-set Typesetter Ltd., Hong Kong
Printed in Great Britain by The Alden Press

ISBN 0 412 58140 X

A catalogue record for this book is available from the British Library

Library of Congress Catalog Card Number: 94-70975

∞ Printed on permanent acid-free text paper, manufactured in accordance with ANSI/NISO Z39.48-1992 and ANSI/NISO Z39.48-1984 (Permanence of Paper).

To the twins, Daniel and Christopher

Contents

Preface to the first edition

A technological book is written and published for one of two reasons: it either renders some other book in the same field obsolete or breaks new ground in the sense that a gap is filled. The present book aims to do the latter. On my return from industry to an academic career, I started writing this book because I had seen that a gap existed. Although a great deal of information appeared in the published literature about various technical aspects of advanced manufacturing technology (AMT), surprisingly little had been written about the systems context within which the sophisticated hardware and software of AMT are utilized to increase efficiency. Therefore, I have attempted in this book to show how structured approaches in the design and evaluation of modern manufacturing plant may be adopted, with the objective of improving the performance of the factory as a whole. I hope this book will be a contribution to the newly recognized, multidisciplinary engineering function known as manufacturing systems engineering. The text has been designed specifically to demonstrate the systems aspects of modern manufacturing operations, including: systems concepts of manufacturing operation; manufacturing systems modelling and evaluation; and the structured design of manufacturing systems.

One of the major difficulties associated with writing a text of this nature stems from the diversity of the topics involved. I have attempted to solve this problem by adopting an overall framework into which the relevant topics are fitted. This, I hope, will provide a panoramic view of the subject areas. However, the reader may find that some of the topics mentioned are only treated at a superficial level, or even omitted altogether. This is inevitable due to the broad-based nature of this book. Fortunately, most of these topics have been discussed in detail in a number of texts, which are listed as further reading materials at the end of the relevant chapter.

Although the text has been written over a relatively short time, the book builds on my work and research experience of manufacturing industry over some eight to ten years. I would like to thank Professor Ray Wild for his help and encouragement, without which the completion of this book would have been impossible. In addition, I am indebted to all of my colleagues in the Department of Manufacturing and Engineering Systems at Brunel University, from whom I have received tremendous help in many ways. In particular, I wish to thank Dr P.H. Lowe and Professor N. Slack for their invaluable advice; Mr S. Harrington for reading the draft and for making suggestions; Dr R.J. Grieve, Mr F. Schmid, Mr H. Maylor and Professor M. Sarhadi for many inspirational discussions. Sincere appreciation is also extended to the contributions made by my former students, Mr D. Harrison and Miss K.R. Matthews.

Grateful thanks are due to a number of special friends: to Mrs K. Cawsey of Cheltenham, who taught me how to use the English language 'properly', indeed, without whose teaching I would not have been in a position to start writing this book; to Mr E. Thompson and Dr D.H.R. Price, whose help during the early stages of my professional career is greatly appreciated.

I would also like to thank my editor at Chapman & Hall, Mr M. Hammond, for his professionalism and effort in putting this book together. His deadlines have taught me some new aspects of the 'just-in-time' concept! The criticisms and comments received on the manuscript of this book have been very helpful and constructive. I would like to thank the reviewers, especially Professor D.J. Williams of Loughborough University of Technology and Professor A.S. Carrie of the University of Strathclyde.

Finally, I thank my family for their continuous support and encouragement.

B. Wu

Preface to the second edition

When the first edition of *Manufacturing Systems Design and Analysis* was published some two years ago, I had hoped that it would serve the purpose of contributing to the then newly recognized manufacturing systems engineering. To date the text seems to have fulfilled this objective well. It has attracted its fair amount of attention internationally, and constructive comments about its structure and contents have been gratefully received from many colleagues, on both sides of the Atlantic who have adopted it for teaching and research in this area.

It is a frequently quoted fact that today's manufacturing functions are becoming more and more interdisciplinary, with new concepts and techniques being continuously and rapidly introduced. I have included some of these in this second edition of the book including, for example, object-oriented perspectives of manufacturing operations and up-dated materials on techniques such as computer simulation. In particular, the design framework of manufacturing systems as presented in Chapter 7 is now a much enhanced version: techniques related both to physical configurations and information integration are introduced; a comprehensive case study of manufacturing systems redesign is presented and a mini-project is outlined. The mini-project is designed to help the reader bring together various techniques of manufacturing systems analysis through the application of the framework in a systems redesign situation. This has been a very popular exercise among my students who find it interesting, stimulating and rewarding.

I am grateful to a number of highly respected experts and colleagues who have helped me in my recent work in this field. Special thanks are due to Professor David J. Williams of Loughborough University of Technology for his continued interest in my work; to Dr John Parnaby, Dr Keith Oldham and Mr Alan Hyman of Lucas Applied Technology Ltd, and Professor James Powell of Salford University for their support of our research in computer aided manufacturing systems design; and to Professor Colin L. Moodie of Purdue University, USA and Professor Alan S. Carrie of the University of Strathclyde for their encouragement.

A number of my colleagues and students have contributed directly or indirectly to the preparation of this edition. I would like to thank Alan Moyse, Niki Costas, Janet McDonnell, Stefan Fritz, Roger Hull and Keith Hodges for proof reading the manuscript. Also, I wish to express my gratitude to Lucas Industries plc and Hobsons Publishing plc for permission to reproduce the Fordhouse case study. I am also indebted to all the colleagues who have reviewed the book and made criticisms and suggestions.

B. Wu 1994

Part One
Systems

1 Overview of manufacturing systems analysis in the technological age

1.1 Introduction

This chapter is an introduction to the background and scope of manufacturing systems engineering (MSE), a recently recognized multidisciplinary engineering function. It is the result of a new approach to manufacturing modernization and represents the core of this book.

Unlike traditional forms of engineering concerned with production, MSE adopts a systems approach to the design and operation of modern manufacturing systems. It incorporates the new manufacturing technologies and techniques into the manufacturing processes, so that **manufacturing systems** can efficiently support the wider company objectives.

Understanding the context and problems of modern manufacturing is the first step to appreciating the need and potential role of MSE. In this first chapter, therefore, some real-life cases will be presented, at both corporate and operational levels, to highlight the key problem areas that are being faced by today's engineers and manufacturers: the impact of accelerating technology and the impact of increasing competition. The first half of the chapter will provide those with no prior knowledge of the structure and operation of manufacturing systems with an understanding and appreciation of the background and development of manufacturing. The second half will provide an overview of the potential of manufacturing systems analysis and the interrelationships of the topics to be covered.

After studying this chapter, the reader should be able to:

- explain the nature and context of manufacturing;
- outline the need for and importance of manufacturing industry;
- highlight the characteristics of new developments in manufacturing technology development;
- describe the environmental pressures on today's manufacturing firms;
- have a preliminary appreciation of the potentials and limitations of the newly developed manufacturing technologies;
- recognize the need for a systems approach in manufacturing modernization; and
- outline the content of manufacturing systems analysis, and the interrelationships of the topics involved.

1.2 Manufacturing industry in context

To understand manufacturing systems, it is first essential to appreciate the context of manufacturing, that is, the nature of manufacturing and its economic and social significance. This section does not provide a full account of these topics, which are discussed extensively elsewhere. However, the context of manufacturing will be briefly considered in an overall sense, to highlight what is involved in manufacturing and how it affects our everyday lives. It is hoped that this brief discussion will help the reader to appreciate the importance of the topic. More detailed description and analysis of the structures and functions of manufacturing systems will be given later in the book.

1.2.1 Manufacturing defined

Manufacturing is the organized activity devoted to the transformation of raw materials into marketable goods. In the terminology of economics, these marketable goods are known as *economic goods;* they cannot be obtained without expenditure. This is in contrast to *free goods,* which are available in unlimited quantities at no cost. Manufacturing industry is also called a *secondary* industry, because this is the sector of a nation's economy that is concerned with the processing of raw materials supplied by the *primary* industry (agriculture, forestry, fishing, mining, extraction of minerals, and so on) into the end products. It is one of the most basic and important functions of human activity in modern industrial societies.

A *manufacturing system* usually employs a series of value-adding *manufacturing processes* to convert the raw materials into more useful forms and eventually into finished products. For example, in the process of manufacturing a machine-tool, pieces of metal are machined into parts; various parts and other items are assembled into sub-assemblies; iron ore is smelted and converted into castings; steel plates are cut into various sizes and shapes and then welded into fabrications; and finally, these are assembled into the machine-tools.

The outputs from one manufacturing system may be utilized as the inputs to another. For example, in the United States, approximately 20 per cent of steel production and about 60 per cent of rubber products are used by the automotive manufacturing industry. The automotive industry is also one of the major consumers of machine-tools, paints and glass, and so on. A manufacturing system is, therefore, a typical input–output system which produces outputs (economic goods) through activities of transformation from inputs (which are also called *factors of manufacturing*). This is why many authors in this field have adopted an input–output approach to represent a manufacturing system in an overall sense. Figure 1.1 gives three such examples.

Although the systems in general may be represented in a homogeneous manner, as illustrated in Fig. 1.1, the actual manufacturing activities are in fact highly diversified – more than 450 separate manufacturing industries have been identified with their products classified into about 20 major groups. These belong to two principal categories: *consumer goods* and *capital goods.* Some examples of typical modern manufacturing industries are listed in Table 1.1.

It is interesting to note that even a university system is analogous to a manufacturing organization, where fresh students are taken into the system and through a series of educational processes transformed into highly valuable professionals. The lectures, seminars, tutorials, and laboratory work are all designed to carry out the transformation process – equivalent to the manufacturing processes involved in an actual manufacturing system.

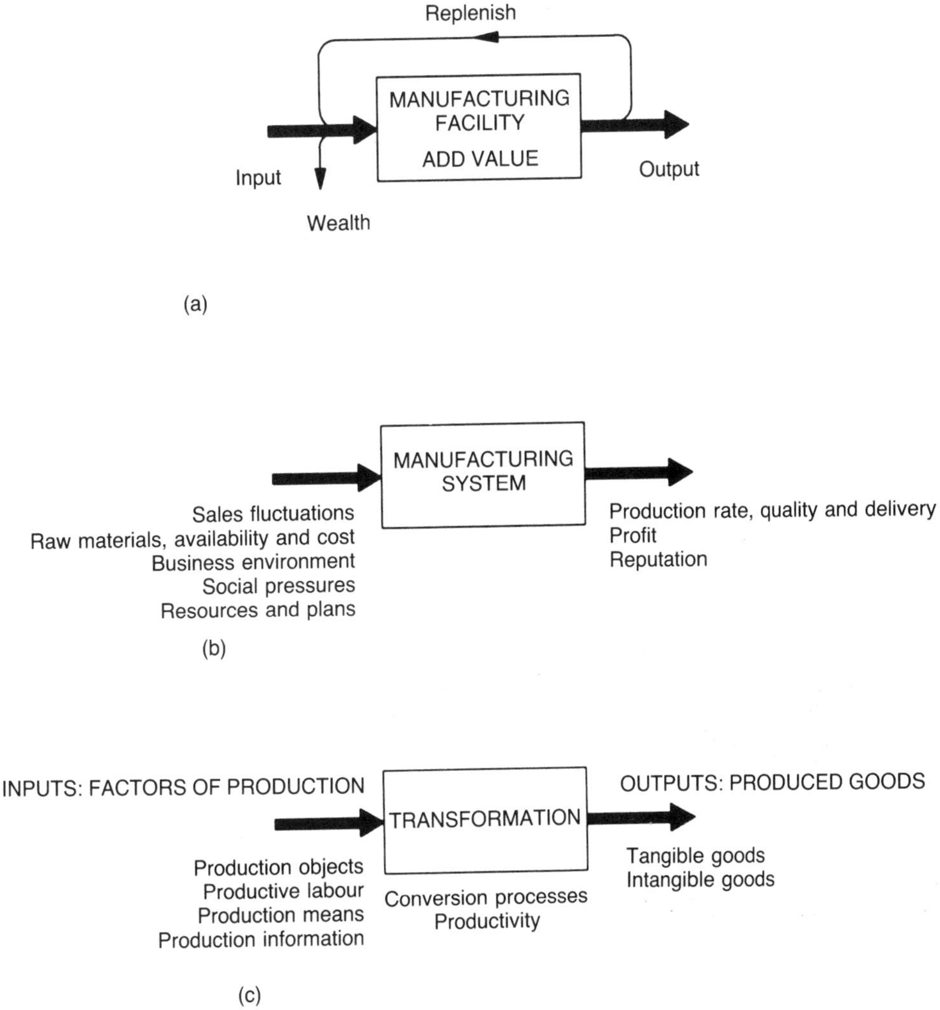

Fig. 1.1 Input–output models of manufacturing. (a) Wealth creation by manufacturing (Lupton, 1986), (b) Overall view of the manufacturing system (Parnaby, 1979) and (c) Basic Meaning of manufacturing (Hitomi, 1979).

The quality of the transformation process is continuously monitored by tests, assessments and examinations. In the end, for those who have passed the final exams and other assessments, a degree certificate is issued to indicate that the individual concerned has successfully completed the transformation process – analogous to the 'pass' certificate of a product. The difference is, of course, that the manufacturing factors here are the students rather than the actual materials involved in a conventional manufacturing system, and these students themselves play a vital part in the success of the transformation process. Once they have graduated, you cannot put a price on them – they are *invaluable* to society.

Table 1.1 Typical modern manufacturing industries

Industry	*Product category*
Aerospace	Capital
Shipbuilding	Capital
Machine-tool manufacturing	Capital
Automotive	Consumer and capital
Electronics	Consumer and capital
Computer manufacturing	Consumer and capital
Computer software	Consumer and capital
Metal, coal, oil	Consumer and capital
Chemical industry	Consumer and capital
Textile industry	Consumer
Leather and fur	Consumer
Clothing and footwear	Consumer
Toy-making	Consumer
Wood and timber production	Consumer
Paper, printing and publishing	Consumer
Building materials	Consumer and capital
Furniture industry	Consumer
Food processing	Consumer
Drink and tobacco	Consumer

1.2.2 Economic and social significance of manufacturing

In an industrialized country, manufacturing industries may be viewed as the backbone of the nation's economy, because it is mainly through their activities that the real wealth is created. It has been estimated that in such a country on average about a quarter of the population is involved in some form of manufacturing activity, and the rest of the population benefits from the products (Harrington, 1984). In the United Kingdom, for example, manufacturing industries (process industries, plus capital and consumer products manufacturers) generate approximately 30 per cent of the nation's wealth and employ 32 per cent of the working population. Furthermore, the jobs of half of those employed in non-manufacturing sectors of the country depend on the close links that exist between these sectors and the manufacturing industries.

To any industrialized country, manufacturing is important internally as well as externally. The internal significance factors are continued employment, quality of life, and the creation and preservation of skills. The external factors are national defence, and the nation's position and strength in world affairs.

The internal significance of manufacturing in a society may be demonstrated by Fig. 1.2. In this figure the various aspects of a social structure (whether based on capitalism or socialism) are represented by a pyramid. The upper structure of the society, representing the quality aspects of life, must be built up on the economic base of the society. Since the height of a pyramid is determined by the size and strength of its base, the quality of life (the height of the pyramid) depends upon the economic strength (the base of the pyramid). The greater the economic strength, the higher the quality of life may reach. The important thing is that the strength of manufacturing in an industrialized society to a great extent determines the strength and scale of its economic base.

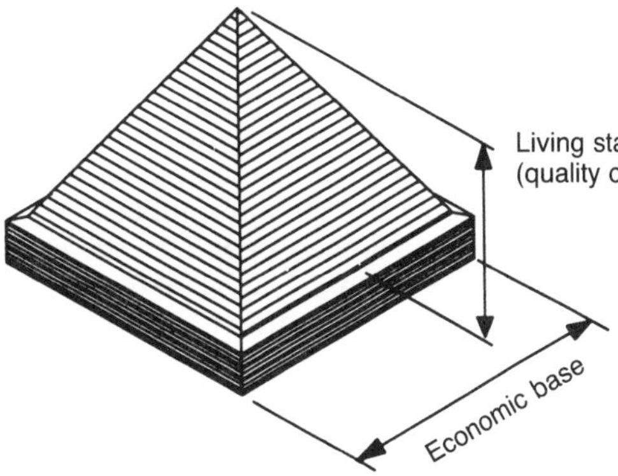

Fig. 1.2 Relationship between upper structure and economic base.

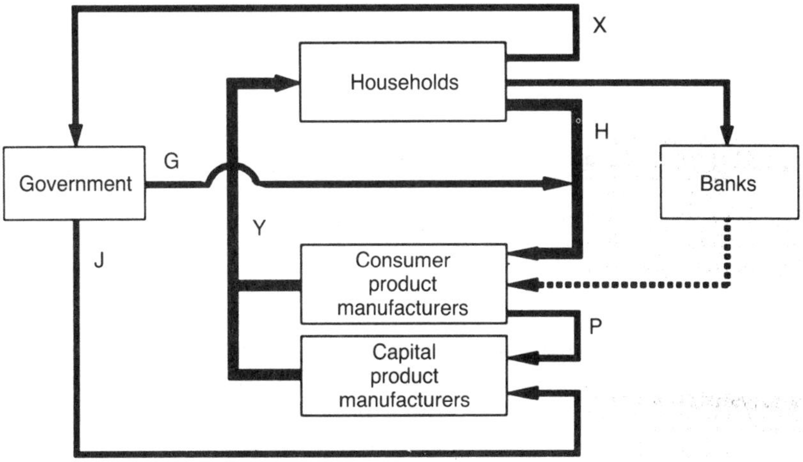

Fig. 1.3 Simplified Keynesian macroeconomic model illustrating the important role of manufacturing firms. Key: Y = National income; X = Government income; J = Government demand for capital products; G = Government demand for consumer products; P = Consumer product manufacturers demand for capital products; and H = Consumer demand for consumer products.

It is not surprising, therefore, that for over two centuries many prominent people have stressed the importance of manufacturing and attempted to evaluate its effects on the macroeconomy. The famous Keynesian macroeconomic model provides a good illustration (Fig. 1.3). This model was popularized by John Maynard Keynes (1883–1946), an English economist best known for his revolutionary economic theories and regarded by many as the most influential economist of the twentieth century. The important role that manufacturing plays in the macroeconomy is clearly demonstrated in this model. Kaldor (1966), for example, summarized the relationship very well: 'Fast rates of economic growth are associated with fast rate of growth of the "secondary" sector of the economy – mainly the manufacturing sector'.

Examples are not hard to find to justify the relationship shown in Fig. 1.2. The effects of manufacturing on human life can be clearly demonstrated by the automotive industry. Before the advent of the motor vehicle, people were restricted to their local areas because the methods of travelling available to them were costly and inconvenient. In the industrialized countries nowadays, the general use of automobiles, together with much improved road networks, has allowed people to travel fast and easily. In the good (?) old days you had to spend a night in Wycombe on your way from London to Oxford, because the horse-drawn coach limited the travelling distance to around 40 miles a day. Today this journey takes only about an hour on the M40 motorway. The effect of this on living habits and social customs has been very profound and consequently the gap between urban and rural life has been significantly narrowed.

Another aspect of the effect of manufacturing industry on the quality of life is shown by Japanese successes. So far, the Japanese have not only improved their general standard of living but also succeeded in producing world-class scientists, economists, managers, engineers, athletes and artists – partly because the success of their manufacturing industries has created enduring wealth and enabled the country to afford the extensive training needed for the talented.

To summarize, the development, prosperity, welfare and living standards of a nation depend to a great extent upon the success of its manufacturing industries. Hence the internal importance of manufacturing to the people of a society cannot be overstressed.

The external significance of manufacturing industry is well summarized by Hall (1987):

> National defence depends on production. High-tech or low-tech, weapons must be produced. Nations with little production capability must use scarce resources to buy them from others, and what they buy is seldom the latest model. A nation that is able to design weapons, but not produce them with quality and quantity, also has a major weakness. . . .
>
> During World War II, production capability was obviously a very big factor. How much this has continued . . . is subject to debate, but there is less question that political weight in the world is substantially affected by manufacturing capability.

1.3 The new environment

In section 1.2, the context of manufacturing was outlined briefly in terms of its nature and its economic and social value. It is clear that for an industrial nation such as the United Kingdom, it is of vital importance continuously to improve the performance of manufacturing industry.

Both the means and structures of manufacturing and the environment within which manufacturing systems operate have changed radically. For modern manufacturing industry to be successful in terms of both immediate profit and long-term competitiveness in today's business environment it is essential for everyone involved to have a sound understanding of these changes.

1.3.1 History of manufacturing development: the three revolutions

Looking back at history, and at the present state of manufacturing industry, it is clear that there have been three revolutions corresponding to the three stages of industrial development.

The Neolithic Revolution: the age of craftsmanship

Manufacturing industry has its origins in the neolithic period (also called the New Stone Age, between the age of chipped-stone tools and the early period of metal tools about 10 000–20 000 years ago) when men became herdsmen and cultivators. Archaeological investigation has shown that neolithic man had techniques for grinding corn, baking clay and weaving textiles. In fact new techniques were developed for both primary and secondary industries. Trade in manufactured products also developed between different communities. This period is regarded by many as the age of the Neolithic Revolution – the revolution which marked the beginning of the age of craftsmanship. In this stage of industrial development, each individual craftsman was responsible for a total production process, from raw material to finished product, and manufacturing activities were entirely powered by human muscle – this is reflected by the fact that the word 'manufacturing' itself derives from two Latin words meaning literally 'hand-making'. As technology progressed, more sophisticated tools were developed. Animal, water and wind power were gradually employed in certain areas, but the basic structure of craft-based production remained unchanged.

The Industrial Revolution: the age of mechanization

The age of craftsmanship lasted for centuries until the Industrial Revolution took place in England and other countries in the eighteenth century. The changes in production techniques from the sixteenth to the eighteenth century laid the foundations of modern industrial production – changes from an agriculture- and craftsman-based economy to one dominated by industry and machine manufacture.

The Industrial Revolution marked the beginning of the age of mechanization. In a machine-dominated factory, the worker obtained new and distinctive skills and his profession changed – from that of craftsman working with hand-tools to that of machine operator subject to factory discipline.

The later stages of the period of mechanization in the early years of the twentieth century were characterized by the analytical study of the production process and the development of scientific management. As the production process became more complex, the total process was analysed and subdivided into a number of simpler production functions. Workers were carefully but rather narrowly trained to operate their own tools and specialized machines with much improved efficiency. The car assembly line at Ford's automobile factory was a classical example of such a production organization – a manufacturing system particularly suited for mass production. Until recently, 'mechanization' has been the key word in manufacturing industries. The concept of mass production was characterized by certain factors such as:

- a stable and 'monochromatic' market;
- long product life cycles and production runs;
- stabilized engineering designs;
- industrial engineering based on breaking a job down into its parts; and
- repetitive operations carried out by each worker.

The New Industrial Revolution: the age of information and automation

The manufacturing industries are now in the midst of a new technological revolution. This revolution, characterized by the increasing application of computers for both information processing and automatic control, has led to significant changes in the techniques of manufacturing. During the first two stages of industrial development, the key issues related to either hand-tool creation and refinement, or machine innovation and power utilization, that is to say, the emphasis was on the development and application of 'hardware'. The latest revolution in manufacturing is different – it has been brought about by advances in computer and information technologies which put more emphasis on information and control than on hardware refinement. The similarity is that just as tools and machines increased the physical abilities of man, so the application of computer and information technologies are heightening his mental abilities.

The third age of industrial development has been accompanied by the evolution of powerful and cheap data-processing systems. This process is taking place in the majority of today's manufacturing organizations.

This development of advanced manufacturing technology (AMT) has attracted much attention. Much recent interest in improving production efficiency has focused upon the development and application of new manufacturing techniques in the following areas:

programmable equipment (the 'hardware'):
- computer-controlled work centres (CNC, etc.),
- robotics and other automation schemes,
- automatic inspection equipment, etc.

the computer-aided/computer-integrated system (the 'software'):
- computer-aided design (CAD),
- computer-aided manufacturing (CAM),
- computer-aided production management (CAPM),
- flexible manufacturing systems (FMS),
- computer integrated manufacturing (CIM),
- artificial intelligence (AI) in manufacturing, etc.

The hardware system is concerned with the actual handling and processing of production materials on the shop floor, while the software system is concerned with the handling and processing of manufacturing and management information, and thus the planning and control of the manufacturing systems. If properly implemented, significant improvements in production efficiency and flexibility are possible. A typical example is the inventory-orientated manufacturing system under material and resource planning (MRP II) control. In successful applications, prompt and accurate information, combined with the computer power and the flexibility provided by sophisticated equipment, can perform planning, control and manufacturing activities in a formal and highly organized manner. Production materials and resources are allocated to the correct places when they are needed, and production proceeds according to a feasible plan produced by the system. Consequently, work-in-progress and production costs can be reduced, and delivery dates met more frequently.

Although the development and applications of these new manufacturing technologies do seem to hold great promise for improving the performance of various functional areas in manufacturing systems, it is very difficult to establish just how much they can achieve. Both successes and failures are frequently reported in the literature. One thing which is certain, however, is that this current revolution in manufacturing industry is taking place at a much more rapid rate than the previous ones. 'The manufacturing field presently is undergoing changes that would have been quite difficult to predict a decade ago' (Bedworth and Bailey, 1987).

For anyone who has recently been involved in the development or adoption of new manufacturing technologies, the words of the famous nuclear scientist, J. Robert Oppenheimer, will be relevant:

> One thing that is new is the prevalence of newness, the changing scale and scope of change itself, so that the world alters as we walk in it, so that the years of man's life measure not some small growth or rearrangement or moderation of what he learned in childhood, but a great upheaval.

1.3.2 The new business environment of manufacturing: international competition

The new technological environment of manufacturing industry has already been outlined. However, we now live in an age which is characterized not only by those rapidly changing technologies but also by improving customer choice, with ever-increasing demand for variety, and fierce international competition. Gone are the days when a company that had dominated a particular market could count on things continuing that way and could afford to say to the customer, as did Henry Ford: 'You can have any colour as long as it's black!' The market of today demands high-variety, small batch volume products of high quality.

What has happened to British manufacturing industry in recent years illustrates the consequences of tough international competition. England led the Industrial Revolution. Highly innovative and creative, the British pioneered the transformation of society from agricultural to industrial – a remarkable achievement in the modern history of manufacturing – and made Britain one of the leading industrial nations and a major exporter of manufactured goods. However, things have changed considerably under the ever-increasing threat from international competition in recent years: in 1983, for the first time in the two centuries since the Industrial Revolution, Britain became a net importer of finished manufactured goods – cars, motorcycles, cameras, hi-fi and video equipment, domestic appliances, and so on. To a lesser extent, a similar situation is being encountered in the United States. It is common nowadays to open up a book on manufacturing industry published in the United Kingdom or United States to find this kind of statement:

> The American production system has been hailed in the United States and around the world for nearly 100 years because of its productivity, capacity, innovations, and contributions to our amazing standard of living. About 15 years ago, the picture began to change. Our number one position in world industrial capability is not what it used to be on most measures. Comparative proportions of the world's manufacturing, leadership in technological and process innovation, ability to compete in world markets, and a dozen sick industries signal that somehow United States industry may be slipping or at least that we are no longer clearly outstanding (Skinner, 1978).

The British motorcycle industry, and the great names of BSA, Triumph, Norton, Royal Enfield, Matchless, AJS, Sunbeam, etc. declined from the apparent prosperity of the early 1960s to almost total oblivion by 1975 (Smith and Rogers, 1985).

Other countries began rebuilding their industrial capability, and by 1970 it was unmistakable that players other than Americans were beginning to do well in the manufacturing arena. The 1970s were filled with dire warnings of comparatively low productivity growth in American manufacturing, and by 1980 the United States was facing tough international manufacturing competition, especially from the Far East (Hall, 1987).

The case study reported by Goddard (1982) may illustrate the extent of the problems faced by today's manufacturers from countries such as the United Kingdom and the United States. In his paper, an analysis of the different planning tools and philosophies adopted by Western and Japanese production management was given. The different planning and operational aspects of the two systems were compared, and the major advantages of each system given. Some spectacular results were cited. In the fork-lift truck manufacturing industry, the American manufacturers would give a delivery lead time in the range of six to nine months, while the Japanese manufacturers' lead-time quotations were around one month. In terms of customer service, the Japanese manufacturers claimed that 100 per cent of their new trucks were delivered on time; and that 97 per cent of orders for service parts were dispatched the same day. If this performance were consistently achieved, the author concluded that it would be difficult for the American counterparts to do better under their present manufacturing practice.

As a result, academics and practitioners involved in manufacturing have been trying hard to find ways to maintain and increase manufacturing competitiveness. There has been no shortage of statements about how things can go wrong with manufacturing in today's business environment, including:

- failure to invest in new plant and equipment;
- inefficient management practice;
- lack of coherent manufacturing strategy;
- inadequate educational and professional training system;
- lack of awareness of the importance of manufacturing;
- high cost of materials and labour;
- unfair overseas competition; and
- cultural background and social attitudes.

There is probably some truth in many of these.

1.4 Can technology solve all the problems of manufacturing industry?

Not only is the manufacturing process itself becoming more complex and sophisticated as a result of rapid technological changes, but the general environment is changing because of tough international competition. Under these circumstances, efficiency and flexibility are of vital importance to the survival of a manufacturing industry which has to operate under much more complex and difficult conditions than ever before. What is the right way to approach the problem? This is the crucial question everyone involved in manufacturing should be asking.

The evidence from successful applications of AMT seems to suggest that at least part of the key to improving productivity lies in technological modernization, that is, the adoption of new manufacturing technologies. But can technology solve all the problems of manufacturing industry? This section illustrates, using real-life examples where appropriate, that for a modern manufacturing system to be successful, a much broader and deeper base of system planning and design is required.

Although it has often been suggested that manufacturing industries can become more productive by employing new and more sophisticated equipment and techniques, surveys (see, for example, Miller *et al.*, 1981; Morecroft, 1983) have shown that in many cases the most significant factor is still the effectiveness of the administration system. Poor planning and control, material shortages, inefficient capacity allocation, lack of understanding of the nature of the organization, and thus lack of formal guidance rules to aid management decision-making, may all result in low productivity, even though the investment in new equipment may have improved local effectiveness.

This can be illustrated by three examples, all directly based on the working experience of this author as a manufacturing analyst in the mining equipment manufacturing industry. Section 1.4.1 illustrates the international environment within which a company must nowadays operate and compete. It compares, at the corporate level, a particular company and its major overseas competitor as far as equipment sophistication and systems performance are concerned. Sections 1.4.2 and 1.4.3 look at some of the manufacturing companies at the operational level. Particular attention is focused on the problems associated with the introduction of AMT into the manufacturing system and on some of the limitations of contemporary technology.

1.4.1 Level of facility sophistication: the deciding factor?

The concept of mechanized underground coalmining is believed to have been developed in Britain. Coalmining equipment is heavy machinery, and a large amount of capital is required to produce it. In the United Kingdom, the home market alone is not big enough to support British manufacturers. Therefore, only about 35–50 per cent of the output is used by British Coal, the rest going to overseas markets. However, international competition in these markets has become very tough, with the German and American manufacturers being the major competitors. The major competing companies are all very well established. Technically there exist no significant differences in their products because the structural and functional aspects of this equipment have long been tested and understood. In order to gain competitiveness in the market, all major manufacturers have adopted well-designed quality assurance systems. The equipment manufactured undergoes carefully planned test and retest processes during various design, development and production stages.

Consequently, as far as the technical and quality aspects are concerned, the equipment produced by these major companies is comparable in terms of competitiveness. What differences there are lie in the price and delivery performances of the competitors. For example, the time taken by the major British manufacturers to deliver an installation, top to bottom, is on average six months, while their German competitor manages to carry out similar manufacturing activities and get an installation with similar quality and performance delivered to the customer in four months. The end effects of this difference can be quite significant. Although it is not possible to assess numerically the consequences of this difference in delivery performances, it is instructive to examine the situation from the customers' point of view.

For the British manufacturers, the biggest and most significant difference comes from the international market. The major customers in the international market in mining equipment include the United States, former USSR, India, Mexico, Hungary, Australia, New Zealand, and China (the People's Republic). Among these customers, China is the second largest coal-producing country in the world and, not surprisingly, one of the biggest equipment purchasers (in the late 1970s, one single order from the Chinese Ministry of Coal Industry (CMCI), to one of the British firms was worth nearly £80 million). Therefore, the view of CMCI planning officials is quite representative: to maximize their production level, the new equipment must be in operation at the coal face as quickly as possible. Assuming there is little technical and quality difference between the equipment of the different suppliers, those able to deliver earlier will be more favoured. Other customers, most noticeably American buyers, have the same attitude. Consequently, this very often puts the German competitors in a stronger position in the market.

Examination of the production practices of the companies reveals that this difference in system efficiency is not due to advantages in manufacturing equipment. The differences in the manufacturing facilities utilized by the major British and German counterparts are not substantial. The British manufacturers are very well equipped with advanced manufacturing equipment: CAD/CAM systems, CNC machines, dedicated special purpose automatic machines, and welding robots have become commonplace in their workshops. One of the British manufacturers even has the advantage of possessing one of the most advanced computer-controlled material and component stores in Europe. From this it can be seen that, with manufacturing facilities comparable in their level of sophistication, a company can still manage to maintain a stronger position through better management practices and a more systemic approach than its competitors.

This real-life example illustrates two common characteristics of modern manufacturing industry. First, if there is little technical difference between the products put on the market by competitors, so that product price and delivery delays are the key variables, then the success or failure of a company competing in that market will to a large extent be dependent upon the organization and production management aspects of the company, that is to say, upon how efficiently the company is run. Therefore, the engineers of today's manufacturing industry should realize that narrow engineering excellence is just one part of the much larger setting of their work.

Second, industrial competition in the new technological environment has shown that international and company competitiveness cannot be gained just by investing in new and sophisticated technologies. Part of the key to improving productivity must be the systematic use of appropriate approaches, techniques and technologies to modernize existing manufacturing systems or design new ones, together with good production management in operating them. In most cases being interested in new technology is not enough, because unless a systems approach is adopted, the use of new technology will not always point a company in the right direction.

1.4.2 AMT operation and overall system performance

This case study is based on the results of a project carried out with the management of a company to evaluate the operation of its manufacturing system. The results are derived from a series of experiments using a large-scale computer simulation model specially developed for the company (this case study also illustrates the usefulness of computer modelling and simulation in MSE, which will be one of the major topics dealt with later in this book). The conclusions drawn from this study highlight some important issues concerning the application of AMT in general.

Recently the company installed a flexible manufacturing cell (FMC) which is dedicated to the manufacturing of one of the most important and complex parts of its products. To evaluate the impact of this sophisticated manufacturing facility on the overall systems performance, a series of simulation experiments were carried out. In particular, the overall work-in-progress (WIP) situation of the company was estimated under various operating conditions. The experiments were so designed that a wide range of contract due-date environment and production control policy possibilities would be covered (Wu and Price, 1989). Some of the representative results are presented graphically in Fig. 1.4–1.6 (in these figures FSFS, FCFS, SI, SI2, and so on, designate production decision rules – see Appendix A for detailed explanation).

The company's decision to deploy an FMC dedicated to the manufacturing of the complicated parts has obviously provided an effective and economic tool, as indicated by the very low material WIP of the brass blocks. Although no doubt an efficient parts-producing tool, the experimental results suggest that the true impact of such an FMC should be related to the overall efficiency of the system. It can be seen that the WIP levels of the assembled valves were the highest in the system. The reason may be visualized as follows.

When works orders were issued to the valve shop, which employs the highly efficient FMC facility, valve parts would be quickly produced and subsequently assembled. However, since the assembly process of the final products could not start until all necessary parts were ready, the valves had

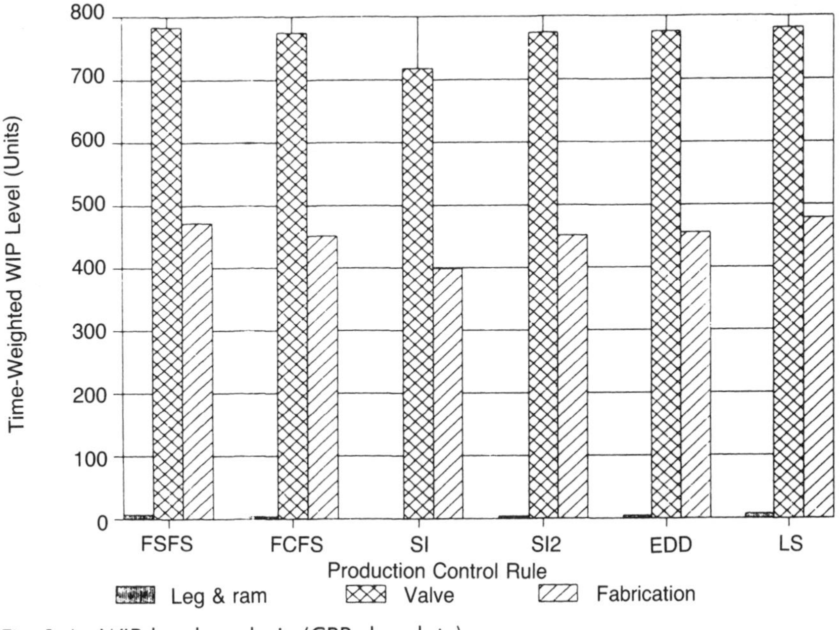

Fig. 1.4 WIP level analysis (CPB due date).

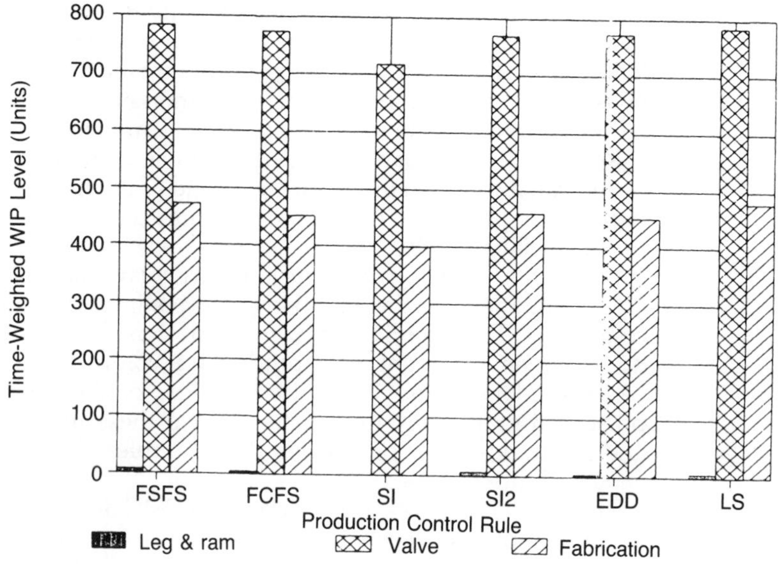

Fig. 1.5 WIP level analysis (CON due date).

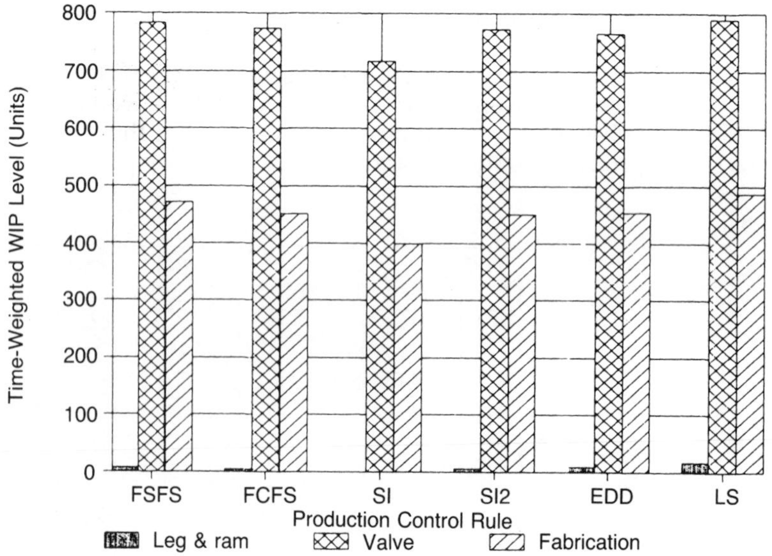

Fig. 1.6 WIP level analysis (CUB due date).

to be 'staged' in a WIP pool, waiting for other parts of the products to be ready. This is then reflected by the high valve WIP levels in the figures. In this case, working capital could be tied up, not in materials or semi-processed parts, but in the assemblies. Therefore, it should be remembered that the high-efficiency processing power of high-technology equipment such as an FMC will not necessarily result in high efficiency in production costs for a complex manufacturing system, unless it is properly incorporated into the system. The simulation results indicate that the potential of this particular

FMC could be further realized by tighter control. More precise timing, together with the high efficiency of the FMC, would result in lower time-related inventory levels of the parts concerned and thus improve the overall WIP cost situation in this production group.

This observation is obviously of general concern with the introduction of sophisticated manufacturing equipment into a large manufacturing system. In general the equipment should not be left as a 'technical island'; it should be a well-incorporated element of the total process. Otherwise, the localized high efficiency will not necessarily bring in overall system benefits.

1.4.3 Operations bottlenecks

It has been said that in high-technology industries such as aerospace, automobiles and electronics, the successful implementation of AMT is a prerequisite for survival. In other industries AMT should represent a significant competitive weapon. However, technology does have its limitations: while there is a great deal of discussion of the development of intelligent factories such as CIM which will require less human involvement, there are few examples in real operation.

For instance, it is common to see a manufacturing company with a high level of automation in its machining and process areas but with the assembly processes still being carried out entirely manually because technology is not yet available for every application. It is well known that manually operated systems, though flexible, do not perform with the same efficiency and consistency as mechanized or automated systems.

An example of this situation is given in Fig. 1.7, which illustrates the production flow situation of a particular manufacturing company. The graph shows the various production stages for one particular batch of jobs. The batch shown arrived at the middle of week no. 20 and consisted of 93 units. From the start of finish, this batch took approximately 27 weeks to go through all the production processes. One of the significant sources of delay is revealed by this graph – the manually dependent final assembly operation took relatively long to complete compared with the other operations which to a large extent have been automated.

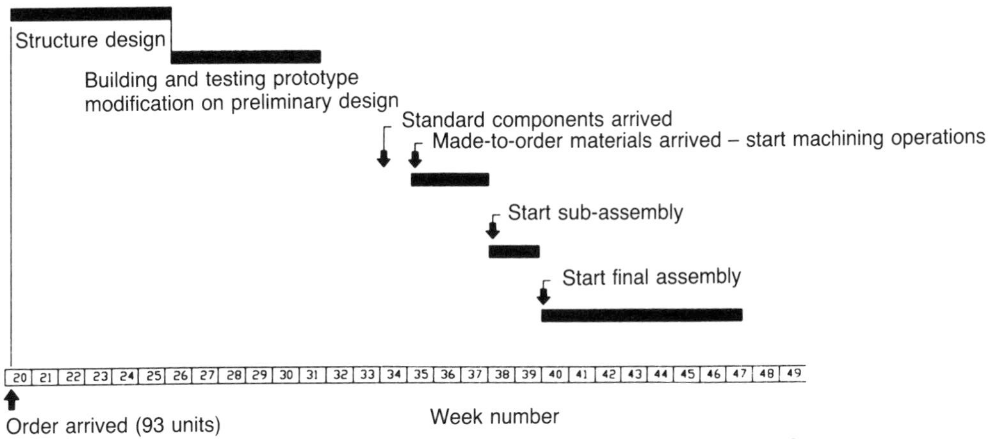

Fig. 1.7 Sample order-flow report.

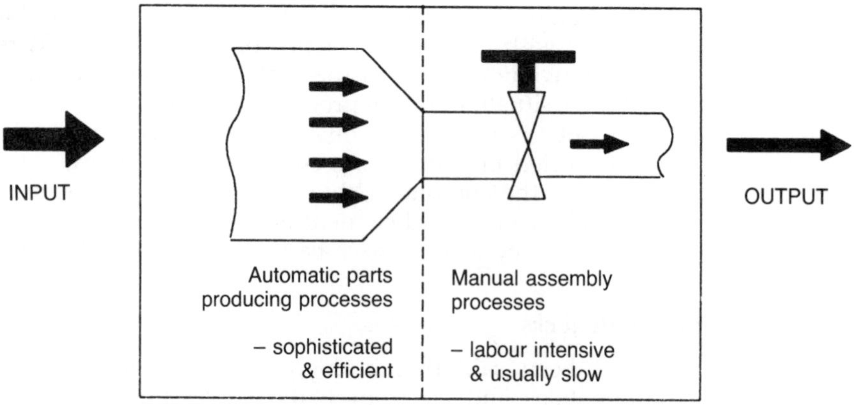

Fig. 1.8 Production-process bottleneck.

The present and the foreseeable future situation of many manufacturing firms is therefore illustrated in Fig. 1.8. This shows why it has been said that industrial assembly will be the centre of automation in coming years. From the published literature it is clear that this area is receiving more and more attention. Concerned with the tendencies of technical, organizational and social development in this area, for example, a number of West German institutional and organization bodies carried out a large-scale collaborative study of German industry in 1983 (Seitz and Volkholz, 1984). Investigations of 355 firms and 43 case studies provided representative data on strategies of rationalization and on automation and human factors of assembly work. Two of the conclusions drawn from this project are worthy of note: first, that the 'traditional' strategies of rationalization and the flexibility of manpower in the field of assembly cannot be easily substituted; and second, that under these circumstances it is obvious that in order to match the traditional assembly system with the sophistication of the parts-producing facilities, strategies of rationalization of work organization must be regarded as an important factor. These observations are certainly in agreement with what we have seen in the particular case study quoted here.

1.5 The systemic approach: the role of manufacturing systems engineering in the modernization of manufacturing

Having read through section 1.4, it may not seem surprising now that, after an extensive 'stock-taking' and review of research in production/operations management, Voss (1984) concluded: 'If UK manufacturing companies are to regain competitiveness, better management of production operations is required. The academic community has a strong part to play in this in identifying, developing, and transferring good management practice.'

This statement can be applied in a broader sense, that is, in the new technological environment; this is generally true for other systemic approaches and methodologies. It has recently been realized that the potential benefits offered by the new technologies with respect to industrial efficiency will not be fully gained unless there is a corresponding systematic investment in research into the design, evaluation, organization and management

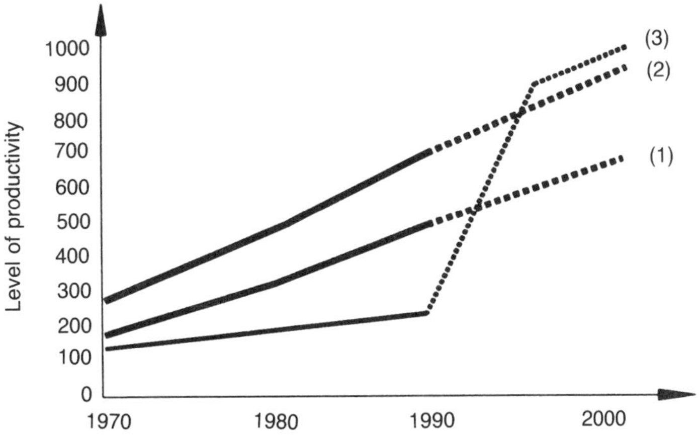

Fig. 1.9 The requirement of a step improvement. Key: 1 = Performance of West German manufacturing industries; 2 = Performance of Japanese manufacturing industries; and 3 = The 'step improvement' required by the manufacturing industries in other industrial countries.

aspects associated with manufacturing systems (Miller, 1981; Towill, 1982, 1984; Voss, 1983; Huang *et al.*, 1985; Smith *et al.*, 1985). Most state-of-the-art technologies, as listed previously, when applied on their own, offer localized benefit only. Even a more systematic approach such as MRP is not always successful. Huang *et al.* (1985) pointed out that despite the increased attention to such computerized systems in the research literature, MRP has frequently proved to be unsuccessful on implementation. For example, it has been reported that less than half of those companies that have employed MRP found it cost-effective and were satisfied with it (Sirny, 1977; Shaw and Regentz, 1980; Miller, 1981; Morecroft, 1983). It is believed that one factor related to this unsuccessful implementation is that the functions and underlying mechanism of the relevant manufacturing systems in question have not been sufficiently analysed and understood by the people concerned in manufacturing industry, as commented by Miller *et al.* (1981):

> Systems allow the manipulation of data and provide massive output for managing a manufacturing operation. Considerable descriptive information is available on how companies use control systems, although from an objective scientific viewpoint, little guidance exists on how to choose controllable factors.

Therefore, although the time has come for manufacturing industries to improve their productivity rapidly in order to achieve and maintain competitiveness – a strategic 'step improvement', as it is called by some academics (see Fig. 1.9) – there is no panacea. As illustrated by the examples given, no single manufacturing technique, even advanced, can necessarily guarantee such a step improvement. The key issues of attaining manufacturing excellence seem to include the adoption of the appropriate methodology, as well as technology, as stated by Don Swann (1984): 'An unsophisticated technique used properly will yield better results than a sophisticated technique used poorly.' It is the realization of this situation that has recently led to the recognition of the potential role of manufacturing systems engineering in modernization of manufacturing industry.

1.5.1 The role of manufacturing systems engineering

Increased competition and new technology are transforming both the factory and the jobs of the manager and professional engineer involved in manufacturing. In a modern factory, capital investment is substantially larger than in an antiquated one. The volume of product is only loosely related to the number of productive employees. The cost of production is strongly dependent upon the cost of raw materials, capital investment and equipment maintenance, rather than on the cost of labour. These changes are increasing the demands on everyone involved in manufacturing and expanding the complexity and risk of their decisions, as indicated recently by Parnaby and Donnovan (1987):

> The modern manufacturing system embraces a wide range of applied methodologies and technologies, processes a variety of materials, and the rates of change in these and in factory and business architecture are considerable. To design and operate the modern manufacturing system requires engineers with multidisciplinary skills.

Until recently the traditional study of production engineering was based on specialized and isolated segments. Very rarely was a manufacturing system viewed from a systems perspective as an integrated combination of processes, machine systems, people, organizational structures, information flows, control systems and computers, designed and operated in order properly to support a coherent manufacturing strategy. This function-orientated style probably has its root in the history of industrial evolution, as commented by Harrington (1984):

> The result of the last two centuries of evolution and expansion in manufacturing is a structure of great complexity – a fractionated and divided structure, with each little part trying to protect its turf and optimise its own performance. This of course has led to suboptimisation of the whole. It has also led to communications barriers between the units at every level, horizontally and vertically.
>
> The result is that today people see manufacturing, not as a single continuum, but as an aggregation of factions, and not always cooperative factions. There are differences in the training, education, and even in the cultural background of the people in these compartments. There has been a loss of perspective on the part of many. This is not good.
>
> Basically these were the result of the division of the original one-man craftsman structure as the scope of the enterprise grew. Skills were specialised and divided. Authorities were specialised and divided.

This traditional view of manufacturing has resulted in many manufacturing systems which are based on a diversity of relatively small and specialized functional units. When the market was only demanding a high volume and low variety of medium-quality products, corporations based on this approach operated reasonably, because their inherent lack of flexibility and other weaknesses were concealed. However, it has now become apparent that the structure of many manufacturing systems designed in this way can no longer cope with the rapid technological change and the new market demand, because this style of functionalism has inhibited a systems approach, encouraging decisions to be made on a short-term and localized basis, as illustrated by the case studies in section 1.4. This traditional approach of production engineering is thus no longer universally appropriate.

The relatively new multidisciplinary engineering function called manufacturing systems engineering, on the other hand, adopts a broader viewpoint than the traditional disciplines of production engineering, and studies the manufacturing process as a whole. It is, therefore, quite different to the existing forms of engineering concerned with production. Manufacturing systems analysts with an MSE background are still required to have the knowledge and skills related to the control, programming and monitoring of single machines and machine systems. In addition, however, they must also have a thorough understanding of systems approaches to the design and operation of modern manufacturing systems, so that they can correctly incorporate the advanced manufacturing techniques into manufacturing systems which efficiently support the wider objectives of the company.

This recently recognized engineering function is the result of a new wave in approaches to manufacturing – the trend toward simultaneous and overlapping development and adoption of manufacturing strategy, manufacturing systems design and analysis, manufacturing technology and manufacturing management.

As a result, some educational and professional institutions have identified the need for manufacturing systems analysts who would be trained at a professional level on a broad-based, multidisciplinary engineering course (the Department of Manufacturing and Engineering Systems of Brunel University, for example, was among the first educational institutions in the UK to do so). Although a relatively new engineering function, the applications of the approaches and methodologies of MSE in rejuvenating manufacturing industry have gained significant successes and much publicity.

1.5.2 The manufacturing systems approach: the context and structure of this book

There is a great deal in the published literature about the various aspects and processes of AMT, but surprisingly little about the systems context within which sophisticated computer systems and hardware are integrated to provide manufacturers with flexibility and efficiency. This book seeks to address this situation by showing how to adopt structured, systematic approaches in the design and evaluation of modern manufacturing plant, with the purpose of optimizing the performance of the whole factory (the three cases quoted earlier in section 1.4 in fact involved some typical examples of the application of manufacturing systems techniques).

This book does not discuss detailed technical aspects of various task functions within a manufacturing system – these are covered extensively by other books in the field. It aims to give the reader an overall framework of manufacturing systems design and evaluation, with particular emphasis on systems analysis, systems design and systems methodology.

The main topics presented are illustrated in Fig. 1.10. As can be seen, the term 'manufacturing systems engineering' contains within itself a number of other key terms, which relate to the main study areas covered:

1. *Systems*. Part I of this book gives an introduction to systems theory to enable the development of a systems approach for the study of manufacturing. The basic concepts and ideas of systems study will be discussed, together with their use in a production environment. This part of the book covers the following topics:
   (a) the basic concepts of systems;
   (b) the 'hard' systems approach;

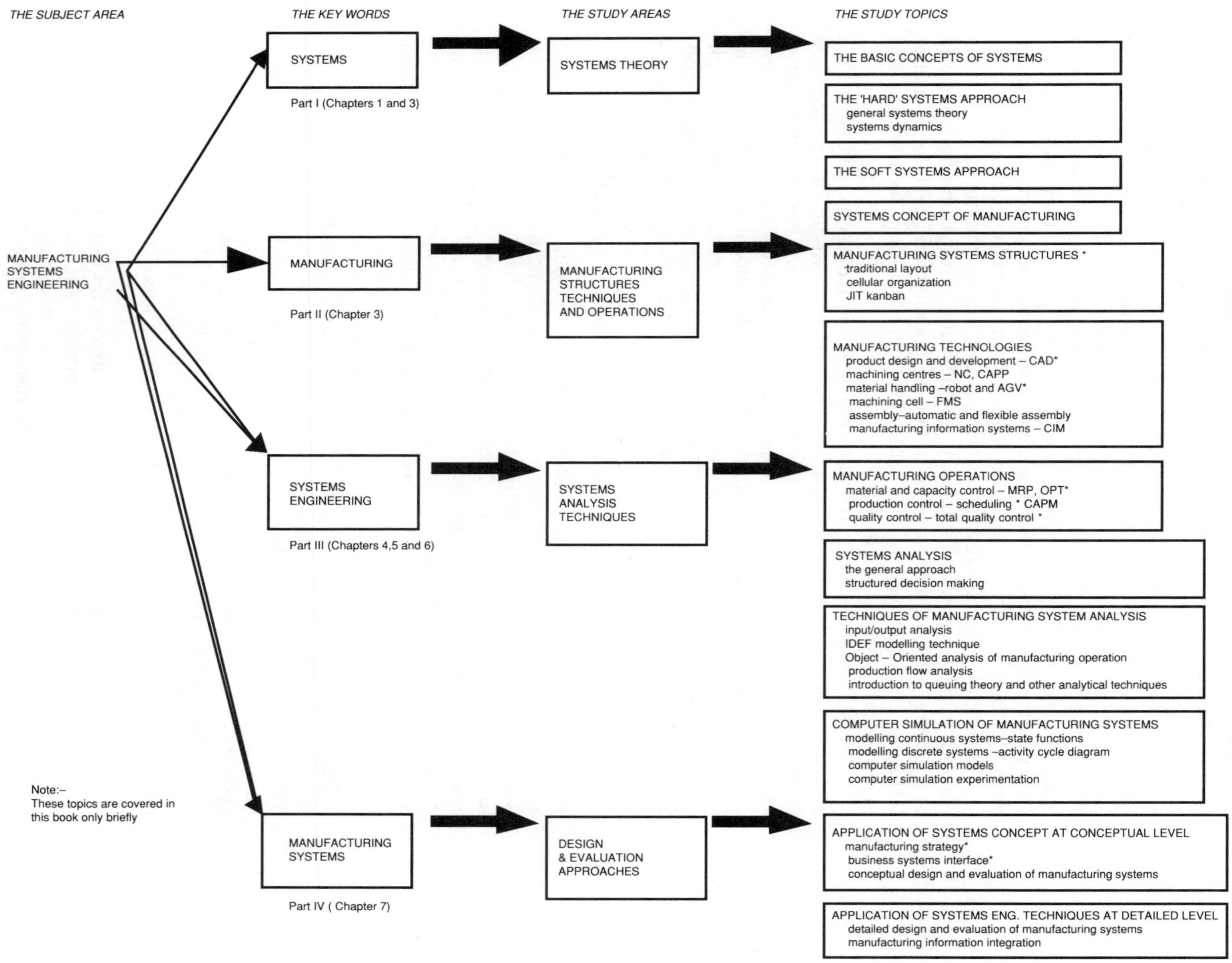

Fig. 1.10 Modular structure of manufacturing systems analysis.

   (c) the 'soft' systems approach;
   (d) a systems concept of manufacturing.
2. *Manufacturing (structures, technologies and operations).* Knowledge of the basic types of manufacturing, the technical characteristics of various elements of advanced technology and the operations of a manufacturing system are obviously prerequisites for the design of a manufacturing system. Part II introduces the reader to the principles of the structure and operation of a manufacturing system and the technologies that are being applied in discrete manufacturing, including:
   (a) manufacturing system structures:
      (i) traditional layout;
      (ii) cellular organization;
      (iii) JIT/kanban;
   (b) manufacturing technologies:
      (i) product design and development – CAD;
      (ii) machining centres – NC, CAPP;
      (iii) material handling – robot and AGV;
      (iv) machining cell – FMS;
      (v) assembly – automatic and flexible assembly;
      (vi) manufacturing information systems – CIM;
   (c) manufacturing operations:
      (i) material and capacity control – MRP, OPT
      (ii) production control – scheduling
      (iii) quality control – principles and techniques.
3. *Systems engineering.* Part III is designed as an introduction to the techniques of systems analysis, which can be used as a general framework to tackle various problems associated with a manufacturing systems design and evaluation project:
   (a) systems analysis:
      (i) the general approach;
      (ii) structure decision-making;
   (b) techniques of manufacturing systems analysis:
      (i) input–output analysis;
      (ii) IDEF modelling technique;
      (iii) object-oriented analysis of manufacturing operation;
      (iv) production flow analysis;
      (v) introduction to queuing theory and other analytical techniques;
   (c) computer simulation of manufacturing systems:
      (i) modelling of continuous systems – state function;
      (ii) modelling of discrete systems – activity cycle diagrams;
      (iii) computer simulation: – programming approaches, simulation languages, and simulation and expert systems.
4. *Manufacturing systems (design and evaluation).* The systems concept, the analytical methodologies, and the technical aspects of manufacturing elements are brought into an overall framework for manufacturing systems analysis in Part IV. A manufacturing systems analyst should be involved not only with the topics covered in Parts I to III, but also with wider business criteria. To achieve competitiveness, a manufacturing company must have a coherent manufacturing strategy which corresponds to its market and matches its corporate strategy. The right choice of manufacturing system for a particular application largely depends on the manufacturing task which the firm has set itself. This part of the book indicates a framework of approaches and techniques which can be used

to analyse and develop manufacturing systems to support business objectives:
(a) manufacturing systems design and evaluation – application of systems concept at conceptual level:
  (i) manufacturing strategy;
  (ii) business systems interface;
  (iii) conceptual design and evaluation methodology;
(b) manufacturing systems design and evaluation – application of techniques of systems engineering at detailed level:
  (i) detailed design and evaluation methodology;
  (ii) manufacturing information integration.

1.6 Conclusion

In this chapter, we have given an overview of manufacturing systems engineering, including its background and general context. Based on the material presented in the previous sections, it is clear that we may conclude our discussion so far with the following motto:

TECHNOLOGY + METHODOLOGY + INFORMATION INTEGRATION = A BETTER CHANCE TO SUCCEED

Further reading

Bedworth, D.D. and Bailey, J.E. (1987) *Integrated Production Control Systems* (2nd edn), John Wiley.
Don Swann, P.E. (1984) Execution is the Key to Success of any System for Manufacturing Material Flow Control, *Industrial Engineer*, October.
Ettlie, J.E. (1988) *Taking Charge of Manufacturing*, Jossey-Bass.
Hall, R.W. (1987) *Attaining Manufacturing Excellence*, Dow Jones-Irwin.
Hitomi, K. (1979) *Manufacturing Systems Engineering*, Taylor & Francis Ltd.
Miller, J.G. (1981) Fit Production Systems to the Task, *Harvard Business Review*, **59** (1), pp. 145–54.
Miller, J.G., *et al.* (1981) Production/Operations Management Agenda for the 80s, *Decision Sciences*, **12**, p. 547.
Milner, D.A. and Vasiliou, V.C. (1986) *Computer-Aided Engineering for Manufacture*, Kogan Page.
Parnaby, J. (1981) Concept of a Manufacturing System, in *Systems Behaviour* (3rd edn), (ed. Open Systems Group) Open University Press.
Towill, D.R. (1984) Information Technology in Engineering Production and Production Management, in *The Management Implications of New Information Technology* (Ed. N. Piercy), Croom Helm Ltd., pp. 15–47.

Questions

1. Study Fig. 1.1 carefully, and then answer the following questions:
   (a) What are the similarities among these manufacturing system models?
   (b) What are the differences? In your opinion, what are the reasons for these differences?
2. Consider a manufacturing system that you are familiar with. Identify the

inputs to and the outputs from the system and then illustrate the overall system in a way similar to that shown in Fig. 1.1.

3. Based on the model given in Fig. 1.3, how did Keynes view the position of manufacturing industries in the macroeconomy of a nation?
4. In your own words, describe how and why manufacturing is important to a nation. Describe the economic and social significance of the manufacturing industry you have analysed in Question 2.
5. What are the main features of the three stages of industrial development in manufacturing? How were the first two significant from a socioeconomic viewpoint, and what in your opinion will be the socioeconomic significance of ongoing developments?
6. What are the lessons to be learnt from each of the case studies in section 1.4? According to Ettlie (1988): 'Reports of failures appear from time to time in the literature and in private, but they are not really useful to study.' Do you agree?
7. Summarize the lessons you have learnt from this chapter by writing a short essay entitled 'Can technology solve all the problems of manufacturing industry?'.
8. Recently the need for a multidisciplinary engineering function called manufacturing systems engineering (MSE) has been recognized. What are the problems and background which have caused such a new wave? What are the objectives and contents of MSE? As a manufacturing systems engineer, how do you see your role in today's manufacturing industry?

2 Systems concepts

2.1 Introduction

Chapter 1 has given some indication of the unprecedented problems and opportunities facing manufacturing industry today. The problems are both complex and interdependent, so that solutions cannot be sought in isolation. Parallel with these problems are the evolution processes of technological development. It can be argued that manufacturing industries are now leaving one technological age which is characterized by machines, and are in transition to the age of systems.

Problem-solving in this systems age will frequently be a team task, with each team member not only being an expert in a particular field but also possessing understanding of the interfaces between different areas, together with an awareness of external influences such as economic, political, social and technological factors. To cope with this, it is necessary for those involved to adopt a systems attitude to their work. The purpose of this chapter is to help build such an attitude by introducing the concepts of systems, in order to provide a basis for our study and analysis of manufacturing systems. Two particular questions will be addressed. First, what are systems? And second, what are the properties of systems? Although there are many general introductions to systems concepts, this chapter is designed to provide a broad survey of relevant systems ideas, with particular reference to the organizational and operational aspects of manufacturing systems. Although it is not always possible clearly to define and subdivide this particular subject area, it is true that 'systems thinking' has been founded upon some key ideas – those of *systems*, *control* and *communication* (Checkland, 1981). These ideas will be used together as a structural framework for this chapter, into which various systems concepts and systems characteristics are fitted and discussed.

Section 2.2 introduces some basic systems concepts, including a definition of systems, systems perspective, the types and structural properties of systems. Section 2.3 deals with the control and communication aspects of systems. The principles, structures, means and problems of these two operational aspects of systems thinking are presented in a format considered appropriate to manufacturing.

Having introduced a relatively wide range of systems concepts, these will be 'brought to life' in section 2.4 by means of the following:

- A prototype system model, based on the structural properties of the particular type of systems which is to be the main concern of this book. This prototype model may be used to help us understand and analyse real systems involved in a particular problematic situation.

- A structured procedure to identify and describe the systems content of the situation.
- A set of preconditions necessary for the effective control and operation of the systems concerned.

After studying this chapter the reader should be able to:

- define and/or explain the following basic systems concepts: component, objective, wider system, emergent property, relationship, hierarchy, environment, system perspective, process, boundary, and sub-system;
- use the technique of input–output analysis to describe the structure and contents of a system of concern;
- identify and contrast the following types of systems:
  physical/conceptual
  static/dynamic
  open/closed
  continuous/discrete
  stochastic/deterministic;
- discuss in general terms the principles, structure and problems of control, and give examples of feedback controls which exist in manufacturing systems;
- explain the basic principles of information theory, discuss the structure, requirements and problems of communication within organizations, and indicate their implications in the area of advanced manufacturing systems;
- apply the structured procedure for system identification and description to a problematic situation to analyse and express the content of the situation;
- use the prototype system model as an archetypal norm so as to check the content of a problematic situation; and
- outline and explain, from a systems viewpoint, the necessary conditions for the effective operation of manufacturing organizations.

2.2 Basic systems concepts

Systems are encountered everywhere in the universe. Despite their diversity, however, all systems have some characteristics in common. This has led to the development of *systems thinking* – an attempt to explain the fundamental structure and nature of systems in a scientific way. As a result of this, systems concepts have been applied in many of the fundamental fields of science and most branches of engineering.

2.2.1 Functional perspective and systems perspective

When faced with a complex situation, an analyst will first attempt to reduce the complexity of the problems involved, so that they are of manageable size and can be dealt with effectively. The traditional analytical approach has been based upon a functional perspective, breaking down problems and then analysing individual functions. On the other hand, the fundamental idea of systems is to recognize and analyse a situation within an overall perspective – in terms of 'system'. The purpose of the systems approach is to simplify the problematic situation through a systems perspective. Unlike the functional approach, the systems approach encourages the analyst

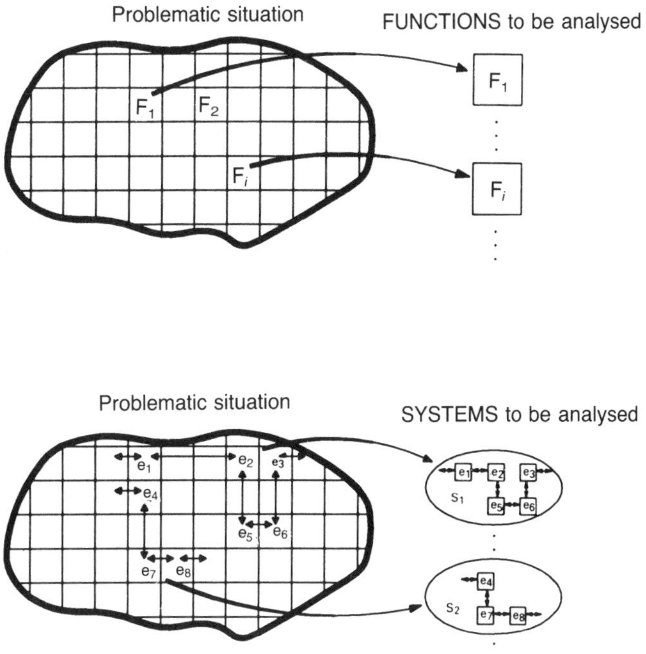

Fig. 2.1 Difference between functional and systems approaches. (a) Simplifying complexity through reduction and (b) Simplifying complexity through a systems approach.

to consider activities in their entirety, utilizing systems concepts such as *objectives*, *relationships* and *transformation*. The difference between the traditional and systems approaches is illustrated in Fig. 2.1. This shows why these approaches are also referred to as *reductionism* and *holism*, respectively.

What is a *system*? When expressed in a formal manner, the definition and characteristics of systems may appear to be abstract and not easily understood – however, it will not be so hard to comprehend the basic systems concepts when they are related to real-life examples. From the schematic representations of manufacturing systems given in Chapter 1, a system can be viewed as a transformation process which converts a set of inputs into a set of outputs, as shown in Fig. 2.2. The inputs and outputs of a system are the main interfaces between the system and the outside world. The process is the totality of systems elements including objects and relationships.

A systems view of a public house provides an everyday example. From a customer's viewpoint, a pub could be regarded as an 'enjoyment-generating system', as illustrated in Fig. 2.3. This system converts money, leisure time and the company of friends (the inputs) into enjoyment (the output) through the sale of beer and other refreshments, and the provision of an atmosphere conducive to conversation and the playing of games (the process). So, a system is an ongoing process – any process that changes the state of one or more objects involved in it. It consists of a number of different elements. The property and behaviour of these elements contribute to the property and behaviour of the system as a whole and in an organized manner. We

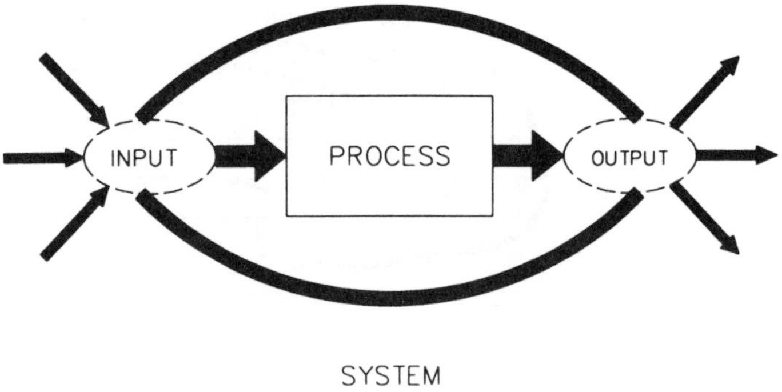

Fig. 2.2 Schematic representation of a system.

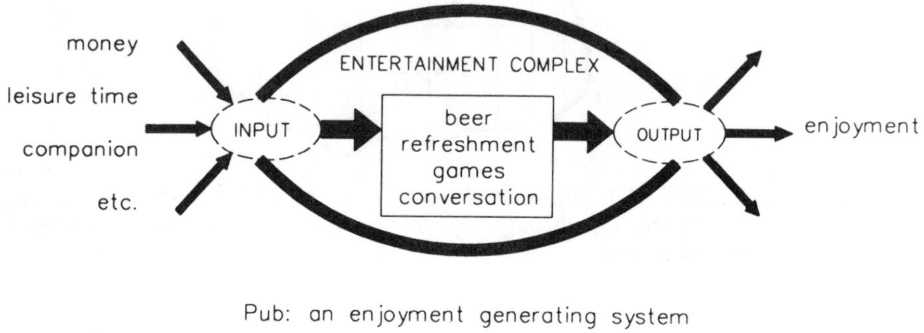

Fig. 2.3 A pub as viewed by the customer.

will explore the common foundations of systems through a formal definition of a system and discuss its characteristics. For our particular purpose a system will be defined as *a collection of components – for example people and/or machines – which are interrelated in an organized way and work together towards the accomplishment of a certain logical and purposeful end*. This definition implies that a system must have the following features:

- *An assembly of components*. These components are the structural, operating, and flow parts of the system which can be individually identified. System components can also be identified as input, process, output, feedback control, and restriction. The input and output of a system may sometimes also be referred to as the cause and effect, respectively.
- *Components connected in an organized manner*. This indicates that the relationships between system components are important. Each component is related directly or indirectly to every other component in the system and affected by them. Without relationships there will be no systems. The concept of relationship constitutes one of the fundamental differences between the doctrines of holism and reductionism. The relationships between the individual parts and the effects they have on each other are considered important by holism – hence a whole cannot be analysed without considering these interactions.

- *A logical objective or purpose*. This implies that the systems we are concerned with are rational. That is, a logical objective must be explicitly identified. The purposeful action of a system which transforms system input into system output should, therefore, conform to the 'principle of causality', that is, for every output or effect exhibited by a rational system, there exists a definite set of inputs of causes that influenced and produced that output. This implies that identical inputs to a rational system should always be likely to produce identical outputs. Manufacturing systems are rational systems. A fortune-telling system using a crystal ball, for example, is regarded by modern science as an irrational system, because its inputs have no rational influence on its outputs.
- *Components which work together towards the common objective*. It is the totality of the components which, together with their attributes and relationships, constitutes a particular system and provides the output for each given set of inputs.

This definition and its associated concepts provide a basis for the systems approach to manufacturing design and analysis adopted in this book. Later in this chapter, a step-by-step approach to assembling, testing and describing a system will be presented.

2.2.2 Hierarchical structure of systems

Having considered the basic constituent of a system, we must now examine both the position of the system within the hierarchy of systems, and what actually lies within the boundary of the system itself.

System boundary, wider system and external environment
To make the study of systems practical and manageable, it is first necessary for the components or elements of a subject system to be enclosed within a *system boundary* and thus effectively isolated from the remaining universe of all other systems and factors. The relationships between a subject system, its

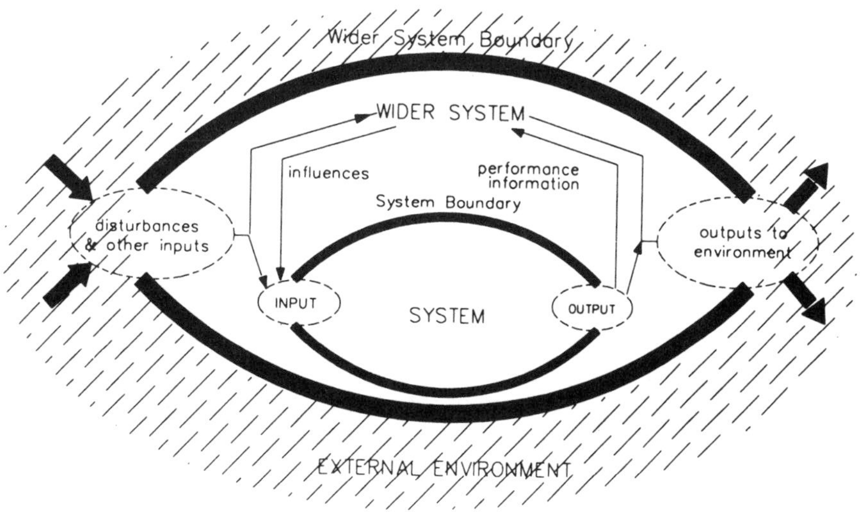

Fig. 2.4 System boundary, wider system boundary and environment.

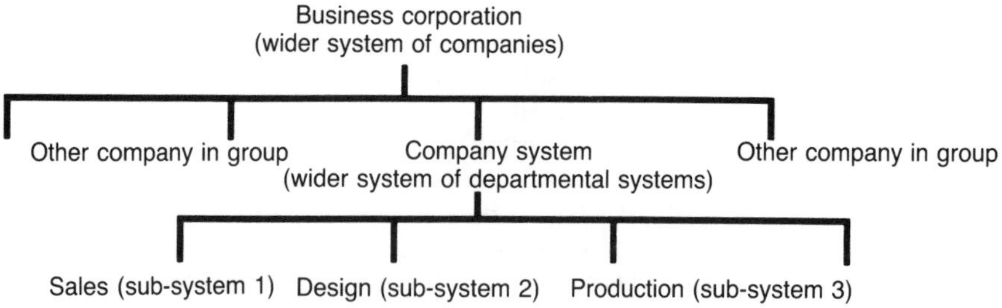

Fig. 2.5 The wider system and sub-systems of a manufacturing company.

wider system and the external environment are as shown in Fig. 2.4. This diagram implies that systems are hierarchical in nature, for the system at one level can be a sub-system or even a component of higher systems. The system which contains the subject system is sometimes called its *wider system*. For instance, normally a number of systems can be identified within a manufacturing company at a departmental level. It is apparent that all of the systems at this particular level must operate within the company system, which is one level up in the hierarchy and hence the wider system of the departmental systems. The company itself can be a system within the wider system of a business corporation, as shown in Fig. 2.5. A wider system influences its constituent system by laying out its operational goals and checking its performance, as well as supporting its operation. Everything that remains outside the boundaries of the system is considered to be the *external environment*. System boundaries are in fact a familiar concept. For example, every country in the world must define its national system by establishing its land, sea and air territorial boundaries. Everything within these boundaries is regarded as 'national', that is, included in the system; everything outside the boundaries is referred to as 'foreign', that is, environmental. Inputs and outputs of the system act as the interfaces between the system and the external environment, as illustrated in Fig. 2.4. One example of an interface between a system and its environment is the customs of a country, through which materials and people wishing to enter the country must pass. Environmental factors also affect the system through disturbances and restrictions.

Sub-systems, system components and emergent property

A systems approach to problem-solving usually involves looking at a situation from the top down rather than from the bottom up. Attention is first focused on the system under study as a 'black box' that interacts with its wider system and environment. Lower levels of system elements are then considered, that is, how the smaller boxes are actually connected and co-operate to accomplish the common system objective. Therefore, having considered the hierarchy of systems within which a subject system must reside, it is necessary to look at the hierarchical structure within the system itself, and at some of the properties of these smaller black boxes.

Depending on the number of hierarchical levels involved, a subject system may be further divided into sub-systems and components, each of which will receive inputs and transform them into outputs. For example, within the departmental system shown in Fig. 2.5, each task team may be considered as a sub-system. The relationship between a sub-system and the system is

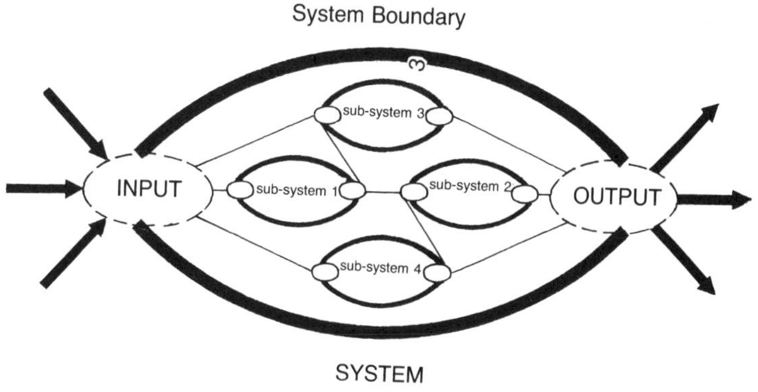

Fig. 2.6 System hierarchy: system and sub-systems.

equivalent to that between the system and its wider system. That is, a sub-system can be a total system itself, consisting of all the components, attributes and relationships necessary to achieve the particular objective which its wider system has set for it. The wider system of a sub-system can either be the subject system itself, or even another sub-system which is one level higher in the structure, as illustrated in Fig. 2.6. Thus each of the task teams of a production department (a sub-system) may have its own shop-floor space (structural components), parts (flow components), equipment and the associated production procedures (process components), all of which are integrated and work towards its operational goals under the control of the department (system).

Individual components are at the lowest level within a subject system which in practice will not be subdivided any further for the purpose of a particular analysis task. If we consider the activities of a milling cell as a sub-system of the production department of a manufacturing company, for example, for the purpose of functional analysis the individual machines and operators are likely to be identified as system components. In this context, it is unlikely that we will be concerned with the biological system that must operate within the body of each operator. Though extremely complex and important when looked at from other systems viewpoints – by biologists and doctors for example – there is obviously no need for these to concern us. The properties of sub-systems or system components are known as their *attributes*. Each component in the system may possess one or more attributes, which are the characteristic information about the component. Attributes may assume many forms – for example, shape, colour, volume and dimension. The attributes of components may be changed as a result of a system process – for example, a piece of raw material transformed into a part by a machining operation. The totality of these attributes determines the system state.

An important concept which is closely related to the hierarchical structure of systems is that of *emergent properties*. The idea of emergence is based on the observation that each level in the system hierarchy is characterized by certain properties which cannot be exhibited by the lower levels. A system is therefore more than the sum of its parts. For example, as a whole a manufacturing organization has characteristics and behaviour patterns which cannot be exhibited by any of its individual departments. This concept has been used as a primary argument by holists to support their claim that the reductionist approach needs to be supplemented by a form of thinking based on wholes and their properties. The argument is that the properties and behaviours of a system cannot be entirely predicted even from the fullest knowledge of each part's performance in isolation, since each of these parts on its own does not exhibit the emergent properties of the whole. As system analysts involved with engineering projects, one lesson we can learn from this is that we should always be aware of the nature and the overall objectives of the situation concerned.

2.2.3 System types in a manufacturing environment

The classification of systems by type will help establish accurately the characteristics of the system under investigation. This is important for the subsequent tasks of modelling, analysis and evaluation. The various types of system which we are likely to come across, together with their basic characteristics, will therefore be outlined. First of all, systems can be either natural or man-made. A significant difference between natural and man-made systems is that, while all man-made systems have explicitly defined objectives, it is usually difficult for us to identify the objective of a natural system. For example, the question of 'the purpose of life' has been one of the liveliest subjects of debate for thousands of years. Although different answers to the question have been offered by different people, depending on their culture and religious background, no single answer has been universally accepted. As manufacturing analysts, our main concern will fortunately be with man-made systems. Investigations into the reason for the existence of the solar system or the purpose of life on Earth, though undoubtedly interesting, are outside the scope of this book. It should be pointed out, however, that the introduction of significant man-made systems, if not properly designed and implemented, may severely disturb the equilibrium of the natural world, with possible catastrophic effects. This problem is growing acute as technology becomes more pervasive. One of many examples of this phenomenon is the much-publicized 'greenhouse effect' and the prospect of 'global warming'. Although the credibility of this theory is being debated, the world at last seems to be recognizing the importance of a systems viewpoint when designing and commissioning a large-scale project. The possible catastrophic effects which the man-made system might have on its environment cannot be ignored. A socially responsible designer or analyst involved in large-scale manufacturing systems should always bear this in mind.

Within a manufacturing environment, systems can be categorized by classifying them according to the following criteria:

physical/conceptual;
static/dynamic;
open/closed;
continuous/discrete;
stochastic/deterministic.

Thus we may find a conceptual system which is also static and deterministic, but we cannot have a system that is both continuous and discrete (though occasionally we may encounter a system that contains, say, both continuous and discrete parts – such a system is called *hybrid*).

The first category separates systems in physical terms. *Physical* systems consist of real objects such as machinery and equipment. *Conceptual* systems have no physical existence; they are normally represented in terms of diagrams, charts, mathematical equations and verbal descriptions to express ideas or concepts. Conceptualizing the physical system of a manufacturing operation, whether it is yet in existence or not, is a very important exercise in our particular area of study (this is also called *modelling*), which will be discussed in detail in Chapters 3–7.

The second category classifies systems according to the activities which may or may not be involved. *Static* systems are defined as having structure without activity, that is to say, the system process will not evolve over time. *Dynamic* systems combine structural components with activity so that the states of these systems will usually change continuously over time. Some of the systems models which we will use are static in nature. Typical examples of this kind are Monte Carlo models, which represent a system at a particular point in time, in accordance with a certain logic pattern. These models incorporate probabilistic elements to derive output statistics for the system. Many of the financial analysis models, such as the various types based on spreadsheets, also belong to this category. However, the majority of the systems which we are concerned with are dynamic. A typical example of this type is given by the system of an inventory organization that combines store buildings, inventories, operators, and procedures. Many activities will be involved and the system variables will change with respect to time, as illustrated in Fig. 2.7.

The third category separates systems in terms of their interactions with the environment. An *open* system interacts with its environment, whereas *closed* systems are self-contained under normal conditions. The distinction between open and closed systems is very important in the study of manufacturing systems. Many of the traditional concepts in organization and management theory consider an organization such as a production system as a self-contained, closed system. Closed systems, however, have an inherent tendency to move towards a high degree of disorganization – to undergo an increase in what is called *entropy*, a term borrowed from

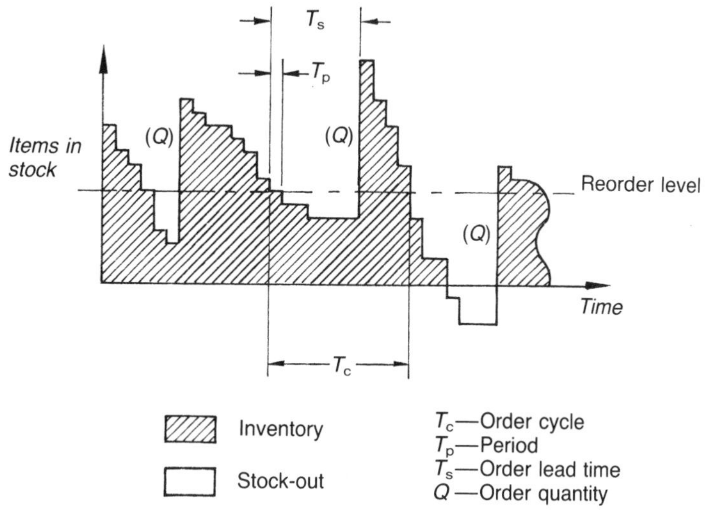

Fig. 2.7 Illustration of an inventory system as an example of a dynamic model.

thermodynamics. In systems theory increased entropy means increased disorganization (Checkland, 1981). Closed systems tend towards a chaotic state in which no further energy transformation or work is possible. In contrast, open systems maintain a dynamic relationship with their environments and, in doing so, offset the process of entropy. It has been realized, therefore, that an organization should be considered from an open-system viewpoint. For example, most production organizations are in fact man-made open systems which, through their business and production activities, continuously exchange their factors of production and produced goods with their environment. An open system must have the ability to adapt to its environment by altering the structure and processes of its internal components. Like natural open systems, therefore, a production organization must retain a high degree of adaptability in order to survive in a competitive business environment. As will be shown later in this book, how to design and operate a production system so that it will be able to cope effectively with environmental changes is one of the most important issues of manufacturing systems engineering.

Dynamic systems can be further classified, using our fourth systems category, into *continuous* and *discrete* systems, according to how the system variables change over time. Continuous systems are those in which the state variables may change continuously over time, while discrete systems are those in which system variables change in a stepwise fashion. The difference between these two types of system is shown in Fig. 2.8. In theory, the inputs of a continuous system may vary continuously by infinitely small intervals and the system responds to these inputs with continuous outputs, as exemplified by the water level in a reservoir. Typically, a continuous system may be described by abstraction, decomposition or aggregation, using differential and/or algebraic equations which express the relationships existing within the system. For the mathematically inclined, the situation may be generalized as follows. A system could be represented by the vector of its state variables $\mathbf{X} \in \mathbb{R}^n$,

$$\mathbf{X} = \begin{bmatrix} X_1 \\ X_2 \\ \vdots \\ X_n \end{bmatrix} \tag{2.1}$$

where the state variables X_i, $i = 1, \ldots, n$ change in value continuously with time t. The relationships of the rates of change of the state variables with

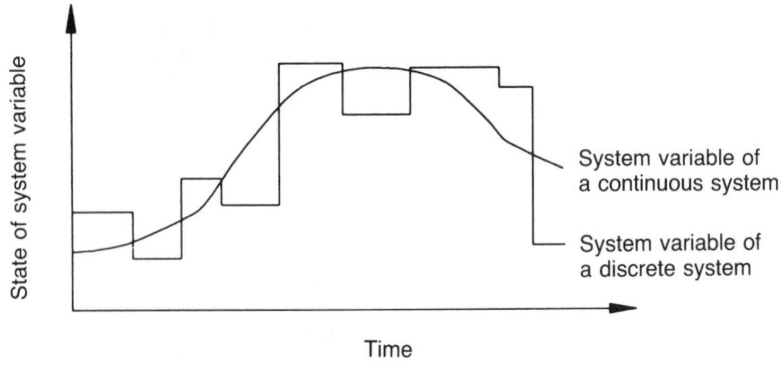

Fig. 2.8 Difference between continuous and discrete systems variables.

respect to time would be given (for space-independent systems only) in the following form:

$$\begin{aligned} \mathrm{d}\mathbf{X}/\mathrm{d}t &= \mathrm{f}(\mathbf{X}, \mathbf{u}) \\ \mathbf{X}(0) &= \mathbf{X}^0 \\ \mathrm{g}(\mathbf{X}, \mathbf{u}) &= \mathbf{0} \end{aligned} \tag{2.2}$$

where $\mathbf{u}(t) \in \mathbb{R}^m$ represents the input vector; **f** is a vector-valued function describing the system relationships; and **g** defines the boundary conditions. By definition these are inherently dynamic. If the system equation describing **f** is simple, it may be solved analytically to give the values of **X** with respect to t as a function of $\mathbf{X}^0$, that is:

$$\begin{aligned} \mathbf{X} &= \mathbf{X}(t) \\ \mathbf{X}(0) &= \mathbf{X}^0 \end{aligned} \tag{2.3}$$

For most continuous systems, however, analytical solutions of the system equations are not possible, and the integration of the differential equations has to be performed numerically on a digital computer. Examples of the application of this type of simulation in management studies include *system dynamics*, which will be discussed later in this book. Continuous simulations are of interest to economists, in modelling the behaviour of economic systems, and planners in studying the behaviour of a system at the aggregate level over a long time-horizon. Researches in operations management are usually concerned with modelling manufacturing systems using a discrete approach.

In contrast to continuous systems, which may be expressed in an abstract form by differential equations because of their continuously changing nature, in discrete systems the state variables change at sudden distinct events. Discrete-event phenomena are inherent in a manufacturing system – as exemplified by the flow of material and parts between different work-stations, and the flow of information among various departments. This type of system is, therefore, more relevant to our study. A discrete modelling approach is more frequently used in manufacturing systems studies in order to describe the detailed properties of such systems. However, in contrast to the modelling of continuous systems, there is no standard way of describing discrete-event systems. The behaviour of discrete systems can be analysed by using a methodology called *discrete-event simulation*. Instead of formulating an intermediate mathematical model, a computer simulation model is directly developed using a programming language – either a special-purpose simulation language or a general-purpose computer language. Quite often the simulation languages are different in the ways they construct a simulation model and in how they handle the simulation processes. Dependent on the particular simulation language or programming approach adopted, the model structures produced will be different from each other. Consequently, the description of discrete-event simulation models is closely related to the particular 'world view' of a special language, and generalization is difficult (computer discrete simulation is one of the major topics of this book, which will be described in detail in Part Three).

In some cases, it is clearly not easy to classify a system as either continuous or discrete. A production organization, for example, may be regarded as a continuous system at the aggregated level but as a discrete system at the disaggregated level. In other cases, we may find that the discrete and continuous parts are combined together, such as in a steel mill where the production operations are carried out in a discrete fashion, but the temperature of the ingot obviously changes in a continuous manner. The combined discrete and continuous systems are, as we have said, called hybrid systems. In these cases the choice of whether to describe the system under study by discrete or continuous or even hybrid approaches must depend upon the objectives of the investigation, as we will see later in this book.

Finally, systems can be either *deterministic* or *stochastic*. Deterministic systems are those which exhibit unique and direct cause and effect relationships between the inputs and outputs, and between the initial conditions and the final state. Thus, for any given input, a deterministic system will always respond with the same output. An example of this type is a clockwork mechanism that has the characteristic of taking fully predetermined actions by design. On the other hand, many systems are characterized by random properties. These are called stochastic systems. Although any stochastic system with a rational nature should also obey the principle of causality, its input, process and output can only be analysed in statistical terms. For example, in most cases it is impossible for a production organization to predict exactly when customer orders will come, or the exact number of products that will be ordered.

The analytic processes for stochastic systems are more complicated than those for deterministic systems. For example, while, given a set of inputs, a deterministic system will produce a unique set of system outputs which can be measured and taken as a true system response, a stochastic system will yield only estimates of the true system response because the outputs themselves are random variables. As a result, when studying stochastic systems, experiments will often need to be run several times in order to *estimate* the true system response and to minimize the estimated variance. The nature of the system under study therefore determines the approaches to problem-solving. For example, under deterministic conditions, a deterministic model can be used to assess alternative policies by asking 'what if?' questions. Different sets of inputs may be used for the model, and then the corresponding sets of outputs are related to these input changes. In contrast, statistical distributions of the input variables will have to be used in the stochastic situation, with the objective of, say, estimating the distribution of the output variables of the system.

2.2.4 Input–output analysis

Having studied the basic structural properties of systems, we are now in a good position to look at a tool that can be utilized extensively during a design–analysis cycle to describe the structure and contents of a system of concern. This is the technique of input–output analysis.

Basic ideas of input–output analysis

Input–output analysis is based on the fact that the simplistic form of a system is composed of three basic components – input, process and output. When faced with a 'black box' situation, in order to complete an analysis of the system (or sub-system) involved and identify the constituents which would fall under each heading, the analyst should first determine what are the system's *outputs*. He should then examine the *process* to determine what *inputs* are required for the transformation to take place.

For example, consider the elements of an assembly activity in a manufacturing situation. The outputs from this process will normally be passed assemblies and rejected assemblies, as shown below:

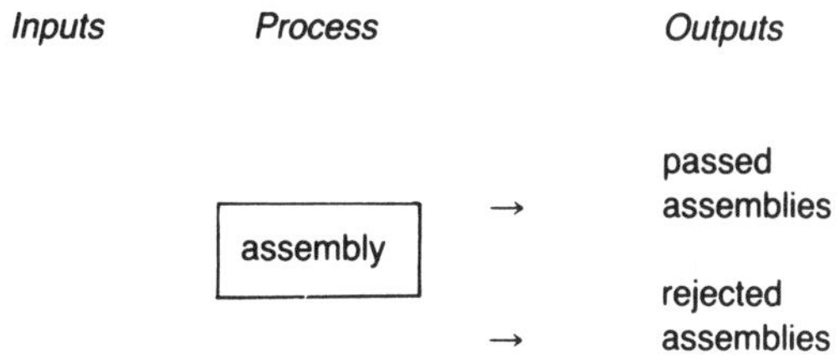

Since the process is an assembly activity, where parts are to be assembled together and then tested, it may then be possible to identify the inputs required, as follows:

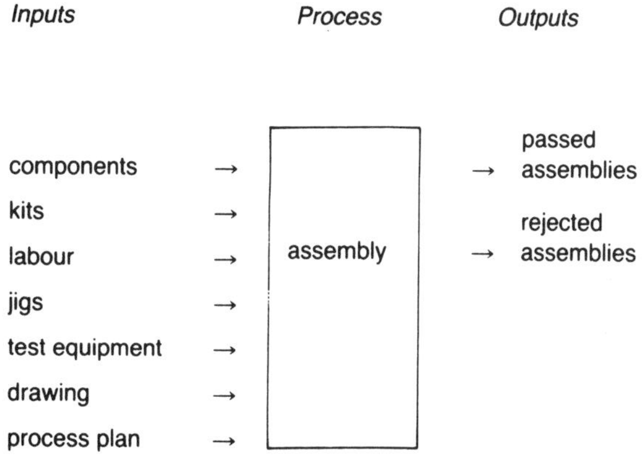

Classification of inputs and outputs

As shown above, the input–output analysis of a system or sub-system should commence with an examination of the outputs required. The difficulty tends to lie in ensuring that all the outputs are identified – this is not as straightforward as it might first appear. However, this activity may be eased by grouping the outputs under two headings: *basic* and *complementary*. Basic outputs comprise the goods or services to be provided to a subsequent customer or user or to another sub-system in the total system process. These might include raw materials, parts, components, final assemblies, drawings, and process layouts. Complementary outputs comprise information generated to record and evaluate the process. These might include records of scrap materials, production progress reports, and reports of production problems.

Inputs can be grouped in an analogous manner to aid the analysis activity and to provide a means of verification to ensure that all of the relevant items have been addressed. Basic inputs comprise the goods or services required to produce the basic outputs. Complementary inputs comprise the resources and supporting information required for the process to be executed, and can therefore be subdivided into *resources*, which include the hardware (plant and equipment) and people required to carry out the process, and *supports*, which consists of the instructions and constraints governing the process.

2.3 Communication and control

This section discusses the operational processes within systems. Manufacturing systems are open systems. For an open system to support its hierarchy and achieve a steady state under changing environmental influences, there must be a set of operational processes within it which regulate or control other operational processes through communication of information. These two important systems concepts will now be described.

2.3.1 Principles of control

The principles of control are closely associated with the concept of automation. The first theoretical analysis of a control function was conducted by the nineteenth-century physicist, James C. Maxwell, who developed a mathematical model of the fly-ball governor of steam engines. However, control engineering and control theory did not become an established professional function and academic discipline until after the Second World War. The term *cybernetics*, conceptualizing the mechanisms of feedback and self-regulation, was first used in the 1940s. Taking a narrow view, this is equivalent to the theory of servo-mechanisms in engineering; in a broader context, cybernetics may be regarded as a science of control applicable to many of the complex processes in nature, society and business organizations. Cybernetics has made an important contribution by identifying the two fundamental characteristics of all control systems.

The first of these is the distinction between the activating power and the information signal. In general, the value of the controlled system variable is changed by an actuator. However, the actuator is powered by an autonomous source which is not attached to the controlling signal. This ensures that the operation of varying the controlled system variable does not distort the information signal.

The second is that principles of control mechanics are fundamentally identical for all control systems. During a control action, the rate at which energy is fed to the actuator to effect the variation action is directly related to the variation between the actual and desired values of the controlled system variable – this is the principle of *feedback* control.

Feedback control

Feedback, in system terms, is the sub-system function that compares the output with a given criterion in order to reduce the deviation between the actual state of a system and the desired state, when the system is subjected to unpredictable disturbances from its environment. Therefore, system feedback takes place whenever information about any of a system's outputs is used to correct its operation. The control system mentioned previously, James Watt's fly-ball governor, is a typical example of feedback control.

The essential components of a typical feedback control operation are illustrated in Fig. 2.9. They are:

- a controlled system parameter or condition
- a monitoring function which measures the current state of the controlled condition
- a decision-making function which compares the current state of condition with the desired condition
- an actuator that alters the system state so as to reach the desired state.

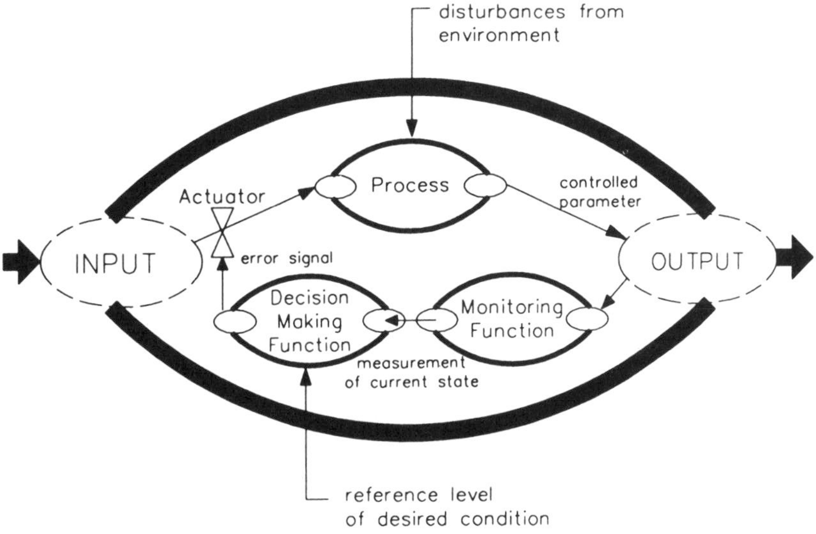

Fig. 2.9* Essential components of feedback control.

Figure 2.9 shows that information about the state of the controlled condition (system output) is obtained through a monitoring function, using a measuring device which in practice can be either a physical sensor or a human inspector. The information is then sent through the feedback path to the decision-making function which produces an error signal from the difference between the output level and the control reference level. This error signal signifies the need for control action and the amount of correction required. Control is then achieved by using this error signal to drive the actuator in such a manner that the level of system output will be directed to the level of reference. A control system which possesses the fundamental components and characteristics shown in Fig. 2.9 is called a *closed-loop* control system.

In any system there may exist two different types of feedback – *intrinsic* and *extrinsic* – as illustrated in Fig. 2.10. Intrinsic feedback is also called *integrated* feedback and occurs when feedback and control are within the system's boundary and undertaken by elements belonging to the system. Extrinsic feedback involves feedback and control outside the system boundary and thus involves entities which belong either to the wider system or the environment. In general, the presence of intrinsic feedback is not readily apparent to an outside observer.

This is one of the reasons why in many system studies it is first necessary to treat the system under investigation as a 'black box'. The behaviour of the system is observed from the external environment through the inputs and outputs crossing the system boundary, and at this stage the system analyst is not usually in a position to study how the system actually processes the inputs to produce its output response. Identification of intrinsic feedback is only possible when the system can be analysed from inside and the source of feedback can be isolated. The distinction between intrinsic and extrinsic controls is illustrated in Fig. 2.11, where a learner driver attempts to pass a driving test.

One of the most important concepts in control theory is that of *stability*. In any control system, the degree of stability is usually the most important

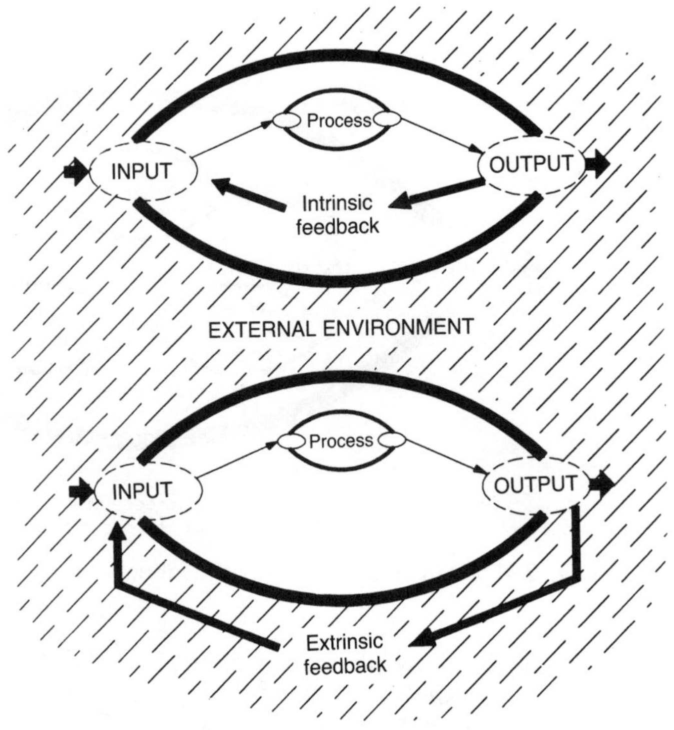

Fig. 2.10 Intrinsic and extrinsic feedback.

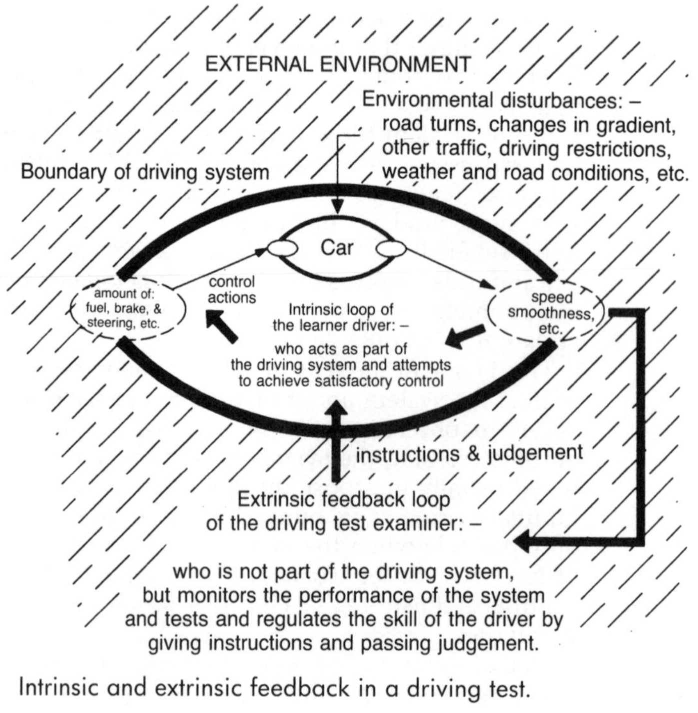

Fig. 2.11 Intrinsic and extrinsic feedback in a driving test.

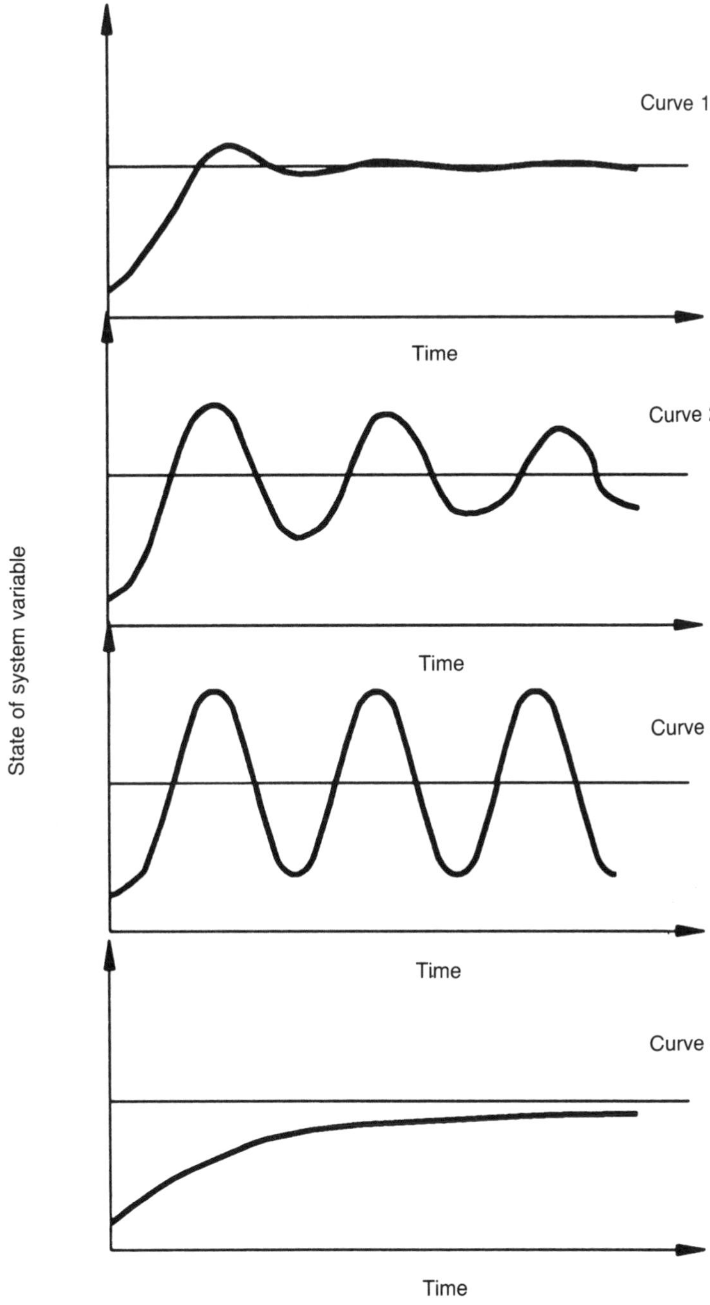

Fig. 2.12 Responses of a controlled system.

objective. The stability of a control system is determined to a large extent by its response to a sudden disturbance from its environment – a sudden increase or decrease in the demand for products from a production system, for example. There are a number of ways in which a system may respond to this disturbance from a stable initial condition, as shown in Fig. 2.12.

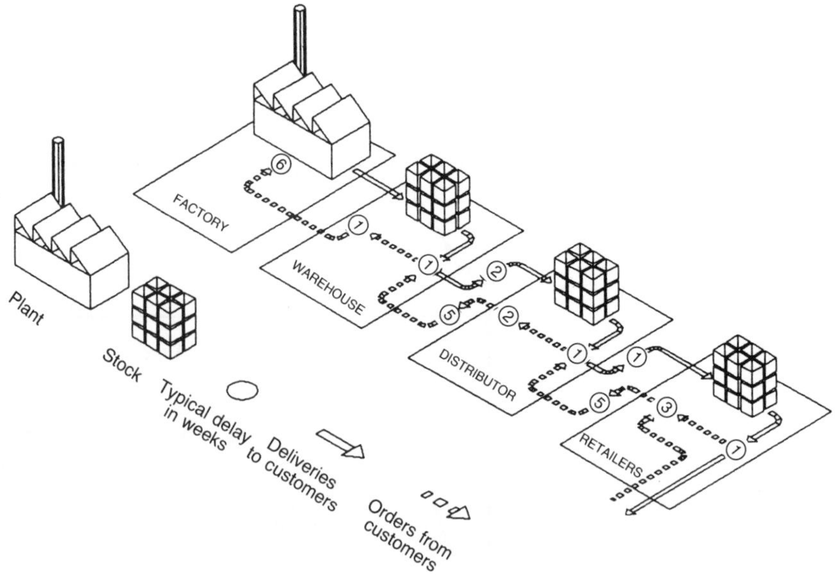

Fig. 2.13 Organization of a production-distribution system (based on Forrester, 1961).

Curve 1 of Fig. 2.12 usually represents the ideal condition where the system is under effective control. When a new target state is set, the system responds quickly and stably by changing its state to that new level and achieving a new state of steady operation.

Curves 2 and 3 represent system instabilities due to various problems inherent in the system concerned. The likely causes of these undesirable response behaviours include: time delays and inadequate simplification.

It is usual for large systems to have intrinsic time delays. For example, it takes time for a workpiece to wait and then be processed by a machine in a machining shop. This tends to slow down the system's dynamic response to input, with the result that it overcorrects itself when in fact only a small change is required. An early systems study carried out by Forrester (1961) provides a good example of undesirable effects such time delays may have on a production organization. He developed a systems dynamics model of a production–distribution system to explore the possible improvements in performance that could result from particular management policies relating to the production, storage and distribution functions of a company. The model is a structural representation of the organization, as shown in Fig. 2.13. By simulating the dynamic behaviour of the system, the researcher correctly predicted the effect of changes in customer orders on the production of different firms in an industry. Because of the inherent time-delay factors shown in Fig. 2.13, the system would produce late and exaggerated responses to customers' changes in demand, resulting in high production costs and misused resources. This is shown in Fig. 2.14. To improve performance, the control system model must take account of the dynamic characteristics of the processes being controlled. For instance, the relevant parameters of the control system should be adjusted, for instance, by reducing certain delays. Addition of extra communication lines could also be considered if necessary.

Inadequate simplification is basically due to the lack of real understanding

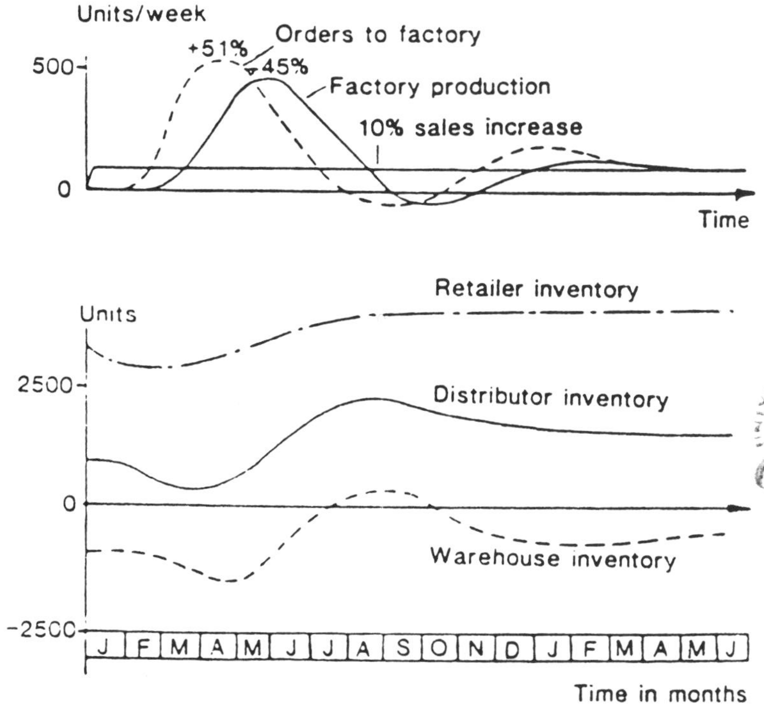

Fig. 2.14 Dynamic response of a production-distribution system (based on Forrester, 1961).

of the processes being controlled. Non-linearity, for instance, is inherent in many systems. Yet our understanding of the behaviour of non-linear systems is very limited – there is no general method for dealing with such systems. As a result, most control systems are designed and implemented in a linear fashion. Linearization may lead to over-simplification and hence to an inadequate control system. When this is the case, unexpected system behaviours may occur.

Curve 4 of Fig. 2.12 shows a system response which is over-damped and sluggish in its response to change. This represents another case where the system cannot satisfy the requirements put upon it. For example, customer orders would not be processed promptly by a production system with such a response characteristic and low output would result. In practice, this may be the result of a lack of ability or inefficiency in communication.

Feedback structure in production

Feedback control systems are not limited to the field of engineering. It is also common to find feedback structures in economics, biology and business organizations. In engineering applications, a control system is usually operated by electricity, by mechanical devices, by pneumatic or hydraulic power. In a business organization, control is accomplished through meetings, procedures and managerial actions.

Many managerial actions are based on the principles of feedback control. Some production systems, for example, may be conceptually viewed as feedback systems. Feedback models for production were initially proposed by several authors who attempted to explain and solve production problems by using a continuous dynamic response model. The basic concept of this approach is easy to illustrate. When represented by a continuous and linear dynamic system, the state equations of a production system may in general be expressed as a set of first-order differential equations, reflecting the feedback structures of the system, and given below as expression (2.4). (Note that this is a linear version of equation (2.2) and a general mathematical statement of the block diagram given in Fig. 2.9. It is again intended for the mathematically inclined to illustrate the basic mechanisms underlying a feedback structure. Thorough understanding of the analytical procedure involved is not necessary and the reader could skip this if so wished.)

$$\dot{\mathbf{X}} = \mathbf{AX} + \mathbf{Bd} \tag{2.4}$$

where $\mathbf{X}$ is an N-vector (the state variables); $\mathbf{A}$ is an $N \times N$ matrix (the transfer matrix); $\mathbf{B}$ is an $N \times M$ matrix (the distribution matrix), with $M \leqslant N$; $\mathbf{d}$ is an M-vector (the control vector). The analytical solution of this state function is easily obtainable (see, for example, Richards, 1979). It may be expressed as follows:

$$\begin{aligned} \mathbf{X} &= \exp(\mathbf{A}t)\,\mathbf{X}(0) + \int_0^t \exp[\mathbf{A}(t-\tau)]\mathbf{Bd}(t)\mathrm{d}\tau \\ &= C_1\exp(\alpha_1 t)\mathbf{u}_1 + C_2\exp(\alpha_2 t)\mathbf{u}_2 + \ldots + C_n\exp(\alpha_n t)\mathbf{u}_n + \ldots \end{aligned} \tag{2.5}$$

where α_i and $\mathbf{u}_i$ are the eigenvalues and eigenvectors, respectively, of $\mathbf{A}$. Each $\exp(\alpha t)$ may be referred to as a 'mode' of the system. For example, $\alpha_i > 0$ could be associated with growth behaviour; $\alpha_i < 0$ with decline behaviour; and α_i as a complex number, with oscillatory behaviour. It is the combination of these modes that governs the dynamic behaviour of the production system. The objective of such a model would be to find the values of the control parameters, or the structure of the control systems, so as to obtain satisfactory dynamic responses for the system under study.

A simple example will illustrate the feedback concepts of control and stability discussed so far. Figure 2.15 gives a simple schematic representation of such a response model of a production process. The purpose of the model is to stabilize production activity in make-to-stock firms by regulating the response to demand fluctuations. Equation (2.6) is a simplified mathematical statement of the model:

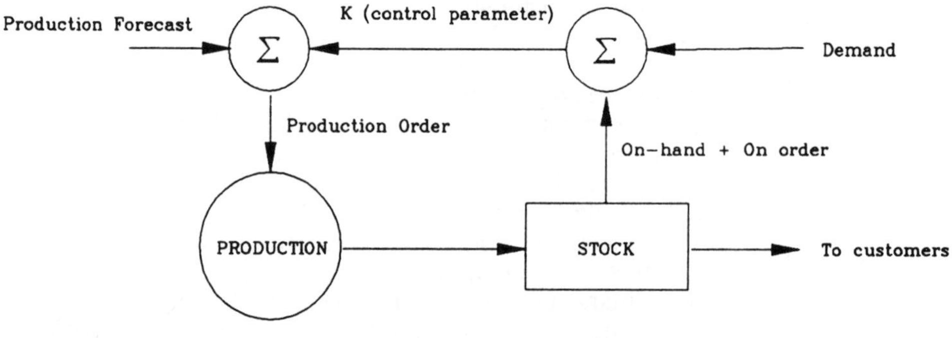

Fig. 2.15 Magee's (1958) response model of a make-to-stock production system.

$$\text{Production level} = \text{Plan or forecast of production} + K[\delta(\text{Demand}) + \delta(\text{Inventory}) + \delta(\text{Delivery})] \quad (2.6)$$

where: K is a control parameter ($0 < K < 1$), and $\delta(\ )$ is the difference between planned and actual demand, inventory or delivery. It is not difficult to appreciate how this simple control mechanism works. Inventory on hand and on order is matched against demand to determine how much additional production should be programmed. However, only a fraction of the difference is used. The combination of this fraction difference and the predetermined production forecast makes up the master schedule which is issued for production. The control parameter K should be chosen such that the system would respond quickly to the changes in demand, but in the meantime maintain as stable a production as possible. Production control as outlined above is only one example of the many feedback controls used in manufacturing industry. Table 2.1 lists some such controls.

Feedforward control

The concept of feedback control can be contrasted with that of *feedforward*, or *open-loop* control. In a feedforward control system no information from the

Table 2.1 Examples of controls in manufacturing

Types of control	*Control objectives*	*Parameters controlled*	*Feedback*	*Mechanism of control*
Production control	To achieve efficient production according to production plan	Job batches, working capacity, material inventory	Production progress reports, number and completion time of products produced in given periods, list of late jobs, list of material shortage	Production planning, material planning, capacity planning, production scheduling order issuing
Quality control	To maintain the expected quality level	Level of inspection standard, level of assurance policies, quality performance indicators	Quality performance report, conformance or defective percentage	Inspection quality standard, quality assurance procedures
Purchase control	To purchase items with adequate quality for production when required and at the right price	Supplier choice, order issuing time, order quantities	Supplier performance reports, goods reception reports	Supplier selection procedures, material requirements
Production cost control	To keep costs at the acceptable level	Overhead, direct costs, indirect costs, etc.	Accountant reports, production cost reports production time study reports	Standard production costs

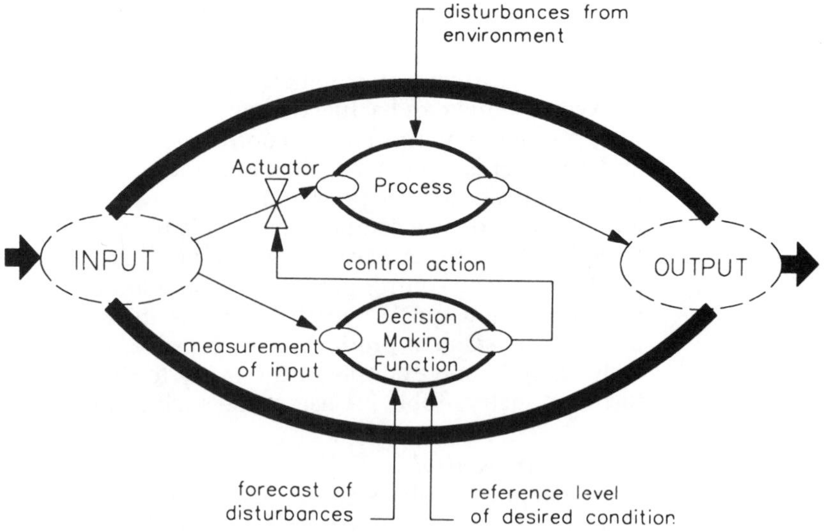

Fig. 2.16 Feedforward control.

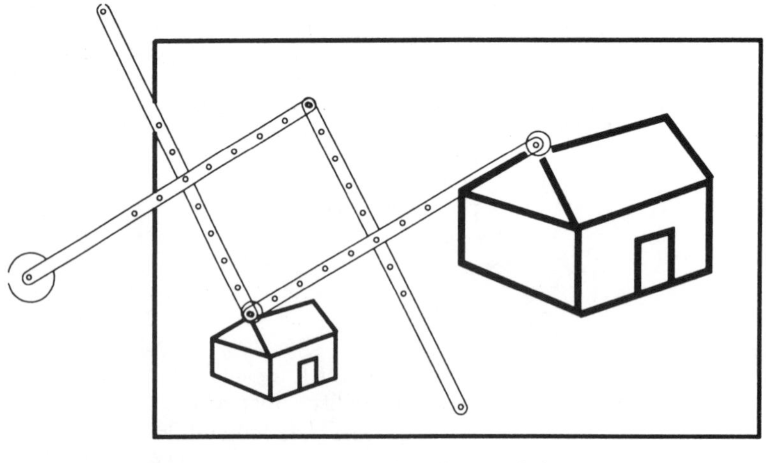

Fig. 2.17 Pantograph: an example of a feedforward device.

system process is used to correct the system operation (Fig. 2.16). Early examples of feedforward control include some of the machine-tools developed in the nineteenth century, in which the cutting tool was guided by following the shape of a model. Another example of feedforward control is the mechanism of a drawing device called a pantograph which was used for copying drawings and maps to a different scale (Fig. 2.17).

Until now the emphasis has been on the study of feedback control, which involves reacting to events after they have happened. Recently, attention has switched to feedforward control with the realization that some disturbances may be predictable and thus possibly measured in advance. If this is the case, then feedforward control may be a useful way of compensating for or cancelling the undesirable effects of these disturbances upon the system output before they actually materialize. For example, if a forecast of

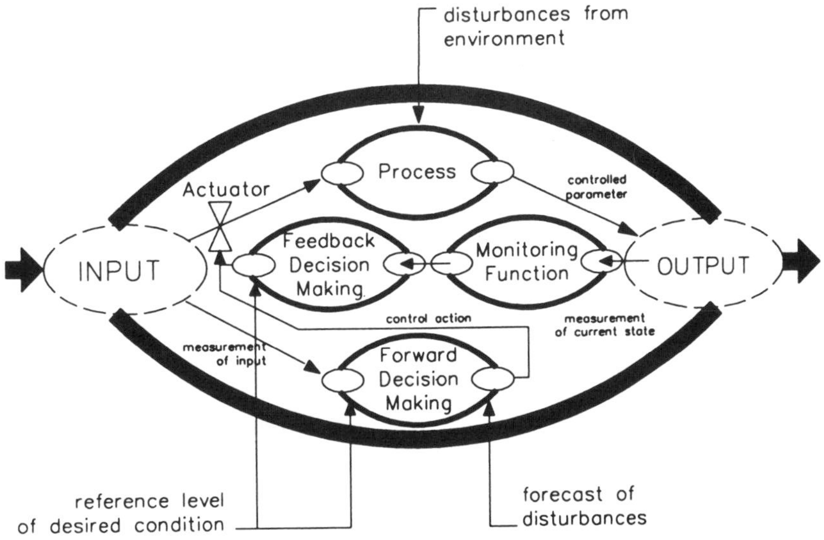

Fig. 2.18 A combination of feedforward and feedback controls.

bad weather is received in the morning before we leave home for work, we may make a decision to take an umbrella and put on more clothes. If the forecast turns out to be right, this foresight will prevent us from getting wet or a cold.

The major difficulties with effective feedforward control are, however, the requirements that firstly the relevant environmental disturbances can be forecast and secondly the decision-making sub-system is capable of taking the right actions to eliminate the undesirable effects. As a result, the analysis and design needed for this kind of control system are demanding tasks. Furthermore, it is not always possible for feedforward control to achieve high accuracy. It is often necessary, therefore, for feedforward control to work together with feedback control so that the sub-system of feedforward control minimizes the effect of measurable disturbances, while the sub-system of feedback control corrects the error caused by unmeasurable disturbances and at the same time increases the overall control accuracy by compensating for the limitations of the feedforward part (Fig. 2.18).

Hierarchy of control systems

Feedback control may appear at more than one level. A higher-level control governs the lower levels by monitoring their overall performance and setting the desired reference levels for them. It is apparent that the concept of multi-level control is closely related to the concept of system hierarchy.

The structure of a management control hierarchy is a good illustration of this more sophisticated control concept, as shown in Fig. 2.19. This shows that the activities of an organization are regulated by a hierarchy which may consist of several levels of control. At the *business* level, control of the organization is exercised by the government or by regulatory agencies (laws and regulations, tax on income), the bank (bank loan, interest payments), the stockholders (dividend payments, shareholders' committee). Note that at this level the effects of the competitors, the trade union (welfare of workforce), the local communities (environmental protection), and so on,

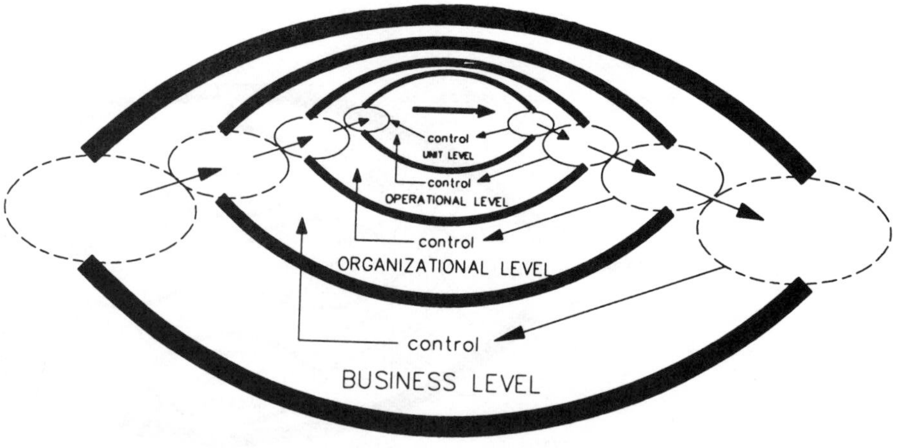

Fig. 2.19 Control hierarchy of a business organization.

are exercised through environmental influence rather than control because these do not directly regulate the business. At the *organizational* level, control is carried out through management decisions on the level of production, product prices, the level of employment, the level of investment, and so on. These decisions have to be made on the basis of the performance of the organization in relation to the business target which is set for the organization. At the *operational* level, the departmental managers are responsible for the fulfilment of the operational tasks necessary to satisfy the organizational requirements. At the *unit* level, each individual process is controlled by a designed physical and/or human-operator approach.

In relation to the concept of levels of control, we should also appreciate the complexity of feedback structures that exist within an organization. For example, the information feedback system that exists within a manufacturing organization is very complex in its structure. This is illustrated by Fig. 2.20, which is a simplified feedback structure of a make-to-inventory production system. It is often easier to tackle a complex control system through a hierarchical approach, that is, to study a system first at its aggregated level, using a model with a manageable number of control functions, and then to decompose each of these into its lower levels. Input–output analysis is one of the techniques for such analytical approaches. Other techniques available will be discussed later in this book.

In addition, the distinction between human control and physical control is of particular importance for our study of manufacturing systems. A manufacturing operation will always comprise both managerial and production process controls. Managerial control, as exemplified by the three higher levels in the organization control hierarchy of Fig. 2.19, can be described as *human activity control*, involving the intervention of human beings. Production process control is a lower-level control which is characterized by physical elements such as mechanical, pneumatic, hydraulic, numerical, and automatic control devices. These may be termed *physical activity control*. It has been argued that the higher levels in the control hierarchy, that is, the managerial control functions, are not simply more complex control systems of the same kind as physical activity control. The differences are analogous to those between a human operator and a robot. The recognition of such a

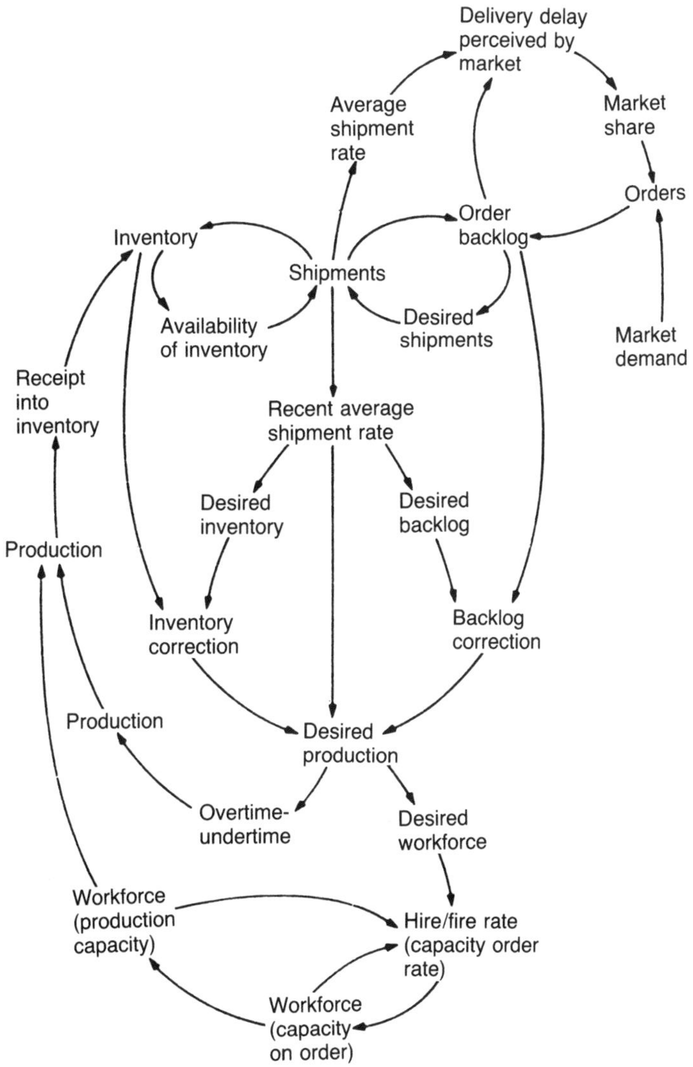

Fig. 2.20 Feedback structure of a make-to-inventory production system (Richardson and Pugh, 1981).

distinction has led to the development of the so-called *soft* systems approach which attempts to cope with the problems inherent in human activity systems. The traditional *hard* systems approach has been successful in applications which involve physical activity problems.

2.3.2 Principles of communication

Effective communication is one of the prerequisites for successful control. As can be seen in Fig. 2.21, the whole control process depends on the flow of information among the three basic functions. First, various pieces of information are needed to enable decisions to be made. Second, the output of the decision-making function will again be information in the form of instructions

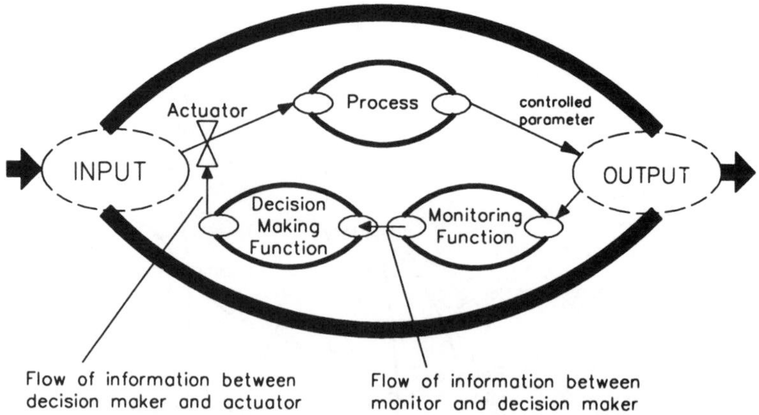

Fig. 2.21 Information flow between components of a control system.

or constraints. These must be channelled to the intended destination to initiate logical control actions. Poor communication within a control system at the various stages of the feedback loop will have serious consequences. If there is poor communication between monitor and decision-maker, erroneous information is received by the decision-maker, making adequate decision-making impossible. Poor communication between decision-maker and actuator makes it impossible for the instructions from the controller to be carried out correctly, even if the monitoring and decision-making functions are working properly and producing the right set of instructions.

Information theory

Information theory provides a theoretical foundation for the design and analysis of communication systems. It deals with the conditions and parameters which affect the transmission and processing of information in a communication system. Since its initial introduction in the late 1940s, information theory has developed rapidly and found widespread applications in many areas.

As suggested earlier by the block diagrams of control systems, a communications function must have an information source which generates messages out of a body of information, and an information receiver to which the messages are sent. In information theory, a number of key parameters are defined for a communication system which are measured and treated mathematically. We consider three such parameters.

The first is the amount of information contained in a message. This is called the *entropy* of the message and measures the degree of order or non-randomness. If a communication system has an information body which consists of a set of N possible messages, and the possibility for message i to occur during transmission is given by p_i, then the average amount of information in a message of this set is given by the entropy H, defined as follows:

$$H = \Sigma p_i \log_2(1/p_i) = -\Sigma p_i \log_2(p_i) \tag{2.7}$$

This measure in fact represents the amount of information in terms of decisions. Put simply, what H tells us is that the size of the information body concerned is such that on average we would have to ask at least H yes/no questions in order to achieve unique identification. In other words, the set will probably contain 2^H different possibilities, so that H binary digits will be required if all the possibilities in the set are to be encoded into 0s and 1s in the most efficient way. H can therefore be interpreted as associated with the construction of codes. This is why such a measure of information has been given an intuitively rational and convenient form – that is, it is measured in bits per message. (Other interpretations of the information function in equation (2.7) also exist.) This measure of information happens to be similar to that of entropy as defined in the flow of energy or the measure of randomness or disorder, but with opposite sign. Hence this 'negative entropy' or 'information', as represented by H, can be regarded as a measure of order or of organization.

The second parameter is the rate at which information is produced at the information source, denoted I and usually measured in messages per second.

The third parameter is the capacity of a channel for handling information, denoted C and usually measured in bits per second.

The fundamental theorem of information theory states that for a noiseless channel, the following relation exists among these parameters:

$$I < C/H$$

That is, the average rate at which information can be transmitted through the channel cannot exceed C/H messages per second. Based on these and other important quantities yielded by the mathematical characterization of a generalized communication system, numerous approaches to encoding and decoding for transmission of information have been developed. These are all intended to reduce transmitting error to minimal levels.

Although information theory was originally applied in the context of telecommunications engineering, it has found applications in other areas. One example is the assessment of performance of human operators in complex systems. In such studies, a human operator is usually modelled as a 'human information processing system', consisting of five sub-systems (sensory transducer, perceptual mechanisms, central processor, effector, and peripheral organs) with different types of information transmitted among them, as shown in Fig. 2.22. The performance of an operator may be assessed for the purpose of design or analysis by estimating the rates of information handling at various points in the system. For example, by using experimental data and applying some of the theorem of information, it has

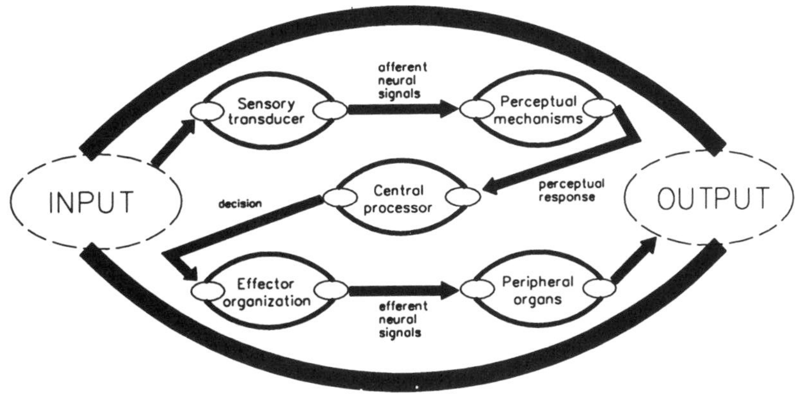

Fig. 2.22 Components of the human information system.

been estimated that the maximum capacity of information transmission through the human eye is approximately 10^9 bits per second. This value, interestingly enough, is just above the input rate from a television display screen and below that from a cinema screen.

Communication in manufacturing organizations

Communication in manufacturing organizations can take place in many ways and through different means. We will now examine some of the most frequent cases to illustrate the structure, means and types of communications occurring in manufacturing organizations.

Communications within manufacturing organizations may be classified first according to the directions in which information flows. *Vertical* communication includes downward and upward information flows; these often correspond to the two portions of the control loop in a control system, as shown in Fig. 2.21. Communications can also take place along *horizontal* paths based on department-to-department and person-to-person levels. If we view the vertical communications as being basically associated with the control functions, then horizontal communications are mostly concerned with the detailed actions of actual input–output transformation. Figure 2.23 illustrates the operational structure of an actual manufacturing company. The important horizontal communications between different departments are clearly shown. Vertical communications of a reporting or instructional nature can also exist within each functional sub-system. Communication systems, therefore, can also have a hierarchical structure, the same as organizational systems and control systems. The detailed informational and functional structure of a typical manufacturing system will be examined later in Part Two of this book.

Communications within a manufacturing organization may also be classified according to how they are carried out, that is, whether they are based on *human* or *physical* activity. Human-activity communication takes place in the higher levels of the management hierarchy and is usually associated with the human activity control functions. This type of communication is based on a unique human heritage called 'language' – the character that most distinguishes us from other animals.

Certain conditions must hold for this type of communication to be effective. First, a commonly understood and accepted language must be adopted. On the surface this may appear to be obvious, but in fact such a language is a very important precondition for effective communication within an organization. This is particularly true if the organization concerned is international or has international dealings, as with many manufacturing groups today.

Second, a common vocabulary must be used within the organization. Organizations will almost certainly have different sets of terms which facilitate their own communication processes. Thus terms like CNC, MRP, FMS and CIM will all have certain meanings for us (hopefully). However, it is very unlikely that they will make sense to people involved in the other professions, such as ballet dancing. This situation is also observed within individual manufacturing systems. To avoid ambiguity, therefore, it is necessary for many organizations to provide specific definitions for the special terms used within their particular business environment as part of their organizational standards.

Finally, there should be a well-defined grammar and a set of rules by which the meaning of a message can be assessed. To illustrate this aspect of communication, let us examine how this may be done in English. For

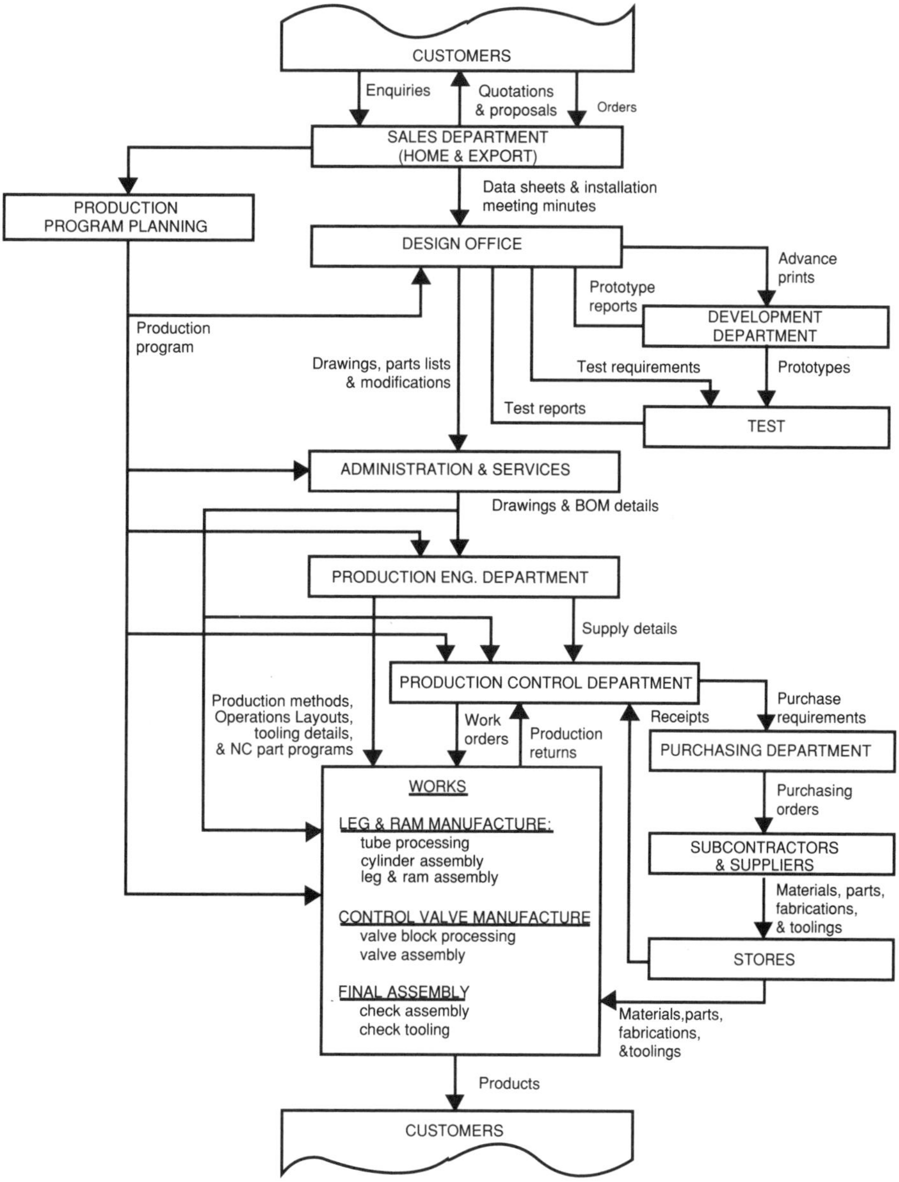

Fig. 2.23 Simplified activity flow chart of a company.

example, according to English grammar, we may use an adverb–adjective–noun pattern to form a phrase to describe a subject. However, the same subject described may be modified by the preceding words so as to produce phrases with quite different meanings. We might even be able quantitatively to measure the communication impact of such phrases. It could be possible to establish a numerical assignment for all adjectives and adverbs, as exemplified in the table below, so that those words with positive values are usually associated with favourable or desirable attributes, and vice versa. This numerical assignment could then be used to assess the impact on a noun modified by these modifiers by algebraic multiplication processes, as shown below:

1. very good student : $1.25 \times 3.09 = 3.86$
2. quite nice student : $1.05 \times 2.62 = 2.75$
3. rather ordinary student : $0.84 \times (-0.67) = -0.56$
4. extremely bad student : $1.45 \times (-2.59) = -3.76$

Adverbs		*Adjectives*	
slightly	0.54	evil	−2.64
somewhat	0.66	bad	−2.59
rather	0.84	wicked	−2.54
pretty	0.88	immoral	−2.48
(no adverb)	1.00	inferior	−2.46
quite	1.05	comtemptible	−2.20
decidedly	1.16	disgusting	−2.14
very	1.25	average	−0.79
unusually	1.28	ordinary	−0.67
extremely	1.45	charming	+2.39
		lovable	+2.43
		nice	+2.62
		pleasant	+2.80
		good	+3.09
		admirable	+3.12

It can be seen that the above phrases shift from one extreme at (1), which is a very desirable modifier of 'student' with a value of 3.86, to the other extreme at (4), which is a very undesirable description with a value of −3.76. Of course, in real life we do not necessarily have to assess the messages communicated through conversation and writing by using a numerical method such as the one shown above, because we have become used to evaluating the impact of such communications intuitively and automatically. However, consciously or unconsciously, we do have to assess them according to certain rules. The example given above illustrates one of the ways in which this might be done.

The above preconditions for effective communication can all be related to the principles of information theory, that is, to increase the capacity of the transmission channel and/or reduce the effect of disturbance and transmission errors. Take, for example, the need for a commonly understood and accepted language. Imagine what would happen in a meeting in a multinational corporation attended by associates from various countries who expressed themselves in different languages. The only way for the meeting to proceed would be to communicate through interpreters. Not only would the overall capacity of communication be severely limited, but also the accuracy of communication would be dependent on the skill and knowledge of the interpreters. Misunderstandings could easily occur as a result of mistakes in translation. Realizing the importance of this aspect of communication, many UK universities and technical institutions are emphasizing the importance of language teaching and learning. Some have actually initiated engineering courses with a foreign language component (usually French or German).

So far we have considered various aspects of communication that are

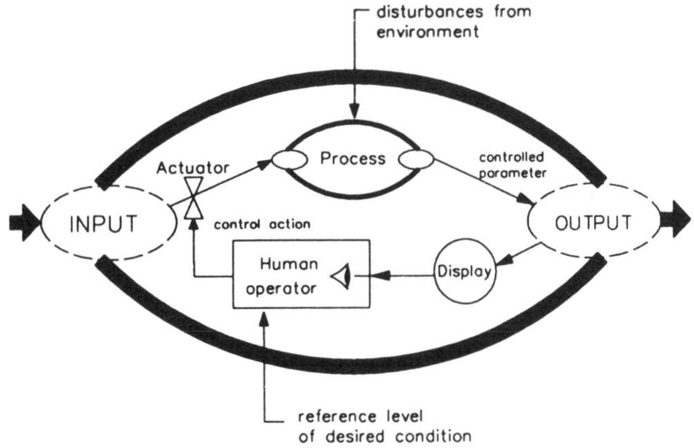

Fig. 2.24 Schematic representation of a human–machine communication process.

likely to occur at the higher levels of the control hierarchy. Within a manufacturing organization, however, a great deal of communication takes place between human operators and machines and even among machines. This kind of communication is more likely to occur at the lower levels of the control hierarchy, that is, at the physical activity level concerned with the operational aspects of the system. Examples of human–machine communication are a CNC machine operator reading the *x*- and *y*-coordinates of a cutting tool, and a production manager checking the progress of certain jobs from the visual display unit (VDU) of a computer system. Examples of machine–machine communication are a CAM computer which downloads numerical control data to a CNC machine through a direct communication line, the central computer system controlling the movement of an automatically guided vehicle (AGV) in an FMS cell through wired or wireless communication links, and computers connected by a network which 'talk' to each another by sending and receiving files.

The usual reason for the information input taken from instrument readings or from a VDU in a human–machine communication is to monitor the state of some variable within a process. The decision may then be taken to allow control to be maintained. A schematic, simplified block diagram of this process is shown in Fig. 2.24.

In human–machine communication, the operator should be efficiently integrated into the overall communication system with the necessary degree of accuracy and flexibility. Their needs must therefore be carefully analysed and accounted for. For example, many computer-based information-processing systems require the cooperation of human operators to achieve satisfactory operation. This implies that the means of presenting instructions to and receiving instructions from the human operator should be rapid, easy and error-proof. When we use a personal computer, it is clear that the input–output facilities such as mouse, tablet, window, pull down menu and yes/no questions are all based precisely on this idea. Because of the noise sources which may be introduced by human or machine, the design and implementation of a human–machine communication system which effectively integrates human and machine components and maintains a smooth and error-free operation is not always an easy task. The application of a manufacturing resource planning (MRP) system illustrates this situation well. MRP can be regarded as a typical human–machine communication system. Operators are required continuously to collect production information and input it into a computer system. These data are then processed by the system according to certain logical procedures so as to provide the operator with plans of material and capacity requirements. To date, MRP is one of the best-publicized approaches for dealing with the planning of multistage production systems. However, very often MRP fails to live up to expectations and many cases of failure have been reported. Although the logic of MRP itself is clear and straightforward, the calculation processes are easy and simple, and the software may be error-free and work well, there are many problems with other aspects that will cause difficulties in the implementation of MRP. For instance, O'Grady (1986) has given the following five reasons for the failure of MRP in practice:

- *Poor master production schedule.* The master production schedule acts as a driver of the MRP system. Any errors in the master production schedule will inevitably result in inaccurate output from the MRP system.
- *Poor stock status file.* The stock levels of particular components are used to calculate the net requirements. Errors in stock-level recording result in errors in net requirements.
- *Inaccurate lead times.* The lead times are used as a fixed entity to give the time-phasing inherent to MRP operation. In practice, however, lead times are likely to vary and the use of a fixed lead time may result in difficulties.
- *Inaccurate bill of materials* (BOM). The enormous task of inputting the BOM file for most manufacturers usually means that some errors in the file are likely to occur. Although these start-up errors are likely to be rectified over a period of time, the necessity of frequent changes to the BOM in a fast-changing technological environment often means that errors occur.
- *Out-of-date information.* In many MRP implementations data are recorded on paper and then entered manually via a keyboard into the computer files. The information that is on the computer may therefore be out of date. This problem can be reduced by the use of an effective shop-floor data recording system.

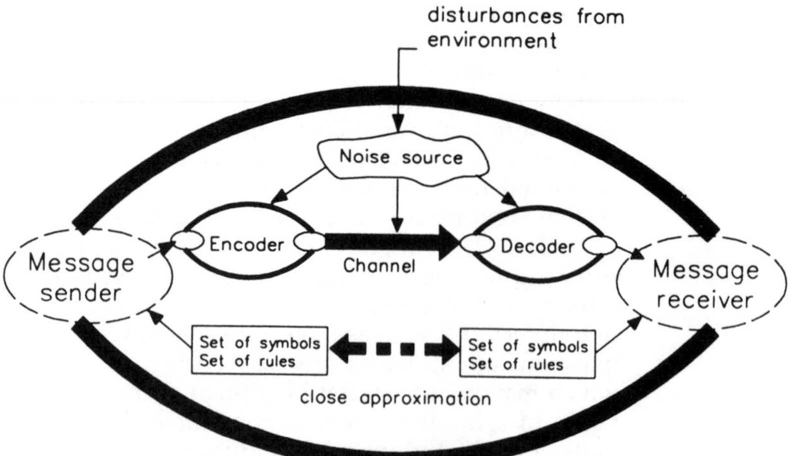

Fig. 2.25 Conceptual model of a communication system.

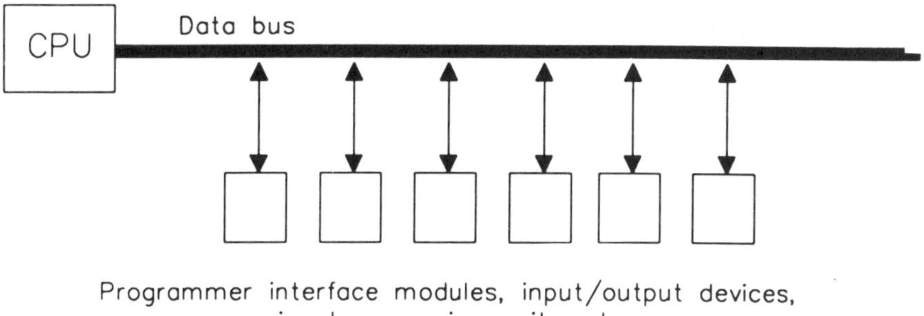

Fig. 2.26 Programmable logic controller (Williams, 1988).

Obviously, for such a human–machine communication system to work successfully, noise sources such as those listed above must be kept to the lowest possible levels.

The basic principles of information flow in machine–machine communication are summarized by the model shown in Fig. 2.25. Although the principles remain the same, the actual methods and details of technology may vary a great deal, ranging from telecommunications through wireless signal transmission and computer information networks, to programmable logic controllers (PLCs). A PLC, for example, is essentially a purpose-built microprocessor that is designed to handle a large amount of input and output, either digital or analog, and execute logic control over a number of cooperating devices, as shown in Fig. 2.26. PLCs have found a wide range of applications in manufacturing industry and are used in many forms of automation.

Advanced technologies have changed the face of many manufacturing organizations. Indeed, it may be said that the whole area of advanced manufacturing technology has been developed around the basic idea of data processing, information transmission, and communication integration. Communication integration is being introduced along both the vertical and horizontal dimensions within manufacturing organizations for planning, coordinating, monitoring and controlling the totality of manufacturing activities. The result is known as computer integrated manufacture (CIM). One of the most prominent features of CIM is the integration of manufacturing information systems through computer networks such as Manufacturing Automation Protocol (MAP), Technical Office Protocol (TOP) or local area networks (LANs). As a result of communications integration, the basic features of the CIM approach are, first, that manufacturing information is integrated by communication systems and by the use of shared databases; second, that manufacturing activities are integrated physically through computer networks; and third, that manufacturing functions are integrated to achieve a coherent approach to the coordination and control of the total operation.

In summary, therefore, the design and analysis of a manufacturing system will inevitably concern the structure and means of communication within the system. Whatever structure and means are chosen, it is always necessary to analyse the situation and design a system in such a way that noise and interruptions introduced into the communication process are kept to a minimum.

2.4 Description and use of a prototype system model

Equipped with a range of basic concepts involved in systems thinking – those of system structure, control and communication – we are now in a position to develop a prototype system model which may be used to help understand the real system involved in a particular problematic situation. In association with this prototype model, we can then formulate a structured procedure to identify and describe the actual systems content, and examine the conditions necessary for the effective operation and control of manufacturing organizations.

2.4.1 A prototype system model

To develop such a prototype system model, we must recall the systems concepts which have been presented in this chapter. Let us start from the top of the systems hierarchy. It is clear that the overall structure of this prototype system model should be similar to Fig. 2.4 – in general a system will reside within its wider system, which is in turn surrounded by the environment. Through our previous discussions on system hierarchy, we know that within the system boundary there may exist a number of sub-systems and/or components, but up to now we have not indicated what these should be. Our examination of the concepts of control, however, has put us in a position to explore further what characterizes the inside of a system. Using our knowledge of the concepts of control and communication, we are able to identify three sub-systems, together with the necessary channels for communication among them, which must be present inside the boundary if the system is to function properly (see Fig. 2.9). These three sub-systems should correspond to the essential components which are necessary for an effective controlled operation as outlined in section 2.3.1:

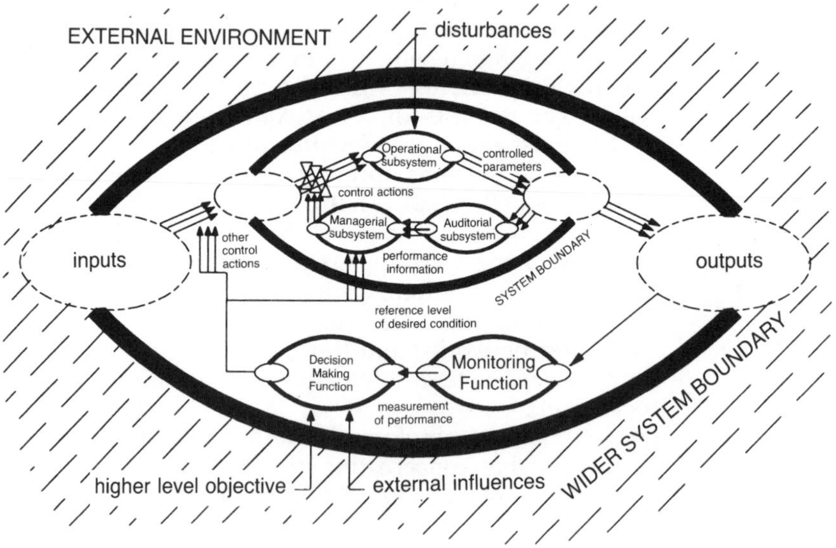

Fig. 2.27 A prototype system model.

- an *auditorial* sub-system which monitors system performances, and reports these to
- a *managerial* sub-system that is responsible for goal-setting and decision-making, thus exercising control over
- an *operational* sub-system consisting of the rest of the system components which actually carry out the specific transformation function of a particular system (turning system inputs into system outputs).

If we now update the hierarchical representation of system given in Fig. 2.4 accordingly, a prototype system model is obtained as presented in Fig. 2.27. This prototype system model tells us, at least at a conceptual level, what a properly designed and implemented system should look like. By using this model as a structural framework, we can make certain that during a system analysis project the important issues will be looked into, as shown later in section 2.4.3. When applying this prototype model, the hierarchical nature of systems should always be borne in mind. That is, the prototype model can be applied at any of the levels in the systems hierarchy.

2.4.2 A structured procedure for system identification

In association with the prototype system model, system identification and description is one of the essential skills which any system analyst must learn. First of all, it must be realized that to analyse the contents of a situation in terms of a system, the system recognized must have a purpose. Different people, depending on their involvement, will have different views on what the system in question does. For example, the systems view of the owner of a pub is very likely to be different from his customers' view (Fig. 2.3). The owner would probably see his pub as a 'profit-generating system' (Fig. 2.28) which converts his investment into profit. Again, the barstaff's view of the situation would be different from those of the owner and the customer. The staff will see the pub as a place where a living is to be earned, hence as a 'wage-generating system' which converts their time and effort into a wage packet (Fig. 2.29).

This indicates that the inputs and outputs to the same situation should be confined according to a particular viewpoint. For instance, although the choice of drinking companion is important to a particular customer, it is unlikely to matter to the landlord or the barstaff. When assembling a system, therefore, the analyst should first of all be very careful about the viewpoint

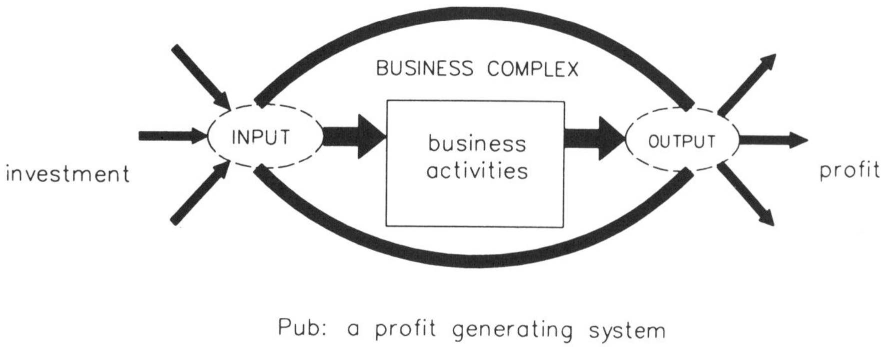

Fig. 2.28 A pub as viewed by the owner.

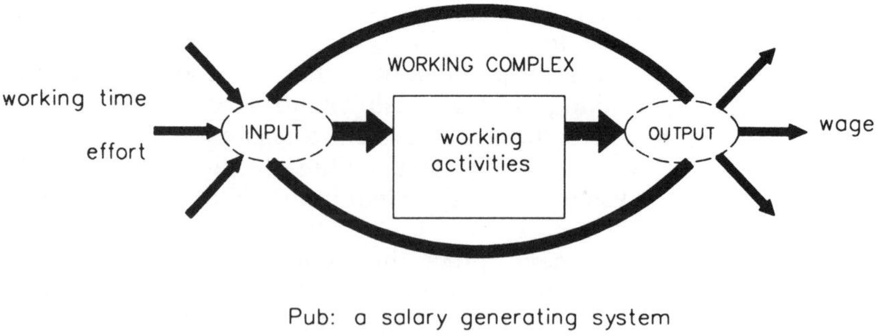

Fig. 2.29 A pub as viewed by barstaff.

he or she takes. A structured procedure which subdivides the process can help in practice to solve the difficulties in putting the pieces of a system together. We now describe one such analysis technique. This technique consists of five steps which we will call the 5W (what–why–whether–which–what) procedure:

Step 1: *Appreciation of a situation*. The first step is for the analyst to be aware of certain problematic situations and the activities or processes which may be involved. The questions associated with this initial stage will be of the '*What* is going on out there?' type.

Step 2: *Identification of the need for analysis*. Next we should check to see if there is any reason to study the situation. Before we go any further, we need a positive answer to a question such as '*Why* should I be interested?'.

Step 3: *Identification of the existence of systems*. Having observed a problematical situation and identified the need for commitment, it is only natural at this stage that the analyst should find out *whether* or not the contents of the situation may be described in systems terms. The answer to this question can be found by applying the formal system definition to the activities and/or processes involved, with particular attention directed to the essential system characteristics. A useful set of guidelines for this is given in section 2.2.1.

Setp 4: *Identification of systems of interest*. Next the analyst should separate out *which* systems are of interest, and identify the nature of these systems. Systems of interest should be separated from each other and from the rest of the situation contents, so that at a later stage they may be treated individually. This must be done by selecting, arranging and connecting together those components for the particular systems. The prototype system model presented in section 2.4.1 may be used to facilitate this assembling process – as a structural framework into which bits and pieces may be inserted. In fact two 'which?' questions will be involved in this step. In parallel with the separating and arranging processes which are intended to identify which groups of components constitute our systems of interest, another 'which?' question should be asked in order to clarify the nature of these systems. The answer to this second question should identify a system with an appropriate title, which usually gives a clear indication of its main function. It is important to appreciate that this step involves iteration of the two 'which?' questions, as shown below:

Which group of components? ↔ Which type/title?

Step 5: *Description of systems of interest*. Having acquired an adequate understanding of the structure, process, and behaviour of the systems involved through steps 1–4, the analyst is now in a position to describe the situation in systems terms to express *what* he or she is actually concerned with. The actual method used for describing systems does not really matter, as long as it is capable of expressing the important features conforming the system definition. Frequently diagrams such as that of the prototype system model are used for this purpose. Specially developed tools, such as input–output analysis (section 2.2.4), the $IDEF_0$ methodology (section 3.2), the object-oriented approach (Chapter 4) and the GRAI approach (section 7.2.1), are particularly useful when manufacturing situations are concerned (these will be described in detail in the following chapters).

It is apparent that this 5W procedure for system identification and description has been designed in close relation to the important system features discussed earlier in this chapter and summarized by the prototype system model.

2.4.3 Preconditions for effective operation of organizations

Similar to the 5W procedure, the purpose of examining these preconditions is to assist the analyst in understanding and analysing complex situations which exist in many manufacturing organizations. A checklist of such preconditions may be used, for example, to decide what sorts of action should be taken to improve the situation, by investigating the problems involved through 'What and where are they now?' and 'What and where should they be?' types of question.

When the systems concepts are applied to investigate the conditions necessary for the effective operation of a manufacturing organization, a checklist of the following kind may be developed:

From overall systems viewpoint (from the top down to the lower levels of the system hierarchy):

1. *Astute intelligence gathering*. The continuing awareness of what is happening in the wider business environment is one of the most important prerequisites for effective operation. Sufficient consideration must continuously be given to the influence of environmental factors such as changes in government policies and institutional regulations, economic and political climate, as well as customer requirements and technological development.
2. *Ability to cope with disturbances*. Sufficient resources and their flexible utilization are necessary to cope with both foreseeable and unforeseeable environmental disturbances.
3. *Coherent organizational and operational strategies*. The objectives or goals of the organization should be defined clearly and comprehensibly, in line with the statements of purpose supplied from its wider system. This applies equally down the organizational tree. That is, the strategies and

policies adopted at various levels of the organization must be coherent and in line with the overall organizational objectives.

4. *Adequate overall structure*. All the necessary parts as defined by the prototype system model must be in existence within the system boundary.
5. *Adequate sub-system structures*. The necessary sub-systems themselves should be designed and implemented properly so that they perform their intended tasks adequately. In particular, resources must be distributed rationally among the transformation sub-system and the related auditorial and managerial sub-systems in accordance with the overall business strategies, so as to achieve a balance between system performance parameters such as cost, quality and quantity.
6. *Effective communication*. This can in fact be considered as a precondition for many of the other preconditions listed here. Internally, the various sub-systems must have an effective means of communicating information in order to achieve effective control and policy implementation. Externally, close links must be established between the organization and its customers, its suppliers and any other relevant bodies to increase competitiveness.

From a more detailed viewpoint (the control and communication aspects of organization):

7. *Adequate understanding of the transformation processes*. To design and implement a control mechanism, the process to be controlled must be understood to a required level of technical detail including its inputs, outputs, flows, states, behaviour, and so on. Such a level of understanding can be obtained through analysis activities such as observation, modelling and experimentation. (A major part of the effort involved in systems engineering has been devoted to the development of analytic methodologies and modelling techniques which are used mainly for understanding and predicting the behaviour of various types of process. Some of these techniques will be discussed in later chapters of this book.)
8. *Adequate measurement of the transformation processes*. It is necessary to be able to measure relevant process parameters in an adequate manner. The key issues associated with this particular aspect include the choice of measurement, the measurability of chosen parameters, the frequency of and accuracy of measurement, and so on. Take the frequency of measurement: many of the control systems encountered in manufacturing operations use discrete process control through sampling – for instance, through periodic meetings. In this case the measurement of the controlled parameter must be taken within a suitable interval. Too long an interval may result in out-of-date information being used for decision-making, whereas too short a sampling interval can be prohibitively costly.
9. *Adequate communication channels*. If not properly designed and implemented, noise introduced into the communication channels can result in poor communication, which may at the very least have the same effect as poor measurement.
10. *Appropriate managerial sub-system*. The importance of this point is illustrated by the recently accepted view in business circles that a business organization can only be as efficient as its management is capable of making it. In systems terms, the managerial sub-systems must be capable of making the right decisions for the particular processes being controlled.

2.5 Conclusion

This chapter has covered a range of relevant systems ideas which provide the basis for our study of manufacturing systems design and analysis. It should be realized that, according to Checkland's (1981) map of the activities within the systems movement, our study of systems ideas is mainly concerned with the 'Problem-solving application of systems thinking in real-world problems' rather than the 'Theoretical development of systems thinking'. In later chapters, therefore, our attention will be focused on the ideas and techniques of the 'hard' and 'soft' systems approaches, both of which are indended to deal with real-world problems.

Further reading

Blanchard, B.S. (1981) *Systems Engineering and Analysis*, Prentice Hall.
Checkland, P.B. (1981) Science and the systems movement, in *Systems Behaviour* (3rd edn), Harper & Row.
O'Grady, P. (1986) *Controlling Automated Manufacturing Systems*, Kogan Page.
Sandquist, G.M. (1985) *Introduction to System Science*, Prentice Hall.

Questions

1. Describe some collections of elements which conform to the definition of a system given in this book. Identify their system components and system objectives. Give examples of collections of elements which do not conform to the definition and outline the reasons why they cannot be regarded as systems within the context of our study.
2. What is the wider system of a manufacturing company? How does this influence company operations? What are the environmental influences and their likely effects on the company? What are the emergent properties of a manufacturing company which its individual departments do not possess?
3. What are the possible effects of the Channel Tunnel project on its environment? As manufacturing systems engineers, what might we learn from this?
4. What are the differences between open and closed systems? Why is the distinction between these two types of system important to our study? Give examples of each. Also identify and contrast a static and a dynamic system; a continuous and a discrete system; and a deterministic and a stochastic system.
5. What is meant by 'control' and why is it needed? Summarize the basic principles of feedback control.
6. 'Delay' is usually an undesirable situation in our work and everyday lives. Outline the reasons for this in terms of control.
7. Give an example of feedback control existing in an organization with which you are familiar (refer to Fig. 2.9).
8. Discuss the role of sales forecasting in a production organization.
9. What are the fundamental differences between the management control system and the process control system of a plant?
10. Consider, using an example from manufacturing, how inadequate communication may contribute to a system's failure.

11. Identify the key communications in your organization. What are the sources of noise which may exist in this system? Also consider how the development of communications technology has affected the structure and operation of the organization.
12. Use the prototype system model of Fig. 2.27 to expand the system which you described in Question 7. Identify the influential environmental factors, the wider system (if any) and the three necessary sub-systems; describe the operational sub-system; indicate what sorts of data are obtained and how these are transmitted between the sub-systems; and indicate how such information is used by the managerial sub-system and how control is achieved.
13. Review sections 2.2.1–2.2.3. Take each of the five steps of the 5W procedure in turn, and find the related system features which are outlined in these sections.
14. Choose a situation from your work environment which contains problems and assemble the possible system(s) involved by applying the 5W procedure. Outline the answers to the questions associated with each of the steps.
15. The following checklist has been provided by one policy studies institute for the important issues of managing innovation in a manufacturing organization. An innovative business should:
    - understand the importance of innovation;
    - have a deliberate policy of research and development;
    - show resolve and determination in the development of ideas;
    - demonstrate accurate market assessment;
    - show aggressive investment planning;
    - have a good labour relations policy;
    - have an integrated decision-making system.

    Take each of the criteria on this checklist in turn and find the item in the precondition list (section 2.4.3) to match it.

Part Two
Manufacturing

3 A conceptual model of a manufacturing system

3.1 Introduction

This chapter provides a structured survey of the most important aspects of a modern manufacturing system, and develops a systems model for an important type of manufacturing organization. Reading through this chapter will be like taking a 'guided tour' through a manufacturing system. We will follow the route that a product takes as it goes through the system, from customer inquiry to final delivery. On the journey through this 'Manufacturing Park' we will visit various interesting places (or, in more professional terms, the functional areas) where advanced manufacturing technology (AMT) is being used to increase productivity. This chapter is a guidebook, with 'maps' to help you to find your way through the rather complicated structure of operations and anecdotal information to assist you to understand the basic concepts involved in various manufacturing functional areas.

The map, a conceptual model of the manufacturing system, will be presented in a systems description technique known as $IDEF_0$. This technique provides a useful way of specifying the information and organization structure within a complex manufacturing system. The method, the features and the advantages and disadvantages of this particular systems description technique will be reviewed in section 3.2.

Next, a particular type of manufacturing system, the order-handling-manufacturing system (OHMS), is identified, defined and analysed to reveal the general context of a manufacturing organization. OHMS is an ideal modelling subject for our purpose for a number of reasons. First, it demonstrates many of the key issues faced by the managers and engineers involved in contemporary manufacturing industry, such as the adoption of AMT in the areas of production planning and control, product design and development, and production-shop activities. Second, with continuously improving customer choice and the ever-increasing demand for variety, its modes of operation are becoming predominant in many industries.

A formal systems definition of OHMS will be developed step by step through interrelation and decomposition, using $IDEF_0$. To aid the understanding process, the structure and operation of a real OHMS company will be examined to illustrate the key features of the manufacturing system.

Although this chapter is not intended to provide a complete description of AMT, brief introductions to the relevant topics will be given where appropriate, enough for the reader to appreciate the basic techniques, the scope of these techniques and, more importantly, the roles they play within the overall manufacturing framework.

In summary, therefore, this chapter is designed to serve three main purposes: to illustrate how systems concepts such as hierarchy, relation, decomposition and control can be applied to describe and analyse the system of a manufacturing organization, by using a systems definition technique such as $IDEF_0$; to develop an overall manufacturing framework to which various manufacturing functions are fitted and related, so as to show what is involved; and to highlight the principles of relevant AMT topics and the problem areas involved in the structural, technical and operational aspects of manufacturing.

After studying this chapter, the reader should be able to:

- Develop a systems structure of a particular manufacturing organization, including its functions, its material and information flow together with its relationships and control structures.
- Identify the role and operations of various functions within the overall framework of manufacturing, including: sales and contract; production planning; product design and development; material and capacity requirements planning; production resource acquisition; production control; and production-shop activities.
- Understand the principles of the systems approach to production resource acquisition and the techniques of manufacturing planning and control, including: aggregate planning approaches, disaggregate planning approaches, and integrated planning approaches.
- Understand the principles and techniques of various AMT areas, including: factory layout; the machining centre and numerical control; robots and AGVs; machining cells and FMS; automatic and flexible assembly; CAD/CAM, CAE, CAPP, and CAPM; computer integrated manufacturing; and quality control.
- Identify the important problem areas and the associated key issues in modern manufacturing systems.

3.2 The $IDEF_0$ system definition technique

Early research into AMT revealed a need for a medium that would assist communication between systems analysts. As a result, methodologies were developed in an attempt to quantify and communicate system concepts. IDEF is one of the more successful outcomes of these attempts.

IDEF is one of the system description techniques mentioned in Chapter 2. The term 'IDEF' stands for 'Integrated computer-aided manufacturing DEFinition', or simply 'Integrated DEFinition'. It was developed by the US Air Force to describe the information and organization structure of a complex manufacturing system. To date, the technique consists of five basic levels: $IDEF_0$, $IDEF_1$, $IDEF_2$, $IDEF_3$ and $IDEF_4$. $IDEF_0$ is a technique that can be used to specify completely the functional relationships of any manufacturing environment. $IDEF_1$ is used to describe the relationship between data items in the environment, such that a relational database model may be specified. $IDEF_2$ is a simulation technique that can be used to investigate a system's dynamic behaviour. $IDEF_3$ is for process design, and the latest addition, $IDEF_4$, is an object-oriented approach to manufacturing software development. These techniques may be used independently. In this study, the $IDEF_0$ technique will be used to describe the functional relationships and specify the flow pattern of information and materials within a manufacturing operation.

$IDEF_0$ is a methodology for the static functional specification of a manufacturing system. It is used to produce a function model which is a structured representation of the functions of a manufacturing system and the flow paths of information and objects which interrelate those functions. It is by nature a 'top-down' approach. This type of approach exposes one new level of detail at a time, that is, it begins the description process by modelling the system as a whole at the highest level and then decomposing this model level by level to describe each of the sub-systems within the system hierarchy. In fact we have used such a top-down approach several times in Chapter 2, during our study of systems properties.

3.2.1 The $IDEF_0$ function block

The basic element of an $IDEF_0$ model is called a *function block*, as shown in Fig. 3.1. In such a model, the individual function blocks are linked together through the inputs, the outputs, the mechanism and the controls. The nature of each of these links can be specified: they can be either physical objects or information. When an input is utilized to create an output, a function will be actuated. The performance of the function is carried out through a mechanism and under the guidance of the control.

The inputs to a function entering the function block from the left are usually (but not necessarily) 'consumed' by the function to produce outputs. Raw materials are typical examples of these. The mechanism, represented by an arrow entering the function block from below, indicates the resources which are required to carry out the transformation process – such as tools, equipment and operators. All resources shown must be used as means to achieve the function. They only become part of the output if they are passed on by the function to support other functions. Finally, the controls which enter from the top of the block only influence the transformation process

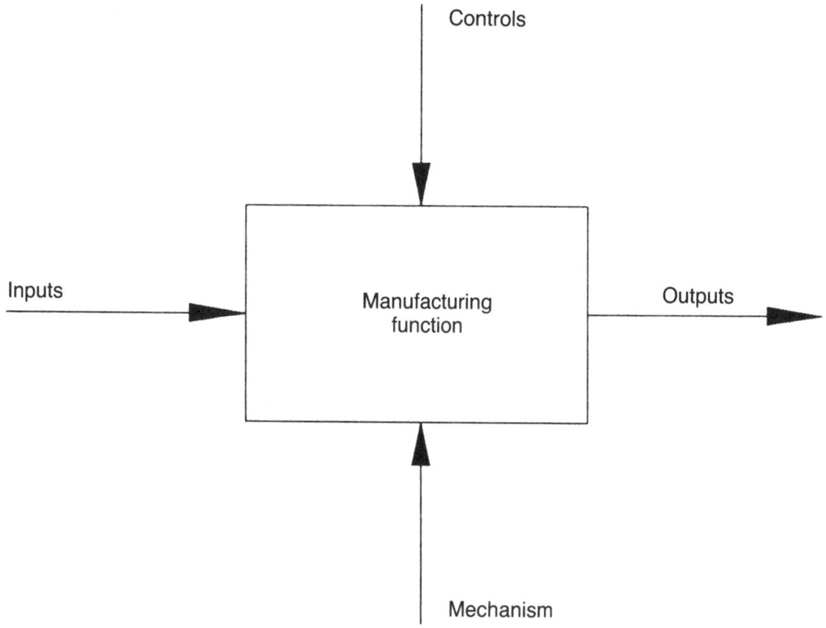

Fig. 3.1 Basic building block of an $IDEF_0$ model: the function.

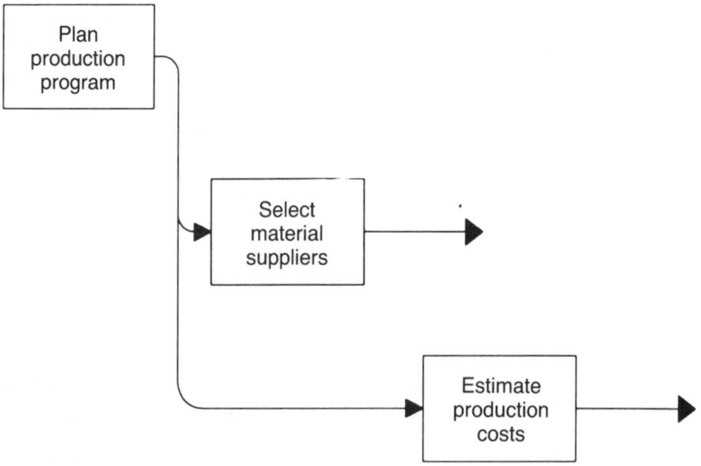

Fig. 3.2 Example of parallel IDEF functions.

and they will not be consumed or processed themselves. It is important to remember that all inputs, controls and resources must be used by a function.

One important point concerning any $IDEF_0$ model is that it is a *static* representation of the system, indicating functional relationships, but not necessarily specifying any dynamic aspects within it. Unlike a flowchart, for example, the flow paths linking function blocks in an $IDEF_0$ model do not indicate a specific sequence evolving over time. The input is not necessarily assumed to precede the particular function and more than one function in the model could be activated simultaneously, as shown by the example given in Fig. 3.2. Also, the frequency of function actualization depends on the particular system – a function may occur many times a second or only once a month.

3.2.2 Function decomposition

An $IDEF_0$ model can be expanded to any level of detail. A function block in the system can be decomposed into more detailed function blocks further down the structure hierarchy. The highest-level function block describes the main purpose of the subject system and the lower-level function blocks describe the supporting sub-systems which exist to serve the upper levels. At the uppermost level, a function block is usually labelled as function A_0, which represents the overall system objectives and system boundary. If A_0 consists of, say, four sub-functions, then these will be called A_1, A_2, A_3, and A_4. Each of these sub-functions, together with their associated inputs, outputs, controls and resources, may themselves be decomposed into the next level in the hierarchy. The sub-function blocks at the next level will be denoted A_{11}, A_{12}, . . . , A_{21}, A_{22}, . . . , A_{31}, A_{32}, . . . , and A_{41}, A_{42}, . . . , as shown in Fig. 3.3.

Although in theory an $IDEF_0$ model could be decomposed into lower levels any number of times, for convenience and clarity the level of decomposition should always be limited in practice to a manageable number. As a rule, it is recommended that each function at a given level should be only decomposed into between three and six sub-functions at the next level. On

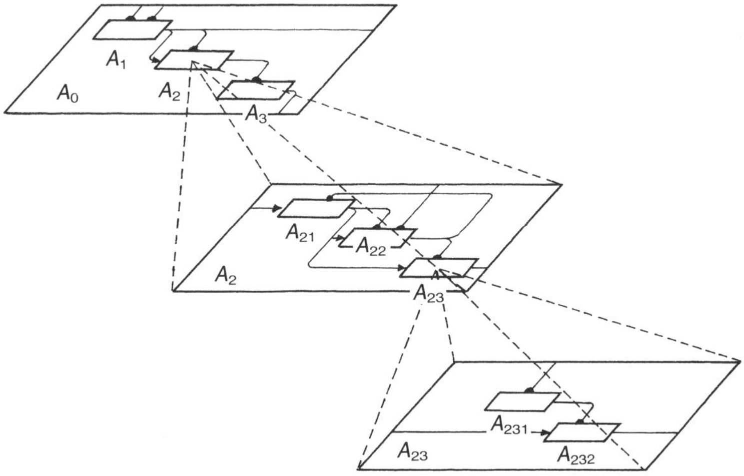

Fig. 3.3 IDEF function levels.

the other hand, if a function cannot be subdivided into at least three lower-level sub-functions, then the particular decomposition should not take place.

The presentation of these sub-functions indicates what is inside the original function and how these sub-functions are interrelated by the object and/or information flow paths. Each decomposition process, therefore, exposes a certain amount of new detail to a clearly bounded part of the model, providing a structured method for organizing lower levels of detail. Finally, a complete $IDEF_0$ model provides a comprehensive description of the functions and their relationships within the manufacturing environment being modelled, as shown in Fig. 3.3. This technique, therefore, provides a clear specification methodology for total system description at any desired level of detail.

3.2.3 Advantages and disadvantages of $IDEF_0$

The IDEF methodology has been utilized for many types of systems design situation and gained wide recognition as a powerful systems description tool, particularly in the field of manufacturing systems engineering. However, of all the currently available tools for system description, none can be said to be the best for all purposes. When choosing the tool for a particular investigation, the advantages and disadvantages of the options should first be examined. This, of course, applies to $IDEF_0$.

The major advantages of $IDEF_0$ include the following:

- It permits an effective, standardized systems communication method whereby system analysts can properly communicate their concepts.
- It permits a system to be described in as complete a level of detail as desired.
- It provides a mechanism for decomposing a function into a number of smaller sub-functions and verifying that the inputs and outputs of the function match those of its sub-functions. This allows many individuals to work on different aspects of the total system and yet be consistent in terms of final system integration.

- It has the potential to be used as an industry standard for manufacturing system design.

These features of $IDEF_0$ are very important in contemporary manufacturing industry, where substantial efforts are being focused on total system integration. Without a language which is understood by all the system analysts working in different parts of a company, communication would be more likely to be subject to failure (section 2.3.2), making successful design and implementation of a total integrated system much more difficult to achieve.

However, as a standardization technique, $IDEF_0$ has a number of disadvantages in terms of learning time involved, cumbersomeness, ambiguity of function specification and, perhaps most significantly, its static nature. Such a hierarchy-of-functions model does not explicitly represent the conditions or sequences of processing – these are usually described in the text of a function's descriptions, but not shown by the model itself. Consequently, a great deal of manual effort and interpretation may be required to identify the appropriate functions which should process a specific input and to verify their consistency.

Nevertheless, the $IDEF_0$ technique is a useful and flexible system description tool, because it assists the systems analyst to understand the functional and information structure within a complex system during investigation. Although initially intended for manufacturing systems definition, it can be used as a general-purpose tool and applied to many subject areas. Note, however, that although $IDEF_0$ is flexible enough to fit into many existing system analysis methods, it is itself only an aid for the analyst. The quality of the models developed is dependent upon the skill of the modellers. Like any other techniques, competence in using this method can only be gained through practice.

Having examined the basics of the $IDEF_0$ method, let us now turn our attention to the description of the systems structure of a manufacturing organization.

3.3 Structure of OHMS

In this section, we will use $IDEF_0$ to develop a conceptual model for a particular type of manufacturing system – the order-handling-manufacturing system (OHMS). The OHMS type of manufacturing system provides an ideal choice to illustrate the structure, context and problems involved in discrete manufacturing. This is because, from customer enquiry to production delivery, its activity chain highlights almost every aspect of a secondary manufacturing process.

This section will serve a number of purposes in our study of manufacturing systems: to illustrate how to use a technique ($IDEF_0$ in our case) to describe a complex situation; to develop a systems concept of manufacturing so as to understand the basic structure and functions of manufacturing systems; to identify some of the key problem areas within a manufacturing system in today's industrial environment; and to establish a bench-mark from which our study of manufacturing systems will evolve in later chapters.

3.3.1 Classification of manufacturing systems

Although the word 'manufacturing' generally refers to the activity of transforming raw materials into marketable goods, the variations involved are almost unlimited, as pointed out in Chapter 1. This is due to many influential variables such as: the type of product involved (consumer good, capital good, etc.); the nature of the product being manufactured (mechanical, electrical, biological, chemical, etc.); the technology involved (from simple to complex and highly sophisticated); the unit value of the product (from a few pence to millions of pounds); the nature of the market concerned in terms of size, maturity, competitiveness, and so on; and the method of manufacture being employed. These factors, together with many others, all influence the type and nature of the manufacturing system involved.

There have been many attempts to classify manufacturing systems according to a general framework. For example, they have traditionally been classified according to the method of manufacture being employed – for example, job (including project) production, batch production, continuous or flowline production, process production.

Shipbuilding and civil construction are typical examples of project production, where the product remains stationary throughout the production process, and the work crews, equipment and materials arrive at the project sites for assembly. The production tasks are usually distinct and performed infrequently – often only once.

An intermittent batch production system refers to the case where products are manufactured in batches. In this production system, the products may follow the same production sequence through a shop (a flowshop). Alternatively, the system may have the characteristics of a jobshop. In the latter

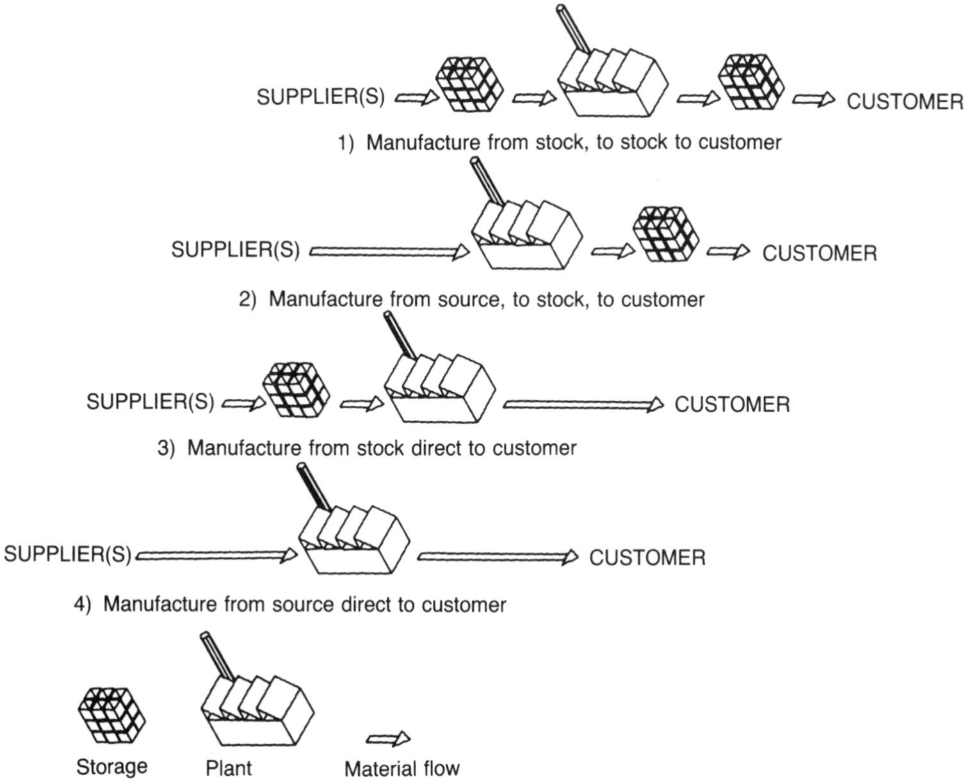

Fig. 3.4 Basic types of production system (based on Wild, 1989).

case, general-purpose plant and flexible workforces are usually employed for the production process.

Finally, continuous and process production systems refer to the manufacture of a small range of products in large quantities, giving scope for mass production. Specialized machinery and workers are employed for this type of production and the process is usually capital-intensive. Examples in this category include car assembly lines and oil refineries.

Production systems may also be classified according to their operating structures. Wild (1989) has defined four basic types of manufacturing to identify the nature of manufacturing systems: type 1, make from stock, to stock, to customer; type 2, make from source, to stock, to customer; type 3, make from stock direct to customer; and type 4, make from source direct to customer. These four basic manufacturing structures are shown schematically in Fig. 3.4. This classification system shows how each of the four different manufacturing types will handle the system inputs and provide for the future outputs. The first two types are also referred to as *make-to-stock*, the last two as *make-to-order* manufacturing systems.

The application of computer-aided production planning and control systems also has significant effects on the definition of production systems. Barder and Hollier (1986) identified six types of production system within the engineering batch manufacturing sector according to product complexity. The authors claimed that this classification can be used as the basis for assessing the effectiveness of production control systems. The characteristics of each production system are summarized as follows:

Type 1 production systems are characterized by technologically complex products with long lead times, where a large indirect workforce is required to support manufacturing operations (ratio of direct to indirect workers less than 1.5) and a very high proportion of the manufacturing cost is attributed to overheads. The percentage of components made in-house is low and, although the product cost attributable to bought-out items is only average, the value of stocks is very high, due to the need for a buffer against supplier lead times. Relatively few products are manufactured and these only to customer order. The large number of new shop orders per production period suggests small batch sizes. Typical companies in this category are the manufacturers of aircraft or complex electrical goods such as radar equipment.

Type 2 systems are characterized by a fairly high product complexity and a complexity of manufacturing operations similar to that of the type 1 systems. However, a greater number of products is manufactured. There is a similar dependence on suppliers, but the ratio of direct to indirect workers is greater than in type 1 systems (typically 1.5) because of shorter lead times and less complex products. The manufacture of machine-tools and switchgear are typical examples of this group.

Type 3 systems are characterized by a low complexity of product and of manufacturing operations. Several different products are manufactured, a relatively large proportion being for stock. Simple plants are involved in the production processes. The size of the workforce is small, with a high ratio of direct to indirect workers (greater than 1.5). Batches are large and throughput lead times short. Typical examples include the manufacture of low-complexity electrical equipment.

Type 4 systems are characterized by a low complexity of product and a very low level of complexity of manufacturing operations. Companies in this group manufacture a wide range of products, mainly to customer order.

Throughput times are short. Very few components are bought in and direct labour is a major element of product costs. Typical companies in this category will include small-size general engineering firms.

Type 5 systems are characterized by a higher product complexity and a higher complexity of manufacturing operations than those of types 3 and 4, but considerably lower than those of types 1 and 2. A fairly wide range of products is produced. A relatively wide range of companies could be classified into this group, for example, companies concerned with toolmaking.

Type 6 systems are characterized by a very low complexity of product and a low complexity of manufacturing operations. Companies in this group are usually large in workforce size and have a very high ratio of direct to indirect wokers (typically 2.5). Again a fairly wide range of products is produced, mainly for stock. Manufacturers of industrial batteries, industrial fasteners, and industrial seals are typical examples.

3.3.2 Overview of OHMS

Having examined the different types of manufacturing system, a sub-set can now be defined which will be studied more closely. This sub-set of manufacturing systems will be constrained by the company definitions of Wild's (1989) type 4 together with Barber and Hollier's (1986) type 2. That is, it will consist of those intermittent order-handling systems characterized by the production of customized capital equipment.

This particular type of manufacturing system, OHMS, involves make-to-order production operations and customer-tailored end products in the form of capital equipment. An OHMS operation demonstrates many of the key issues faced by the managers and engineers involved in today's manufacturing industry, such as production planning and control, product design and development and production-shop activities. Furthermore, with continuously improving customer choice and the ever-increasing demand for variety, these modes of operation are becoming predominant in certain industries. It is also evident from practical experience that production planning and control procedures can be more difficult to carry out when jobs are produced to order rather than for stock, because the operation is complicated by inherent sources of uncertainty. Nevertheless, proper production planning and control of the production operations of this type of manufacturing system are of the utmost importance, since they often involve a high level of investment. Inadequate management may lead to costly and ineffective use of resources.

A formal definition of OHMS will now be given using $IDEF_0$. We will refer to the structure and operation of an actual OHMS company to illustrate the key features of the OHMS operation. Even if the reader is not involved in an OHMS organization, the material to be presented should still be highly relevant, since it will highlight many of the problem areas which are typical in manufacturing.

The objective of an OHMS operation may be defined as the manufacture and delivery of goods of proper design and quantity to customers' specifications, with an appropriate guarantee of product quality and prompt delivery at an acceptable cost.

In order to achieve its objective, an OHMS organization must ensure that personnel with the appropriate skills, together with the necessary tools and materials, must be made available in the right place at the right time.

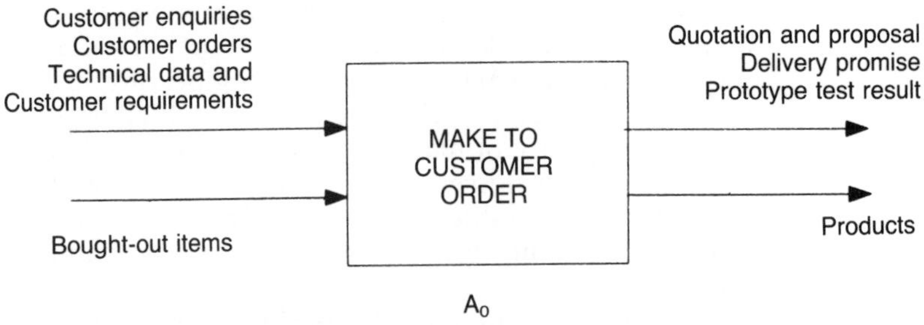

Fig. 3.5 Simplified overview of an OHMS operation.

Furthermore, all activities must planned, scheduled and carefully monitored and controlled to ensure that specific deadlines are met.

Following the $IDEF_0$ methodology, Fig. 3.5 provides an overview of an OHMS operation. Since we are mainly concerned with direct manufacturing activities, Fig. 3.5 represents a very simplified summary of a manufacturing organization. Indirect (though relevant) constraints such as financial, personnel, social and political factors in the operation will not be included in our conceptual model. Therefore, from customer enquiry to product delivery, the production functions throughout the organization may be grouped into the following four areas:

1. *Formulating the production plan*, involving the sales department, costing control, design office, and production planning department.
2. *Designing and developing the product to order*, involving the design office, development and testing departments.
3. *Gathering production resources*, involving the purchasing and stock control departments.
4. *Producing products*, including parts-producing activities, sub-assembly and final assembly operations. These involve the machine shops, assembly lines and the production control department.

The relationships between these functions are as illustrated in Fig. 3.6, which is the expansion of the OHMS model to its first level of decomposition:

A_0 MAKE TO CUSTOMER ORDER
- A_1 FORMULATE PRODUCTION PLAN
- A_2 DESIGN AND DEVELOP PRODUCT TO ORDER
- A_3 GATHER PRODUCTION RESOURCES
- A_4 PRODUCE PRODUCTS

Since an OHMS operation specializes in customer-tailored production, its production plan at the company level (or the *master production plan*) is to a large extent dictated by the sales contracts between the company and its customers. Contracting is, therefore, the first key function in its activity chain. Its effectiveness is vital. In today's extremely competitive market, the task of the sales department for the majority of manufacturers is very clear – to win enough orders so that the plant is sufficiently loaded. Whether an order can be won will depend largely on the overall effectiveness of a firm's performance. It has long been identified, for example, that the incoming rate of customer orders may in general be influenced by the delivery perfor-

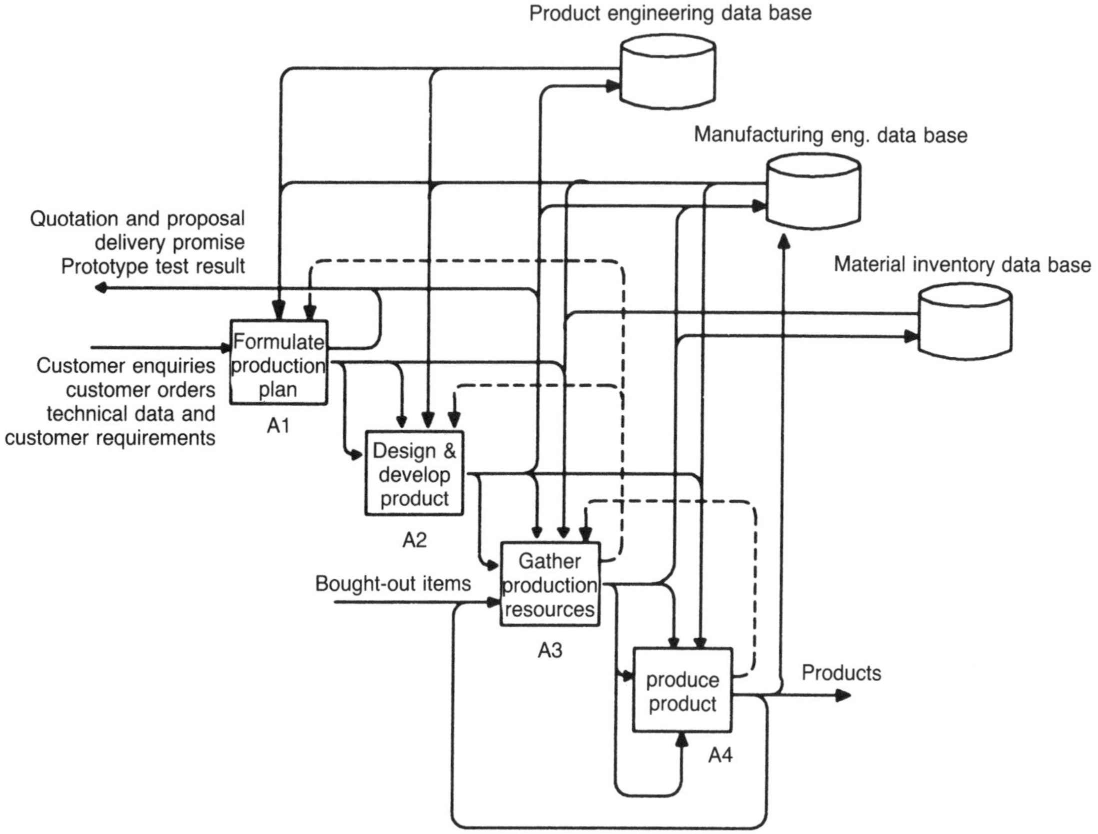

- - - -▸ Problem feedback/modification requirement

——▸ Functional information/material flow, e.g.: –
technical specification
production programme
drawings
production/modification instruction
status report

material
fabrication
assembly
tooling, etc.

Product eng. data base – design & technical information i.e.:
design standard, material specification,
drawing & parts list, etc.

Manufacturing eng. data base – production & cost information i.e.:
standard layout, methods & tooling information,
production lead time, production cost, & production capacity, etc.

Material inventory data base – material & supply information i.e.:
material status, material orders, material costs,
supply lead times, subcontractor & suppliers inf., etc.

Fig. 3.6 Simplified $IDEF_0$ model of an OHMS operation at level A_0.

mance of the manufacturer, the product price, and product quality (for example, Forrester, 1975).

The 'tailoring' nature of the production process inherent in OHMS is one of the most important features of this sub-set. Significant design/redesign, development and testing activities are involved to meet a customer's particular requirement in the handling of each individual contract. OHMS organizations need to cope with non-standard elements, frequent change of design, inconsistencies in operation lead times, and frequent late modifications required after the work orders have been issued to the work shops. This makes the development and implementation of a formal (computer-aided) control system for such an organization a very difficult task. Consequently this tailoring feature represents a considerable problem in today's market conditions. Despite the successful development and introduction of formal production control methodologies in many manufacturing systems (especially those of make-to-stock type), many companies in the OHMS subset still rely on informal, or partially informal, operating systems because of the difficulties outlined. Informal systems do not guarantee effective operations and can break down easily in practice. Practitioners often referred to them as 'panic' systems.

Another feature of the operation of OHMS is that the availability of necessary raw materials and bought-out items (for example, semi-processed materials, components, sub-assemblies and fabrications) is important if the company is to provide acceptable delivery lead times. A tailored product means that tailored specifications of product materials and/or other bought-out items are likely to be required. This will always involve a longer than usual supply delay. In order to shorten the overall lead time and so become more competitive, some companies have adopted the policy of purchasing the key materials and/or other specially required bought-out items for a possible contract some time before the order is actually confirmed. Since these items can take a substantial amount of working capital, very high risks have to be taken by the management.

Only after all these pre-shop-production activities have been carried out can the production processes themselves take place in the workshops. The workshop activities can be similar to a make-to-stock, batch production, jobshop or flow-shop. However, differences do exist. The most significant feature of the make-to-order production operation in question is the low commonality of the products involved. Utilization of the workshops in such systems tends to be low because of the frequent small failures and difficulties that occur during various stages of the production process, due to the 'learning curve' effect and/or design errors. These failures and difficulties are one of the major causes for frequent design modification requests and for the inconsistency of production lead times.

Therefore an OHMS activity chain will involve both pre-shop-production and post-shop-production, as well as the general workshop functions, as follows:

Pre-shop-production functions:
- design and development of product
- material and bought-out items acquisition

Shop-production functions:
- workshop activity – parts producing

Post-shop-production functions:
- sub-assemblies and product assembly
- product testing

Having examined the basic structure and functions of an OHMS operation, it may helpful to examine an actual company. CLD Equipment Ltd is a UK-based capital equipment manufacturer, whose organizational and operating structure conforms to the typical characteristics of an OHMS organization, as described in Box 3.1.

Box 3.1 CLD Equipment Ltd – An Example of an OHMS Operation

General profile

The company, established in 1948, is a member of a well-known UK industrial group which employs a total of 17 000 personnel in the UK and in overseas subsidiaries. The group consists of 45 companies with a total floor area of 325 000 m^2, and had a total turnover of £463 million in 1985. It has extensive interests in aerospace and defence equipment, industrial hydraulics and electronic equipment, and also in the mining industry.

CLD Equipment itself, which specializes in the production of heavy-duty, power-operated roof-support systems for underground coalmining, is the largest of the companies within the Mining Division of the industrial group. The company employs approximately 1500 people. Together with its two much smaller sister companies, also based in the UK, which often work in collaboration with it on the same contracts, the company achieved a total turnover of £154 million (£71 million from overseas markets) in 1986.

To understand the nature of the company's products and market, it is necessary to give a brief description of the techniques used for underground coalmining. In modern coalmining industries all over the world, three basic techniques are widely used underground to cut the coal from the coalface and at the same time to guarantee the safety of the personnel and equipment. These coal-cutting techniques are known as the *room the pillar* method, the *shortwall mining* method and the *longwall mining* method. With the shortwall and longwall methods, coal is extracted along

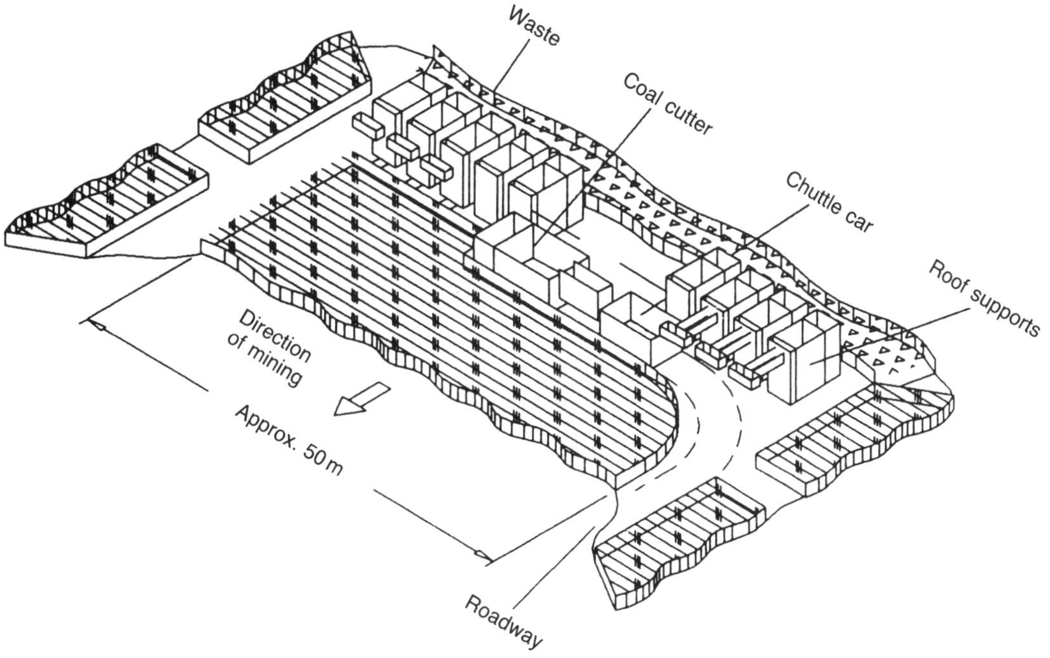

Fig. 3.7 Shortwall mining.

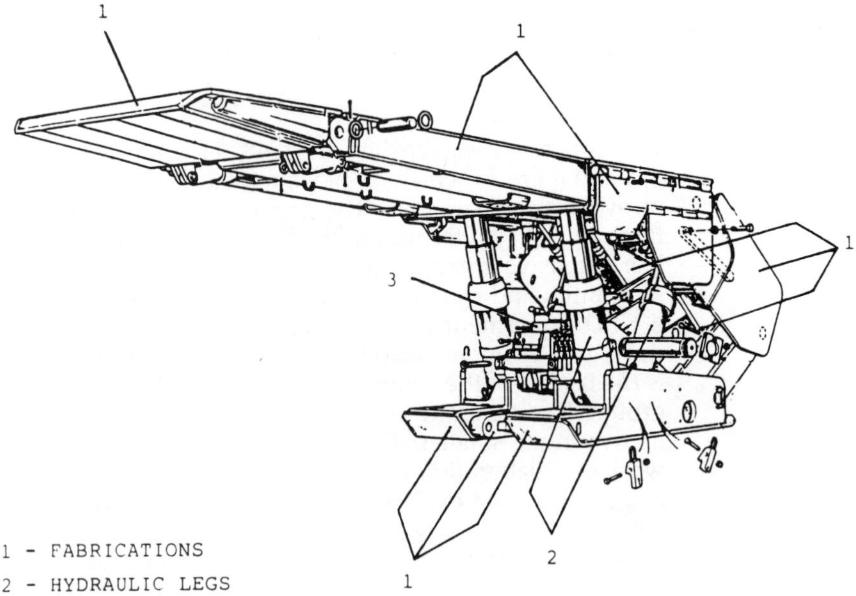

Fig. 3.8 Structure of a hydraulic support unit.

a coalface whose roof is supported by a series of specially designed self-advancing, hydraulically powered roof supports, as shown in Fig. 3.7. A typical modern face layout will consist of 150–200 such roof supports. As a result, longwall installations are very expensive – the cost of a typical installation of hydraulic supports on a 200 m long, 2 m thick coalface will be of the order of £2 000 000. The company was formed in 1948 to market the then newly developed hydraulic support system. Since then, it has developed a comprehensive range of hydraulic supports for both short-wall and longwall mining.

Company manufacturing operations

The structure of a typical hydraulic support unit can be seen in Fig. 3.8. A support unit is constructed from three major parts – the welded fabrications, the hydraulically powered legs and rams, and the hydraulic valve control system. A large proportion of the hydraulic components are standard items bought out to meet the requirements of the company. The manufacture of fabrications is carried out by sub-contractors. The company itself concentrates on the following key manufacturing activities:

1. the design, development and testing of the equipment for each new installation (two prototypes are built and tested for each contract);
2. the manufacture of the hydraulic legs and rams for each installation, including machining, welding, plating and polishing of the steel cylinders and assembly and testing of hydraulic legs and rams;
3. the manufacture of the valve systems for hydraulic control, including machining and processing of valve blocks and assembly and testing of valves;
4. the final assembly and testing of the products.

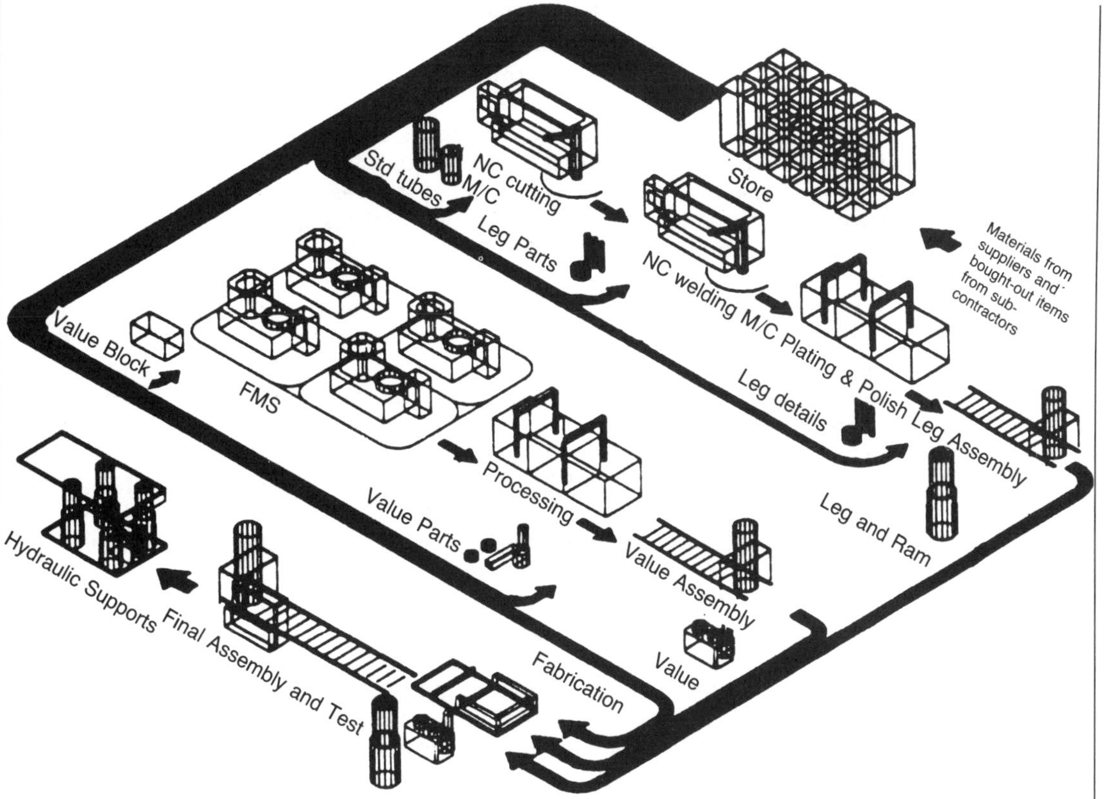

Fig. 3.9 Flow pattern of components, and company structure.

The flow pattern of the components and the structure of the company are shown graphically in Fig. 3.9. The production management of the company always attempts to carry out manufacturing activities in an organized manner. If everything always went well, then each of these activities would take a more or less fixed amount of time to complete, and the whole manufacturing process would be a fabric composed of the above separate, yet interdependent functions. However, in real life things are not so simple. Usually there are many orders for different installations going through the system at the same time, each with its own target delivery date. Consequently, hundreds of manufacturing activities are happening simultaneously, many interrelated. The occurrence or non-occurrence of any one thing could affect many other things, or even the whole system. It is easy to appreciate how challenging it is for the management to run the organization smoothly.

Company manufacturing facilities

Since the competition in this market is very fierce, both from domestic and overseas rivals, the ability of the company to react quickly to engineering changes and market demand is essential. With an average delivery lead time of five or six months on orders worth £2 million, the company simply cannot afford to waste time in its production schedule. In order to increase its ability to react quickly and reduce lead times, it has made significant investments in the latest manufacturing equipment such as CNC machines, dedicated robotics, CAM, CAD and even FMS.

The company's CAD facility offers the ability to create two-dimensional engineering drawings and three-dimensional models, giving a professional presentation of company designs. As a result, both the response speed to customer enquiries and the quality of design quotations have been increased. The use of this CAD system assures customers that roof-support designs developed by the company take account of their special mining and operational requirements. It also facilitates the process of prototype design modification if this is required. The CAM system installed in its production engineering department is used for a wide range of tasks, including process planning and NC part programming, as well as jig and tool design. In addition, a bill-of-material (BOM) system has been developed in-house to control the key materials and bought-out items.

The majority of the leg-manufacturing activities are carried out with NC machines, dedicated automatic machines and robots. In addition, an FMS has recently been installed in the company. The system costs nearly £4 million, and has been fully operational since early 1986. This FMS is used for machining the very complex brass blocks of hydraulic valves. It comprises four CNC machining centres connected by an AGV material-handling system. The principal benefits of this system to the company depend on the direct and indirect cost savings, reduction in lead times, rapid response to production program changes, and real-time management information on the manufacturing of valve blocks.

From the above description it can be seen that the company has allocated a substantial amount of capital to deploying the most sophisticated technology available, resulting in higher efficiency and flexibility in some of its activity areas. However, most of these benefits are localized. Many managers and engineers in the company refer to these sophisticated systems as 'technology island', because their full potential has not yet been utilized.

3.3.3 A_1: Formulating the production plan

Having developed an overall picture of the OHMS manufacturing system, let us now examine in more detail what is involved in each of its functions.

Planning for production is the first of the four parts of the manufacturing activity, shown as function block A_1 in Fig. 3.6. For an OHMS operation this planning consists of two major tasks, as listed below and shown in Fig. 3.10 (although it is recommended that in real applications an IDEF function block should be broken down into between three and six subfunction blocks, for illustrative purposes we will not necessarily follow this rule):

A_1 FORMULATE PRODUCTION PLAN
- A_{11} Sale and Contract
- A_{12} Plan Production Programme

A_{11}: Sale and contract

For OHMS manufacturing companies, the sales procedure is usually as follows:

1. A potential customer calls the company to make an enquiry, specifying technical as well as other particular requirements.

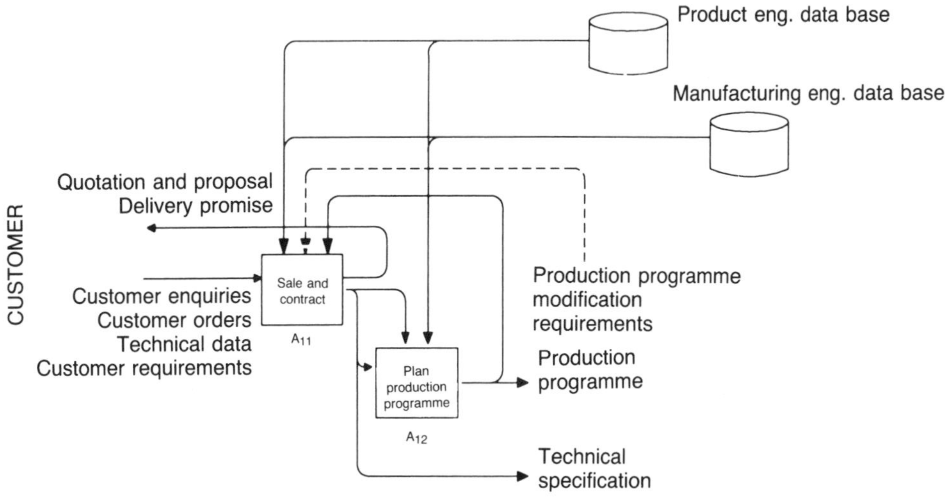

Fig. 3.10 A_1: Formulating the production plan.

2. In response to this customer enquiry, the manufacturer will prepare a proposal (quotation) including technical specifications, price and delivery date.
3. The potential customer will carefully examine the proposal in terms of quality, price and delivery date, probably comparing it with alternative proposals provided by competing companies.
4. If not satisfied, the customer will either disregard the proposal and approach another company, or he will contact the manufacturer again and ask for modifications.
5. Through a number of iterations of stages 3 and 4, a mutually acceptable agreement may finally be reached, at which point the customer issues an order for the agreed equipment.

It is important to recognize that, from a systems point of view, the customers act as the wider system to the manufacturer. The company must take note of its customers' objectives in order to maintain a reasonable relationship between itself and its wider system. It is the task of the sales department to satisfy the requirements of the wider system and in doing so to secure the inflow of resources by receiving orders from customers.

For the sales department to operate efficiently, necessary information about current workload (production schedule), as well as technical and production information, should be readily available. This information may be obtained by accessing the company management information system, as shown in Fig. 3.10. In reality, demands cannot always be met and not all orders can be accepted. The sales department must balance demand against its technical, capacity and cost information to determine the sales plan which best satisfies the requirements of both customers and company.

A_{12}: Plan production programme

The objective of this block, also known as *aggregate planning* or *master scheduling*, is to produce a production programme at the aggregate level, so as to meet the total customer demand over a relatively long period of time (often three months to a year). In the make-to-stock manufacturing environment, such a production schedule is mainly based on forecast, while in make-to-order companies the orders already received from customers to a large extent determine future workload, although some forecasting may still be necessary.

The major output of this function is called a *master production schedule*. This expresses the overall plan in terms of specific end products together with their technical specifications and promised delivery date. Relevant information about the type of product, the number of units of product involved and the due date associated with a contract should all be provided in such a production programme. In reality, however, things will not always happen this way. As a result, feedback links must exist between this function block and the following function blocks. This is included in Fig. 3.10 and labelled as 'Production programme modification requirements'. Requests for programme modification may arise for a number of reasons, such as material acquisition problems, capacity acquisition problems, or production problems.

The importance of adequate production planning and control cannot be over-emphasized. This is reflected by the fact that production planning and control have been the major topics dealt with in production and operations management. Many publications are available in this area. A survey of the techniques used to aid the decision-making processes for production planning and control is presented in Appendix A.

3.3.4 A_2: Designing and developing the product to order

Following production planning, product design and development is the next manufacturing function in the system. This function can be broken down into two sub-function blocks, with the second broken down further into three activities:

A_2 DESIGN AND DEVELOP PRODUCT TO ORDER
- A_{21} Control Design and Development Process
- A_{22} Develop Prototype

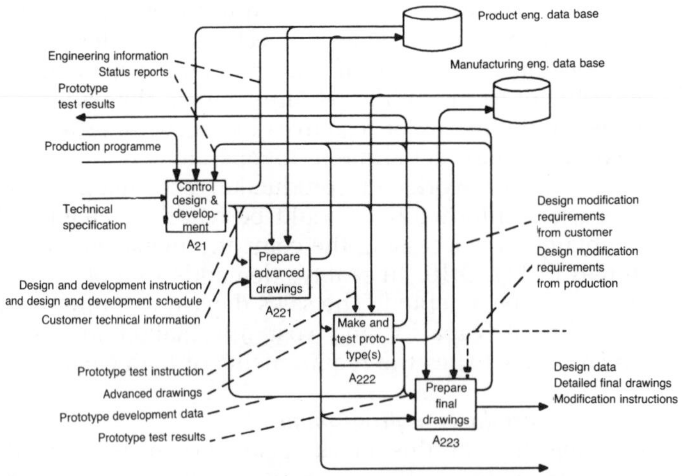

Fig. 3.11 A_2: Designing and developing the product to order.

A_{221} prepare advanced drawings
A_{222} make and test prototype(s)
A_{223} prepare final drawings and part lists

These functions are described, using the $IDEF_0$ format, in Fig. 3.11. They can now be examined in more detail.

A_{21}: *Controlling the design and development process*

Administration of design project

As shown in Fig. 3.11, following the established production programme, the next major task of an OHMS operation is to produce a product design which is tailored to specific customer needs. By nature the tasks of design and development are project-specific and as such they should be properly managed. At this point, it is necessary to review some of the systems concepts introduced in Chapter 2. It was shown that the first requirement of any efficient operation is an efficient administrative function, which will first set goals and then monitor the performance of the operational sub-system. When applied to the present case, this principle tells us that for efficient project management there should exist a project administrative function (shown here as function block A_{21}, which can, of course, be further broken down to expose more detail of its structure if required) which has the following two major responsibilities.

Production of detailed instructions for design and development

Under the guidance of all the relevant information provided and from the product specification and customers' requirements, this function should create a detailed design and development schedule for each individual production order contained in the master production schedule. These detailed procedures become the work plans of the design teams concerned and also act as milestones against which the performance of the design teams will be measured. When compiling such work plans, the manager or planner should first attempt to establish the extent and nature of the design and development work necessary by asking questions such as:

- What is the difference between this and the previous design?
- Should the work be based on a totally new design, or can it be simply a modification of a previous concept?
- How difficult is the task involved?
- What is the amount of work involved?
- Which task team is suitable (or available) for such a project?

Following these, a clear and precise description of the desired product should be created. If this fails, customers' requirements will not be satisfied by the resulting product. However, there are also certain other important considerations (of which the selection which follows is not exhaustive) that affect detailed procedure planning of this kind.

Serial dependency implies that the occurrence of certain activities is dependent on the status of certain other activities. For instance, before various tests can be carried out on a prototype, the prototype must be made available. This means that all the activities needed to produce a prototype, such as the preliminary design and prototype-production activities, must be completed first.

Resource dependency means that the capacity of the available resource will to a large extent determine the duration of project activities. The capacity of the resource will affect activity durations in two different ways. First, it determines the length of the processing time required for each activity; second, different activities may have to compete for the same resources, making it necessary to add a certain amount of waiting time to the total throughput time.

Production time considerations are concerned with maintaining the overall production schedule. By nature, product design is a time-consuming business. Measures should be taken so that the throughput production time of an order is kept to a minimum. This may need interactions of the design function with other manufacturing functions. For instance, some of the bought-out items needed for production may have long lead times. Orders for these may have to be issued early – perhaps even before the whole design process is complete – so that the overall production lead time can be kept as short as possible. Figure 3.12, which illustrates the relationships of the design functions and their associated production activities in CLD Equipment Ltd, provides a good example of this consideration. As can be seen, the lead times of the big steel tubes used by the company for hydraulic leg and ram manufacturing are exceptionally long – on average 36 weeks. If orders for these steel tubes were issued when the final designs of leg and ram became ready, then unacceptably long lead times for the end equipment would result, because the delay caused by the legs and rams would halt the whole production process. In order to avoid this, the company has adopted a steel tube acquisition practice in which a provisional order is issued to the tube manufacturers as soon as the necessary specifications become available. With these provisional orders the suppliers (also OHMS manufacturers in this case – an example of what is known as a *supplier chain*) will only be given the specifications on the diameters of the tubes and the approximate quantity required. Based on this information, the suppliers can carry out their production preparation process and start producing steel tubes to the specified diameters. The specifications of tube lengths will become available only when the final designs of legs and rams are ready. At this point specific tube cut instructions can be given to the suppliers and the tubes are then cut to the required lengths. As can be seen from Fig. 3.12, this practice cuts the throughput production time of the company by approximately nine weeks.

Monitoring project progress and updating work schedule

Once the detailed procedure for a project is planned and issued to the appropriate team, the administrative function should oversee the progress of the project, closely monitoring performance. This can be accomplished through regular review meetings, or by continuously receiving progress reports from the design team. Unforeseen problems should be quickly detected and promptly dealt with. Detailed schedules should be updated in time to overcome the practical difficulties. However, attempts should always be made to modify the plans in such a manner that the new versions will still comply with the master production schedule.

In summary, administration (or management) is the first of the functions which make up product design and development. As in any other operation, management plays a vital role in the successful completion of the tasks involved. The discussion has highlighted the importance of information processing and communication. As shown in Fig. 3.11, a management information system (MIS) should be made available with data bases containing

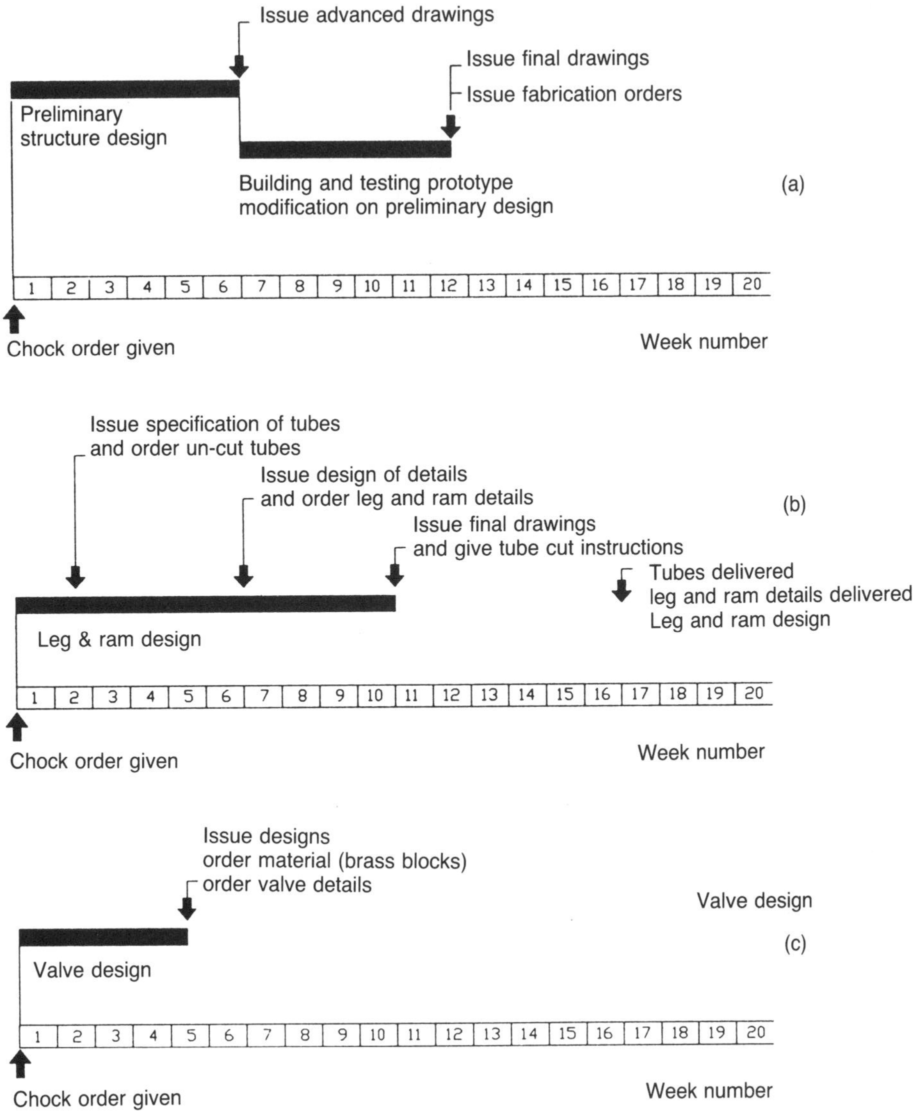

Fig. 3.12 The design function and its associated activities. (a) Structure design, (b) Leg and ram design and (c) Valve design.

relevant information, readily accessible by the key personnel and easy to analyse for the purpose of making decisions. The maintenance of such a usable MIS depends on the availability of appropriate computer hardware, software, communication channels, as well as on the competence of the systems analyst. Figure 3.11 also illustrates that the provision of a proper feedback mechanism is also important in ensuring a successful design function.

The fundamentals of effective project management and the tools available are reviewed in Box 3.2.

Box 3.2 Project management

It is clear that, by nature, a product design function involves project-based activities – clearly defined activities which are planned for a finite duration and carried out by a combination of resources pulled together to achieve a specific goal. Project management is an art (or science?) that is closely related to the ideas of the systems approach and is concerned with the efficient execution of project-based functions. Once the project objective and project tasks are defined, the elements of the process may be identified as follows:

- assign resources to accomplish the tasks (*goal-setting*)
- check task progress (*monitoring*)
- analyse and re-schedule the remaining tasks (*control*)

These should be done in such a way that the project goal is achieved in the shortest possible time and at minimum cost.

The main techniques used to plan and control complex projects are the program evaluation review technique (PERT) and critical path analysis (CPA). Similar to a Gantt chart (an example of which is shown in Fig. 3.12), a PERT chart is a visual scheduling aid which shows the sequence and timing of project tasks. However, PERT is a more precise method for scheduling project activities. It not only illustrates the interrelationships of the tasks required to accomplish the goals of a project, but also reveals, as the project progresses, which task chain within the network is the critical path – a potential bottleneck. PERT highlights those activities to which special attention should be paid so they do not delay the entire project. Figure 3.13 gives an example of such a PERT network with its current critical path highlighted.

The techniques used in modern project management are the result of evolutionary changes over many years. In the early days, project man-

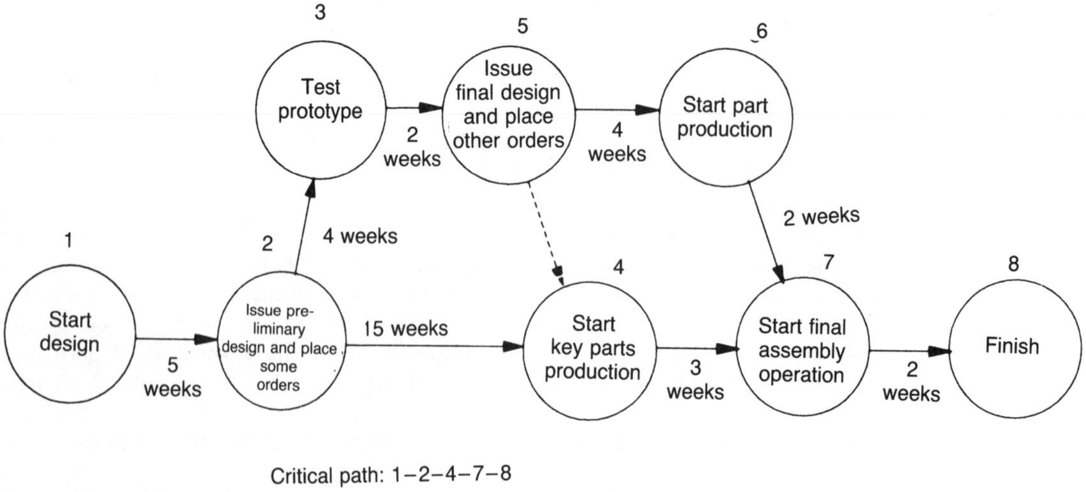

Fig. 3.13 Example of a PERT network.

agers relied entirely on drawing Gantt or PERT charts using a drawing board and ruler, or at best with a specially designed template. Analysis of critical path and resource allocation was also carried out by using hand calculation methods. The more complex the project, the more difficult the planning and control and hence the greater the potential for error. As in many other areas, however, the advent of computers has revolutionized the tools available to tackle the complex tasks involved in project planning and control. Nowadays, specially developed software, with sophisticated analysis procedures and graphic aids, allows complex projects to be managed with an ease and efficiency unmatched by the pencil-and-paper methods. What is more, such power can easily be brought to your desk – PC-based project management software is commercially available at a cost of no more than a few hundred pounds.

These computerized PERT programs not only allow a large number of tasks to be handled with ease, but also provide useful facilities such as interactive graphics, flexible network creating and editing, project data base development, automatic task status display, flexible resource scheduling, variable resource work hours, real-time scheduling with resource calendars, optimal resource utilization with 'what-if' analysis, resolution of conflicts through schedule staggering, resource under- or over-usage analysis over multiple projects, resolution of resource conflicts across multiple projects, resource cost control, advanced report generating, and data transformation to and from other software packages.

There are a number of PC-based project management packages available (SuperProject Plus, for example, supplied by CA Computer Associates Ltd). Demonstration disks can usually be obtained free of charge from the suppliers, allowing one to explore their usefulness. They will give a good demonstration of some of the features listed above and perhaps be enough to whet one's appetite.

A_{22}: Developing the prototype

Once the design project of a production order is assigned to the appropriate term, work may start as dictated by the schedule. As shown in Fig. 3.11, the main inputs to this function are the product specifications, transmitted from the sales function and through design management. The main output from it will be the complete final design of a product, tested and approved to be both functionally and characteristically satisfactory according to the specifications. The final design will provide a complete parts list, together with detailed drawings and specifications for every part, sub-assembly and final assembly involved. Also included should be any other relevant information such as production and test instructions.

In reality, few design tasks start completely from scratch. Most of the design processes involved in a manufacturing company will be based on familiar ground. For example, from prior development experience, a company may have assembled a library of existing designs. A new product development task may only involve modifications on some of these existing designs, or it may be modelled on these designs. Box 3.3, for example, illustrates the range of designs which has been developed over the years by CLD Equipment Ltd. This again highlights the importance of an efficient MIS – since a manufacturer's collected prior experience is one of its most valuable assets, the means to store, retrieve and analyse this experience must be made available.

Box 3.3 Product Range of CLD Equipment Ltd

As an OHMS organization, the company adopts a 'make-to-customer-requirements' policy. A new or modified installation is designed, developed, tested and then manufactured to each customer contract, such that all the products manufactured are tailored to suit a specific operating environment. Although each installation is unique in its detailed design, the equipment may be classified, according to structure, into the following types: frame supports, chock supports, and shield supports of four basic types – chock shield support (four legs), two-leg shield support, four-leg shield support, and revlem support (see Fig. 3.14).

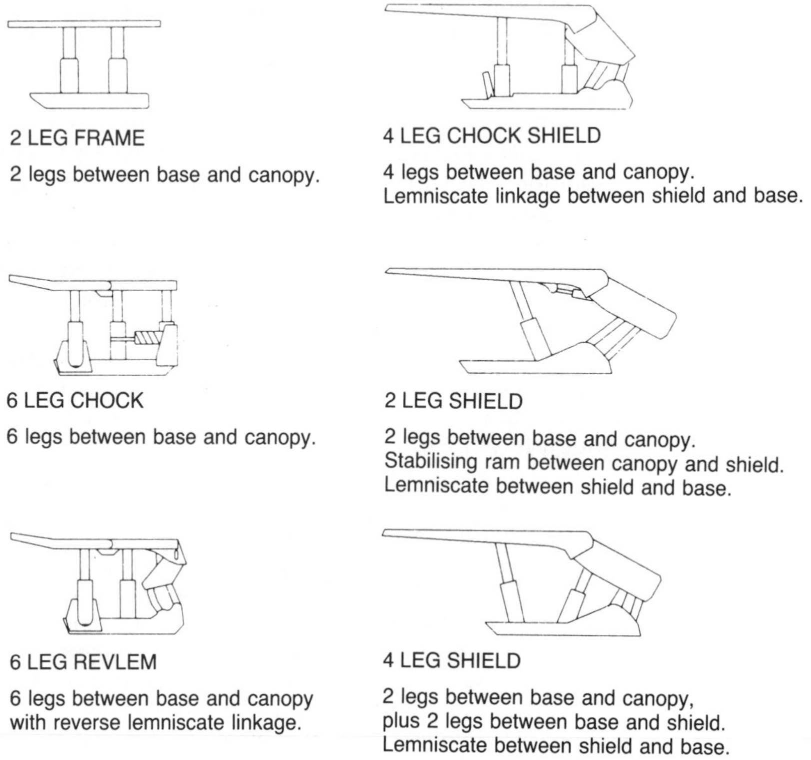

2 LEG FRAME
2 legs between base and canopy.
4 LEG CHOCK SHIELD
4 legs between base and canopy.
Lemniscate linkage between shield and base.
6 LEG CHOCK
6 legs between base and canopy.
2 LEG SHIELD
2 legs between base and canopy.
Stabilising ram between canopy and shield.
Lemniscate between shield and base.
6 LEG REVLEM
6 legs between base and canopy
with reverse lemniscate linkage.
4 LEG SHIELD
2 legs between base and canopy,
plus 2 legs between base and shield.
Lemniscate between shield and base.

Fig. 3.14 Types of powered roof support.

Generally, within the design process of function block A_{22} (all of the activities involved in this block are centred around a product prototype) three main activities can be identified:

- *preliminary design:* preparing outline and detailed design proposals;
- *product development:* developing prototype(s), and preparing and executing validation studies;
- *final design:* preparing finished design in a form suitable for use in manufacturing.

These are shown in Fig. 3.11 as function blocks A_{221}, A_{222} and A_{223}, respectively. In reality, a product design process may not flow in an orderly manner through the above three stages as shown in Fig. 3.11. Loops may exist between any two of the three functions. For example, the production development group may request the preliminary design group to reconfigure the product; the final design may ask the preliminary design to modify certain design features. Frequent modification requests on product design are a common feature in modern manufacturing industry, particularly with OHMS operations. Thus no design can ever be regarded as 'final'.

A_{221}: Preparing advanced drawings

This function takes the definition and specification of a production order and generates a broad solution in the form of a *preliminary design* (which some companies may call an *advanced design*). Often a choice must be made between a number of schemes. It is clear, therefore, this function itself can be further broken down to expose the more detailed activities involved, such as:

- *Analysis*: analysing the design requirements provided and establishing crucial issues; proposing a course of action.
- *Synthesis*: finding possible solutions for the product specifications and building up complete outline designs.
- *Evaluation*: evaluating the accuracy with which alternative designs fulfil product specifications and other requirements and then selecting the final choice from these alternatives.
- *Detailing*: converting the outline design into a detailed form suitable for use in building a prototype.

Again, considerable iteration may take place within the above cycle before a detailed preliminary design emerges. In fact, this function is where the most important design decisions are made – customer requirements, prior experience, production methods and cost aspects must all be brought together. Consequently, this is where the greatest demand is made on the skills of the design team.

During this early stage of the design process, the design team is faced with the task of defining product layout, performing design analysis using various evaluation techniques, selecting materials and components, and so on. In order to help the designer solve numerous problems and to turn out an effective product in as short a time as possible, design aids of every kind should be made available. One of the most important of these is a product engineering data base, providing technical information such as design standards and engineering drawing practices together with various types of design data based on past experience – material specification, parts and components data and previous drawings, and so on. Of course, data will be retrieved from, as well as sent to, such a data base, as shown in Fig. 3.11.

A_{222}: Making and testing prototype(s)

The end product at this stage – the preliminary design presented in a form suitable for prototype building – leaves significant scope for further improvements. A preliminary design may contain many features which are expected to function in a certain manner, but whether these features will in reality function as expected remains to be seen. The structural and functional properties of product design may be predicted to a certain extent by using available analysis techniques. The strength of steel structures, for instance, can be analysed by the theoretical methods such as static or dynamic mechanics, or numerical techniques such as finite element analysis. Nevertheless, it is frequently necessary to carry out tests on the technical and functional suitability of a design. If this is the case, then prototypes and models based on the preliminary design will have to be built for the purpose of experimentation. Many OHMS manufacturers dedicate a function-layout development workshop, with general-purpose manufacturing facilities, to prototype production.

Once the prototype is made available, various types of test may need to be carried out on it to evaluate the extent to which the preliminary design meets the production definition and specification. Test design requires a full understanding of the product under investigation, as well as knowledge in other specialized areas. The aim of testing is to reveal the true characteristics of the prototype. Depending on the type of product and the type of information required about the prototype, this may be achieved by such procedures as strength tests (pressure, tensile, fatigue, etc.), environmental tests (dust, temperature, humidity, vibration, etc.), and functional tests (mechanical, electrical, hydraulic, etc.).

Extreme testing conditions are often used to make certain that the product will function properly over the full range of its operating specifications. A certain safety factor is often built into these testing conditions. For example, a safety factor of 1.5–2.0 is used by CLD Equipment Ltd when carrying out fatigue tests on its prototypes. That is, the prototype is subjected to a stress level 1.5–2.0 times the maximum operating specification. Consequently, often more than one prototype will have to be produced because some of these tests can be destructive. Reliability and maintainability of the product are often the main concern here. In general, a product must be designed and manufactured in such a way that it will give failure-free operation throughout a specified period of time under the specified working conditions. This quality aspect is known as the *reliability* of the product. On the other hand, *maintainability* is concerned with the ease with which the product may be repaired when a failure does occur. Under certain circumstances, *fail-safe* design may be required. This refers to a design of product which gives it the ability to continue to function even if one or more parts fail during its operation. The control system of a nuclear power station is designed with such a requirement. Passenger jet aircraft are designed in such a way that if any of the engines fails in flight, the rest of the engines will be able to keep the aircraft going until a safe landing place is found.

Apart from tests that can be carried out in-house, the prototypes may also need to be sent out for certain kinds of test to be conducted. For example, special testing equipment may be required which the company does not possess itself; prototypes may have to be sent to specialists. Or the performance of the prototype may need to be evaluated under real operating conditions – this type of test is called a *field test*; it is frequently necessary when the product is intended to be used in hostile environments. Or again, the equipment may be designed to work in conjunction with equipment made by other manufacturers – a *compatibility study* may need to be carried out to see if this is possible; this type of test requires close cooperation from the companies concerned.

Figure 3.11 shows that there is a feedback from the test function to the preliminary function, where the product goes 'back to the drawing board'. Problems revealed by prototype testing must be rectified through a number of such iterations. Also, the customer must be kept informed about the progress of the development process and provided with reports of product performance, as illustrated. Sometimes final approval from the customer may be required before the manufacturing function proceeds. Based on the information contained in the test reports, a customer may decide to ask for further improvements or modifications to the design.

At this stage (and the next stage) the design team should also keep an eye on the production side of the problem. A good designer should be equipped with a sound understanding of the producibility of the product in relation to the company's manufacturing facilities, as well as the relevant production technology that is currently available. The adoption of automatic and computer-controlled manufacturing facilities has demanded that much greater consideration, at the design stage of a product, be given to producibility. Although sophisticated, these facilities, whether NC tools or robots, cannot compete with human operators in terms of adaptability. 'Design-for-manufacture' has, therefore, become a new concept for engineering designers which is receiving more and more attention. Furthermore, it is obviously undesirable to put forward a product design whose production depends on a manufacturing technique which the company does not have. On the other hand, the possibility of an alternative and more efficient way to manufacture and hence to design a product must be continuously explored. Since design-for-manufacture is such an important and relevant topic in our study of manufacturing systems analysis, this will be discussed in more detail later in this section.

A_{223}: Preparing final drawings and parts lists

By now a great deal of communication of problems and instructions will have taken place between various functions. As a result, a large number of modifications and rectifications will also have been carried out on the preliminary design. All the activities undertaken so far have been aimed at allowing a design of good quality to take shape – and a sound design of good quality is exactly what the final design function must accomplish. The aim of this function is to carry out refinement of the design elements (design layout, preliminary part and assembly drawings, test reports and various feedback information, and so on), so as to produce a detailed, complete version of the product design in a form understandable to those who will make the product. The most widely used form of communication is the engineering drawing. These drawings have to communicate precise instructions and should, therefore, be error-free and subject strictly to the standard rules, codes and conventions. (Remember the prerequisites for effective communication listed in Chapter 2?) To aid this crucial communication process, sometimes a complete three-dimensional representation has to be produced if the product concerned is complex in its construction, as exemplified by Fig. 3.15, which shows a parts list and an exploded three-dimensional view to indicate the structure of a hydraulic control valve. This type of design communication is particularly useful when a design has to be presented to a client.

It is clear that a considerable amount of time and effort will be required to carry out this process. Nowadays computer-aided facilities play an important role in the tasks involved in this function. They are being used not only as analytical tools to evaluate the design characteristics of a part, but also to reduce the tedium of the skilled and patient work required by both preliminary and detailed design tasks. In addition, the quality of the end re-

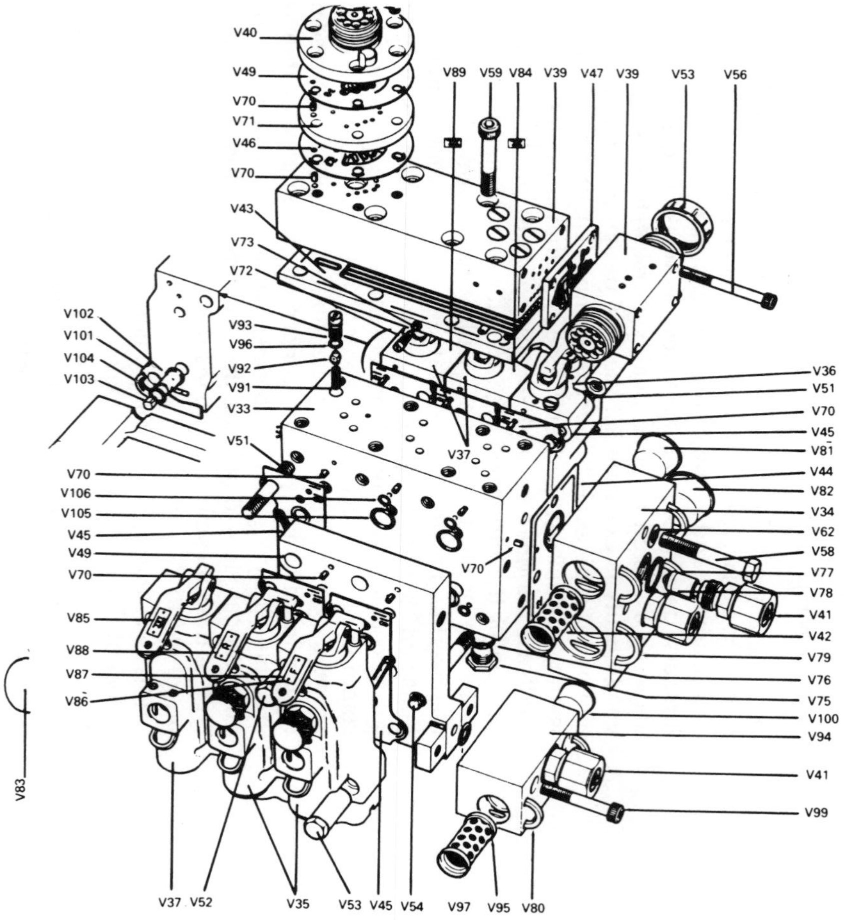

Fig. 3.15 Structure of a hydraulic control valve.

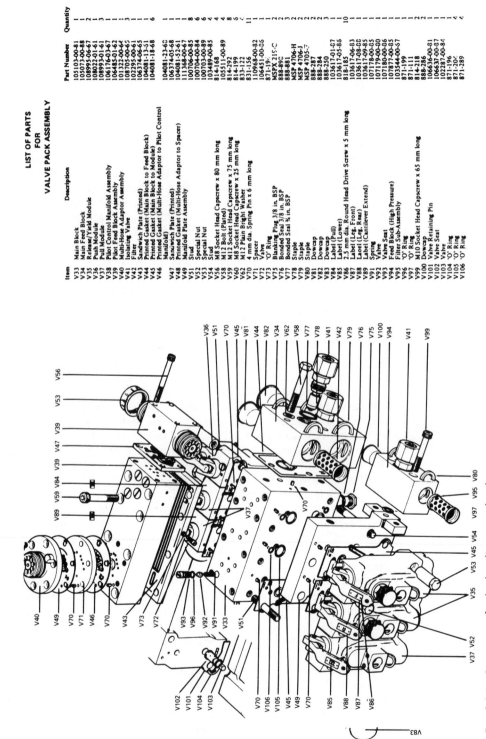

LIST OF PARTS
FOR
VALVE PACK ASSEMBLY

Item	Description	Part Number	Quantity
V33	Main Block	105103-00-81	1
V34	Main Feed Block	105073-00-88	1
V35	Release/Yield Module	108995-06-67	2
V36	Push Module	108022-01-61	1
V37	Pull Module	108993-01-61	3
V38	Pilot Control Manifold Assembly	106176-03-67	1
V39	Pilot Feed Block Assembly	106485-01-62	1
V40	Multi-Hose Adaptor Assembly	101322-00-64	1
V41	Isolating Valve	108705-00-65	3
V42	Filter	102295-00-61	1
V43	Sandwich Plate (Printed)	106374-06-65	1
V44	Printed Gasket (Main Block to Feed Block)	104081-13-51	1
V45	Printed Gasket (Main Block to Module)	104081-18-68	6
V46	Printed Gasket (Multi-Hose Adaptor to Pilot Control Manifold)	104081-23-60	1
V47	Sandwich Plate (Printed)	106374-05-68	1
V48	Printed Gasket (Multi-Hose Adaptor to Spacer)	104081-52-61	1
V49	Manifold Plate Assembly	111368-00-67	1
V51	Stud	100706-00-85	8
V52	Special Nut	100704-00-84	6
V53	Special Nut	100703-00-89	6
V54	Stud	103489-00-85	4
V56	M8 Socket Head Capscrew x 80 mm long	814-168	4
V58	M12 Bolt (Plated)	105511-00-89	4
V59	M12 Socket Head Capscrew x 75 mm long	814-292	8
V60	M8 Socket Head Capscrew x 25 mm long	814-199	6
V62	M12 Plain Bright Washer	833-122	4
V70	4 mm dia. Spring Pin x 6 mm long	831-156	11
V71	Spacer	110968-00-82	1
V72	Valve	106451-00-86	1
V73	'O' Ring	871-194	1
V75	Blanking Plug 3/8 in. BSP	MSPX 219-C	2
V76	Bonded Seal 3/8 in. BSP	888-890	2
V77	Bonded Seal ¾ in. BSP	888-881	3
V78	Staple	MSP 4706-H	2
V79	Staple	MSP 4706-J	2
V80	Staple	MSP 4705-F	2
V81	Dowcap	888-287	2
V82	Dowcap	888-284	2
V83	Dowcap	888-250	3
V84	Label (Pull)	103617-01-87	1
V85	Label (Lower)	103617-05-86	1
V86	2.5 mm dia. Round Head Drive Screw x 5 mm long	818-185	10
V87	Label (Leg, Front)	103617-06-83	1
V88	Laoel (Leg, Rear)	103617-08-88	1
V89	Label (Cantilever Extend)	103617-09-85	1
V91	Spring	107178-00-85	1
V92	Valve	107179-00-80	1
V93	Valve Seat	107180-00-86	1
V94	Feed Block (High Pressure)	107877-00-85	1
V95	Filter Sub-Assembly	103544-00-67	1
V96	'O' Ring	871-199	1
V97	'O' Ring	871-111	1
V99	M10 Socket Head Capscrew x 65 mm long	814-218	2
V100	Dowcap	888-286	2
V101	Valve Retaining Pin	106636-00-81	1
V102	Valve Seat	106637-00-87	1
V103	Valve	102287-00-84	1
V104	'O' Ring	871-196	1
V105	'O' Ring	871-204	4
V106	'O' Ring	871-289	4

sults is being significantly increased. Computer-aided design (CAD) is the longest-established computer tool used in manufacturing. It has been used beneficially by many manufacturers in all three functional areas of design. The basic features of this computer-based design tool are examined in Box 3.4.

It should be noted that each of the three design functions must send a status report back to the control function, in order periodically to update the current project progress and highlight any problems. Also, design modification requests may come either from the shop floor in the light of production problems, or from the customer. These communication processes are presented by the information flow channels shown in Fig. 3.11.

Box 3.4 Fundamentals of CAD

Computers are utilized extensively by engineering designers in generating design configurations, drawings and graphic displays. Although this is the area with which CAD is most commonly associated, in fact CAD incorporates all the design activities that involve the use of a computer, including modelling, analysis and evaluation, as well as design documentation and communication.

The development of CAD has provided designers with a powerful and flexible tool. A designer can now prepare and evaluate in hours a preliminary design that would have taken days or even weeks using traditional manual methods. A CAD system not only increases the productivity of the designer but also improves the quality of the design, for the following reasons:

- With the capability provided, more complete design analysis can be carried out and more alternatives evaluated, resulting in less risk and uncertainty.
- With the ability quickly to store and retrieve design data, the amount of effort involved in various stages of the design process is significantly reduced. More attention can thus be paid to the quality aspect of the design so that there are fewer opportunities for errors to occur.
- With the flexibility provided, design changes can be more readily incorporated, allowing design errors to be more easily rectified and modification requests more easily accommodated, resulting in a much more rapid response.
- CAD gives a much more accurate and professional presentation of a design, resulting in better documentation and communication.

If the general process of product design can be summarized by the three design functions outlined earlier – preliminary design, product development and final design – then a true CAD system should assist the full range of activities involved:

1. *Analysis:* including preliminary product conceptualization and assessment of technical and economic feasibilities.
2. *Synthesis and evaluation:* finding possible solutions for the product specifications and building up complete outline designs; analysis of

both static and dynamic properties (structural, functional, power transmission, etc.).

3. *Optimizing:* evaluating the accuracy with which alternative designs fulfil product specifications and other requirements and then selecting the best choice from the alternatives.
4. *Presentation:* producing detailed drawings, including parts and components lists, material and other specifications.

Up to now, although CAD has been beneficially used in areas 2–4, most systems do not yet support such a complete range of design activities. The use of the majority of CAD systems is still restricted to area 4.

A CAD system is usually a combination of computer hardware and specially written software. Depending on the application, there are various installation configurations, from large and powerful mainframe-based set-ups to inexpensive and flexible PC-based systems. However, the basic elements are similar for all CAD systems. Figure 3.16 illustrates the configuration of a typical PC-based CAD system. Like any other computer application, a CAD system comprises input devices (in this case keyboard, electronic tablet, mouse, light-pen and joystick), a processing unit (computer CPU plus software), a data-storage unit (magnetic tape storage, hard or floppy disk drive), a display unit (a graphic terminal) and some output devices (plotter, graphic printer). During a design process, the designer will communicate with the system by putting parameter values and commands into the computer through the input devices. These data are needed to define the geometry and other properties of the design. They will then be used by the processing unit to create a mathematical model of the object and communicated back to the designer by means of an interactive image of the object on the display unit. The designer can then modify or update the design by issuing new parameter values and/or commands to the system. Hard copy of the design can be obtained through printers or plotters. The storage device connected to the system is used for keeping the design data and the system software.

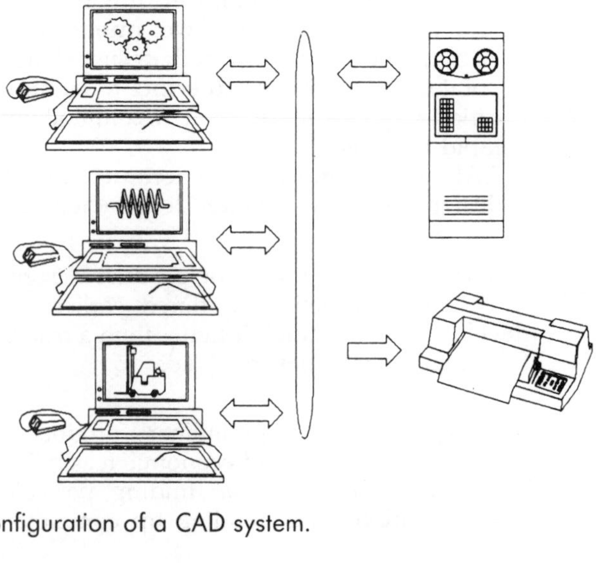

Fig. 3.16 Configuration of a CAD system.

Software for CAD can be divided into three basic categories according to its modelling abilities. Computer-aided drafting, using *two-dimensional* (2D) systems, is the starting point from which computer-aided design evolved. These systems provide computer-aided drafting ability to create and modify two-dimensional engineering drawings. The majority of CAD activities are carried out using 2D systems, with the objective of increasing drafting productivity and drawing quality. Such systems allow single entities such as points, lines, arcs and circles, as well as text, to be manipulated in two dimensions: they can be added together, copied, moved, rotated, mirrored, magnified, and deleted. Various viewing options are provided so that, for example, any part of a drawing can be 'zoomed' to reveal its details. Special features are also provided to accommodate the particular requirements of engineering drawings, including automatic filleting, chamfering, hatching and dimensioning.

A drawing or part of a drawing created using these systems is saved as a file on magnetic tape or disk, and can be retrieved for later use. In this way, a library of digitized drawings of standard parts, components and previous designs can easily be built up. This can be used to increase productivity by avoiding duplicated effort later. Sometimes drawing libraries of standard parts are even provided by the software suppliers themselves. For example, AutoDesk supplies a series of such libraries which can be used with its PC-based CAD system AutoCad, covering a wide range of application areas in various engineering disciplines including mechanical, electrical, electronic, structural, civil, architectural, and process.

Two-and-a-half-dimensional systems are basically similar to 2D systems but with the additional ability to create 3D images by extending an entity from its 2D base. The principle of this approach is illustrated in Fig. 3.17. The top view of an entity is first presented in the usual manner as a two-dimensional orthographic drawing. This is then given a 'thickness' along the third dimension to define the shape and an 'elevation' to specify the position of the entity in the 3D space. According to the shape and position definition given by the value of these two parameters, a 3D image of the entity is formed when extended along the third dimension. In addition, this 3D image may be viewed from any angle chosen by the

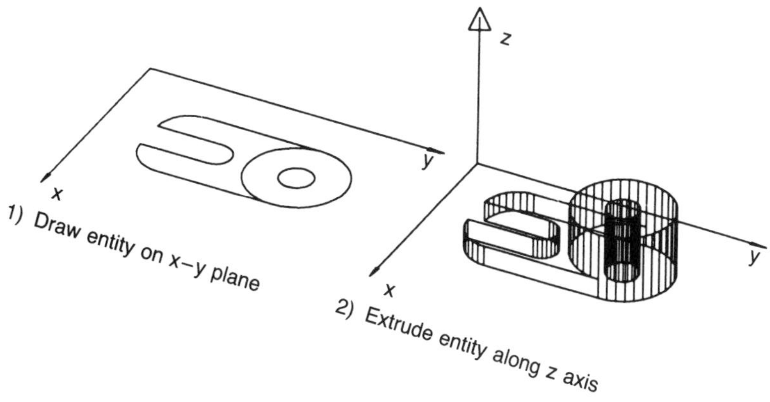

Fig. 3.17 Construction of a 2½D image.

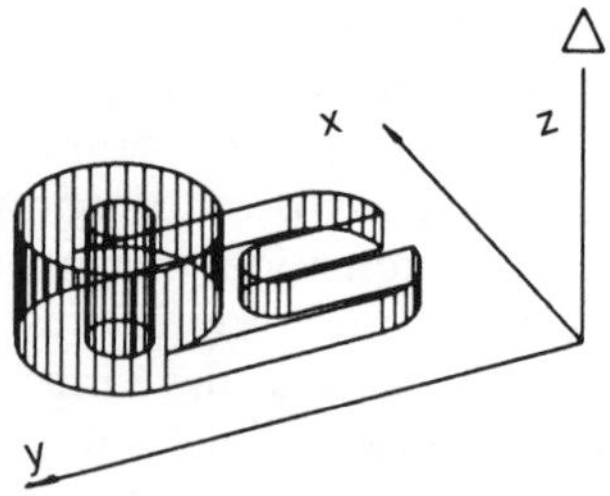

Fig. 3.18 The image constructed in Fig. 3.17 viewed from a different angle.

designer, as shown in Fig. 3.18. There are a number of widely used $2\frac{1}{2}$D CAD systems, with AutoCad being a typical example.

This additional feature provides the designer with the ability to analyse engineering designs interactively to a certain degree. The spatial relationship between a number of basic objects, for instance, may be assessed by using this approach. However, the use of a $2\frac{1}{2}$D system in this type of application is rather limited because, first, it does not provide a realistic representation of a 3D object, and second, it is very difficult to model complex shapes.

Three-dimensional systems are the true electronic graphic systems which are capable of accurately presenting a 3D object in the form of either a wireframe or a solid model. This approach is generally referred to as 3D geometric modelling. CAD systems with 3D geometric modelling capabilities can be used in all of the four CAD application areas listed.

With geometric modelling, a 3D image of an object can be represented either by a wireframe model, which uses interconnecting lines to depict the shape of the object, or by a solid model which is a vision of the object as it would appear in the 3D space. Wireframe modelling systems demand less computer power because an object is represented only by the outline of its shape. When the object concerned has complicated geometries, a wireframe model can become confusing because details of the object may not be clearly visible. Solid modelling systems provide more accurate and clear representation of an object, but they usually need the support of more sophisticated computer hardware. Although there exist a number of PC-based 3D modelling systems, the slow speed associated with their model generating and display processes severely limits their application values.

To facilitate the engineering analysis processes, some systems also provide the ability of mass property analysis, component interference checking, finite element analysis, and even colour animation. These allow structural (component shape characteristics, configuration of various parts of an assembly, operation of moving mechanisms, and so on), strength (static and dynamic), as well as other (heat transfer, for example) types of analysis to be performed during the design stage.

Design-for-manufacture

The concept of design-for-manufacture is an important topic. Consideration, during the design stage, of producibility is the first step to guaranteeing the smooth production of a product. It is of importance to the efficient operation of the overall manufacturing system used. A design which proceeds without taking design-for-manufacture into consideration is likely to cause problems during later production operations. It is not infrequent, for example, for a minor design fault to halt the whole production operation, and this can be extremely costly to a company. It must be remembered that a design defect at the drawing stage can be corrected with a pencil and rubber – or nowadays with a few key strokes. If such a flaw remains undetected until production, however, the resulting production delays, together with the subsequent modifications to tooling, jigs, fixtures, process set-ups, and so on, will certainly have a much more significant and negative impact on the whole manufacturing operation.

To highlight this important issue, Box 3.5 gives part of an actual report produced by the production engineering department of CLD Equipment Ltd. It indicates clearly some of the production problems caused during the final assembly stage of the product, when the design-for-manufacture aspect is not carefully considered during product design. The effects of this in terms of production cost are also clearly shown.

Box 3.5 An example of production problems caused by design: extract from the report of an investigation into the problems of Chock Assembly

Summary

This report is a study of the problems in chock assembly. Throughout a particular contract, all the problems, faulty components, etc., were recorded and time studies were taken of good assemblies. Therefore, the actual cost of assembly could be compared with the possible cost of assembly. It has been calculated that because of poor design, bad components, shortages, etc., approximately £22 500 was lost in the assembly of the particular contract investigated.

It is recommended that considerably more effort, particularly from Production Engineering, is put into Chock Assembly in order to realize these savings and possible further savings through improved methods and shop layout.

Objective

This report is the result of an investigation into the assembly of chocks. This contract was chosen to be studied from beginning to end in order to obtain some detailed information on the problems which occur in chock assembly. It has also been the aim to quantify the time and money lost because of the problems occurring.

The contract

This chock is a steep seam unit and with the following specifications:

Chock type:	4L × 280T steep seam chock shield
Quantity req:	176
Target rate:	6 units/day

This contract is considered to be fairly typical and a good example of the variety of problems encountered in chock assembly.

The contract was primarily assembled on line 4. During the contract there were many problems and faults which caused the line to work inefficiently and occasionally to stop completely.

Design faults

1. The first canopy had to have two holes flame-cut to allow the upper shield hinge pin retaining plates to be fitted. The operators had to leave the line while the flame-cutting took place. Three other canopies were modified on site by cutting off the retaining plate housings and rewelding in a new position. The remaining canopies came in correct from the supplier.
2. The first few canopies had a section of fabrication flame-cut to allow four hydraulic legs to be fitted. This was done off the line.
3. On assembly, the shield-securing discs were found to be unsatisfactory and were scrapped off. A new one was designed and made. This was found to be over-size and had to be modified again.
4. The design of the upper shield did not consider assembly requirements. Therefore, the side ram aperture had to be enlarged.
5. On no. 1 chock a valve assembly bracket had to be cut off and a new one designed and fitted at a lower height as per customer's requirements.
6. Another type of assembly bracket had to be reshaped to suit assembly requirements.
7. When the banjo bolts were fitted to the valve block they were too close together, causing the adjacent bolts to foul. This problem continued throughout the whole face.
8. The shape of the lower side shield was changed several times during the contract. Most of them had to be flame-cut and ground – some of them while on the line.
9. Some of the staples were very difficult to fit, due to the bad positioning of the staple holes in the block assemblies.

Cost of Chock Assembly

Actual assembly costs

As a result of all the problems and faults, the cost to assemble the contract was unnecessarily high. The total number of man-hours worked to complete the contract was 6426.4 hrs. The total cost to assemble the contract was therefore:

$$6426.4 \times £16.40 = \underline{£105\,393}$$

Possible assembly cost

Time studies were taken of each station on and off the line. The longest station and, therefore, the controlling cycle was canopy sub-assembly with a target time of 71.75 min.

The total number of man-hours to assemble the contract with a 21 man crew could have been:

$$\begin{aligned}\text{Total} &= (\text{No. in crew}) \times (\text{Hrs worked per day}) \times (\text{No. of days reqd})\\ &= (21) \times (7.8) \times (176 \times 71.75/468)\\ &= 4420\,(\text{hrs})\end{aligned}$$

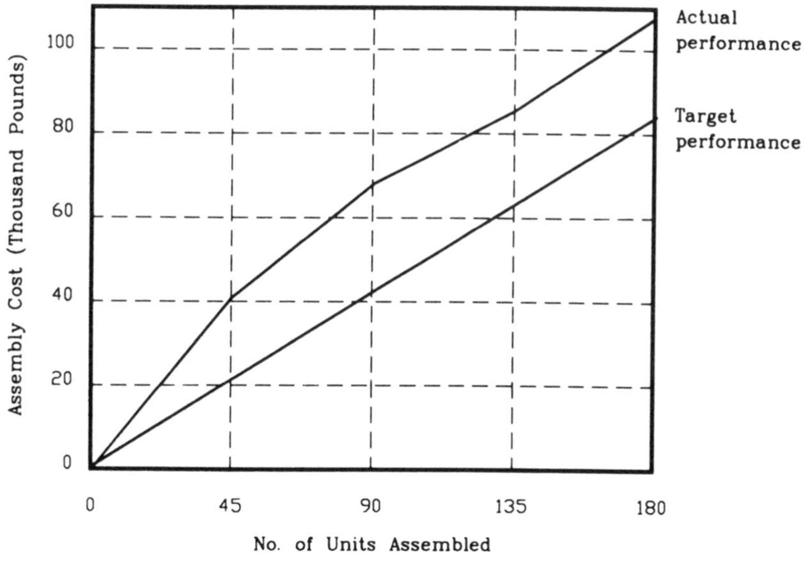

Fig. 3.19 Summary of assembly costs.

Time is also necessary to set up the line and a period of four days with a full crew is usually considered to be sufficient. The total number of man-hours to complete the contract, therefore, should be set at:

$$\text{Total} = 4420 + (21 \times 4 \times 7.8)$$
$$= 5075\,(\text{hrs})$$

Then the total cost of assembly should have been:

$$5075 \times £16.40 = \underline{£83\,230}$$

Therefore, the cost of unexpected modifications and other faults on this particular contract can be valued at:

$$£105\,393 - £83\,230 = \underline{£22\,163}$$

The situation is represented graphically in Fig. 3.19.

It is quite clear that if the designers had had a clearer picture in mind as how to put things together in the final stage of production, then most of the production delays listed in Box 3.5 could have been avoided. This strongly supports the argument that one reason for the many problems experienced during the production process is a lack of collaboration between the production activities and their related manufacturing functions. These problems may have their origins elsewhere and a wider systems approach must, therefore, be adopted in order to trace and solve them. Design, obviously, is one of the most critical interfaces which influence the performance of production. Realization of this has led to an increasing emphasis on design-for-manufacture. For instance, in a more recent philosophy of manufacturing planning and control known as just-in-time (JIT) design-for-manufacture is treated as an essential part of the design function. This is necessary because research has shown that a very high proportion of the early JIT failures were found to be related to design – looking at the problems reported in Box 3.5, this may not come as a surprise.

If a product design is to be 'manufacture-friendly', then the designer must have a sound knowledge of the manufacturing practice concerned. First, he or she should know the ability and limitation of the existing production facilities. Then, during the design process, a clear plan must be developed to show how to produce and process each of the individual parts and how the parts and components are to be put together. The preliminary questions for the designers to ask in this context are therefore:

1. Is the in-house producibility adequate?
2. Do the sub-contractors have adequate producibility for the bought-out items?
3. If the answer to question 1 or 2 is 'no', what can be done about it?

At the next more detailed level of design, consideration must be given to each of the individual production processes, such as part manufacture, product assembly, transportation and storage, and the problem of distribution. There are general design-for-manufacture rules related to each of these areas, which can be applied to reduce product complexity and hence production cost. The following rules, for example, can be applied to assist product assembly processes (Williams, 1988):

Rules for assembly design:

1. Minimize the number of parts. By reducing the number of parts in an assembly, the number of assembly tasks to be carried out is immediately reduced. This reduction also significantly reduces the organizational and transport problems associated with the assembly process.
2. Use standard parts where possible. The use of standard parts – the same fastener throughout an assembly, for example – reduces the amount of tooling required and contributes to the effective reduction of the number of parts in an assembly from an organizational point of view.
3. Ensure that the product has a suitable base part on which to build the assembly.
4. Ensure that the base part has features that will enable it to be readily located in a stable position in the horizontal plane.
5. If possible, design the product so that it can be built up in layers, each part being assembled from above and positively located (so that there is no tendency for it to move under the action of gravitational forces).
6. Try to facilitate assembly by providing chamfers or tapers to help guide and position the parts correctly.
7. Avoid expensive and time-consuming fastening operations, such as screwing and soldering. Use snap fits or other push-together mechanisms where possible.
8. Make modular sub-assemblies.

Rules for part design:

9. Avoid projections, holes or slots that will cause tangling with identical parts when placed in bulk in a feeder or bin. This may be achieved by arranging that the holes or slots are smaller than the projections.
10. Attempt to make the parts symmetrical, to avoid the need for extra orientating devices and the corresponding loss in feeder efficiency (ensure that parts cannot be assembled ambiguously).

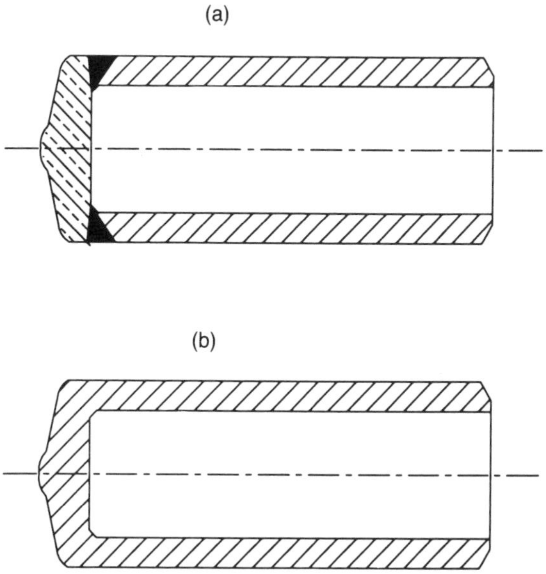

Fig. 3.20 Different production processes. (a) Welding and (b) Extrusion.

11. If symmetry cannot be achieved, exaggerate asymmetrical features to facilitate orientating or alternatively provide corresponding asymmetrical features that can be used to orientate the parts.

Similar design rules also exist to assist metal removal and other production processes. These may be found in relevant textbooks or design manuals.

Designers should also be aware of recent developments in production methods and the implications of these to their particular company – at the very least they should be continuously kept informed by those involved in production. The body of a hydraulic cylinder, for instance, can be made up either by welding a cap on the end of a steel tube, or by an extrusion operation from a single piece of steel, as illustrated in Fig. 3.20. Although functionally there are perhaps no significant differences between these two designs, differences will certainly exist in the production facilities needed and in the production costs. The decision must made as to which of these is the better choice for a particular application. This requires all relevant factors, such as material availability, cost of new facilities, need for new skills and effect on works organization, to be taken into careful consideration.

Finally, the importance of continuous communication between design and production functions should be stressed. Formal and effective links must be set up between these two functions to ensure that in-house producibility and new development are properly considered. This is particularly important in modern manufacturing industry, where advanced manufacturing technology is being developed rapidly. It is good practice, for example, to have a production engineer closely allied with the product design function on a full-time basis.

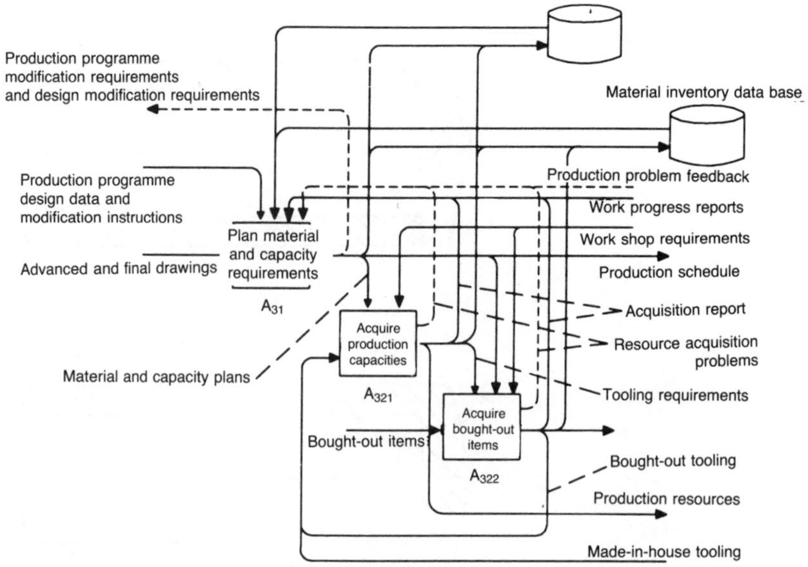

Fig. 3.21 A_3: Gathering production resources.

3.3.5 A_3: Gathering production resources

Having produced a detailed product design, the next function (which runs partly in parallel with the design function) is to gather together all the resources needed to manufacture the required quantity of product (Fig. 3.6), according to the time set by the overall production plan. These resources will include machine capacity, operator capacity and, of course, the materials and tools.

For any type of manufacturing operation this gathering of resources will consist of the following tasks (Fig. 3.21):

- A_3 GATHER PRODUCTION RESOURCES
  - A_{31} Plan Material and Capacity Requirements
  - A_{32} Gather Resources
    - A_{321} acquire production capacities
    - A_{322} acquire materials and bought-out items

A_{31}: Planning material and capacity requirements

This function contains some of the most important activities involved in ensuring an efficient production operation. For many manufacturing systems, this could be the starting point of the activity chain, particularly if products of standardized design are manufactured to stock rather than to customer order. Manufacturing systems designers and analysts should possess a reasonable understanding of the decision-making methods and techniques which are used in this area. Appendix A provides a survey of traditional and current techniques of production planning and control of manufacturing systems. This is intended to serve three purposes: to assist the reader in understanding the manufacturing planning and control functions; to highlight the problem areas to which some of the techniques dealt with later in this book will be related (for example, computer simulation techniques when used to analyse scheduling problems); and as a brief guide for future reference.

The models and techniques included will not be restricted to OHMS operations. Many problems in manufacturing planning and control are not analytically complex, but the variety and quantity of information that has to be tackled is enormous. This is illustrated by the fact that there are many weighty textbooks devoted to this subject area. The survey given in Appendix A is, therefore, by no means exhaustive. However, it does illustrate the basic principles of various approaches. The reader is advised to read through this survey to develop some basic understanding of this extremely important area of manufacturing systems.

A_{32}: *Gathering resources*

The actual acquisition of production resources is the second of the two functions which make up function block A_3. In our illustrative model, for the purpose of simplicity, production resources are represented in a highly aggregated manner by 'bought-out items', referring mainly to materials, components, toolings and other sub-contracted items. Generally speaking, however, these are only part of the production resources, which should include all of the 'factors of production'. They have been classified into the following four categories (Hitomi, 1979):

1. *Production objects*. These are materials on which activities of production are performed; they consist of *primary* materials and *auxiliary* materials. Primary materials are to be converted into products through production processes. These include raw materials, parts and so on. Auxiliary materials are added to the primary materials (as when paint is used) or are supplementary to their production (for example, electricity and lubrication oils consumed in the process of production, and the light and air-conditioning which support the productive effort).
2. *Productive labour*. This is human ability, including the physical and mental ability of an individual worker, by which production activities are performed.
3. *Means of production*. These are the media by which production objects are converted into products with the aid of productive labour. It includes *direct* means of production – the so-called *production facilities* – and *indirect* means of production. Production facilities work directly on raw materials – for example, machines, equipment, apparatus, jigs and tools, power machinery, and so on. Indirect means of production – such as land, roads, buildings and warehouses – are not directly concerned with the productive processes. Production objects are transformed and thus lose some or all of their original properties (shape, colour, and so on), whereas means of production can be utilized repeatedly over a given length of time.
4. *Production information*. This is the knowledge to implement effectively (that is, efficiently and economically) productive processes for manufacturing tangible goods. It includes production *methods*, which are the technical procedures for implementing productive processes. These methods follow objective engineering laws including empirical rules and, in some cases, techniques which are subjective arts gained by training of individuals, relying on experience and intuition.

All of the above must be made available in the right place and at the right time. Because of the different natures of the resources listed, different planning actions are required for their provision at different time-scales. An example of a production resource which needs long-term planning is the people with necessary skills. It has become normal practice for companies to take on school-leavers, sponsor them through university, and provide them with the necessary industrial and professional training. Six years or more may elapse between the start of such sponsorship and the time when the student finally becomes a qualified engineer and begins to contribute to the company. Nevertheless, staffing the company with the necessary skilled people is one of the key issues in building a successful operation, particularly in today's technological settings where the increased use of AMTs require the continuous import of necessary skills to the company.

Detailed discussion of the problems associated with the provision of each of these categories is beyond the scope of this book. We will concentrate only on the key issues associated with bought-out item acquisition, including all those materials and items which are to be included in the final assembly of the product (the production objects) and the tools and jigs which are associated directly with the immediate contracts (some of the production means). Bought-out items can include a wide variety of materials which fall into two main categories: *standard* items and *made-to-order* items.

Standard items include the standard raw materials (steel and/or alloy bars, plates and sheets; plastic and wood products, and so on), finished components (motors, bearings, seals, nuts and bolts, hydraulic/electronic components and modules, and so on) and tools (hammers, spanners, screwdrivers, drills, cutting tools and gauges, and so on) which may be bought off the shelf.

Made-to-order items are the items which are made to order specially for the company. This again includes a wide variety of objects, but the following are the more frequent items which are purchased from suppliers in the specialized fields: semi-processed materials, made-to-design parts, made-to-design sub-assemblies special-purpose jigs and fixtures, castings, forgings and fabrications.

The provision of these items is the responsibility of the purchasing department. The purchasing function of a manufacturing organization usually has the following important features:

- For an average manufacturer it ties up the largest slice of working capital.
- Its efficient operation is critical to productivity and product quality.
- Its complexity requires the cooperation of a number of different groups of people.
- It requires careful systems organization.

The quantity of bought-out items which are included in the delivered product differs from industry to industry and from company to company. Even within the same market, the manufacturers may adopt different procurement strategies. Some companies only manufacture a small percentage of the parts required for their products in-house, leaving a large amount of work to sub-contractors. For example, it is not unusual to find equipment manufacturers using up to 90 per cent bought-out items for their products. On the other hand, some manufacturers may prefer to use in-house facilities all the time for either technical or economic reasons. On average, however,

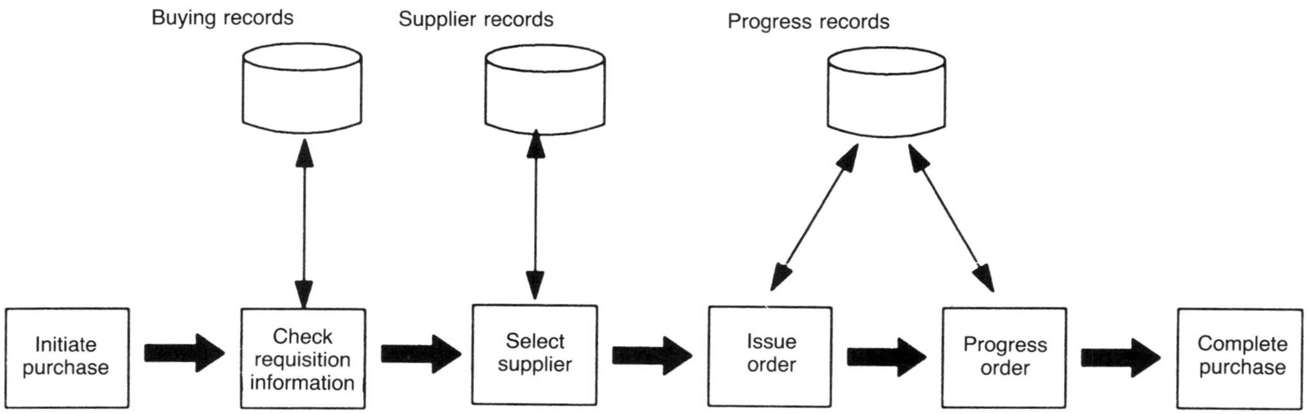

Fig. 3.22 General purchasing procedures.

manufacturers usually spend a large slice of their turnover with outside suppliers and rely heavily on them for materials, parts and other services.

The design and implementation of economical and efficient purchasing procedures should be regarded as one of the most important aspects of manufacturing systems design. An inefficient purchasing function is likely to result in an inefficient business, with possible late deliveries, production hold-ups, high production costs, dissatisfied customers and, eventually, failure to achieve and maintain corporate objectives. Purchasing is a critical manufacturing function which contains a complex set of activities and requires a large number of people with different technical backgrounds who must interact efficiently if failures are to be avoided. Owing to the variety of company practices, however, these is no single method or set of procedures which will suit every purchasing department. The proper procedures will depend very much on the type of product involved and the manufacturing method used.

Generally speaking, it is important that the objective of this manufacturing function should be in line with a coherent corporate objective. For an average manufacturer, for example, this could be: 'To purchase the necessary items needed for production in such a manner that items with adequate quality are obtained and stored at minimum cost to the company and without risking any delay in the production schedules.' If the company concerned is operating in a prestige market, on the other hand, the objective of this function could be: 'To purchase goods of the highest possible standard at a cost which guarantees a certain extra competitive edge.' If properly implemented and operated, a purchasing function should efficiently support the overall corporate strategy. For example, it should in general not allow 'panic orders' to happen.

Procedures for purchase transactions

Although in the real world the actual practice varies a great deal, there is a general pattern of purchasing transactions with a number of associated key issues which may be identified in the purchasing department of an average manufacturer. This is shown graphically in Fig. 3.22. In a manufacturing system, the process of purchasing begins with the notification that, according to the material requirements plan, certain bought-out items are needed to support the production schedule. This notification is normally known as a *purchase requisition*, which serves the purposes of specifying the items needed, authorizing the expenditure of company funds and recording the transaction. A purchase requisition form, whether computerized or manual, should clearly include:

- the job or contract number;
- the date of requisition;
- the goods required;
- the quantity required;
- the date required;
- delivery instructions;
- any other special instructions;
- originator and authorization;
- suggested supplier.

This information will be checked and used for purchase inquiries and purchase orders.

Frequently a purchasing process needs some kind of early warning system. An example of this has been given earlier in this chapter to show how CLD Equipment Ltd uses such an approach to minimize its product throughput time. An early warning system may also be desirable where it is necessary to take a long time to investigate suppliers, negotiate price and terms, and so on. Under these circumstances, an early warning procedure can be used to avoid the situation in which the purchasing department acts too late.

Once a purchase requisition arrives in the purchasing department, it initiates a chain of activities very much like those involved in the sales function (A_{11}), only this time the company is a customer rather than a supplier.

The first step is to choose a number of potential suppliers, using the company records of purchase history and other relevant resources. Of course, these choices should be made in the light of commercial factors such as price, operational factors such as delivery performance, and quality factors such as rejection history. Experienced purchasing staff with sound knowledge of the products involved and their market, as well as of value estimating, are required. In addition, a data base containing records of previous purchases can be of great help (see below).

The next step is to get quotations from the possible suppliers on the shortlist by issuing an enquiry form to each of them, which may include information transferred from the requisition form plus other relevant items, such as:

- description of what is required;
- quantity required;
- date required or period covered, where appropriate;
- location of delivery;
- general conditions of purchase;
- special conditions applicable to this particular transaction;
- terms of payment;
- name, address, phone number, telex and fax numbers of purchasing organization;
- date and reference number.

Having received replies from the suppliers, their quotations should be summarized in a form which facilitates the comparison process (such as a decision matrix), to highlight the reasons for choosing a particular vendor. Based on such information, the best bid in terms of price, quality, delivery promise and service will be given the order.

Timing of goods delivery has become of vital importance in the modern manufacturing environment. Ideally the parts needed should be delivered at the right place just in time for the production process. The purchasing department is responsible not only for placing the orders for bought-out items but also for ensuring that the goods are delivered on time. Therefore, between issuing the order and receiving the goods, the purchasing department should carry out certain 'order progressing' actions to follow up orders when necessary. This can be particularly important if the goods concerned are bottleneck items. A PERT network model, therefore, may be used to expose the orders with critical delivery dates which need to be expedited. Similar information can also be obtained from a computer-based production control system, such as MRP. The actual progress actions may be taken by sending the supplier a reminder by post, telephone, telex or fax, or by personal visit.

When the goods are received, several actions should be taken promptly to complete the purchase transaction. The procedures used for completing the purchase may vary depending on the particular manufacturing system concerned. The type of goods-receiving activities may be affected by the volume, complexity and inspection requirements of the goods involved. The general procedure for receiving deliveries, however, is as follows:

1. Identify the goods delivered against the advice note or packing note by visual examination, to ensure that the right goods in the right quantity have been received. If a wrong delivery is made, the appropriate receiving staff should decide on the spot whether the goods should be accepted. If not, the goods should be sent back right away.
2. If the goods are to be accepted at this stage, then carry out a technical inspection as required by using appropriate measuring and test equipment. Supplies should not be allowed to the production area unless inspected and accepted.
3. If the goods do conform, prepare a goods-received note to notify the users and record the deliveries. (If a computer-based system is used to deal with purchase transactions, it is important that records of goods receipt are entered into the system promptly – ideally in an online manner.)
4. If it is decided that the goods do not conform, then initiate corrective action. This is dealt with in more detail later in this section.

Quality assurance

For the purchasing function of any manufacturer, the three major systems parameters are quality, price and delivery guarantee of goods. There should be a coherent purchasing strategy which is aimed at striking the right balance among these parameters, under guidelines set up by the corporate strategy. For example, if we now apply the systems principles summarized by the prototype system model, then the following activities can be identified which should be carried out in function block A_{322} to ensure that goods of proper quality are procured:

- understand company objectives;
- define quality specifications;
- select suppliers;
- communication;
- inspection; and
- corrective action.

Understand company objectives

Objectives are referred to in the prototype system model as a 'reference level of desired conditions', which in this particular case will be the goal of the purchasing function, set by its wider system – the company. In accordance with the overall corporate strategy, answers to the following questions should be sought:

- How can a quality assurance programme contribute?
- How should this programme be organized?

- What level of quality is appropriate?
- What kinds of procedure are required?
- If higher quality is required, how much will it cost to implement this? What is the implication of this for the overall strategy?

Define quality specifications

Based on the corporate objective, the right quality specifications must be defined for the particular applications. In many cases, defining the right quality level and specifications for the particular application is not primarily a purchasing responsibility. However, the purchasing function still has to make sure that the specification is adequate and up to date and that suitable inspection procedures are properly expressed in the purchase order.

Select suppliers

Another decision-making process which the managerial function must perform is to choose suppliers who are capable and willing to meet the quality specification. Obviously the choice of suppliers should be made based on considerations such as price, technical capability and reputation. An extremely helpful instrument for making this decision is a data base containing records of purchase history, including information such as the names and addresses of suppliers which have been used previously, together with the quantities purchased from each, prices charged, level of quality achieved and the delivery performance. The availability of this data base helps avoid having to start a supplier-selecting exercise from scratch every time a requisition arrives, which obviously wastes resources. In addition, it is the policy of many manufacturers to purchase only from suppliers able to demonstrate that they implement systems designed to control the quality of the goods or services they supply and to provide proof that those supplies will conform to specification. This data base can then be used to provide information about a strong source of supply which consistently meets specified requirements.

The content of this data base should be kept up to date through continuous supplier assessment exercises which may consist of the following steps:

1. Normally, the purchasing department of the buyer initiates a supplier assessment by issuing a request. The purpose of this initial assessment is to establish that the supplier has the means and ability to ensure that items or services are delivered to conform with the specified requirement.
2. The performance of the supplier will then be assessed by the buyer against technical, quality and commercial criteria.
3. Subject to satisfactory performance, the supplier will then be issued with a certificate indicating the types of supply for which approval to tender has been granted.
4. Details of this approved supplier will then be kept in the data base.

The performance of the supplier will be monitored and records updated accordingly. The supplier's performance should be checked on a regular basis. The supplier's quality performance can then be evaluated by a vendor rating formula such as:

$$\text{Vendor Rating} = \frac{\text{No. of batches rejected}}{\text{No. of batches delivered}} \times 100\%$$

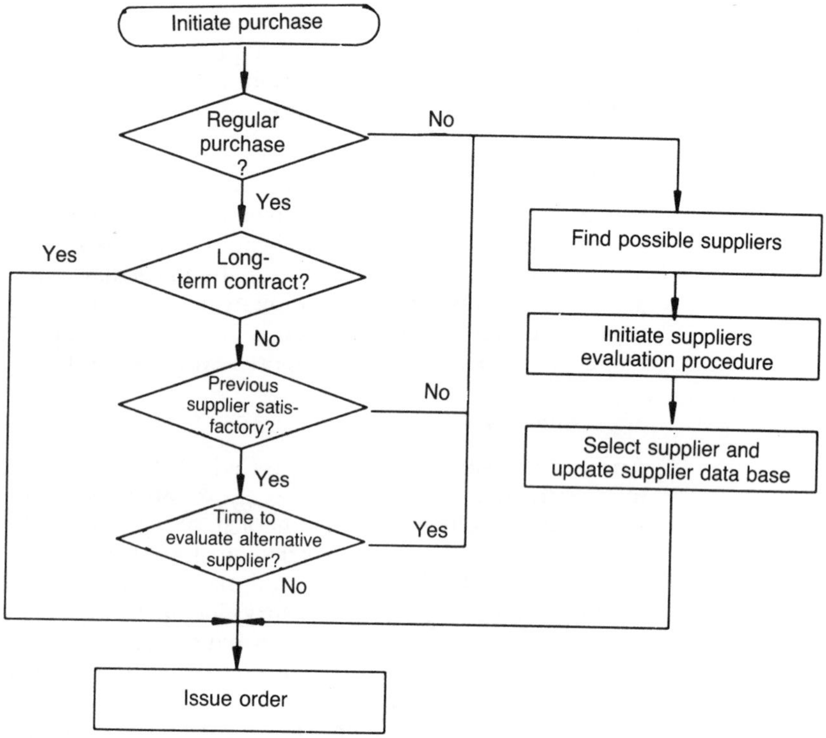

Fig. 3.23 Supplier selection procedure.

For the normal purchasing operation, on receiving a requisition, the purchasing department should establish and maintain an effective supplier selection procedure, such as that shown in Fig. 3.23.

Communication

The chosen suppliers must understand exactly what is expected from their deliveries. When necessary, sheets of technical specifications, engineering drawings or even samples should be included with the purchase order.

The principle of systems thinking tells us that such a problem must not be approached in isolation. We must take note of all those influences which might affect the efficient operation of our own organization. Some of the key issues of the purchaser–supplier relationship are outlined in Box 3.6.

The purchasing function may be considered as behaving as (at least part of) the wider system for the suppliers and sub-contractors, particularly if these vendors are not large in size. Whenever necessary, the purchasing company should try to persuade the supplier to achieve and maintain an acceptable level of quality, both before and during the vendors' production processes. The significance of this is illustrated by the Japanese experience, where the JIT approach has been implemented to produce some spectacular improvements in company performances. It may not come as a surprise that one of the key factors for successful JIT implementation is the close relationship and cooperation between the manufacturer and its vendors.

Specifications and technical instructions regarding the quality of goods must be adequately communicated to the vendors. Sometimes it is even desirable for the purchasing organization to help suppliers to do better by providing assistance and technical advice. This is because when a supplier fails to deliver the right goods at the right time, the purchasing department of the manufacturer itself will also have failed in its function. Therefore, some organizations have devised supplier quality assurance systems for their suppliers to use. Appendix B illustrates a sample document of procedural requirements for supplier quality assurance which is used by such an organization. These companies use such a document both to establish their supplier quality assurance programme and to inform the vendors of their quality expectation. This approach is basically due to the recent recognition that quality is best controlled by those responsible for manufacture and that it is not possible to inspect quality into a product.

Box 3.6 Competition or cooperation: a key issue of the buyer–supplier relationship

Purchasing is a functional area of manufacturing where systems thinking and a systems approach are of vital importance. One aspect of this is that the formulation of strategy and the design of its operational procedures should consider not only the characteristics of the company itself (the internal system parameters such as its size, structure, production practice, product quality and technology used) but also the influences of the external environment (the state of both the buyer's and supplier's markets, the state of the suppliers, and so on).

As a result of some recent successful cases which are based on systems thinking, the buyer–supplier relationship is now receiving increasing attention. Many manufacturers have realized that there is a need to change the nature of their relationships with suppliers. Some of them are in the process of implementing a scheme called Supplier Quality Assurance (SQA), as illustrated by the sample company document presented in Appendix B. The following text examines briefly the key issue that underlies this move.

The nature of the traditional purchaser–supplier relationship is based on competition. That is, the transaction process is driven by the motivation of self-interest from both the purchaser and supplier. Under such conditions, each party will attempt to secure the best deal for themselves without considering the possible effects of this best deal on the other party. In the short term, this relationship can work well because it will stimulate competition between suppliers which forces them to operate more efficiently. In the long term, however, both the buyer and the supplier may suffer due to purchaser–supplier disruption, increased purchasing/selling costs and lack of cooperation.

It has recently been recognized that purchaser and supplier need not have to be in competition with each other – in fact the relationship can be built on cooperation and high trust. That is, purchaser and supplier can work together to compete more effectively in the end-product market, so as to benefit both parties. Compared to the competition relationship, this cooperation-based approach is more systems-orientated. The advantages of this approach include reduced misunderstanding and conflict, less resource waste, better quality control and better delivery performance.

However, it should be noted that not all are convinced that a cooperative relationship between purchaser and buyer is the best approach. Based on the well-tested classical argument of motivation by self-interest, the opponents of this approach fear that suppliers may be encouraged to become less efficient because of lack of competition. Furthermore, there are certain prerequisites for its successful operation, including: integration of planning activities by purchaser and buyer; standardization of parts design; shared new technology and manufacturing methods; and a high degree of trust and intimate cooperation between the two parties.

The most publicized application of this approach in the UK is the case of Marks & Spencer, which has benefited greatly through a close and intimate cooperation with its suppliers, from primary producers to producers of the finished products. In Japan in particular, where the mutual interdependence between purchaser and suppliers is widely recognized, this type of cooperative relationship is very common. Kanban production environments, for example, are established on such a basis. Suppliers deliver goods or services of assured quality to the manufacturer's premises just in time for the production process, greatly increasing production efficiency because of the elimination of both material stock and defective parts. In return, the suppliers gain a secured long-term relationship with the manufacturer and guaranteed demand for their products, allowing them to develop experience-based cost reductions and a high level of quality performance.

Inspection

The fundamental purpose of this function is to check goods received so as to ensure that they conform to the specified requirements. Traditionally, inspection has been the prominent function in quality control practice. Recently there has been a trend to change the emphasis of inspection control to one of audit control. Nevertheless, inspection still remains one of the most important means of ensuring that product quality is being maintained by vendors.

It should be remembered that inspection is not a value-adding function. In addition, it can also be expensive to carry out. Therefore, the general rule for inspection is: do without it unless it is necessary. If inspection must take place, however, then the policy and method used should be adequately devised so that the system ensures that the items or services delivered meet specifications efficiently and economically. The procedure may be based on a 100 per cent technical check, which is relatively expensive to conduct but carries a high quality guarantee. It may also be based on a sampling process so that only some of the goods are inspected to reduce the cost involved but with less statistical confidence. To choose the right practice, therefore, the following questions should first be answered:

- Which order/delivery should be inspected?
- Which feature of the items/services should be inspected?
- Should a 100 per cent inspection be conducted?
- If sampling inspection is appropriate, which method is the most efficient and economical?

There are general rules which can be applied to obtain the answers to the above questions. For example, if destructive inspection is involved then a sampling approach is inevitable; on the other hand, 100 per cent inspection must be applied if the consequences of a defective item being used are disastrous. Some sophisticated analytical models are available which provide more specific solutions to the problem by considering the possible costs involved for a particular inspection scheme. The following simple analysis illustrates the basic principle of these models.

Given the number of items in a batch (N), the proportion of defective items in the batch (p), the cost of inspecting one item in the batch (C_i), and the expected consequential cost of a defective item (C_f), the cost of 100 per cent inspection is given by NC_i, and the consequential cost of expected defectives is pNC_f; 100 per cent inspection is advised if $NC_i < pNC_f$. That is, it will be more economical to conduct 100 per cent inspection if $p > C_i/C_f$. On the other hand, if $p < C_i/C_f$ then it is cheaper not to conduct any inspection.

In the real world, however, what is required is usually something between the two extremes of 100 per cent inspection and no inspection at all. British Standard BS 6001 can be used to obtain the appropriate sampling plan for a particular specification. This is outlined in Box 3.7.

Inspection may also be carried out, not on the items themselves, but on the supplier's quality assurance system – system structure, procedures, documentation, measuring and test equipment, and so on – as outlined in the sample document of quality assurance given in Appendix B.

Box 3.7 Choosing a sampling inspection scheme

A sampling inspection process is carried out by first randomly selecting one or more samples from a batch. These are inspected and the result then determines whether the whole batch should be accepted or rejected. This technique can be conducted in several ways, the following being the most widely used.

Single sampling

This is the simplest sampling procedure, which consists of the following steps:

1. Randomly select n items from the batch under inspection.
2. (a) Accept the batch if only m or fewer defectives are detected; otherwise
   (b) reject the batch.

This is usually referred to as an $n(m/m+1)$ inspection scheme. For example, 100(4/5) describes an inspection scheme which checks the quality of a batch from a sample of 100 items and accepts the batch if there are no more than four non-conforming items in this sample.

Double sampling

This scheme uses up to two samples to determine whether or not to accept a batch in the following manner:

First sample

1. Randomly select n_1 items from the batch under inspection (for the same level of assurance n_1 is smaller than the single sample size n).
2. Accept the batch if only m_1 or fewer defectives are detected, or reject the batch if m_2 or more defectives are detected (for the same level of assurance, m_1 is smaller than m, and m_2 is larger than $m + 1$, m being acceptance number of single sampling); otherwise go to step 3.

Second sample

3. Randomly select n_2 items from the rest of the batch and consider the combined sample of size $n_1 + n_2$.
4. (a) Accept the batch if only m_3 or fewer defectives are detected from the combined sample; otherwise
   (b) reject the batch.

In a similar way, this is referred to as a $n_1(m_1/m_2)$, n_2, $n_1 + n_2(m_3/m_3 + 1)$ inspection scheme.

The purpose of double sampling is to increase the efficiency of the inspection process by using a smaller sample size while retaining similar performance. For the same level of assurance, an inspection process conducted by using an appropriate double sampling scheme will in general need less inspection work than that required by a single sampling scheme, since the second sample is not always necessary. Based on the same idea, a third, fourth, or even fifth sample can be taken during an inspection process. This is known as *multiple sampling*. However, sampling schemes with more than two samples are rarely used in practice, because of the complexity involved in establishing and maintaining such a scheme. In addition, there are various other types of procedure for increasing sampling efficiency. These can be found in the relevant textbooks.

Performance of a sampling scheme

The efficiency of a sampling scheme is demonstrated by what is known as its *performance curve*, an example of which is given in Fig. 3.24. The performance of a scheme is shown by plotting its probability (Q) of accepting a batch against the probability (p) of an item in the batch being defective. Such a curve can be used to reveal the probability of wrongly accepting a batch when it should be rejected, and of wrongly rejecting a batch when it should be accepted. This is explained below.

One of the most important parameters when choosing a sampling scheme is what is known as the acceptable quality level (AQL). AQL is defined as 'the maximum per cent defective (or the maximum number of defects per hundred units) that, for purposes of sampling inspection, can be considered satisfactory as a process average (BS6001: 1973, p. 5).' This is shown as p_a in Fig. 3.24 (which is often taken as the value of p that gives a the value of 0.05 in practice). Now if a batch of delivery has a proportion p_1 non-conforming (that is, if p_1 is the probability of an individual item in the batch being defective), where p_1 lies between 0 and p_a, then it should be accepted. However, at this quality level the sampling scheme represented by this performance curve will on average accept only $100(1 - a)$ per cent of the batches delivered. That is, the scheme will wrongly reject $100a$ per cent of the batches. This is why a is usually

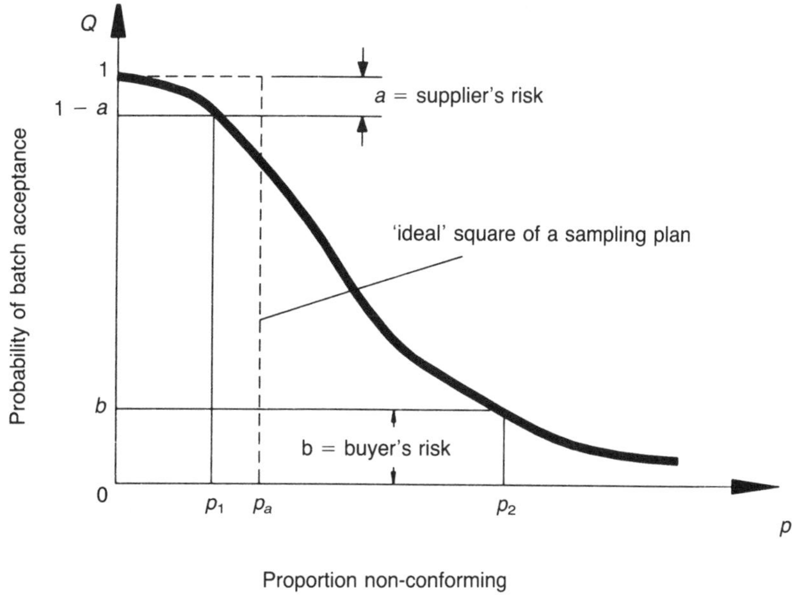

Fig. 3.24 Characteristic sampling curve.

referred to as the *supplier's risk*. Similarly, with this scheme the buyer also has to take certain risk of wrongly accepting a batch when $p = p_2$, shown as b in Fig. 3.24 and known as the *buyer's risk*. Obviously, an ideal inspection scheme should exhibit a 'square' performance curve as shown in the figure. The actual performance curve of a scheme therefore illustrates the extent to which the scheme approximates to the ideal situation.

Use BS 6001 to choose the appropriate sampling scheme

Choosing the appropriate sampling plan for a given performance curve is not always an easy task. For example, for a double sample procedure there are five parameters whose value must be determined in order to describe the scheme. In practice, therefore, it is recommended that the standard procedures such as those provided by BS 6001 be used to facilitate the process.

BS 6001 is widely adopted as the standard for acceptance procedures used in Britain. It is also in line with the major acceptance sampling procedures that are used internationally. With the aid of this standard, the task involved in choosing the appropriate sampling scheme for an application is reduced to the following steps:

1. Decide the inspection level, which is a measure of the 'squareness' of the performance curve. BS 6001 offers seven levels (S-1, S-2, S-3, S-4, I, II, and III in increasing order of squareness), of which level II is recommended for general applications because it provides an adequate compromise between inspection accuracy and sample size.
2. Based on the batch size and inspection level, decide the sample size(s) by using a sample size table provided. BS 6001 lists the appropriate sample sizes for single, double, multiple, as well as some other types of sampling procedures.

3. By following a similar procedure, decide the value for the required quality level.

Average outgoing quality

When dealing with the issues of inspection control, the sample company document defining operational procedures for quality assurance (Appendix B, para. 3.6.2) demands that:

> Sampling procedures used by the vendor are to be in line with BS 6001, General Inspection Level 2, single-sample plan, but with no allowable rejects permitted within the sample. Should a reject be found during sampling, revert to 100% sorting for defective features.

By now we know exactly what is required by the first half of this paragraph. However, understanding the meaning of the rest of the statement requires the introduction of another important concept involved in sampling inspection – that of *average outgoing quality* (AOQ).

If rejected batches are 100 per cent inspected and all non-conforming items replaced, as demanded above, then the expected proportion non-conforming of the items that have passed the inspection is given by pQ – where p is the probability that an item is defective and Q the probability that it has avoided the scrutiny of the inspection process. This is the AOQ. By referring to Fig. 3.24, it is clear that the AOQ of a given sampling scheme, when the defective replacement rule is enforced, is in fact given by the product of the coordinates of its performance curve.

If an AOQ curve is plotted against p it exhibits a shape as illustrated in Fig. 3.25. The graph shows that the AOQ reaches a maximum value of Q_m at p_m, because for p values below p_m the value of AOQ decreases as the quality of the deliveries itself increases; for p values above p_m the AOQ also decreases as the quality of the accepted deliveries again increases as a result of 100 per cent inspection being applied to more

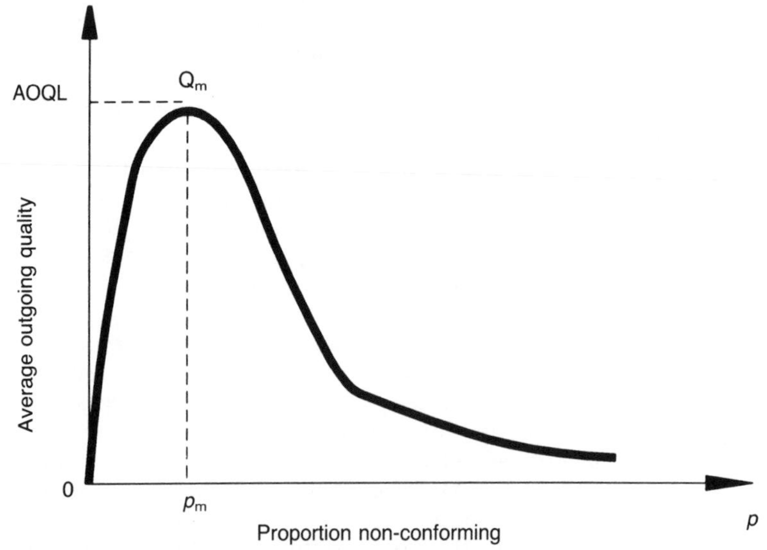

Fig. 3.25 Average outgoing quality curve.

batches. The maximum value that the AOQ is allowed to reach by a sampling scheme is called the average outgoing quality limit (AOQL).

Therefore, the inspection procedure demanded by the sample document in fact intends to implement a long-term guarantee that, irrespective of the actual quality of the supplies, the proportion non-conforming after the inspection function will not exceed a pre-specified level – the AOQL of the chosen inspection scheme.

Corrective action

The purchasing department should establish and maintain prompt and effective corrective procedures for controlling supplies and/or services that do not conform to the specified requests. These procedures should include provision for identification, segregation and disposition as appropriate. These must include: appropriate action upon receipt of non-conforming supplies to ensure only acceptable supplies are processed; defective processes and work operations to help suppliers eliminate potential causes of non-conformance; and analysis procedures for properly granting concessions. As a supplement to the previous set of questions, therefore, the following two additional questions should also be answered:

- What procedures must be used to communicate with suppliers about non-conforming items/services?
- What procedures must be used for dealing with defective items which have been delivered?

Some of the corrective procedures should be agreed with the supplier. For example, rejection of non-conforming items/services will be cause for negotiations with the supplier to ensure adequate replacement. All reject supplies should be clearly identified and promptly removed from the production routes to prevent unauthorized use or mixing with good material or components.

Non-conforming supplies may be subjected to further sorting (as outlined earlier), reworking or even scrapping. Sometimes a concession procedure may be established for items which do not conform to requirements (usually when the items concerned are too expensive to be scrapped). A concession should be applied through a formal procedure. Concessions must be carefully investigated and scrutinized by all parties concerned before being granted. This makes it necessary for a concession application to be submitted for investigation to the technical department for design implications, to the sales department for customer requirements, to an approvals manager for approvals specification and to the quality department for adherence to procedure, concession application control, registration and issue control.

3.3.6 A_4: Producing products

We now come to where the actual transformation processes take place – where metals are cut, parts are manufactured and components are put together to construct the end-product. This is basically where most of the hardware tools of AMT are located. This function block consists of the tasks listed below and shown graphically in Fig. 3.26:

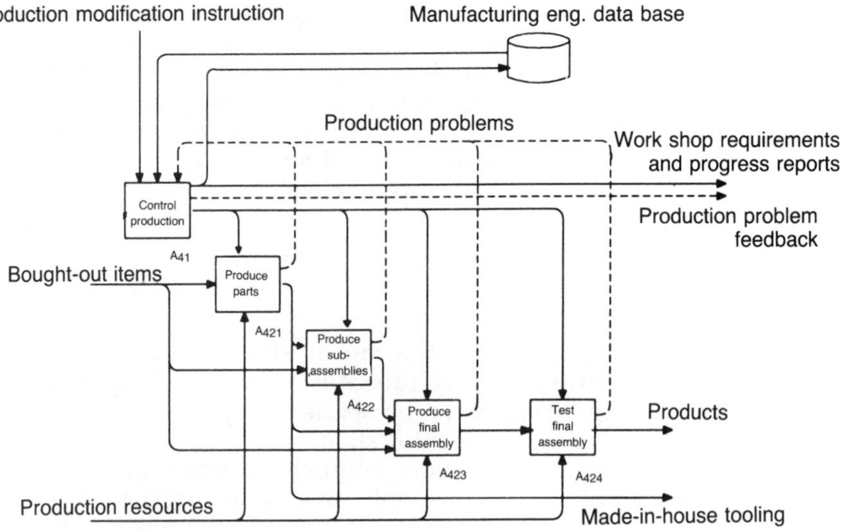

Fig. 3.26 A_4: Producing products.

A_4 PRODUCE PRODUCTS
- A_{41} Control Production Activities
- A_{42} Carry Out Production Activities
  - A_{421} produce parts of products
  - A_{422} produce sub-assemblies of products
  - A_{423} produce final assembly
  - A_{424} test final assembly

***A_{41}* Controlling production activities**

In decomposing the 'produce products' function block, once again control is at the head of the activity chain. We are concerned here with the control of shop-floor activities. Within a manufacturing control system, shop-floor control represents the lowest level of the control hierarchy, as shown in Fig. 3.27. Production control of shop-floor activities is the focal point in the transformation of materials into finished products. It is concerned with the detailed execution of the master production schedule. Its main functions include the determination of a detailed plan of production for each part, component or assembly of the end product and also corrective action when problems arise in the shop. Hence its output comprises shop-activity scheduling, a production process plan, production and corrective instructions, and production status reports.

Shop-activity scheduling and control is the most important function of this section: work issued to a shop must first be planned in such a way that all of the parts, components and sub-assemblies of the end product are produced and delivered when they are needed with the lowest possible production cost and highest possible production efficiency. The problems and difficulties associated with such a planning task, as well as some of the techniques that may be used to aid the planning process, are reviewed in the second part of Appendix A. Some of the other important issues concerning shop production control and processing planning are reviewed in Box 3.8 and Box 3.9, respectively.

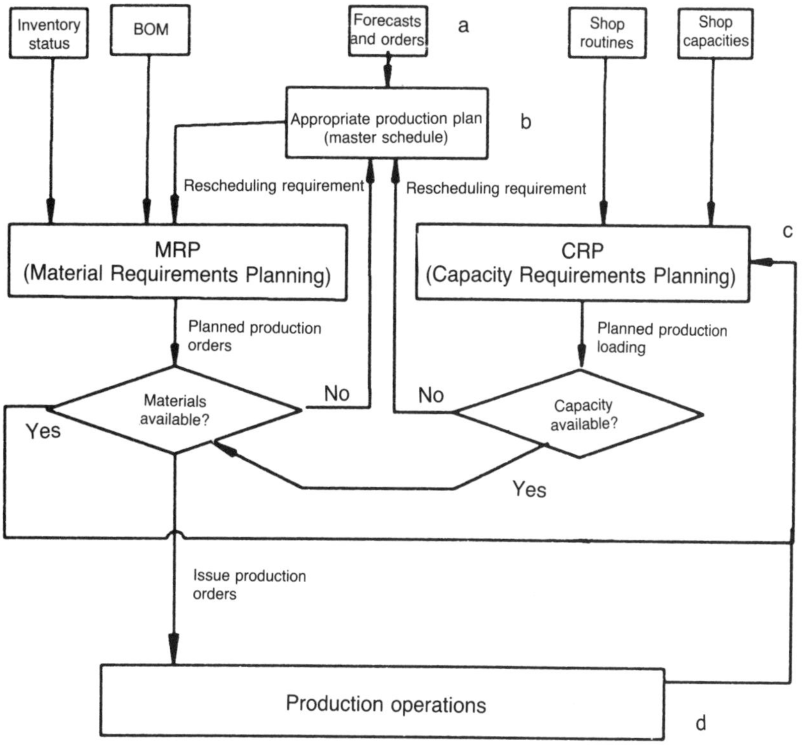

Fig. 3.27 Levels of manufacturing control. Key: a = time horizon – one to two years; b = time horizon – a few months; c = time horizon – weeks or days and d = time horizon – hours or minutes.

Once a work order is issued to the shop, the next task for production control is continuously to monitor the progress of production by tracking the flow of work from task to task, so as to make sure that all the items are processed in time and delivered to the appropriate subsequent stations. Collecting status information on work in progress is one of the key requirements of a shop-floor control system. The ability promptly to update the production plan and to inform higher-level control and customers depends on effective data collection and transfer. Once again, therefore, we see the need for feedback status reports, as illustrated in Fig. 3.26. When problems occur during the production operations, concerning the work schedules or production quality, reports should be sent to the control function so that the necessary action may be taken.

Box 3.8 Centralized or decentralized control: which is best?

So far, we have come across the word 'control' several times as we have moved down the manufacturing systems model from the master planning level to the level of shop-activity control. It is clear that this is because control activities, being system functions, are hierarchical in nature. Within a manufacturing system, in particular, the control hierarchy may be as shown in Fig. 3.27. This diagram reveals that the control functions at different levels must operate in series: the aggregate planning system and the MRP system for example, must provide a shop-floor production plan to the shop scheduling system. However, each level of control has its own time horizon, objective and technical requirements. First, the time horizon of control at each level becomes successively shorter, from several months at the aggregate planning level down to several days or even hours at the shop control level. Second, as indicated in Appendix A, each level of control looks at the manufacturing environment from a different perspective. That is, the objectives of the control functions differ from level to level. At the aggregate planning level, for example, the aim is to plan and control production and capacity so that product cost is minimized to maximize profits. But at the shop-floor level the planning and control function seeks to schedule the shop activity sequence in a way that will meet the production plan efficiently. Finally, each level will need different technologies to support its particular dynamic characteristics and information requirements.

It is important, therefore, that each level in the control hierarchy functions efficiently in relation to the higher-level needs. As shown in our $IDEF_0$ model, at each level control is exercised to meet local objectives, but with higher-level constraints. If any problem occurs that makes meeting higher-level constraints impossible, then feedback loops will result in other sets of constraints being introduced until feasibility is accomplished. This description of the system of control hierarchy is a widely-accepted representation of the control structure which should exist within a manufacturing system if efficient manufacturing is to occur. The advantage of systematically dividing the overall system into levels is that it simplifies the task of system design and analysis by allowing the particular set of problems at each level to be analysed and solved individually.

Although the general requirements are well understood, when it comes to modern practice there seem to exist two very different approaches in the design and implementation of manufacturing control systems: these are known as the *centralized* and the *decentralized* approaches. They will now be reviewed and compared below.

Centralized control

Centralized manufacturing control is based on the observation that production control is an interrelated chain of events, all of which must occur and agree for efficient control of overall production operation. The key to this, it is argued, is the integration of the information system and its data base. In order that the control functions all act dependently they must be 'glued' together by a data base that flows through an integrated information system. As a result, a centralized control system often contains an integrated single system which attempts to run everything within a manufacturing system. Manufacturing resources planning (MRP II), for instance, is a typical example of such a centralized approach which uses a generalized data base for overall systems planning and scheduling. A system based on this approach is sometimes referred to as computer-aided production management (CAPM) or integrated manufacturing control (IMC). Such a system often requires sophisticated computer hardware

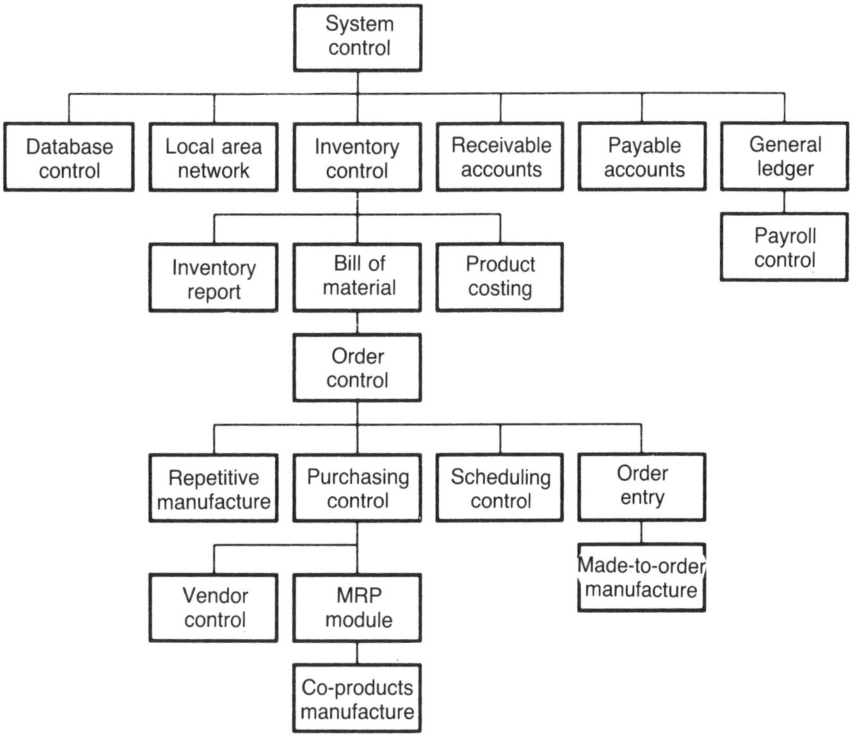

Fig. 3.28 Example of an integrated manufacturing control system.

and software comprising 20–30 software modules designed to carry out different planning and control tasks at different levels, but all working around a single centralized data base. Figure 3.28 illustrates the typical structure of an integrated control system. This is in fact the modular structure of a popular PC-based manufacturing control system called FOURTH SHIFT which consists of a series of software modules arranged around a centralized data base. According to the product brochure:

> The FOURTH SHIFT Manufacturing Software Series is a family of computer applications . . . designed to support decision making in a manufacturing company. FOURTH SHIFT communicates consistent priorities to all people in an organization through its common data base, on-line inquiries and exception-oriented reports. It establishes a framework in which the whole company can work more effectively together.

This illustrates clearly that such a system will attempt to take virtually every aspect of manufacturing systems control into consideration. It has been shown that if successfully implemented, an integrated control system could result in improved productivity in terms of improved output rate, reduced overtime, reduced inventory investment, and improved customer service levels. Stories of success with this type of control system are not hard to find; but they by no means tell the whole story.

Decentralized control

The complexity of manufacturing system control is fundamentally due to the enormous quantity of data required in order properly to support the planning and control functions. In a centralized control system it is common for a manufacturer to have to maintain accurately 5 million individual pieces of data to support a component part scheduling function. The requirement to maintain such a large amount of data in an appropriate, consistent and accurate condition makes it extremely difficult to implement a centralized manufacturing control system successfully.

With a centralized approach, therefore, the system may become overdetailed because far too many information flows are required. As a result prompt gathering and communication of information to each individual area of the manufacturing process becomes difficult to achieve, if not impossible. According to one authority (Parnaby, 1989):

> As a result, all the data sampling time interval requirements, essential to effective control of manufacturing complexity, cannot be met. The personalized requirements of the individual control functions and supervisors cannot be met. The system therefore fails to meet local needs. It becomes too detailed, too complicated, too sophisticated and no one can understand it. The database becomes too involved and too complex, and is run by a committee which is too large, cross-functional and cannot agree.

A decentralized control approach will attempt to simplify the tasks involved by eliminating the complexity. Figure 3.29 illustrates the basic structure of a decentralized control system. As can be seen, with such an approach the entire control system is divided into individual control areas, each of which uses a dedicated local planning and control system. Since the technical and structural characteristics differ from area to area, the area controls will be specially designed to meet local product and local cell needs. In fact, each of these local areas may be viewed as a self-contained manufacturing module with a set of dedicated software tools to support its effective operation. The operations of these manufacturing modules will be coordinated by an operations planning system (such as a MRP II) at the top level, the aim of which is to harmonize the production of the individual cells so as to meet the needs of the whole business operation.

The features and advantages of a decentralized control system are as follows. First, the fundamental requirement of a decentralized system is a

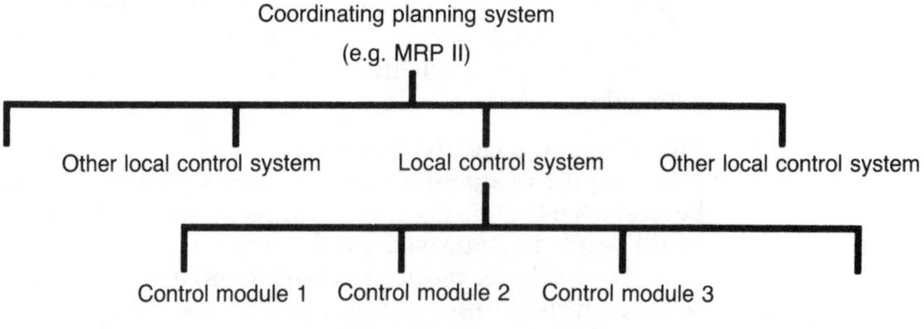

Fig. 3.29 Structure of a decentralized control system.

cellular layout of manufacturing. The cellular layout is closely related to the ideas of group technology (GT), which attempts to bring together and organize common expertise in order to improve productivity. Within a cellular manufacturing environment, manufacturing facilities are divided into production cells. Each cell is designed to produce a family of parts which could be either a set of end products or just part of the total product. This set of parts must be produced by similar machinery, tooling, machine operations, jigs and fixtures and production expertise. Normally a cell is responsible for all the operations needed to transfer raw material into finished parts. The advantages of such a manufacturing layout include reduction of necessary control, reduced material handling, reduced set-up time, reduced tooling, reduced work-in-propress, reduced expediting, increased operator expertise and job satisfaction, and improved human relations.

Second, the top-level control system plays a coordinating role and therefore needs only to produce the overall production plans for the cells, thus significantly simplifying its implementation and operation tasks.

Third, planning and control in each cell can be carried out by a simpler and tailored control system, allowing the local cells to exercise the discretion needed to deal with their particular local requirements.

This approach has also been termed a *human-centred* approach, because if a manufacturing cell is simple enough then it may be possible to use simple manual or PC-based computer systems to control its operation, with a greater degree of human involvement. This can be of tremendous advantage because people are much better than computers when it comes to using intuition and experience and making subjective judgements. This is illustrated by the success of the kanban environment in which scheduling, planning and control functions are carried out by the people who are directly involved in the production operation, without the need for expensive and complicated centralized service. The European Commission has been funding development work in human-centered CIM systems through its Esprit information technology programme. This has resulted in a number of pilot plants being implemented. In the UK, for example, the organizations involved in this development include ITT, Dorman Smith Switchgear and BICC (*Electrical Review*, 1990).

However, the decentralized approach is not without disadvantages. First, a cellular manufacturing layout is known to reduce shop flexibility, with possible undesirable effects on machine utilization, job flow times and job tardiness. Second, its applicability is limited to only a certain type of manufacturing system. Finally, if the formal control link ends at the cell level with a manual activity, then the communication process cannot take place in the same way as with a centralized system. Greater attention, therefore, must be paid to the design and operation of status reporting procedures in order to effectively monitor and assess the cell-level operations.

So, centralized or decentralized: which is the best? Unfortunately, once again, there seems to be no ready answer to the question – as with so many other aspects of AMT, there is no magic formula which will solve all manufacturing control problems. Centralized or decentralized, the use of computer hardware and software cannot *cure* all the problems which are inherent in a manufacturing system – it will only *aid* the decision-making process.

Box 3.9 The role of process planning in a computer integrated manufacturing system

Production process planning is another important manufacturing function which must be carried out by function block A_{41} before the actual production processes can begin. Process planning may be defined as the function of transforming design information into work instructions. It covers the interpretation of drawings, the selection of raw materials in the correct form and shape (note the overlapping of this and the design and planning functions), the selection of the correct tools and production methods and the sequencing of operations. The objective of this function is to produce process plans that enable parts to be manufactured to the correct specification and in the most economical manner.

Process planning uses the information provided by the function of detailed design (that is, function block A_{223}, including part drawings and the bill of materials) and leads a workpiece through the processes until the part is completed. It is an extremely important part of the total manufacturing process – any mistakes here will have a significant and undesirable impact on subsequent production operations. It is also a very tedious function which demands much skill and patience, since a process plan has to be an extensive document that details manufacturing methods and steps, including some or all of the following items: the raw materials required, their shape, form, specification, dimension, and condition; the operations to be carried out, the machining operations used and the sequence of their execution; the tools, jigs and fixtures to be used with each operation; the machining data; process parameters such as temperature and atmosphere; inspection requirements and criteria; the standard time allowed for the operation; and storage and transportation instructions. Included in this document should be a comprehensive set of data which has many uses throughout the organization. Apart from constituting work instructions to the shop floor, the cost-control function will also need the data to estimate machining and labour costs; and the control function itself needs this information to schedule production.

With the traditional manual-based system, process planning relies heavily on the knowledge, skill and experience of a trained production engineer. He or she has to understand fully the production processes and the associated tools that can be used to produce the parts or fabrications. He or she must also know about the alternative ways in which the same jobs can be done, so that the most effective methods and/or technical routes are chosen. Computer-aided process planning (CAPP) systems are currently under development and a few have already been introduced into industry. The main reasons for the development of CAPP are to improve the efficiency of the process planning function; to achieve reliable and consistent process plans; to retain the knowledge of process planners; and to use and integrate company data. To achieve these objectives, a CAPP system needs to provide information for work order routeing, production scheduling and cost estimating; to operate as an integrated planning aid to obtaining relevant data automatically from the data base and generating a set of process plans accordingly; to allow online transactions for continual update of information; and to be structured in such a way that expansion and modification can be easily accomplished.

The key element of a CAPP system is usually a knowledge-based expert system – a rule-driven software system which seeks to emulate the reasoning capacity of an expert in a particular area. One way in which such a system works is to store knowledge as rules which are then proved true or false as the user executes the software. Eventually a conclusion will be reached depending on which rules are proved true. To be truly 'intelligent' a CAPP system should also have optimizing features which are able to utilize the available resources in an efficient manner. To date, development of successful knowledge-based CAPP systems has proved very difficult because of a number of problems, the most significant being extracting expert knowledge from the expert – how can this be done efficiently, quickly, and in the certainty that all the relevant knowledge has been collected and anything irrelevant ignored? Recently this has been attracting much attention.

CAPP can be of two types, *variant* or *generative*. Variant CAPP systems rely on the parts to be processed being similar to each other. This type of system can be used effectively, for example, when GT manufacture has been implemented. They are basically storage, retrieval and edit systems for standard process plans of families of parts and are based on classification and coding (as such they are less 'intelligent' than the generative type outlined below). A huge effort is usually required in the classification, coding and creation of part families. These systems are rather inflexible and still need an experienced planner to edit or modify the plan. Furthermore, with a variant approach the CAPP system becomes less effective when there is high variability of either part design or production method.

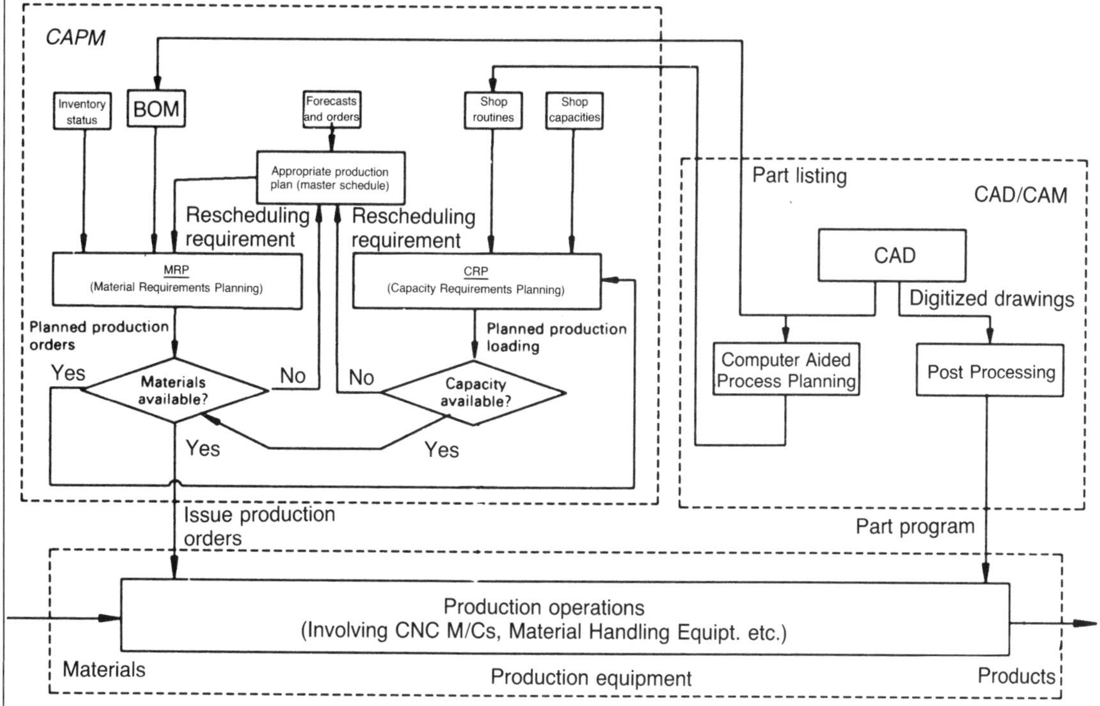

Fig. 3.30 CAPP: an important link within the integrated manufacturing environment.

Generative CAPP systems are principally advisory systems which set out to develop a process plan for each part from scratch, beginning with the part drawing. This approach, therefore, overcomes the problems associated with the variant approach because no previous manufacturing plans are needed. Most systems of this type still rely on interaction with the system user to gain correct information about the part, that is, they can only interact indirectly with CAD. However, recently there have been a number of attempts to develop generative process planning systems that interact directly with a CAD system, but their real application value is still rather limited.

In conclusion, to integrate fully various manufacturing functions we must look to CAPP, as shown in Fig. 3.30. Although automatic production of NC programs is now almost a reality (this is sometimes known as CAD/CAM), it is clear from the literature that CAPP is one element of CIM that has not as yet enjoyed as much success as those in other AMT areas. There is a long way to go before fully practical CAPP systems are available that can be made part of a real CIM system.

A_{42}: Carrying out production activities

Here we are concerned with a sequence of production operations through which components are manufactured to correct design and specification and are gradually made into the end product. In general, a manufacturer (discrete production) will employ four shop activities to accomplish its production operation, which are designated by production blocks $A_{421}-A_{424}$, inclusive. These four blocks perhaps indicate the manufacturing areas where the state-of-art 'island' types of AMT have been most successfully applied (except for the final assembly function, which to a large extent still remains a human-intensive activity). A sound understanding of the techniques and methods involved in these production areas is of vital importance to manufacturing systems engineers. The relevant concepts in these areas will include the various aspects of manufacturing technologies. We cannot, however, cover every aspect of these topics in detail – that would be neither practical, because there is insufficient space in this book, nor necessary, since many of them have been discussed in detail in a number of well-written books in this area (Groover, 1987; Williams, 1988). What is needed here is a knowledge and appreciation of the techniques, the scope and, more importantly, the roles they play within the overall manufacturing framework. For this purpose the relevant topics will be described briefly below.

Factory layout

Traditionally, the physical distribution of manufacturing facilities has been to a large extent dictated by the functional aspects of the operation. That is, in accordance with the characteristics of product identified in section 3.3.1, a number of basic types of facility layout have conventionally been used – project layout, functional layout, product layout and group layout (Wild, 1989).

While to date the above scheme is still very much in force, the more recent approaches to factory layout seem to have paid much more attention to the control aspects of manufacturing systems. Two such approaches will be outlined here, both of which aim to reduce complexity and hence have profound implications on the structure of control systems involved (in practice these two approaches can be interrelated).

The first is called *cellular manufacturing*, and divides a total manufacturing system into business divisions. A business division is basically a self-contained product-group unit designed to serve a specific market. Thus within a manufacturing system one division will deal with those well-established products characterized by high-volume and low-variety production, while another is concerned with new or made-to-order products with their associated low-volume and high-variety. Each of these business divisions is supported by a number of autonomous or near-autonomous cells with their own dedicated workforces and control systems (Choobineh, 1988; Henry C. Co and Araar, 1988). This type of cellular manufacturing is closely related to the concept of decentralized system control, whose advantages were listed in Box 3.8.

The second is a Japanese manufacturing philosophy known as *just-in-time* (JIT) manufacturing, which has attracted worldwide attention (for perhaps two main reasons: the difficulties which have been experienced with the MRP approach and the successes which the Japanese manufacturers have had with their approach). JIT can be viewed as the combination of a simplified structure of manufacturing facilities and an integrated manufacturing planning and control approach, with the ultimate aim of achieving an economic batch quantity of one, zero inventory, reliable and predictable production, and zero disturbance.

The JIT methodology makes the assumption that inventory and work-in-progress (WIP) are used as buffers which disguise inefficiencies and problems that arise in the system. If these problems are viewed as rocks, then the level of such a buffer is equivalent to the level of the water which conceals these rocks underneath the surface. With zero inventory the level of the water will reduce so that system defects are exposed and can than be targeted for improvement.

The implementation of JIT is usually carried out through a system of kanban cards to control materials flow and ensure that stock is kept to a minimum. A work centre trades a card to obtain material and receives one for a finished component. This type of system is nowadays known as a *pull production* system, because a work centre is only authorized to produce when it receives a kanban card from another work centre downstream. No work centre may produce parts just to keep itself occupied. For every part that is being processed in the system, therefore, there must exist an immediate need. The jobs involved are consequently pulled step by step over the various production processes; this contrasts with the approach usually adopted by the conventional control methods (including MRP), where jobs are 'pushed' through the system.

The key consideration in a kanban system is the rate of production rather than the size of the discrete batches. The number of cards on the line is controlled in such a way that build-up of WIP on the line is avoided above the level allowed by the number of kanban cards. The main system parameters here include the number of kanban cards or containers, N; container capacity, C (parts/container); work-centre lead time, T (hrs/container); and the demand rate for parts, R (parts/hr). The relationship between these parameters can be derived by applying one of the basic equations of system dynamics (see section A.2.1 of Appendix A). When viewed as continuous flow process and under the equilibrium condition, the delay that a part experiences in a system is equal to the total number of parts in the system

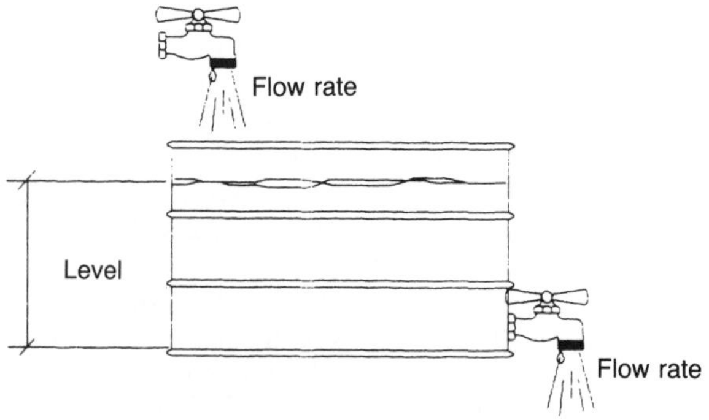

Fig. 3.31 A continuous flow situation.

(or the level, in units) divided by the flow rate of the parts (parts/hr), as illustrated in Fig. 3.31. In a kanban situation the delay is equal to the work-centre lead time, T; WIP is given by NC; and the flow rate is equal to the demand rate for parts, R.

It should be noted that under normal conditions the delay experienced by the parts will be the sum of the work-centre lead time and the waiting time for the next operation. Under the ideal JIT condition, however, the parts should be immediately processed by the subsequent work centre to keep a continuous production flow. Consequently the delay experienced here would be determined by the process lead time only. Applying the relationship outlined above gives:

$$T = \frac{NC}{R}$$

That is, under the specified operational conditions, the number of kanban cards required to keep the WIP level in the system to a minimum is given by:

$$N = \frac{TR}{C}$$

In practice, however, a very small amount of safety stock should be carried between successive operations, so that the formula more generally used to calculate the number of kanban cards required in practice is given by:

$$N = \frac{TR}{C}(1 + s)$$

where s is a factor for safety stock whose value should be set at no more than 10 per cent. Once the number of kanban cards and the capacity of the containers are determined, the following operating rules should be applied during production:

1. Each container must have one and only one kanban card.
2. The parts must always be pulled.
3. No production must be authorized without an unfilled parts container.
4. No more and no fewer parts must be produced than the capacity of the container.

The JIT approach demands organizational and/or technical support from every functional aspect of the company. For example, a closely related concept to the successful implementation of JIT is that of total quality, which aims to achieve satisfactory fulfilment of customer needs at lower production cost, through: appropriate organization and operational structure (for example, cellular layout of operational units); close ties between design and production; continuously improving quality levels (or a company-wide quality improvement with a 'no defects' goal in manufacturing); continual improvement in all functional areas involved; total employee participation; and close ties with vendors.

JIT therefore constitutes a new way of thinking in manufacturing, rather than merely a new technique. In fact, it is an excellent example of a systems approach to problem-solving which seeks to optimize the system as a whole. Such thinking is also being reflected in other areas of development – for example in so-called concurrent engineering.

AMT components and level of automation

Figure 3.32 shows a family tree of the components of AMT (exclusive of CAD/CAM and CAPM) and their relationships to each other in terms of level of automation.

At the lowest level are the building blocks which can be either programmable, such as CNC machines, material-handling/welding robots and automated guided vehicles (AGV); or fixed automatic equipment, such as dedicated assembly machines. From an operational point of view, the two most important features of programmable equipment are flexibility and repeatability. Because such equipment is controlled by a program, it can be easily instructed to accomplish a range of different tasks. In addition, by means of NC, repetitive operations can be performed with consistent accuracy. Therefore, when adopted correctly they will be able to carry out the tasks involved in function blocks A_{421}–A_{424} efficiently and with improved product quality. Although fixed automation can be used to the same effect, due to its lack of flexibility it can only be applied where high-volume and low-variety parts are involved.

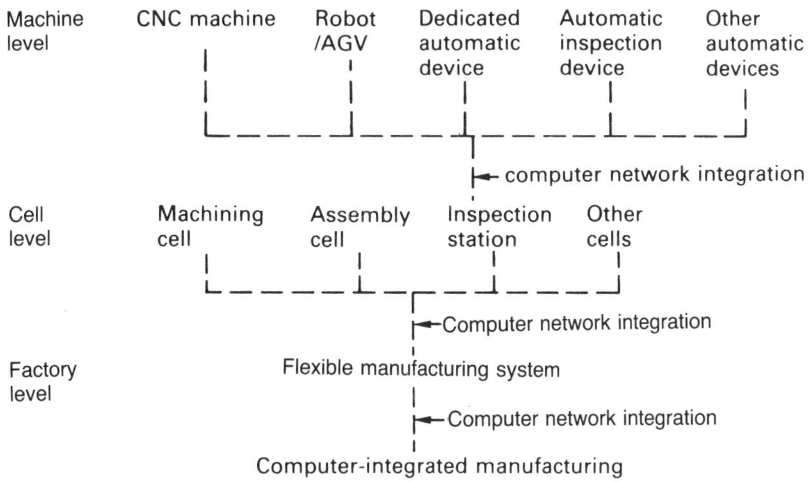

Fig. 3.32 Components of AMT.

At the higher levels of automation, the basic AMT blocks are integrated into flexible cells and systems, and then into a totally integrated environment known as computer integrated manufacturing (CIM), where manufacturing functions are no longer independent entities and thus the versatility offered by AMT is utilized from a systems perspective rather than at the functional levels of $A_{421}-A_{424}$ alone. Integration of individual manufacturing activities is achieved through the use of computer networks, which are a typical example of communication systems where an information sender will be able to communicate to an information receiver through a transmission channel (see section 2.3.2). At present, the main type of network used in this regard is the local area network (LAN), which uses data cables to permit data transmission over up to a kilometre. Several networking standards can be adopted for manufacturing control, among which General Motors' manufacturing automation protocol (MAP) seems to be the most widely accepted (Williams, 1988). More detailed treatment of these individual techniques can be found in the literature cited at the end of this chapter, which provides sufficient materials on the AMT aspect of function block A_4. However, it should be pointed out that one manufacturing area where AMT has not made as much impact as in other areas is the final assembly operation, that is, within function block A_{423}. The reason for this is illustrated by the case study presented in Box 3.10.

Box 3.10 Limitations of AMT: assembly operations. The case of CLD Equipment Ltd

The level of automation in most of the machining and process areas at CLD Equipment Ltd is quite high. However, all the assembly processes are still carried out entirely manually. This is not uncommon in modern industry. Although there are many cases where dedicated automatic or flexible assembly systems have been successfully applied, technology is not yet available for every application and manpower still remains as an important factor in this area (Lotter, 1985). The CLD production management is fully aware that the overall efficiency of the company's system is not only determined by how quickly and flexibly it can produce the individual parts, but also by how efficiently it can assemble them into end products. Having used sophisticated means to improve its design and parts production operations, the management is eager to seek solutions to the problems associated with its traditional manual assembly systems. After an extensive study of all the possibilities, however, it seems that manpower flexibility in the field of assembly cannot be easily substituted and that assembly operations will remain labour-dependent, at least for the foreseeable future. This point can be demonstrated by a method for analysing products for their suitability for assembly automation. The method (Langmoen and Ramsli, 1984) analyses each part in a product with respect to different criteria such as weight, size and tolerance. For each of the criteria, different difficulty levels are established, with a score assigned to each level. The scores indicate how relatively difficult an assembly automation would be, so that:

0 – very easy to automate
1 – easy to automate

2 – possible to automate
4 – difficult to automate
8 – best solved manually

According to this method, each part in a product will be evaluated with respect to a number of criteria. If a part gets 4 or 8 points for any of the criteria, then it should not be considered a candidate for automated assembly. The implication of this measuring system is that the present technology restricts the application of assembly automation within those areas where the products to be assembled are small in size, lightweight and not very complicated. Unfortunately, CLD's products do not belong to this category. When applied, scores of 8 are assigned to all its assembly operations (valve assembly, leg and ram assembly, and support assembly), indicating that the assembly problem of this company at the moment is still 'best solved manually'. Under these circumstances, it is obvious that in order to match the traditional assembly system with the sophistications of the parts-producing facilities, strategies of rationalization on work organization must be regarded as an important factor.

3.3.7 The overall OHMS model

An order-handling-manufacturing system as defined in section 3.3.2 may be described from the uppermost level, A_0, down to the fourth level of decomposition, as illustrated by the complete $IDEF_0$ diagram of the conceptual model (Fig. 3.33). We have seen that the manufacturing functions of such a production system may be represented by the following function blocks:

- A_0 MAKE TO CUSTOMER ORDER
  - A_1 FORMULATE PRODUCTION PLAN
    - A_{11} Sale and Contract
    - A_{12} Plan Production Schedule
  - A_2 DESIGN AND DEVELOP PRODUCT TO ORDER
    - A_{21} Control Design and Development Process
    - A_{22} Develop Prototype
      - A_{221} prepare advanced drawings
      - A_{222} make and test prototype(s)
      - A_{223} prepare final drawings and part lists
  - A_3 GATHER PRODUCTION RESOURCES
    - A_{31} Plan Material and Capacity Requirements
    - A_{32} Gather Resources
      - A_{321} acquire production capacities
      - A_{322} acquire materials and bought-out items
  - A_4 PRODUCE PRODUCTS
    - A_{41} Control Production Activities
    - A_{42} Carry Out Production Activities
      - A_{421} produce parts of products
      - A_{422} produce sub-assemblies of products
      - A_{423} produce final assembly
      - A_{424} test final assembly

This shows clearly the path that a customer order would follow from inquiry to delivery. Obviously, each particular company in the real world will have its own organizational structure and it is usually difficult to generalize the

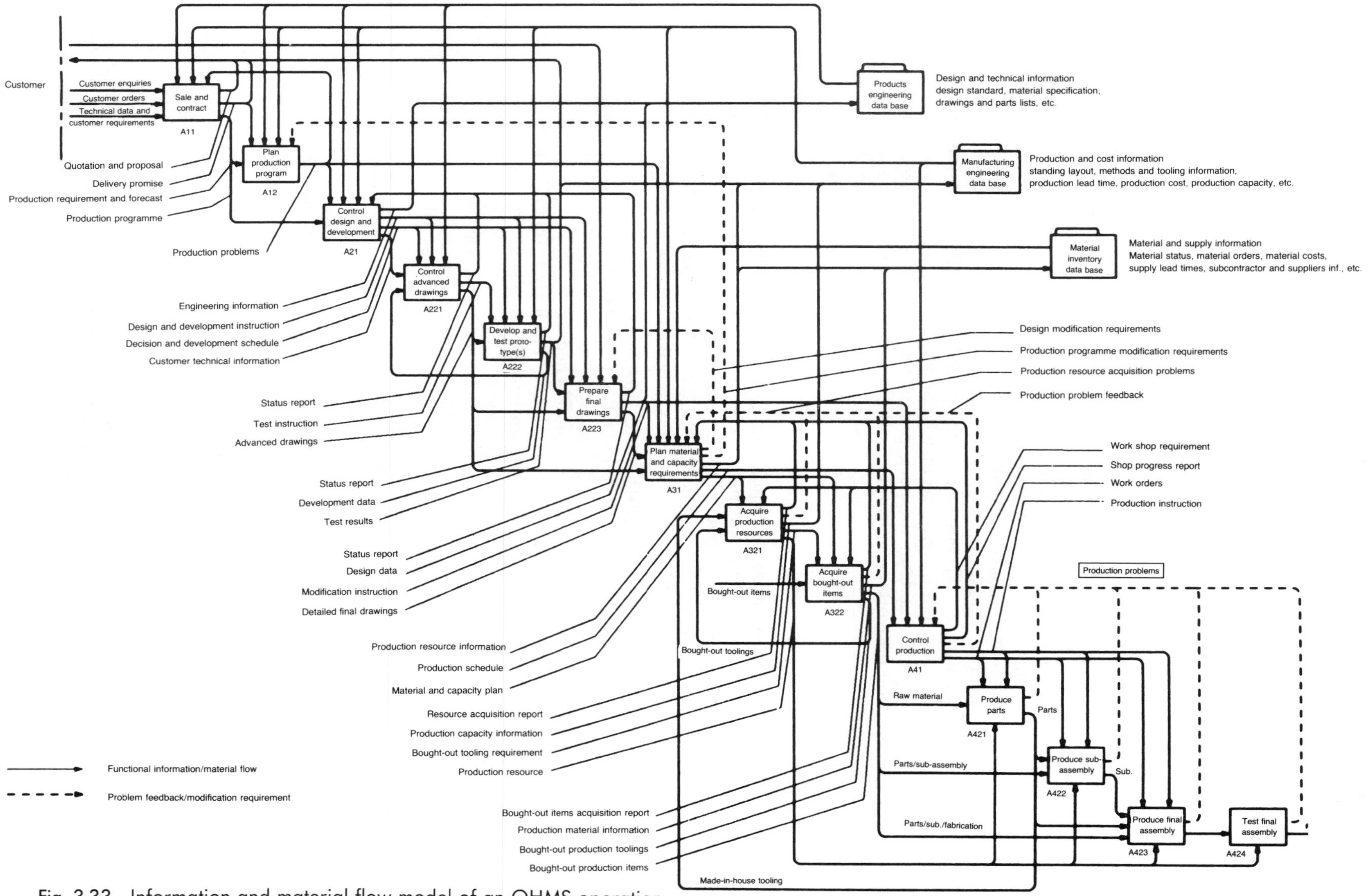

Fig. 3.33 Information and material flow model of an OHMS operation.

operations of different companies using a single model. However, within the system classification given earlier, the companies that fall into this sub-set of manufacturing systems usually exhibit similar functional and informational characteristics. Furthermore, unlike other methodologies (for example the system dynamics approach of Richardson and Pugh, 1981), an $IDEF_0$ diagram provides only a static representation of the system. The time-dependent aspects of a system which are significant in dealing with particular companies are not included in the model. This eliminates one of the major obstacles to generalization. Therefore, it is believed that the model presented in this chapter provides a summary description of the functions and their relationships which are characteristic of those companies that may be classified as belonging to the OHMS sub-set.

3.4 Conclusion

We have now completed our tour of a 'manufacturing park'. One of the implications of a tour is that during the journey we may only visit various places briefly and sometimes we cannot even go to all the places that are of interest. In this chapter we have been able only to see things rather superficially. Nevertheless, this is fine for our purposes. As long as, on completion of this tour, we have developed a mental framework of a manufacturing system showing the structure, the operation, the new techniques, the key issues and the remaining problems involved, then the objective has been met.

The model that has been developed is not entirely conceptual and imaginary. We have compared, where appropriate, the features of this conceptual model with the structure and operational functions of a real company, so as to illustrate the relevant issues through real-life cases. Of course, not all enterprises can be represented by this conceptual model as comfortably as CLD Equipment Ltd. A system model developed for a particular company may be quite different from the one that we have seen. But the important issue is that this particular model has served to illustrate *what* is involved in a manufacturing environment through the structure of a particular type of manufacturing organization, and to show *how* systems concepts such as hierarchy, relation, decomposition and control can be applied to describe and analyse the system of a manufacturing organization.

Having developed a systems concept of manufacturing, we are now in a much better position to understand and adopt the structured approaches and techniques needed for the design and analysis of modern manufacturing systems.

Further reading

Bauer, A., Browden, R., Browne, J. *et al.* (1991) *Shop Floor Control Systems: From Design to Implementation*, Chapman & Hall.

Bignell, V., Dooner, M., Hughes, J. *et al.* (eds) (1985) *Manufacturing Systems*, Basil Blackwell.

Browne, J., Harhen, J. and Shivnan, J. (1988) *Production Management Systems*, Addision-Wesley.

Bullinger, H.-J. and Warneche, H.J. (eds) (1985) *Toward the Factory of the*

Future: Proceedings of the 8th International Conference on Production Research, Springer-Verlag.
Corke, D.K. (1985) *A Guide to CAPM*, Institution of Production Engineers, UK.
Elmasri, R. and Shamkant, B.N. (1989) *Fundamentals of Database Systems*, Benjamin/Cummings.
Groover, M.P. (1987) *Automation, Production Systems, and Computer Integrated Manufacturing*, Prentice Hall.
Hay, E.J. (1988) *The Just-In-Time Breakthrough*, Rath & Strong.
Miller, J.G. (1981) Fit Production Systems to the Task, *Harvard Business Review*, **59** (1), pp. 145–54.
Sartori, L.G. (1988) *Manufacturing Information Systems*, Addison-Wesley.
Smith, W.A. (1985) A *Guide to CAD/CAM* (revised edn), Institution of Production Engineers, UK.
Towill, D.R. (1984) Information Technology in Engineering Production and Production Management, in *The Management Implications of New Information Technology* (ed. N. Piercy), Croom Helm.
Williams, D.J. (1988) *Manufacturing Systems – An Introduction to the Technologies*, Chapman & Hall.

Questions

1. In Chapter 2, a number of systems concepts were introduced, including: system, sub-system, system element, relationship, process, boundary, emergent property, feedback control, and communication. Take each of these in turn, and show how $IDEF_0$ as a system description tool expresses or reflects these systems properties.
2. Draw an $IDEF_0$ model to describe a manufacturing operation that you are concerned with.
3. A manufacturing organization is a typical system within which many systems concepts may be applied. Discuss this in terms of structure, function, relationship and control, etc., and hence identify what is required for its efficient operation.
4. It was claimed in this chapter that: 'Purchasing is a manufacturing function for which a systems approach is of vital importance.' Discuss.
5. What manufacturing functions have benefited from the development and implementation of AMTs? In particular, discuss the role of process planning in a computer-integrated manufacturing system.
6. What parameters, problems, techniques and systems are involved in manufacturing planning and control? Which is best: centralized or decentralised control?

Part Three

Systems Engineering

4 Object-Oriented Perspective of Manufacturing Operation

4.1 Introduction

As discussed in Chapter 1, today's manufacturing companies not only need to produce a wider product range than before, they also have to be able to renew this product range more and more rapidly. In addition, as a result of the impact of information technology on manufacturing, it has been envisaged that a large number of manufacturing organizations in the future will tend to be more coordination-intensive (*Scientific American*, 1991). The differences are summarized in Table 4.1. This means inevitably that manufacturing engineers will have to keep re-designing or modifying existing plants as well as designing new facilities. As a result, these organizations will need not only manufacturing information systems to plan and control the operations of their existing manufacturing structures, but also methodologies and computer-aided tools to help, possibly frequently, restructure their organizational arrangement (either functionally or physically).

However, generally speaking the traditional tools are suitable only for infrequent design projects involving the analysis of hierarchical based manufacturing organizations. There is therefore a need for a systems definition tool (or tools) for the analysis and specification of coordination-intensive manufacturing systems. As with the way in which tools such as $IDEF_0$ have been developed and used for hierarchical manufacturing structures (as illustrated in the previous chapter), such a tool should provide manufacturing systems engineers and managers with an easy-to-use method of specifying the structure and functionality of these systems. Compared with tools such as $IDEF_0$, however, the new approach should be more flexible in its updating process, and should protect the volatile information structure involved.

Due to its easier maintainability, closeness to reality and simplicity, an object-oriented approach seems to have a lot to offer to satisfy the above requirement. The traditional systems analysis methods can be classified as either 'functional decomposition' orientated or 'data' orientated (Coad, 1991). With the functional decomposition method, a system is seen as a set of functions and sub-functions, as clearly illustrated by the approach of $IDEF_0$. The data orientated methods, on the other hand, concentrate on the data structure of a system, making extensive use of dataflow diagrams and entity/relationship concepts. It is only in the recent few years that object-oriented (O-O) methodologies, both for systems analysis and design, have been emerging. There are currently several methodologies available, which in general have been developed by two camps, *revolutionaries* and *synthesists* (Yourdon, 1989). The revolutionaries believe that it is a radically new concept, rendering previously existing methodologies and ways of thinking

Table 4.1 Comparison between traditional and coordination-intensive manufacturing operations (source: Wu, 1993)

Feature	*Traditional manufacturing*	*Coordination intensive manufacturing*
Structure	Hierarchy of functional departments	Collection of autonomous manufacturing units (AMU)
Areas of concern for production management	Capacity, material, activity planning and control within an existing manufacturing structure	Coordinative organizing of flexible manufacturing structures, **plus** planning and control of individual AMUs
Integration approach	Hierarchical integration through centralized planning and control	Coordinative integration through dynamic organization **and** decentralized control
Computers and networks	Large scale, time-sharing computing facilities	Coordinative computer centre **plus** distributed PC networks
Computer-aided production management (CAPM) software tools	Large scale, centralized planning packages aiming to produce detailed production plan down to the lowest level of organizational hierarchy	Integrated Decision Support System for frequent system restructuring at factory level, **plus** factory coordination system and modularized packages for planning and control at AMU level

obsolete. By contrast, the synthesists see the concept as a logical step which can be incorporated into, and enhance, the present methodologies. Whichever way one looks at the situation, however, there is no doubt that the concepts of O-O approach are fundamentally different from traditional structured methods, and require a different way of thinking (Booch, 1989). Although there are different approaches, these share commonalities of the basic O-O concepts in various ways. Section 4.2 will attempt to provide an outline of the major principles involved and, through the use of examples, discuss the potential advantages of this new paradigm.

Examples of the currently available O-O analysis and design methodologies include: Abbot (Abbot, 1983), JSD (Jackson, 1983), ObjectOry (Jacobson, 1987), OOSA (Shlaer and Mellor, 1988), GOOD (Seidewitz, 1989), OOSD (Wasserman *et al.*, 1990), Booch (Booch, 1986, 1991), OOA (Coad and Yourdon, 1991), OMT (Rumbaugh *et al.*, 1991) and HOOD (Graham, 1991). These general O-O methods are briefly reviewed in section 4.3. Based on these, a new approach will be presented: Hierarchical and Object-Oriented Manufacturing Systems Analysis (HOOMA). This method utilizes the useful features of two general methods of O-O analysis, OOA and HOOD, but is developed to support specifically the requirements of system analysis and definition within the manufacturing context. Compared with the traditional modelling methods which are usually static and do not allow modelling of the time-dependent aspects of a manufacturing operation, HOOMA allows temporal logic to be encapsulated in the system objects, and hence dynamic processes can be presented through object interactions. Section 4.4 provides an overview of the HOOMA tools and procedures, and section 4.5, using the products and operational structure of the model OHMS company as an example, illustrates how such an O-O method of system specification can be used in practice to cope with more frequent systems changes.

4.2 Concepts of object orientation

4.2.1 Background

The increasing interest in the object-oriented concept has been led by the advent of O-O programming languages. There have been several object-oriented programming languages available for quite a long time including, for example, *Simula, Smalltalk, CLOS, Object Pascal* and C++. As a result, until the mid 1980s much of the work in the O-O arena emphasized O-O programming and software systems. It is only in the past few years that the technique of O-O analysis has been introduced, and the paradigm extended into the manufacturing area.

The main difference between conventional and O-O methodologies is that the former models a system at the system or sub-system level, whereas with the O-O approach the system is modelled at the level of individual entities within the problem domain. Consequently, with the conventional approach a system is functionally decomposed using functions and processes as the building blocks so that the entities are passive data stores, manipulated by activities or procedures. The associated analysis/design procedure can be thought of as following a top-down process: the analyst starts with a high level abstract description of the function of the overall system and progressively breaks this down and refines it to produce successively less abstract descriptions of the function of smaller and smaller parts of the system. In contrast, decomposition of problem domain with an O-O approach is based on the classification of objects and their relationships with each other, resulting in systems entities which are self-contained in terms of their operations and the corresponding data, and which communicate explicitly with each other. It is not possible to categorize this as fundamentally top-down or bottom-up. The approach can include elements of both approaches: when an analyst reuses an existing object or class, it could be said that a bottom-up approach is being followed; on the other hand when decomposing parts of a system into smaller elements which are then mapped onto objects and classes, the approach is more of a top-down one.

It is now widely recognised that the O-O paradigm provides an excellent approach to manage and express complex entities through data *abstraction, encapsulation* and *inheritance*, which are its main features. With this new perspective, a system consists of a collection of objects which are an encapsulation of data, and whose functionality is obtained by defining a set of messages to which they respond. Through this method of encapsulation, adding a new or deleting an existing kind of object will have little influence on other objects, so that flexibility is enhanced and maintenance eased. In addition, each object belongs to a class. Sub-classes can be declared as a specialization of another class, and they are allowed to inherit the structure and functionality of its super-class. Thus through this mechanism of inheritance new objects can be easily created by using the useful functionality of existing objects wherever possible without having to start from scratch. As a consequence, the main advantages offered by this approach can be summarized as follows:

- *Realistic view.* In comparison to the traditional methods, an O-O approach allows the analyst/designer to identify and manipulate a system abstraction in a more direct way, due to the fact that the real-world consists of physical entities, or objects, rather than 'functions/processes'.

- *Flexibility*. The modularized system structure consists of system units which can be combined to execute the required application, and yet each of these units can be implemented or modified independently without affecting any other modules in the system.
- *Reusability and extensibility*. Such an approach facilitates rapid system design/ development by reusing components already available. Once a system component has been defined, it can be used as the basis for other similar components, hence reducing duplicated effort in system development and modification.

4.2.2 O-O perspective: examples

We will use two examples here to illustrate the basic principles of O-O perspective, the first illustrating the difference between the functional and O-O view points, and the second revealing more details about the structure, makeup and behaviour of objects.

If we look at the structure of a car from a traditional perspective, then it will be conceptualized functionally as shown in Fig. 4.1. That is, this system would be decomposed into four sub-systems each corresponding to a major function that the car is designed to perform. Each of these functions would have associated with it a set of inputs and outputs, as specified by an $IDEF_0$ block. It could be argued that such a presentation of the car appears rather unnatural, for the functions or processes identified here do not physically exist in the real world. They are artificially created in our mind based on our understanding of the operation of the car as a whole. In addition, due to its rigid structure, modifications to this model after its development would be quite difficult and complex.

In contrast, with the O-O perspective, such a system would be conceptualized as shown in Fig. 4.2. Here, real objects and their own actions have been identified, each possessing its own data necessary for the execution of these actions. Instead of a collection of functions, a car in reality will have a dashboard that combines with the pedals to provide an interface with the car. To drive the car, the driver needs to interact only with the controls of the dashboard and the pedals, without necessarily having to possess any knowledge of internal engineering design or functions. As long as the driver is aware of the responsibilities and actions undertaken by these parts, and as long as he or she knows when and how to issue instructions to initiate these actions, he or she will be in a position to control the car properly.

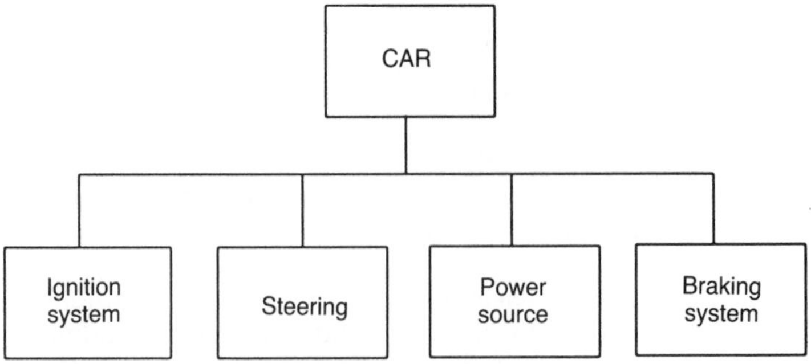

Fig. 4.1 Functional decomposition of a car.

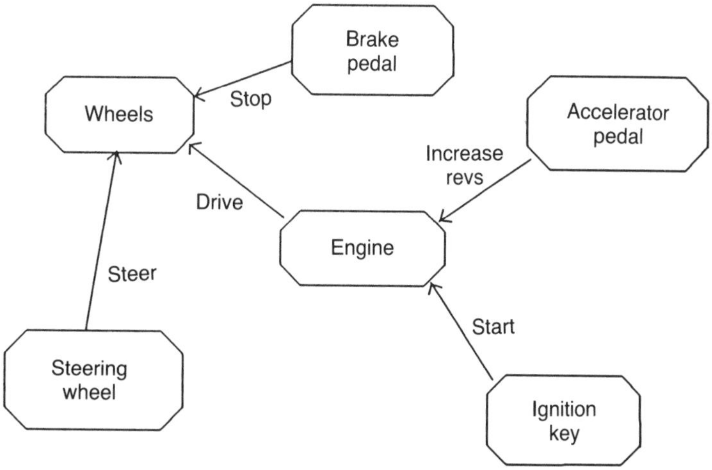

Fig. 4.2 Object-oriented view of a car.

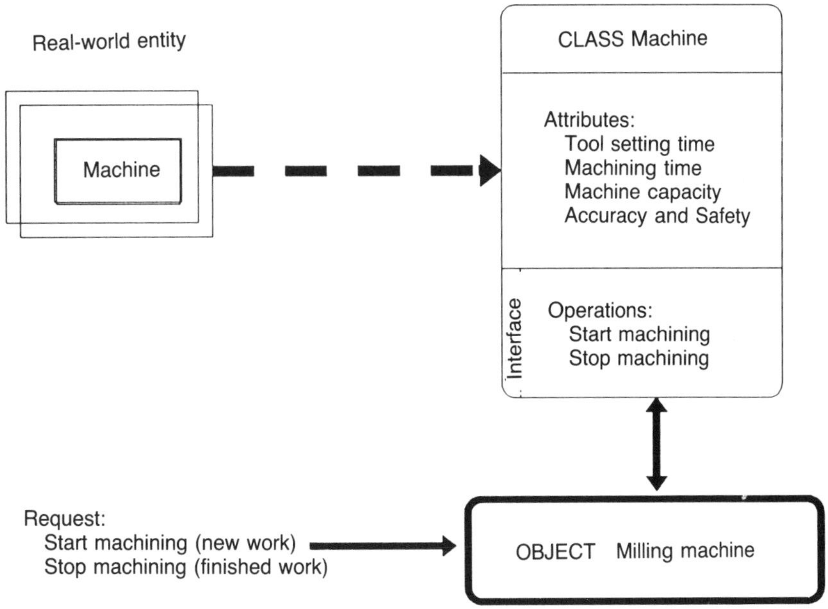

Fig. 4.3 A machine viewed as an object.

More details about the make-up and behaviour of an object are provided by the example shown in Fig. 4.3, which shows a real world entity called Machine as represented through the O-O perspective. As can be seen, further detail of the reality can be modelled by indicating what behaviour will be provided by an object, in terms of a set of *attributes* (variables) and *operations/services*, representing object status and behaviour respectively. The status of an object can be accessed or changed by sending a message to the object to invoke the corresponding services. For instance, in this case the message of 'Start machining' sent to a machine object, once accepted and the required operation initiated, would alter the machine's status from 'idle' to 'busy'.

Objects which share the same structure and behaviour can be grouped into classes. Consider a hierarchy of machines as illustrated in the figure. The Machine class is a generalization of such objects when a plant can have various types of machines. A Milling Machine, on the other hand, is a specialization of Machine, and inherits attributes and services from it. In addition to their inherited properties, a specialization can add new attributes and/or services so that, for example, a Milling Machine will be able to tackle certain specialized jobs which cannot be handled by a less specialized Machine. These form the essence of inheritance hierarchies.

4.2.3 Basic concepts

The great interest in this new paradigm has led to many approaches, resulting in a proliferation of definitions, interpretations and non-standardization of the concept. However, the following is an attempt to formalise certain useful O-O concepts for the purpose of specifying the structure of a manufacturing operation through an O-O perspective.

- *Encapsulation.* This refers to information hiding within an object: *a principle, used when developing an overall program (system) structure, that each component of a program should encapsulate or hide a single design decision . . . The interface to each module is defined in such a way as to reveal as little as possible about its inner working* (Oxford English Dictionary). An object hides details about its implementation (structure and operation). Such hidden information is not accessible by the user of this object.
- *Objects.* Objects are an *abstraction* of the entities in the real world which *encapsulate* information about themselves, such as *attribute* values which determine their state, and a defined set of *operations* on that state which characterizes their behaviour (that is, how they act and react to each other in terms of changes of their state). An object has an identity that denotes its separate position from other objects within a system. Since all the data needed by an object to perform its duties is encapsulated within it, there is no need for any shared data store. Once defined in such a way, objects can be stored and reused for other tasks of system development independently of each other.
- *Attributes and Operation/Services.* Attributes are named properties and hold abstract states of each object, which can only be manipulated by the operation or services of that object. Operations/services characterize the behaviour of an object, and are the only means for accessing, manipulating and modifying its attributes.
- *Messages.* Each object has a visible interface comprising a list of services that it is able to provide (the external view of the object is nothing more than this interface). An object will communicate with another through a request, known as a message, which identifies the operation to be performed by the second object, and in response the 'receiver' object will select and perform the service requested.
- *Classes.* It is important to distinguish between an object and its class. A class is an object that acts as a template, specifying the properties and behaviour for a set of similar objects in terms of:
  - a set of attributes
  - a set of messages defining the external interface

- a set of services invoked by messages

Each object is considered to be an element, or an instance, of a class. Hence, a class is a specification for its instances and thus an abstraction. Classes are related through *sub-class* and *super-class* relationships. A class may have several sub-classes, through which individual procedures and data manipulation tasks are organized into a tree structure. Objects at any level of this tree structure inherit behaviour of higher level objects.

- *Class-&-Object*. A class and the object in that class. Objects are instantiated on the basis of class. Since all instances of a given class have the same structure and behave similarly, instantiation makes it possible to easily reuse the same definition to generate objects. Every object is an instance of one class, but a class may have no instances.
- *Inheritance*. This allows the sharing of behaviour and data between objects through the definition of a sub-class by using the template of its super-class(es). When a sub-class inherits commonalities from the super-class, the relationship is known as inheritance. When a sub-class inherits commonalities from two or more super-classes, it is called multiple inheritance.
- *Generalization-Specialization*. In addition to the inherited structure and behaviour, a sub-class may possess its own specific attributes, services and messages. This is known as *specialization*. Specialization allows the reuse of existing classes by defining new sub-classes which are more specific than the existing abstractions. On the other hand, when two or more classes have an overlapping set of attributes and operations, they can be grouped together to create a new super-class representing the abstraction of those overlapping commonalities. This process is known as *generalization*.

4.3 Object-oriented systems methods

To date the O-O concepts have not only been used in the area of software engineering but are also found in other types of application. In particular, O-O approaches for systems analysis and definition are now considered as a useful alternative to the traditional approaches (Manfredi *et al.*, 1989). A positive benefit of following an O-O approach in this regard is the closeness between abstraction and reality, because organizing a required system around class-&-object actually maps real-world entities into required components.

To start with, any method which aims effectively to describe the structure of a system from an O-O perspective should have a means of representing classes, objects and their relationships. This is usually achieved through the use of a set of graphical tools. Based on the basic O-O characteristics outlined previously, such tools must be able to specify an object clearly in terms of:

- its name and state
- the class to which it belongs
- the action that it provides and its external interface that is visible to other objects.

To model a system, the following general steps are then needed:

- Specify system structure in terms of class and inheritance hierarchies.
- Identify the objects and their attributes.

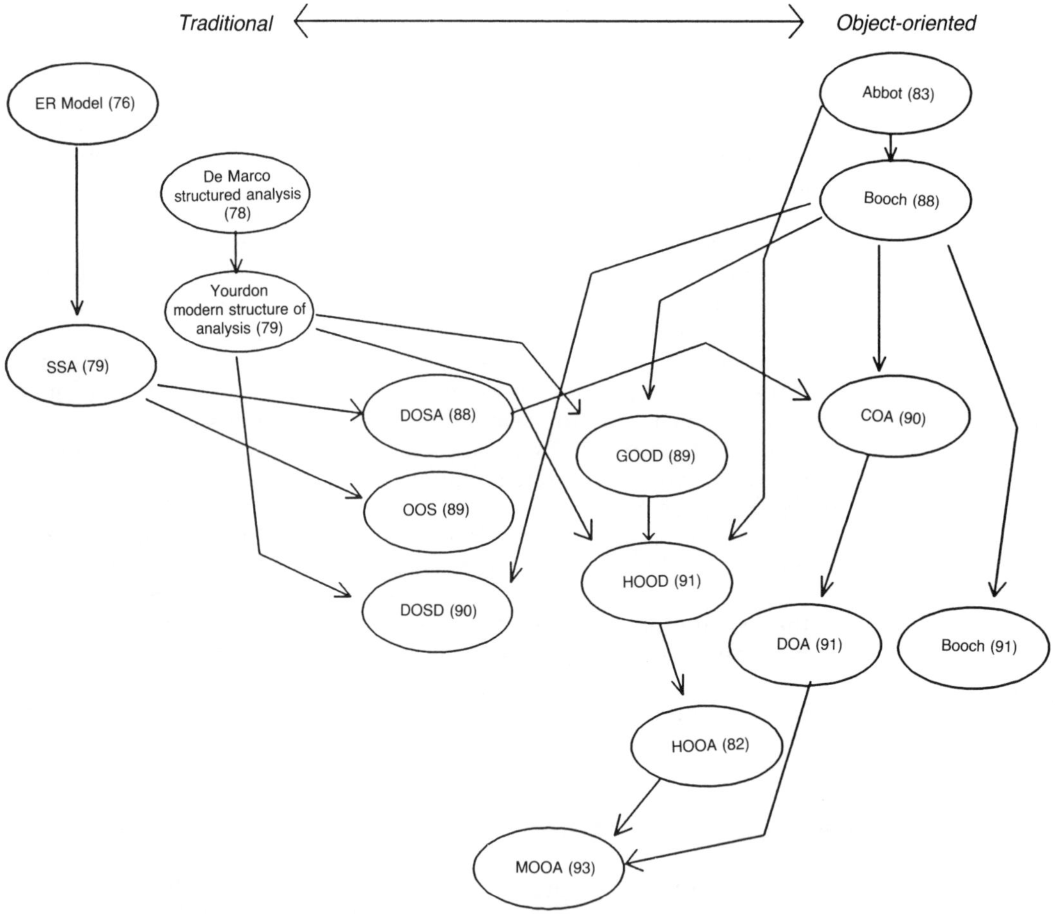

Fig. 4.4 The evolution of object-oriented methodologies.

- Identify the operations provided by and required of each object.
- Establish the visibility of each object in relation to other objects, and hence its interface.
- Establish the communication between objects through message-passing.

To date, several O-O systems design/analysis methodologies have been suggested, the main purpose being to bridge the gap between systems analysis and systems design, and to bring together the functional and data approaches via the use of message-passing between objects. The evolution process of this new development is summarized in Fig. 4.4. Here we briefly review the basic structures of a few typical examples of such methods.

The BOOCH approach (Booch, 1991) is perhaps the most comprehensive O-O system design/analysis methodology to date. This method uses a set of graphical tools to help visualize system structures. Figure 4.5, for example, represents a class with an external visible part indicating its type and operations, and a private part describing its internal details. Arrows are used to show the interdependencies among the objects. Generally this methodology will involve several steps including first the identification of classes and

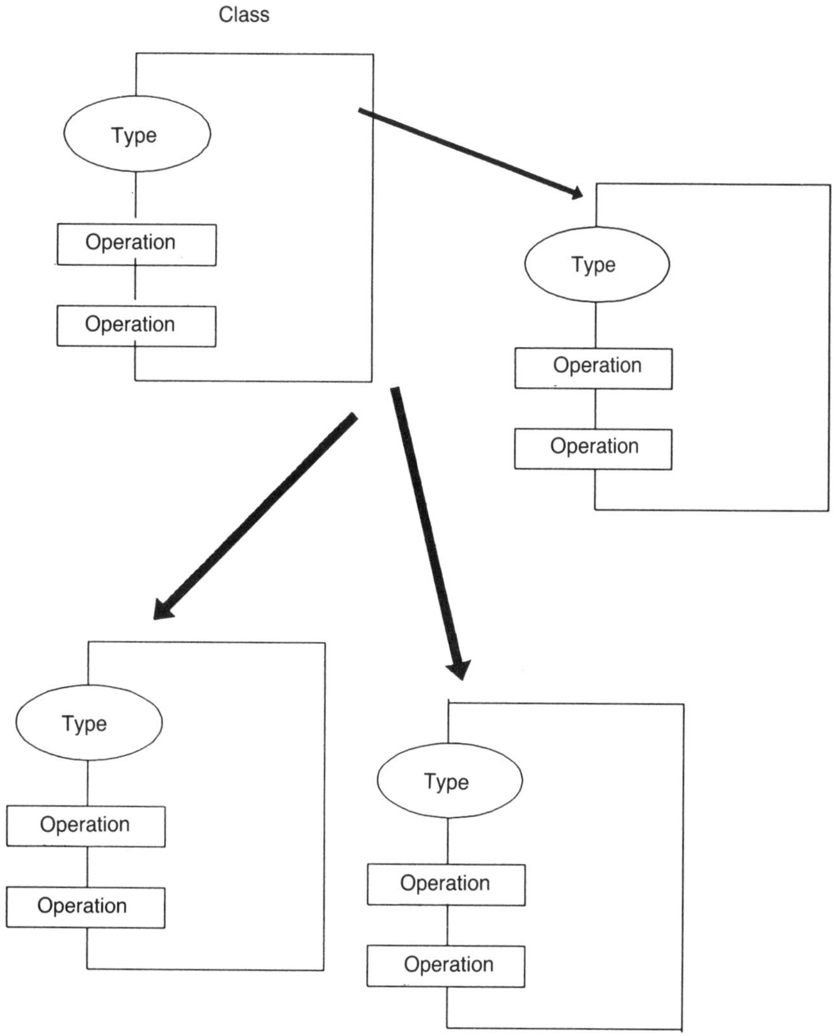

Fig. 4.5 Booch graphical representation.

objects at a high level of abstraction in the application domain, secondly the identification of the semantic of these classes and objects, and finally the identification of relationships among these classes and objects and their interaction within the system.

The GOOD (General Object-Oriented Development; Seidewitz, 1989) and the HOOD (Hierarchical O-O Design) present an interesting approach which is similar in many ways to the traditional structured approach through functional decomposition, and neither seems to consider inheritance hierarchies. The main feature of the GOOD method is the identification of objects at a particular level, and then the decomposition of these objects into sub-objects. In this way the lower-level objects are detailed to meet the specification of the relevant upper-level objects. The hierarchical structure of the system thus formed can then be modelled through a layered collection of object diagrams as shown in Fig. 4.6, where a dotted line separates one layer from

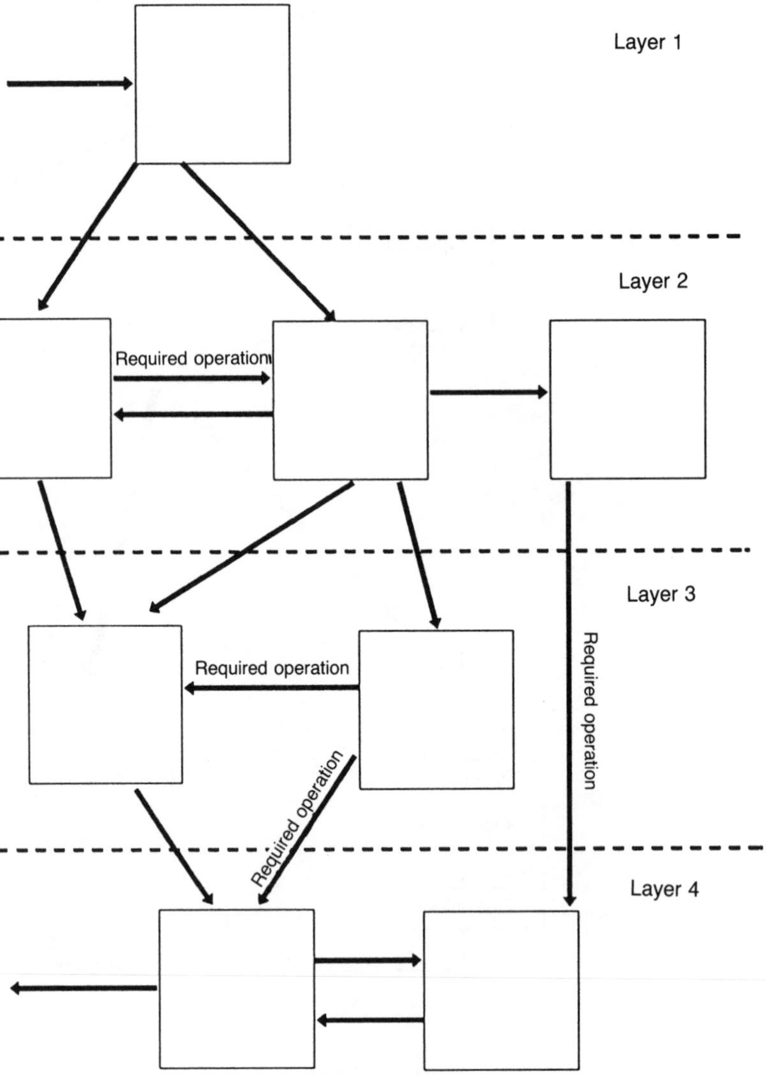

Fig. 4.6 GOOD hierarchical structure of objects.

another and a solid line represents the control flow and inter-dependencies between objects.

The HOOD approach was developed by the European Space Agency (ESA) as a design method for the architectural design phase of an ESA software engineering life cycle. As a design method, HOOD uses a number of steps to decompose an object by involving diagramming techniques from other structured analysis and design methods. A typical HOOD notation is shown in Fig. 4.7, where the arrows represent a 'uses' relation, and the enclosure of a sub-object in the diagram denotes that it is considered as part of the upper object. Through these it is possible to specify how the functionality of an upper-level object is actually implemented through the use of lower-level objects. From a manufacturing point of view, it is important to

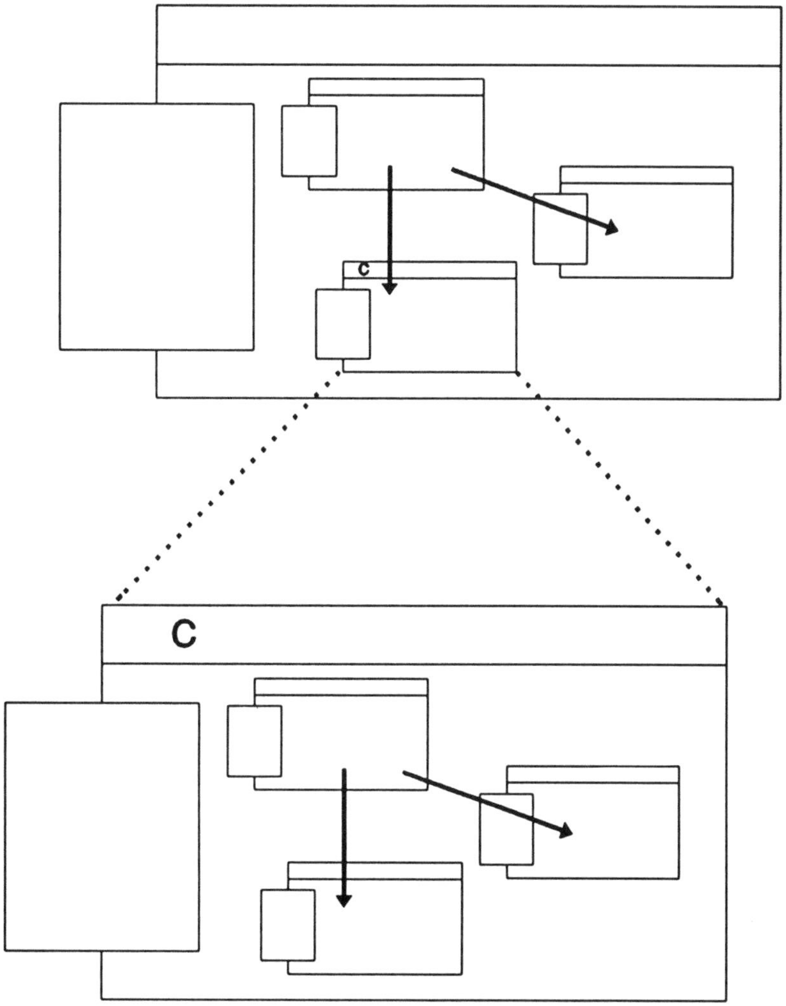

Fig. 4.7 Decomposition of a HOOD object.

note that HOOD identifies two types of object: *passive objects* (e.g. production materials or parts) whose operations are executed immediately when control is passed to them, and *active objects* (e.g. a machine) whose reaction to a request may be delayed depending on the object control structure.

Another attempt to provide the means to support both structured and O-O approaches is OOSD (Object-Oriented Structured Design; Wasserman *et al.*, 1990). This method aims to help the analyst/designer gradually shift from top-down structured development to an O-O approach. OOSD provides a method clearly to show relationships between classes through a combined structured/O-O process. However, it does not represent detailed interactions between objects. There is also a lack of formal procedures and guidelines.

In contrast to the above methodologies which are more design-biased (i.e. more emphasis on system implementation or software coding), some methods are developed specifically for general O-O system specification rather than system implementation. One of these methods is the OOSA

(Object-Oriented System Analysis; Shlaer and Mellor, 1988). To specify the structure and functionality of a system, OOSA uses a combined structure/O-O approach to construct three models:

- Information model specifying objects and their relationships.
- State model defining the behaviour of the objects by using a standard state transition diagram.
- Process model representing states and transitions in the state model.

The graphical notation used by OOSA is mainly taken from entity-relationship diagrams and data flow diagrams from the traditional structured methods. Many of the O-O concepts are not supported by an OOSA presentation.

OOA (Object-Oriented Analysis; Coad and Yourdon, 1991) is perhaps the latest general-purpose O-O method for systems definition. It guides the analyst through the basic analysis steps as outlined at the beginning of this section. OOA also provides a complete set of graphical notations to account for the basic O-O concepts. To start with, OOA divides a system into partitions (called *subjects*) in order to control the amount of a model that an analyst considers at one time. Within a subject box, an individual class/object is then represented by a rectangular box, with its title, attributes and services indicated in three separate sections. The Generalization–Specialization relation is denoted by lines with semicircles at the end of the sub-objects. In addition, a triangle marking represents a Whole-Part Structure. Finally, a bold line represents a message connection between objects, while a thin line shows an instance connection between objects. ('An instance connection is a model of problem domain mapping(s) that one object needs with other objects, in order to fulfil its responsibilities.' (Coad and Yourdon, 1991)) The usefulness of this set of graphical notations for the O-O modelling of manufacturing operations will be illustrated later in this section.

Finally, it is important to point out that the most direct development in O-O design in the manufacturing arena seems to be that of the recent $IDEF_4$ (Mayer *et al.*, 1992). Like $IDEF_0$, $IDEF_4$ is an O-O design method developed under the Information Integration for Concurrent Engineering (IICE) project founded by the US Air Force. However, this is a methodology which aims to provide a comprehensive set of tools and procedures to help design and develop good O-O software systems. The issues related to O-O systems analysis are not specifically addressed.

4.4 Hierarchical and object-oriented manufacturing systems definition

It is evident that none of the methodologies reviewed above has yet achieved the status of being a widely recognized standard comparable to some of the conventional methodologies. In addition, when these are considered for use within the context of manufacturing systems analysis, it becomes clear that there are certain features that need to be catered for. For example, there needs to be some indication of the process flows involved, and both the static (hierarchical) structure and the dynamic interaction of systems entities need to be presented. Therefore, a new approach will be presented here: Hierarchical and Object-Oriented Manufacturing Systems Analysis and Definition (HOOMA), which utilizes the useful features of two general methods of O-O analysis, OOA and HOOD, but is developed to

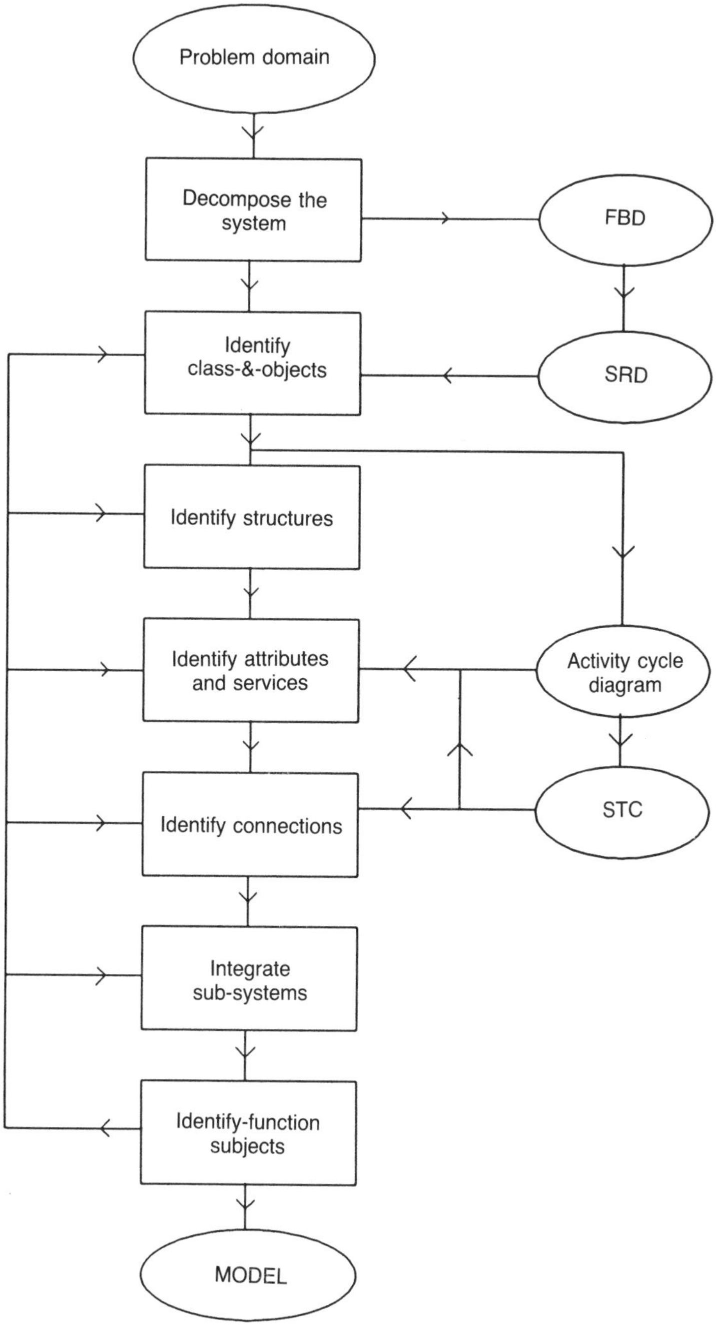

Fig. 4.8 Summary of HOOMA procedure.

support specifically the requirements of system analysis and definition within the manufacturing context (Wu and Fung, 1993). The method aims to bridge the divide between the function-based approach and the pure O-O approaches. Within a coherent framework it not only adopts the general O-O concepts for manufacturing systems specification, but wherever possible also utilizes existing tools provided by conventional approaches. It is hoped that such an approach will be able to provide the user with the best of both worlds, so that both top-down and bottom-up approaches may be followed in practice. In addition, compared with the traditional modelling methods which are usually static and do not allow modelling of the time-dependent aspects of a manufacturing operation, HOOMA allows temporal logic to be encapsulated in the system objects, and hence dynamic processes can be presented through object interactions. The following provides an overview of the HOOMA methods.

The O-O concepts are represented with HOOMA graphically, largely based on the OOA notation but with a number of new additions. It uses several tools and mechanisms in its analysis. The analysis procedure of HOOMA is summarized in Fig. 4.8. The figure shows the steps to be followed, the graphical tools related to these steps and the iterative process involved. A complete conceptual model of a manufacturing operation may require several such iterative cycles.

4.4.1 System decomposition

The decomposition of a system may be achieved in two different ways. First, the system may be tackled bottom-up, i.e. in the true Object-Oriented fashion. One would approach the system and begin to identify Class-&-Objects. Once all the Class-&-Objects have been identified along with their Attributes, Services, Connections and Structures, these small components can then be integrated.

Secondly, decomposition of a system can be based on sub-systems related to functions within the system. This step specifies the overall hierarchical structure of the system, and is very much a top-down approach. This seems to conflict with the pure object-oriented concept, but makes practical sense since a completely bottom-up process based on the pure object-oriented concept is rather difficult to apply within the manufacturing context. Hence it is sensible to adopt a hybrid approach so that the traditional top-down view is employed initially to produce an overall picture of the system, and the bottom-up view can then be applied effectively.

HOOMA approaches the task of system decomposition by using 'Function Subjects' (FS). The basic functional structure of an FS hierarchy can be illustrated effectively with a Function Block Diagram (FBD) as shown in Fig. 4.9. The FSs themselves are represented by a rectangle with the FS name and the main tasks involved. The rectangle encloses the Class-&-Objects belonging to that FS area, within which they can be further decomposed to reveal new FSs. Therefore an upper-level FS may contain certain lower-level FSs, providing an FS hierarchy to guide a user through a large and complex model, as shown in Fig. 4.10 (the dotted lines separate one layer from another, while the solid arrow lines represent the interdependencies between FSs). An upper-level FS is usually implemented by its lower-level FSs, and its tasks accomplished through the execution of the tasks of the lower-level FSs. Such a FS model is also useful as an indication of the overall sequence of operations.

Although decomposition through FS hierarchy is very useful to assist the analyst in specifying the functionality and the static structure of the system, it is important to be aware of the danger of placing too much emphasis on system functions. Due to the inherent conflict between functional and object-oriented approaches, a too detailed decomposition with FS may result in a

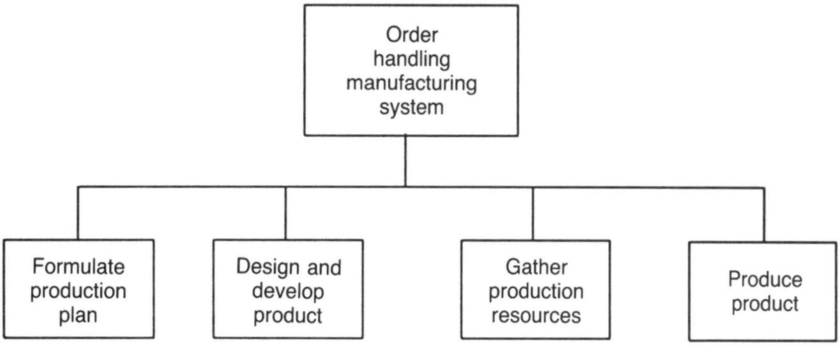

Fig. 4.9 Example of a function block diagram (FBD).

functional biased system without the benefit offered by object-based orientation. A FBD consisting of a maximum of three or four levels of decomposition seems to strike the best balance in this regard.

4.4.2 Identification of Class-&-Objects

Having specified the FS hierarchy, further examination of the sub-systems involved is now possible. HOOMA ultimately decomposes a manufacturing operation into its Class-&-Objects (C&Os). The system's states (variables and information) and operations will be encapsulated within these entities. In addition to the definitions given in the beginning of this chapter, HOOMA classifies C&Os into two further categories: Customer Class-&-Objects (CC&O) and Server Class-&-Objects (SC&O):

Customer: C&Os requiring services (processes) only, from one or more server C&Os. These objects are dynamic because they usually undergo physical transformation and have a limited life-cycle/span within the system.

Server: C&Os that can perform services on certain C&Os. These objects are (relatively) permanent within the system.

The symbol used to represent C&Os is similar to that for OOA, as shown in Fig. 4.11. A class is represented by the bold rectangle and an object by the light rectangle. The symbol contains a label with its class name, attributes and services. Customer C&Os are differentiated from Server C&Os by *italic* labelling. The server C&Os within the manufacturing context are usually based on the abstractions of actual physical equipment, process information, process procedures, human roles and functions, whereas customer Class-&-Objects are the abstractions of data, materials and product parts.

Identification of C&Os is not an easy task, especially in a non-static situation where objects may be created, changed internally and destroyed. The level of detail is therefore very important. This is the first thing to be identified in order to make sure that data is encapsulated at the right level. An FS hierarchy developed previously will provide a valuable insight into initial C&Os. However, to aid in more precise identification, individual FSs may be further developed to reveal their data structure through the use of a Sub-System Relationship Diagram (SRD), which shows how data flows

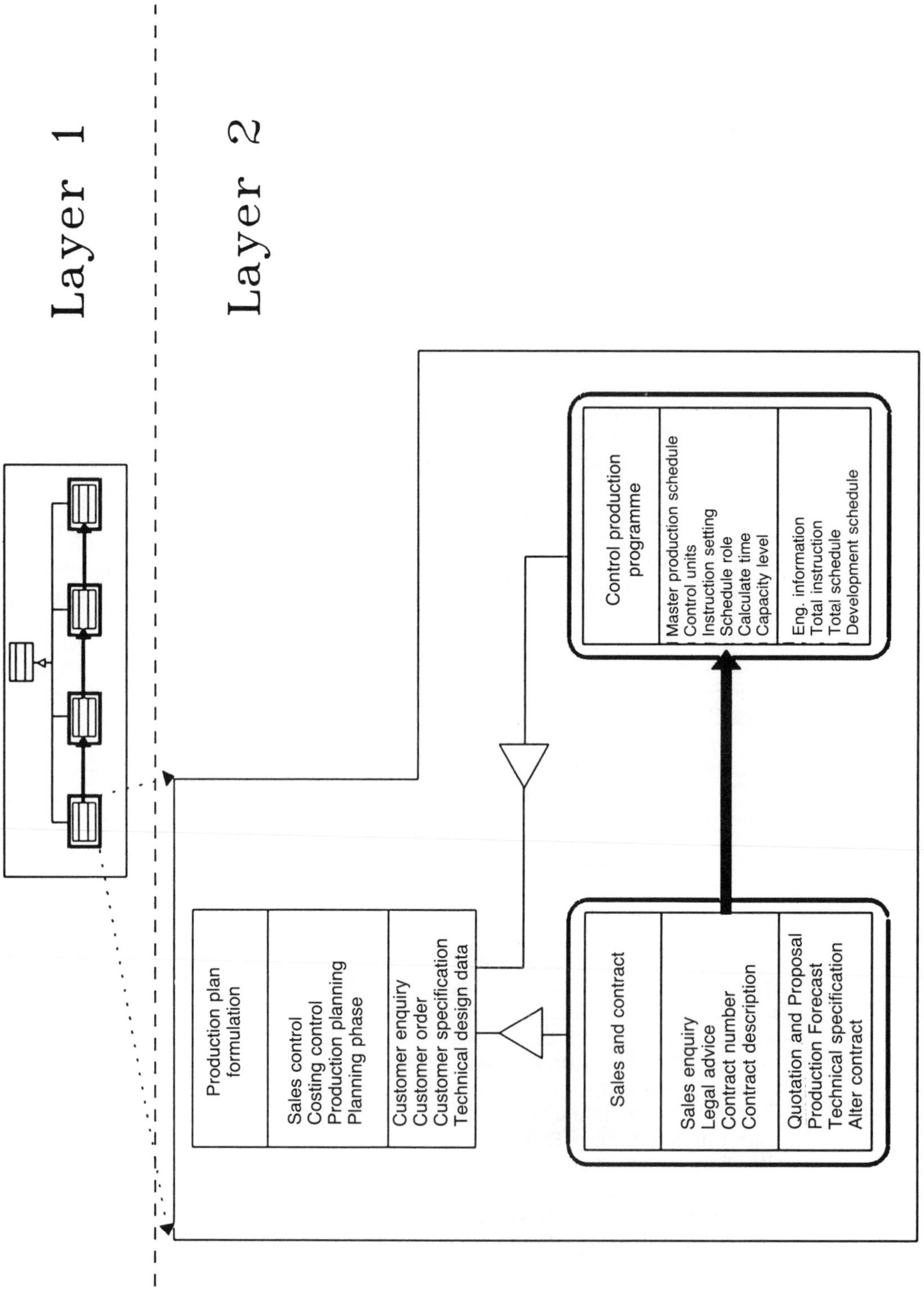

Fig. 4.10 Example of system decomposition through function subjects hierarchy.

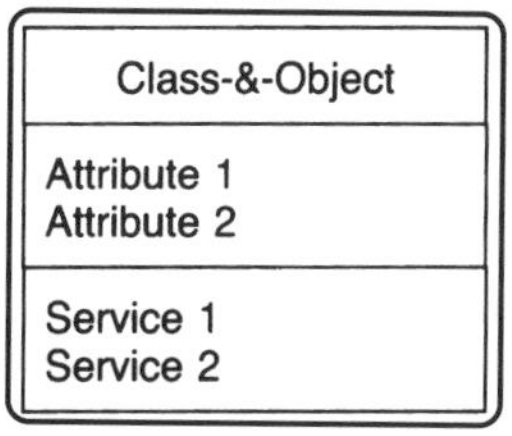

Fig. 4.11 Class-&-Object with attributes and services.

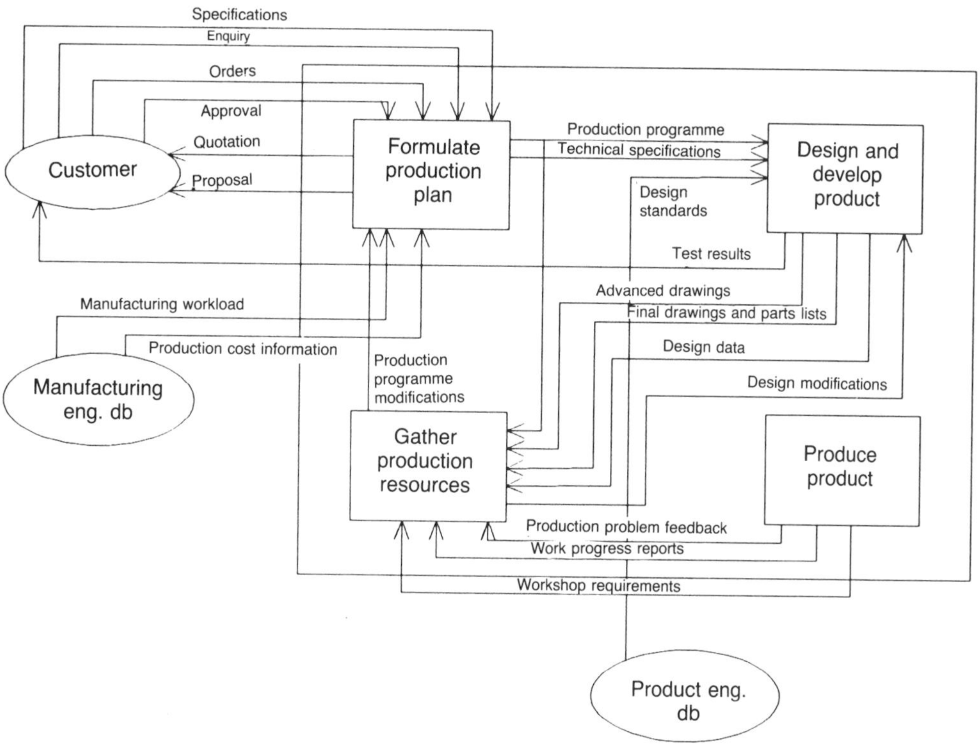

Fig. 4.12 Example of a sub-system relationship diagram (SRD).

around a system FS (Fig. 4.12). The sub-blocks, data and data connections expressed by an SRD will give further clues to potential C&Os, as well as their attributes and services. Although the depth of detail or abstraction will depend on the context of the problem under consideration, the activity of C&O identification should be as thorough as is practically possible. In fact, for those who are familiar with $IDEF_0$, it is not difficult to see that a system model that is functionally decomposed in the traditional way can be used as a viable substitute for the FBD and SRD diagrams at this stage. However, this must be applied carefully in practice so that the analyst does not lose the point of O-O perspective. Again, no more than three levels of functional decomposition are recommended for this purpose.

In relation to the level of detail at which data is to be encapsulated, the process of C&O identification can be helped by asking questions like:

- How does a function block work?
- What entities are responsible for its functioning?
- How do these entities interact?
- What influences it?

If one is faced with a Machining Cell block, the answer to the first three questions could be as follows:

> This machining cell consists of a conveyor, a material-handling robot, a NC machine and an inspection station. Under the control of a central computer, the cell is capable of cutting a number of different parts. A part is put on the conveyor belt that advances automatically until it reaches the machine, when it will be its turn to be machined. Once the machine becomes available, the robot loads this part and, according to its type (read in by a barcode scanner mounted on the conveyor), the machine selects the appropriate NC programme together with the right tools. Following this, the actual cutting process commences. Once all the required operations are complete, the robot unloads the parts. The inspection process which follows is carried out in a similar manner.

This description is very useful to help identify a number of objects, together with some of their attributes:

Parts (barcode_no, description, weight, size, current_stage)
Conveyor_Belt (on/off, speed, length, current_queue_length)
Barcode_Scanner (input)
Robot (weight_limit, size_limit, current_status)
Machine (machine_no, busy/idle, current_programme_no)
Tools (tool_no, current_setting)
NC_Programme (programme_no, description, size)
Inspection_Station (machine_no, busy/idle, current_status)
Computer

Further, answering the last question will enable us to identify some additional, but perhaps not readily apparent objects. For instance, the computer might ask for some of the parts to be diverted to an alternative manufacturing facility if there is a long queue before the cell. This in turn may depend on the current work load of the factory and its scheduling policy. A Cell_Scheduler object may need to be added to the list of objects as a result.

4.4.3 Specification of static structures

The next step is to describe the manufacturing system in terms of the structures of its Class-&-Objects in order to illustrate its static layout. Structures within HOOMA are important to describe the interactions and relationships among the Class-&-Objects that exist within the system, as well as to encapsulate the concept of inheritance (through Generalization-Specialization structures in which the general attributes and services are inherited by the specialization) and aid the readability of the model.

The term 'structure' is used here generally to cover both Generalization-

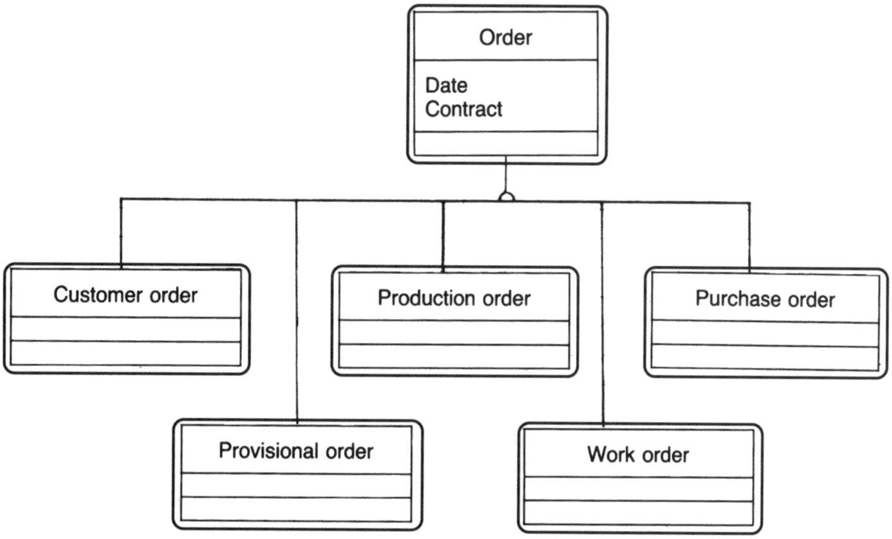

Fig. 4.13 Example of a generalization-specialization structure.

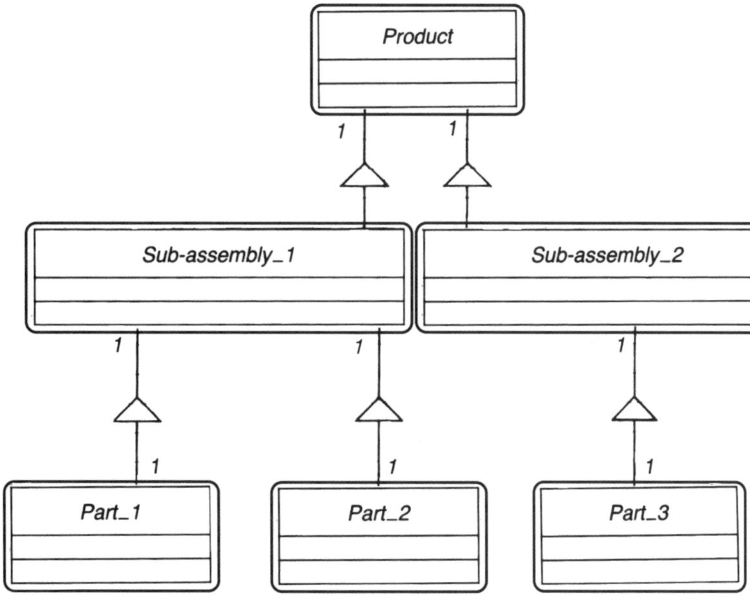

Fig. 4.14 Example of a whole-part structure.

Specialization and Whole-Part relationships of Class-&-Objects. The graphical presentations of these structures are again similar to those used by OOA. The Generalization-Specialization structure notation (Fig. 4.13) is directional, with the Generalization Class at the top connected with the line and semicircle to the Specialization Classes below. The Whole-Part structure (Fig. 4.14) provides an additional element to decompose the structure of a system. It provides a method for displaying the relationship in which a Part object can be part of a Whole object. As shown in Fig. 4.14, a Whole-Part structure has the Whole object at the top connected to Part objects at the bottom. The connecting line has a triangle to distinguish it and indicate its direction. This line is marked with an amount or range indicating the number of parts that a Whole may have, and vice versa, at any given moment. The Whole-Part notation has an extra dimension when it involves customer Whole objects and customer Part objects. The line, which now has *italic* amounts and ranges, often implies a transformation of the Part objects into the Whole object by a service (process) performed on them by a server object. No customer Whole object can exist unless all of its component Part objects have gone through the necessary services and are put together to create it. Therefore a customer Whole Object and its particular component Part objects can not exist at the same instant. For example, Fig. 4.14 shows a Whole-Part structure with a Sub-Assembly_1 object consisting of two Part objects, Part_1 and Part_2. These Part objects may require certain services to be performed on them (e.g. machining or painting) before they can be transformed (e.g. an assembly operation) into a Sub-Assembly_1 object. Consequently, once this is done a Sub-Assembly_1 object will be created and its constituent Part objects destroyed.

In order to specify Generalization-Specialization structures, one should consider a Class-&-Object and identify other Class-&-Objects with similarities in attributes and services. This type of structure can be either hierarchical or lattical, although within the manufacturing context the hierarchy structure is usually more common. Examples of manufacturing entities to be represented by this type of structures include groups of products, groups of specialized machines and cells. The Whole-Part structure, on the other hand, is particularly suited to describe tree-structured entities such as Bill-of-Materials, as shown in Fig. 4.14. Therefore, when seeking this kind of structure, the products and their parts (customer Class-&-Objects within the model) are usually a good place to start.

4.4.4 Specification of dynamic interaction

One of HOOMA's most useful features is its ability to model the dynamic aspects of a manufacturing operation through an O-O perspective. Within its overall framework, dynamic behaviour of individual objects and the interaction between server and customer Class-&-Objects can be effectively described by using two standard tools: specification of object life cycle using State Transition Chart (STC), and modelling of object interactions through Activity Cycle Diagram (ACD; see Chapter 6, Box 6.3).

As far as ACDs are concerned, HOOMA employs them for a number of purposes including, for example:

- To help identify and illustrate the dynamic interaction between customer and server objects.
- To aid the identification of attributes and services.
- To help describe life cycle of objects.

These diagrams provide an ideal means to illustrate the states and interactions of an object throughout its existence within the system. From them, services and attributes may be easily identified and the dynamic and logical interaction between objects clearly specified where necessary.

ACDs are a very useful tool, especially when applied in conjunction with State Transition Charts (STC, Fig. 4.15). State Transition Charts can be used

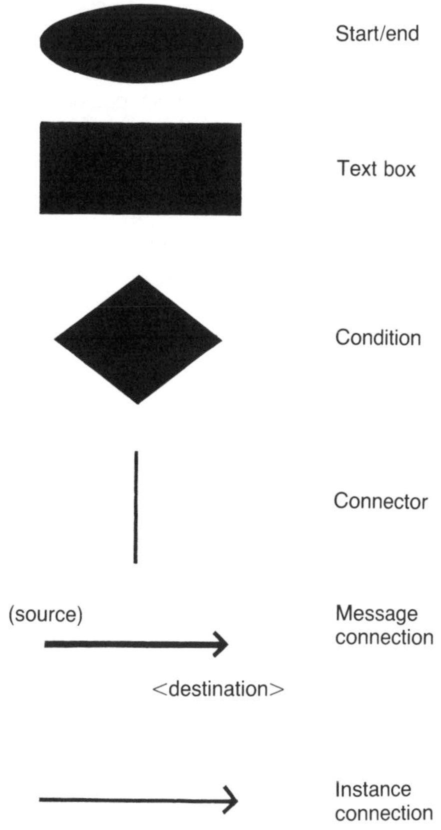

Fig. 4.15 Notation of state transition chart (STC).

to describe precisely the life cycles of customer objects and the operational sequences of server objects, together with their relevant services and connections. For example, as illustrated in Fig. 4.15, associated with each condition block there can be a connection to another object with which the current object is interacting. Through connections like this an STC can provide a detailed life cycle of objects specifying clearly their logical sequences, the timing of connections and the dynamic interactions (together with a relevant ACD where appropriate). Depending on the application, STCs and ACDs can be developed for the whole or parts of the problem domain to specify adequately the dynamic aspects of the system concerned. An example to illustrate HOOMA's usefulness in this regard will be given in the next section.

Based on the objects and the structure of the system as identified in sections 4.4.2 and 4.4.3, it is now possible to further refine the definition of object attributes, and to specify their services/operations step by step following a logical pattern. Since the finalized form of attributes and services must assure both completion and simplification, a number of iterative processes may be necessary to achieve satisfactory results. For instance, the first set of operations can be identified as being the procedures responsible for the change of state of the object, and therefore their identification is related to the development of objects STCs. Similarly, when the operations consist of complicated decision processes or other objects, implying the involvement of dynamic interaction and/or message/instance connection, the use of ACDs and the identification of object connections as outlined in section 4.4.5 and the use of ACDs may provide a very useful insight.

4.4.5 Specification of object connections

There are other relationships and interactions within the system that are not, or cannot be, expressed using the two structures described in section 4.4.3. Two more relationships are needed to provide a more complete picture of the system: Instance Connections and Message Connections (Coad and Yourdon, 1991).

An Instance Connection, which indicates the dependence of one individual object on other object(s) in order to meet its responsibilities, is represented by a line connecting two objects (rather than classes). Each object has an amount or range on its end of the line to reflect its relationship with other objects (Fig. 4.16). For example, assume the objects and structures shown in Fig. 4.13 and Fig. 4.14 are all parts of a particular manufacturing operation. Then in order for the relevant server object (probably an assembly station) to carry out the assembly transformation, usually a certain Work_Order object must exist (in this case the relationship is rather like a control line on the top of a function block in $IDEF_0$). Since one assembly station can be associated with a number of Work-Orders at any time, but a Work-Order is usually connected to one particular operation, here we are likely to have a many-to-one instance connection between the responsible server object and the Work_Order object.

Message Connections indicate communication between objects, relating to the services that an object may require. As shown in Fig. 4.17, at some point a 'sender' sends a message to a 'receiver' to acquire a service, and the receiver receives the message and then takes appropriate action which may return a result.

Having reached this stage following the HOOMA cycle, an outline of these connections is probably already provided by the relevant SRDs. However, this has to be mapped onto the Class-&-Object model with the relevant connections attached to objects. For instance, there should be at least one message connection associated with each service. Instance Connections, on

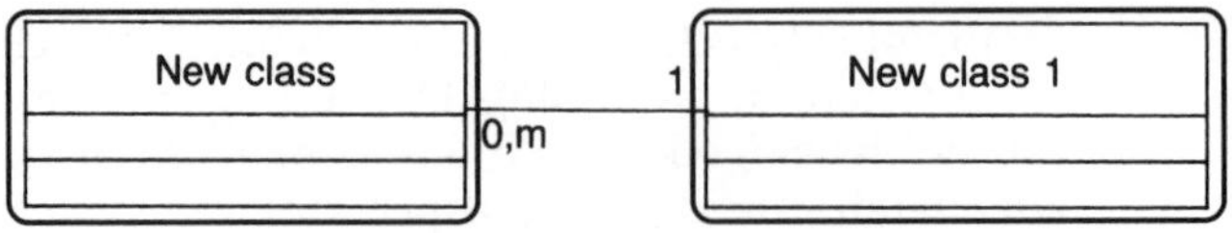

Fig. 4.16 Notation of an instance connection.

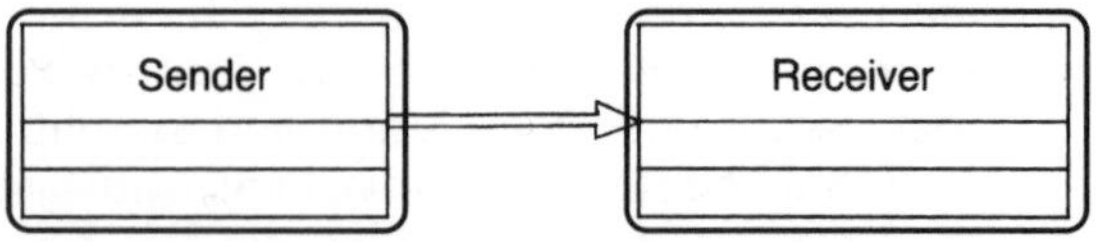

Fig. 4.17 Notation of a message connection.

the other hand, are usually associated with object attributes. Hence they can be identified through examination of relevant attributes. Again, the use of STCs and ACDs will significantly ease the task of identification.

4.4.6 Sub-system integration

Finally, the sub-systems should be integrated into one whole graphic model to convey the necessary information about the problem domain. The divide between sub-systems has already been spanned by earlier Structures and Connections. It is at this point that the Function Subject Layers and the general process flows can be utilized for further integration of the model. For instance, in general the initial (top-level) processes should be placed at the top of the model diagram with subsequent processes below. It is also desirable to have a separate customer Subject Layer consisting of all the customer Class-&-Objects to enhance readability.

4.5 HOOMA presentation of OHMS

As an example, the structure of a typical order-handling manufacturing system (OHMS) will be modelled using HOOMA to illustrate its use. The products and operational structure of the model OHMS company are as summarized by the IDEF_0 model presented in Chapter 3.

4.5.1 Static presentation

As outlined previously, in order to ease the process of analysis, the first step is to decompose the problem domain into smaller and hence more manageable components. Four main Function Subjects can be identified for a typical OHMS operation:

1. Formulate production plan
2. Design and develop product
3. Gather production resources
4. Produce product.

The four sub-systems can then be developed further according to the procedures outlined in section 4.4. In relation to the four-level FBD presentation given in Fig. 4.18, the simplified diagram shown in Fig. 4.19 illustrates the shape of the complete Function Subject hierarchy of this system.

Analysis models can then be conceptualized for each of the FSs identified so as to provide the required O-O presentation of this operation. For example, there are several workshop activities within Produce Product. Together with an SRD to reveal more details (if so desired), the static O-O structure of this sub-system can be developed as shown in Fig. 4.20. The product structure illustrated in Fig. 3.9 is now shown in terms of a CC&O structure. The main 'chock' product is a specialization of the Product Class. The chock assembly consists of two sub-assemblies: the valve sub-assembly and the leg_ram sub-assembly, and some additional parts. These need assembling, hence the instance connection to the Assembly Line Class-&-Object. Similarly, the valve sub-assembly consists of the valve block and valve parts. The valve block requires to be processed by the CNC machine objects belonging to the Flexible Manufacturing Cell, hence the instance connection with the FMS.

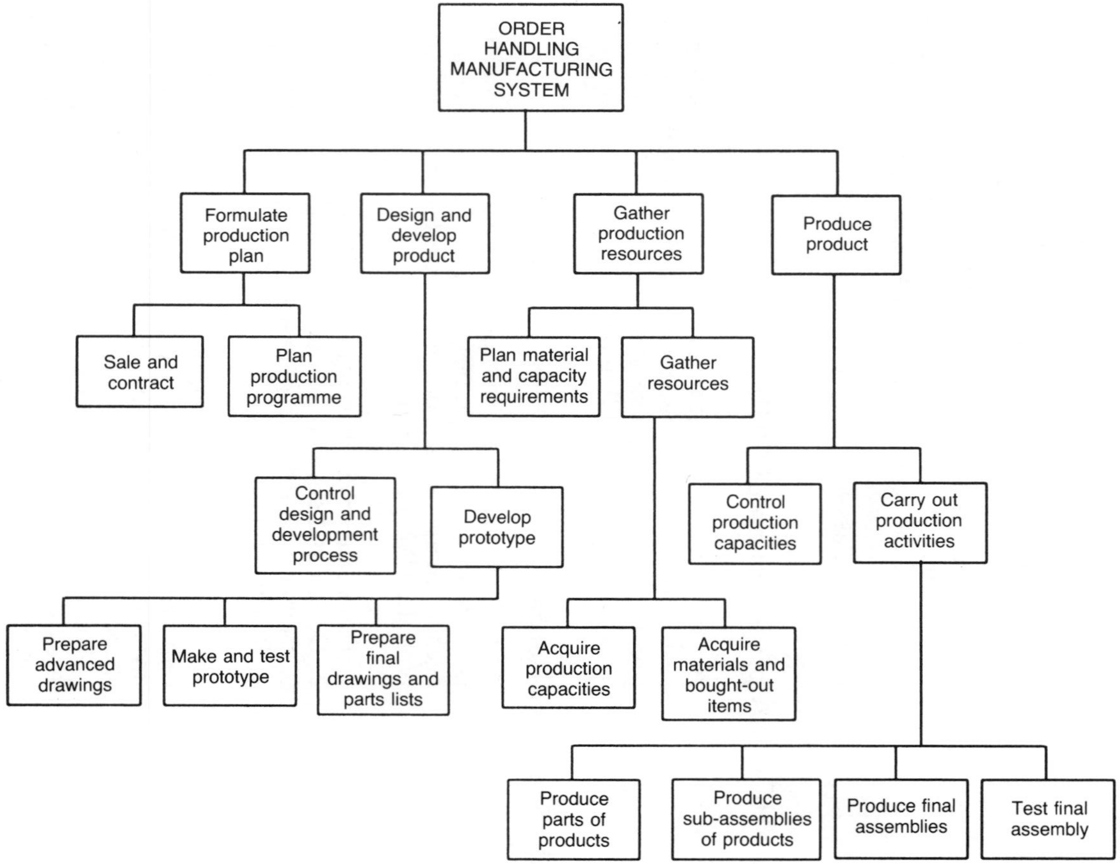

Fig. 4.18 An FBD presentation of an OHMS operation.

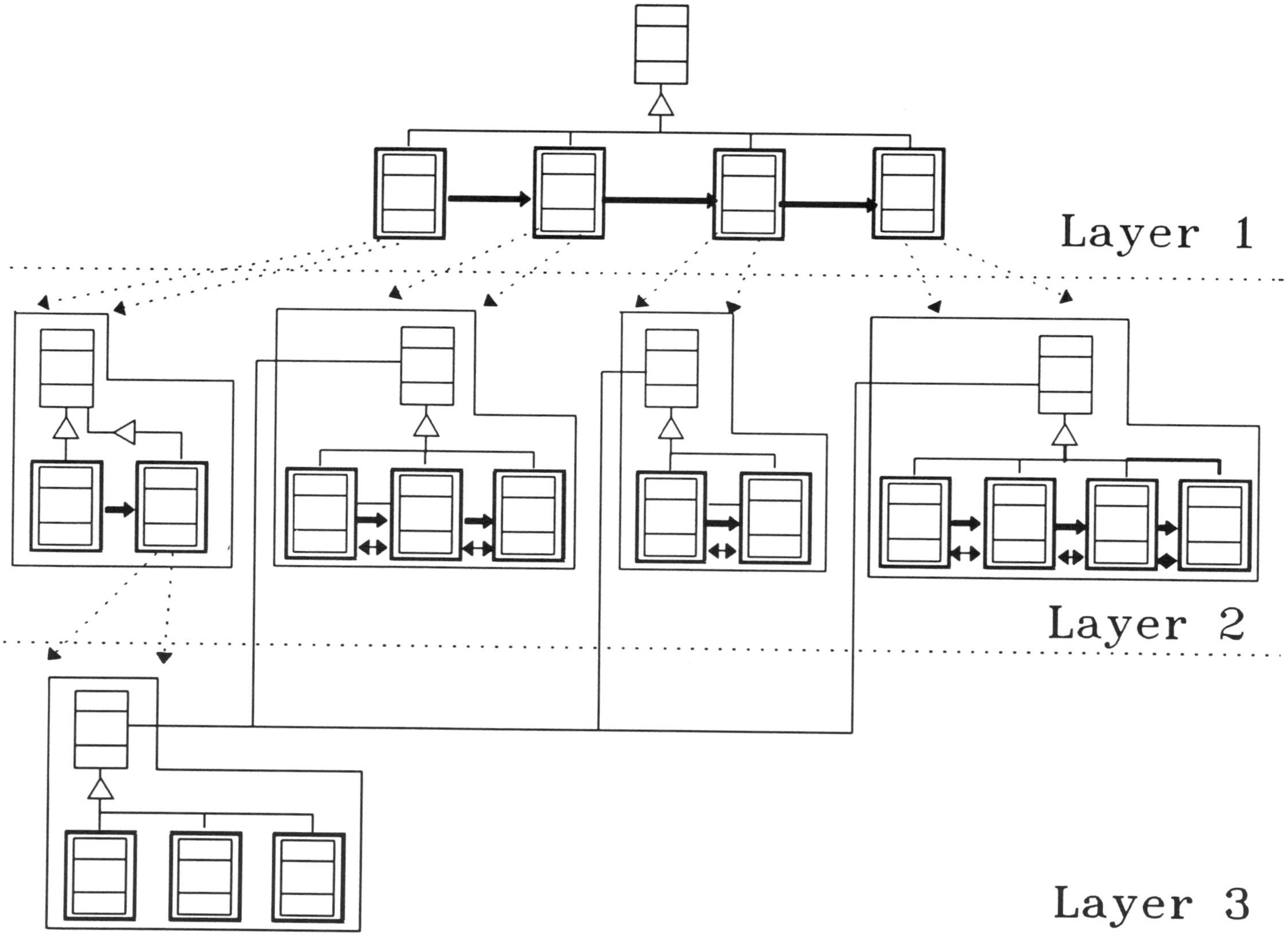

Fig. 4.19 Complete FS hierarchy of an OHMS operation (text omitted)

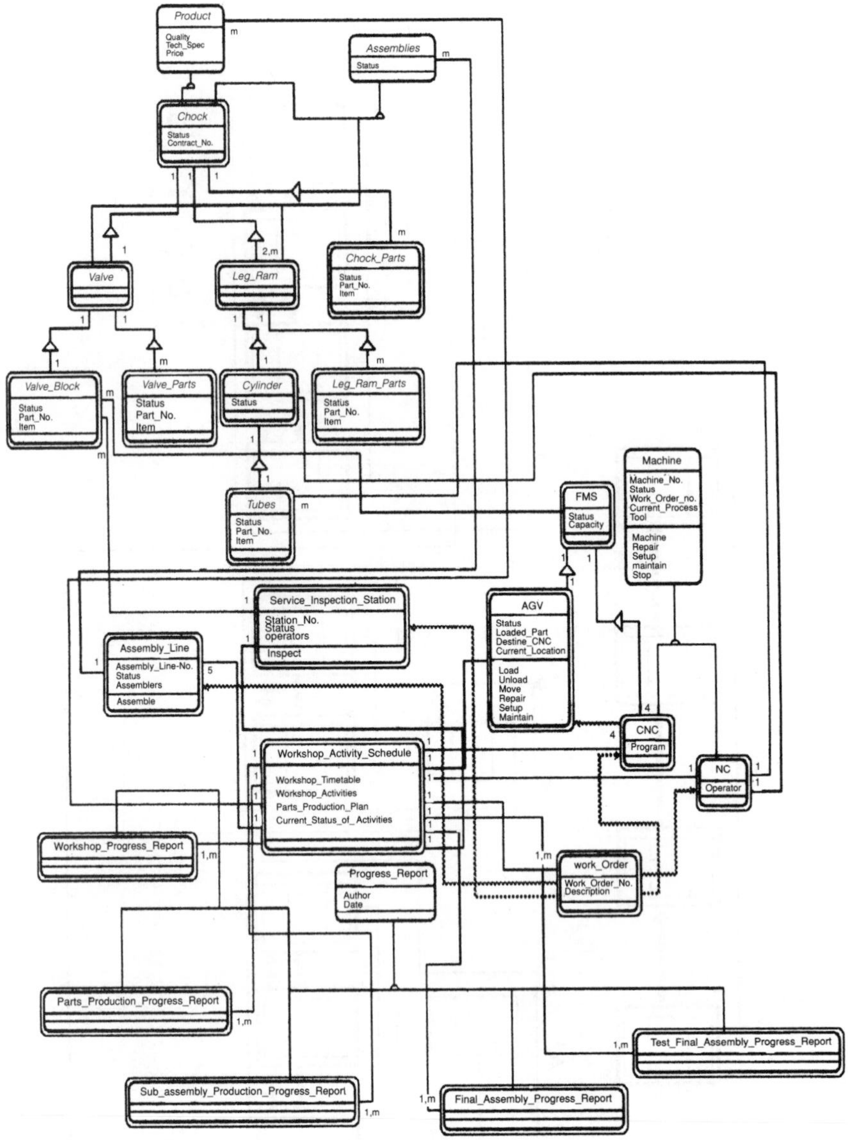

Fig. 4.20 Object-oriented presentation of 'produce product' sub-system.

The valve blocks also undergo inspection prior to the assembly operation. The leg_ram sub-assembly is constructed from a cylinder and leg_ram parts. As shown, the production of the cylinder involves cutting the steel tubes using dedicated NC machines, resulting in an instance connection between the Tube object and the NC object. Finally, a work order must be issued to the relevant objects to authorize any work to be carried out.

4.5.2 Dynamic modelling

Once the static structures of the system are established, the dynamic behaviour of the individual objects and their interactions can be specified

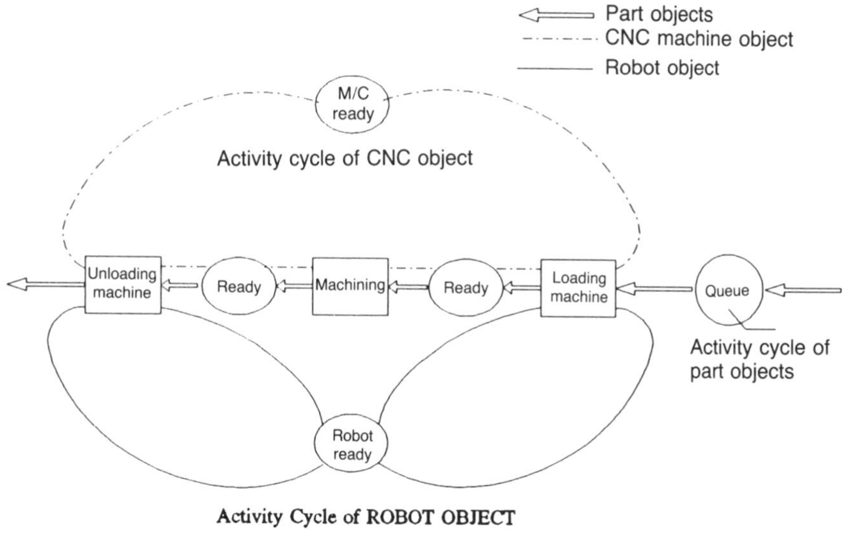

Fig. 4.21 Example ACD illustrating object dynamic interaction.

where appropriate. For instance, the production of valve sub-assemblies uses the FMS. The dynamic interactions among the three objects involved in this process (CNC, AGV/ROBOT and Valve_Block) are clearly revealed by the activity cycle diagram given in Fig. 4.21 (the attributes and services of the CNC and AGV objects are as shown in the sub-system model). A more detailed life cycle of the Valve_Block objects is presented in terms of an STC as shown in Fig. 4.22. STCs for the AGV and the CNC are illustrated in Figs 4.23 and 4.24, respectively.

As can be seen, together these diagrams provide a means to completely specify the dynamic process involving various objects with their relevant attributes, services and connections. For example, the life of a Valve_Block object begins with its instantiation from its Class. Once created, they will wait in the system and request to be loaded onto a CNC. This would involve a message connection to require a 'load' service, hence the destination ⟨X⟩ indicating that the intended receiver of the message in this case is the AGV (Figs 4.22 and 4.23). Generally, this type of message may be sent either by certain controlling objects like a Work_Order or by the customer objects themselves.

Upon reception of the message, the AGV object will identify the source of the message and, if possible (i.e. if ready to take on the task), invoke the appropriate service as requested. Therefore, in order to provide a service a server object will generally require certain attribute conditions to be fulfilled. These attributes may need to be checked with instance connections to other objects involved. In this case, the object's own status is examined for a ready condition. If the attribute holds this value, the AGV will engage itself in the loading operation. The object will then search for an empty CNC to load by inspecting the status attributes of all the CNC objects. Once a CNC is loaded, the AGV object is returned to a ready status. A similar CC&O/SC&O relationship exists between the Valve_Block and the CNC; once a Valve_Block is loaded it will request machining from the CNC. This will involve another message, and this time the destination is the CNC object it is loaded onto.

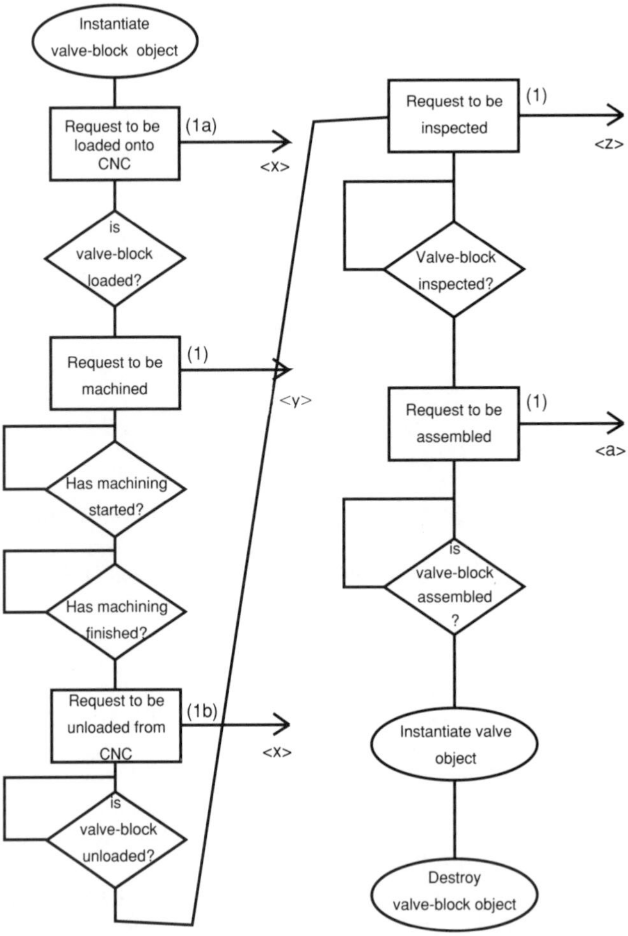

Fig. 4.22 State transition chart of 'valve block' object.

Again the CNC object will check the source of the message and attempt to provide the appropriate service. In this case it will check for the correct tools to be used, and if it finds them the Valve_Block will be processed as necessary.

After the machining operation, according to the logical sequence specified by its STC the Valve_Block object will request to be unloaded. This procedure follows closely after the load CNC request. This will be followed by requests for inspection and assembly. Finally, the life of a Valve_Block object comes to an end when it attains 'assembled' status once it has been transformed into a sub-assembly. When this point is reached, instantiation of the Valve sub-assembly is executed and the Valve_Block object is destroyed. The life cycle of the Valve object can now be modelled along similar lines to a Valve_Block object.

If necessary, a similar exercise can be executed for the other objects to

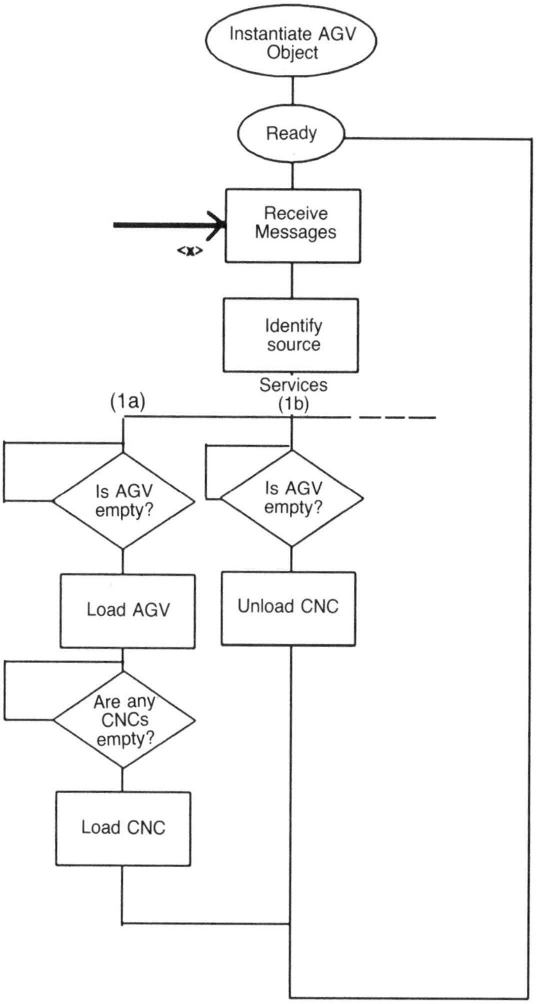

Fig. 4.23 State transition chart of AGV object.

provide the required detail of dynamic modelling. These diagrams and charts will not be shown here, but the above shows clearly the power of these two tools when combined.

4.5.3 Overall model

The overall static HOOMA presentation of the modelled operation is as illustrated in Fig. 4.25. In order to improve readability, this figure does not show the Connections and Structures which must be present within subsystem presentations. However, the Function Subjects have been identified and are shown in the figure.

The overall conceptual model of this OHMS implies the general flow of process involved, from top to bottom. Here the Sales Department is the first stage of the operation, making contact with customers to produce the all-important orders for the OHMS. Once a contract has been signed and the

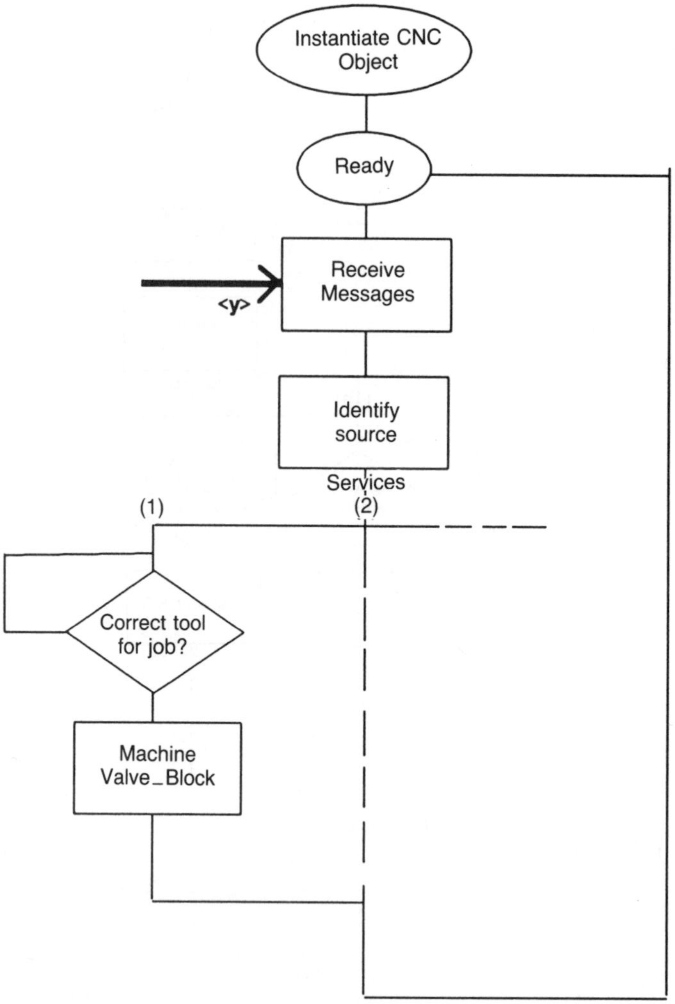

Fig. 4.24 State transition chart of CNC object.

product production plan formulated, the details are forwarded to the Design and Development Departments. With the design finalized, the product parts must be acquired. This will be the responsibility of the Purchasing Department. Acquisition of materials may be from existing inventory, the Stores, or from external suppliers. Purchased materials are delivered either to the workshop area or to the Stores. The workshop, which contains several smaller workstations, is the next and final step in the OHMS operation. The output of the OHMS operation is the product which is at the bottom of the model. The product is contained in its own Function Subject.

If now we compare this HOOMA model and the IDEF$_0$ model of the same manufacturing operation (Fig. 3.33), the differences between the functional and O-O approaches become clear. The IDEF$_0$ model has several layers of functional decomposition, but does not attain the detailed level of entity description of the HOOMA model. In addition, the IDEF$_0$ model is very much static in nature and does not cater for the dynamics involved in

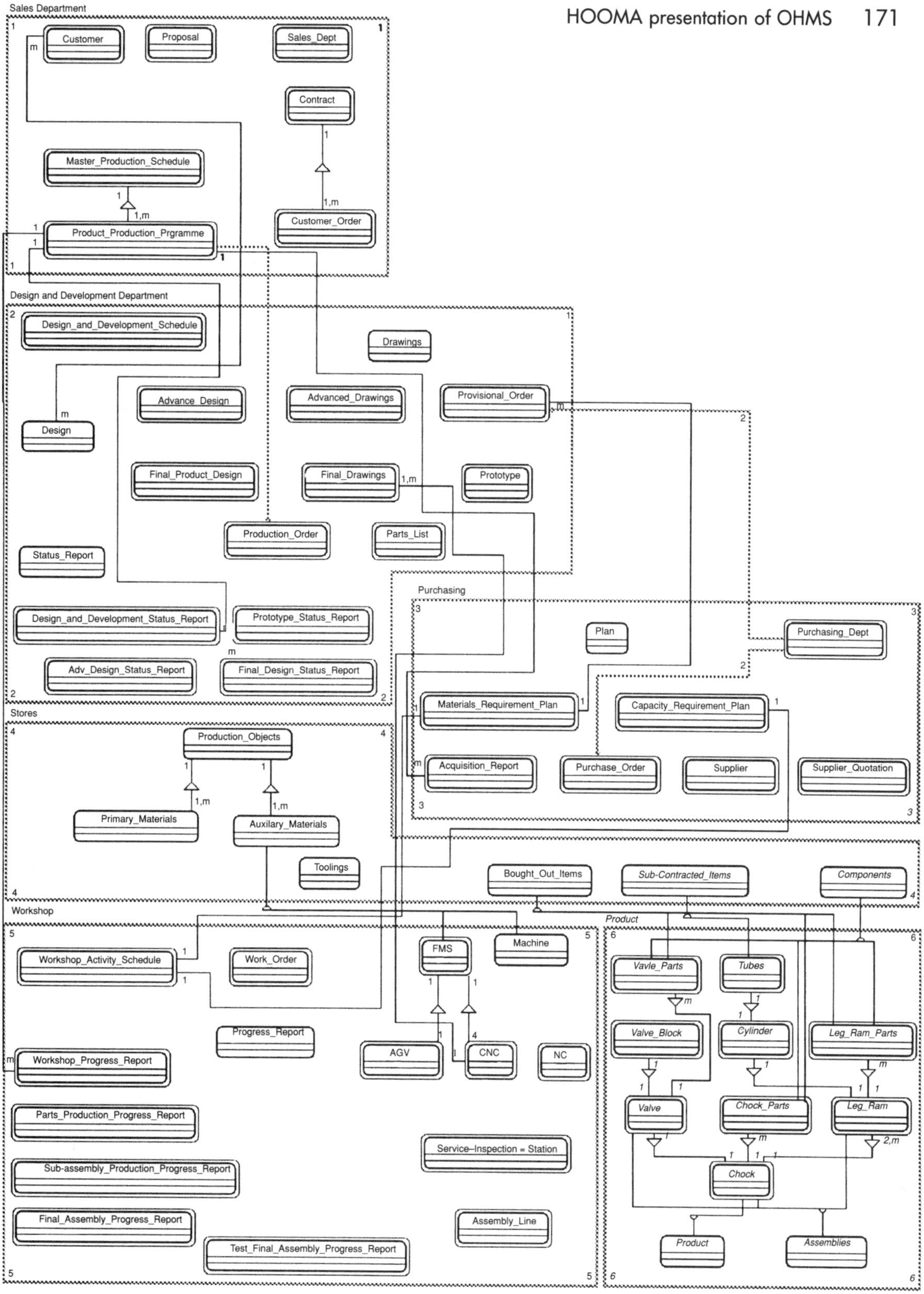

Fig. 4.25 The overall static HOOMA presentation.

manufacturing operations. The HOOMA presentation has a clear advantage here, in being able to represent the system dynamics in terms of object states and activity cycles.

It must be said that graphical representation of the HOOMA model leaves something to be desired when compared with the neatly structured $IDEF_0$ presentation, requiring a little more prerequisite knowledge from the analyst and the reader to be able to make sense of the whole thing. However, this could well be a small price to pay for the true potential such an approach can offer. The example given illustrates the kind of typical CC&O/SC&O relationships that exist within the manufacturing context. In this case the CNC and AGV objects provide a set of pre-fixed services, but go through a cycle of activity that is determined by the life cycle of the Valve_Block object. The life cycle of the Valve_Block, on the other hand, depends mainly on its design and its manufacturing processes, which in turn determine the various logical connections as this temporary object progresses through its transition within the system. Therefore one of the main advantages of the HOOMA approach to modelling manufacturing operations is now apparent: its flexibility. Since the dynamic behaviour of the objects can be individually specified, the resulting system specification will be based on a set of self-contained building blocks. In particular, once the basic operation and behaviour of those SC&Os are specified, they will tend to remain consistent within a system model, and hence can be easily reused for other purposes. For example, suppose now a new type of product is to be introduced to an existing system and hence a new manufacturing operation needs to be specified. With a HOOMA model this can be reduced to the following tasks:

- Specify the structure of the new product utilizing the existing product object structure, and identify the services required to produce the parts and assemblies involved. Draw their STCs accordingly.
- According to the above, identify the reusable SC&Os and specify the new dynamic interaction required through object connections.
- Add any new SC&Os needed and specify their dynamic interactions.

Therefore only a relatively small amount of change will be needed, as against the case with a functional-based model where a large part or even the whole model would need to be altered to cater for such an organisational change.

4.6 Conclusion

When compared with other methods of system specification, the approach suggested in this chapter has certain advantages. With a HOOMA model the flexibility needed due to recurrent organisational changes can be catered for. Class-&-Objects may be added or changed without the necessity of significantly altering an existing model. In addition, HOOMA has the advantage of being able to represent the dynamic aspects of advanced manufacturing systems in terms of object states and interactions.

However, it should be recognized that this methodology is still in its early development stages. Further work is required to develop it to maturity. One area which needs further attention is the end-to-end process modelling. HOOMA does attempt to tackle this, but a specific model may be required for describing global processes end-to-end. The concept of a global process is at odds with the object-oriented ethos, and so it is not surprising that it is lacking from current methodologies. Another area for attention is reuse. Ideally reuse should be incorporated, designed and formalized into the analysis procedure, particularly if the need for frequent systems specification for coordination-intensive manufacturing structures is to be effectively satisfied. This should be more rigorous than the proposed format. Nevertheless, HOOMA and its approach may prove to be the first step towards achieving this goal.

Further reading

Booch, G. (1991) *Object Oriented Design*, Benjamin Cummings.

Coad, P. and Yourdon, E. (1991) *Object-Oriented Analysis* (2nd edn), Prentice Hall.

Graham, I. (1991) *Object Oriented Methods*, Addison-Wesley.

Mayer R.J., *et al.* (1992) *Information Integration for Concurrent Engineering (IICE): IDEF$_4$ Object-Oriented Design Method Report*, AL-TR-1992-0056, Wright-Patterson Air Force Base, Ohio, USA.

Questions

1. Why is there a need for a more flexible tool of systems definition for today's manufacturing operations?
2. Using real-life examples, summarize the basic concepts of object-orientation.
3. Discuss the main differences between an object-oriented approach and a function-based approach. What are the advantages generally associated with object-orientation?
4. What is the relation between a Class and its Objects?
5. Give examples of typical Class-&-Objects from a manufacturing environment. Specify these in terms of attributes, messages, services and state transition.
6. Describe the structure of the Class-&-Objects as specified in question 5.
7. Draw a HOOMA model to describe a manufacturing operation that is familiar to you. This model should specify both the static structure and the dynamic relations of the manufacturing process involved.

5 The general systems approach to problem-solving

5.1 Introduction

According to the 15th edition of the *Encyclopaedia Britannica* (Propaedia, pp. 9–15), the fields of human knowledge and interest may be classified primarily into the following categories:

1.0 Matter and Energy
- 1.1 Atoms: atomic nuclei and elementary particles
- 1.2 Energy, radiation, and the states and transformation of matter
- 1.3 The universe

2.0 The Earth
- 2.1 The Earth's properties, structure, and composition
- 2.2 The Earth's envelope: its atmosphere and hydrosphere
- 2.3 The Earth's surface features
- 2.4 The Earth's history

3.0 Life on Earth
- 3.1 The nature and diversity of living things
- 3.2 The molecular basis of vital processes
- 3.3 The structures and functions of organisms
- 3.4 Behavioural responses of organisms
- 3.5 The biosphere

4.0 Human Life
- 4.1 Stages in the development of human life on Earth
- 4.2 The Human organism: health and disease
- 4.3 Human behaviour and experience

5.0 Human Society
- 5.1 Social groups: peoples and cultures
- 5.2 Social organization and social change
- 5.3 The production, distribution, and utilization of wealth
- 5.4 Politics and government
- 5.5 Law
- 5.6 Education

6.0 Art
- 6.1 Art in general
- 6.2 The particular arts

7.0 Technology
- 7.1 The nature and development of technology
- 7.2 Elements of technology
- 7.3 Major fields of technology

8.0 Religion
- 8.1 Religion in general

- 8.2 The particular religions
- 9.0 The history of mankind
  - 9.1 Peoples and civilizations of ancient Southwest Asia, North Africa, and Europe
  - 9.2 Peoples and civilizations of medieval Europe, North Africa, and Southwest Asia
  - 9.3 Peoples and civilizations of East, Central, South, and Southeast Asia
  - 9.4 Peoples and traditional civilizations of sub-Saharan Africa to 1885
  - 9.5 Peoples and civilizations of pre-Columbian America
  - 9.6 The modern world to 1920
  - 9.7 The world since 1920
- 10.0 The branches of knowledge
  - 10.1 Logic
  - 10.2 Mathematics
  - 10.3 Science
  - 10.4 History and the humanities
  - 10.5 Philosophy

This list not only is an excellent outline of the many fields of human knowledge but also provides a useful guide to the areas where various systems approaches of modelling, analysis and evaluation have been applied both to expand human knowledge and to solve practical problems. Indeed, it has been pointed out that the range of applications based on systems thinking is perhaps as varied and pervasive as the accumulated knowledge of mankind as listed above (Sandquist, 1985) – so much so that a potentially all-embracing systems theory known as general systems theory has been developed, with a society specifically dedicated to its research and development, the Society for General Systems Research, which aims:

> To encourage the development of theoretical systems which are applicable to more than one of the traditional departments of knowledge; to develop theoretical systems of concepts, relationships and models and to investigate the isomorph of concepts, laws, and models in various fields, and to help in useful transfers from one field to another; to encourage the development of adequate theoretical models in the fields which lack them; to minimize the duplication of theoretical effort in different fields; to promote the unity of science through improving and breaking down the barriers of communication among the specialists in various scientific fields.

Looking back on what we have discussed so far in this book, the value of systems thinking in our particular field of interest has now become apparent: it has allowed us to develop an overall framework of manufacturing operations, and to identify the key areas where improvements could be made by applying systems concepts.

The basic systems concepts were discussed in Chapter 2. In addition, .it was pointed out that our main concern with systems ideas would be focused on the systems methodologies and techniques which are to be used for real-world problem solving. More specifically, this implies the application of systems thinking to the design and evaluation of advanced manufacturing systems. This chapter is designed to introduce the methodologies of traditional systems analysis and some of the more recent systems thinking in the area of real-world problem solving. Work which has already been carried out in the development of specific methodologies for the design and evaluation of modern manufacturing systems will be reviewed later.

This chapter introduces some concepts of *systems analysis*, including a general problem-solving framework which has long been recognized as a useful model for structured decision making to solve real-world problems. This problem-solving framework is in fact the basis of all the techniques involved in an important systems discipline known as *systems engineering*. More recently, methodologies based on this particular problem-solving framework have come to be referred to as the *hard* systems approaches. The ideas of a relatively new systems approach, known as the *soft* approach, will also be discussed and the differences between the two approaches compared.

Systems analysis is usually based on a problem-solving methodology. The reason why a methodology is needed for systems design and analysis can be illustrated by looking back at the history and development of this particular analytical discipline. In the 1960s, the first large-scale computer systems were beginning to be developed. The design tasks consisted of defining what the systems would do, specifying their sub-systems and interfaces, and then creating these sub-systems. However, it was rapidly seen that the costs of work carried out after the systems were delivered were unacceptably high and the systems design process had to be reviewed. Analysis revealed that the majority of the problems which were causing the increased expense stemmed from errors which were made during the early analysis and design of the system. These errors could be up to 100 times more expensive to correct when the system was installed than if they had been corrected in the early phases of development. Such an obviously unsatisfactory situation eventually led to the realization that better methods had to be used in the early phases of the development process. In addition, as discussed in Chapter 1, the late 1970s and 1980s saw the onset of a major upheaval in industry all over the world: industry is becoming more competitive, resources are becoming increasingly scarce, customers are demanding more variety, and international markets are forcing prices to fall. Manufacturing firms must, therefore, increasingly react to initiatives from their competitors, or must adopt the right strategy and use new technology to give themselves an operations advantage. This implies that firms are having to restructure more rapidly than at any other time in history. The high technology often implemented under these circumstances is making the planning of such changes very significant. The system components must interact with each other to a higher degree in order to achieve a successful system, as there is now less human intuition to deal with ambiguities. The cost of failing to react appropriately is also much greater today than previously because of the higher cost of capital equipment and the need for rapid response to new product markets, which will be affected by the efficiency of manufacturing systems operations. All of these, therefore, have created a need for a set of guidelines or a methodology which will reduce the chances of mistakes in systems design and thus enhance system effectiveness.

The remainder of this chapter gives an overview of the general systems approach to systems analysis for problem solving or systems design. After studying this chapter, the reader should be able to:

1. understand in general terms the principles, structure and techniques of the hard systems approach to systems design and problem solving, and to use this approach in certain specific problem situations;
2. understand in general terms the principles, structure and techniques of

the soft systems approach to systems design and problem solving, and to become familiar enough with the methodology to apply it where appropriate;
3. explain the fundamental similarities and differences between the two approaches, and be able to choose the most suitable approach according to the particular circumstances;
4. appreciate the relevance of the systems concepts (as outlined in Chapter 2) in relation to these systems methodologies.

5.2 Systems analysis: a model of a problem-solving cycle

5.2.1 Overview

The general methodology of systems analysis has probably been the most widely used systems approach to problem solving. It has been used in many of the subject areas listed in section 5.1, including, of course, the area of manufacturing systems design and analysis. The following is a typical situation where the methodology of systems analysis can help. Imagine that you work as a manufacturing systems analyst in the consultant team of a large manufacturing organization. One day the management of your organization decides that the production planning and control system of one the organization's companies is to be computerized. You are asked to work with the people from the company to design and implement the new system.

The system concerned in this situation is likely to follow a 'systems life cycle' as illustrated in Fig. 5.1. The model shown in this diagram can in fact be viewed as a summary description of the majority of problem-solving situations concerned with the design of manufacturing systems (or their sub-systems). Therefore, systems analysis in general may be viewed as the

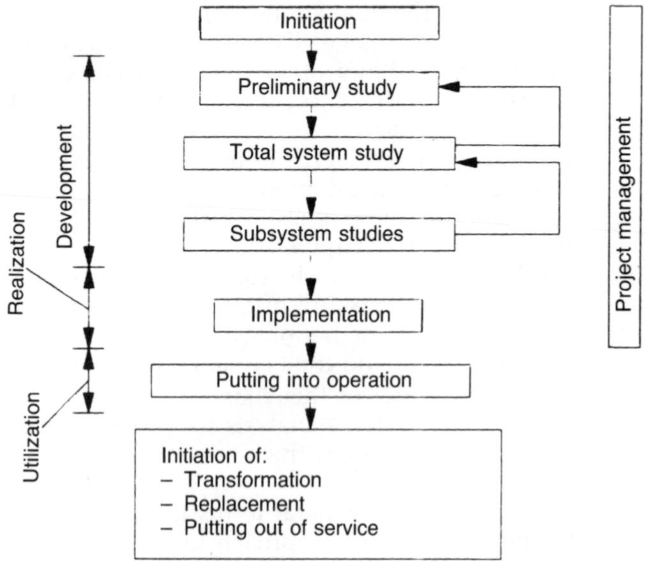

Fig. 5.1 System life-cycle model.

process of studying a system with the objective of modifying and improving it. According to Cleland and King (1983):

> Systems analysis is a scientific process, or methodology, which can best be described in terms of its salient problem-related elements. The process involves:
>
> 1. systematic examination and comparison of those alternative actions which are related to the accomplishment of desired objectives;
> 2. comparison of alternatives on the basis of the costs and the benefits associated with each alternative;
> 3. explicit consideration of risk.

Also appearing in many guises, dependent on specific application, the methodology of systems engineering generally follows a problem-solving cycle which provides a structure to deal with the various problems throughout the system life cycle of Fig. 5.1. The underlying logic and relations of such solution-seeking processes can be illustrated by the situation where one wishes to travel from A to B. Let us now examine what tasks are involved in solving such a problem and what kinds of question one should ask to reach the desired end:

1. *Where are we now?* In order to reach the destination, one must know exactly where one starts the journey. This is why, when planning a journey, we begin by trying to identify our exact location on a map, for there is little point in finding where we want to go without knowing our present location.
2. *Where should we be?* A more obvious question to ask next, having identified the present location, may appear to be 'Where do we want to go?'. However, for such a journey to take place there has to be a good reason – one will not normally spend time and effort travelling without purpose. In searching for the location of the destination, the underlying reason(s) why one should wish to undertake such a journey must first be outlined. Therefore, 'Where should we be?' may be a more appropriate question to ask here. In addition, at this stage the following points will have to be considered:
   (a) What relevant background and environment information is required?
   (b) Is the journey feasible? There may be constraints on distance, time, money and means of transportation – it is not possible, for example, to travel to Mars, or to reach New York from London in one hour – not with today's means of transportation technology anyway.
   (c) What are the most important factors to consider when planning the journey? For example, if we are in a hurry to meet a business appointment then the time taken by the journey is clearly an important factor. On the other hand, if the purpose of the journey is to reach a holiday destination, then financial considerations may become a more important issue.
3. *What are the possible routes and means?* There may be a number of alternative means to reach B from A, travelling by road, rail, sea or air. Each of these may again entail a number of possible routes. In order to make a choice, we must first gather as much information as we possibly can.
4. *Which route to take?* We should now analyse the possible consequences of each choice of route, and choose the best based on the information we have. We must have effective means to carry out this evaluation process –

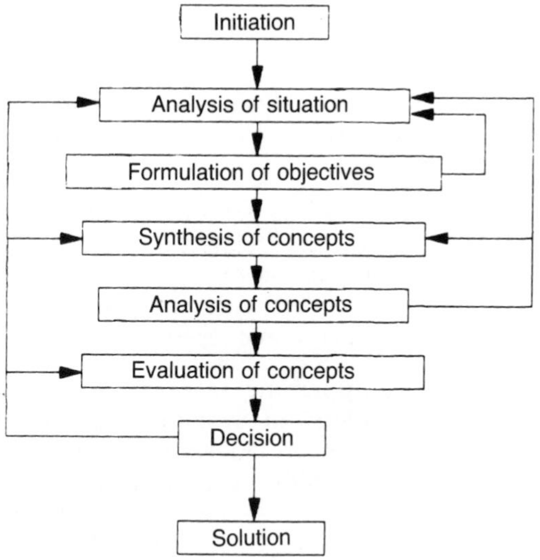

Fig. 5.2 System engineering problem-solving cycle.

perhaps a calculator to sum up the total time or money needed by each of the choices in this particular case.

5. *Start the journey and keep an eye on progress*. According to the journey plan derived from the completion of the previous steps, we can now begin the journey. However, in order successfully to reach B, we have to monitor continuously the progress of the journey. On our way we may have to alter the original journey plan taking into account any change of circumstances, such as delays or cancellations. We may even have to change the location of the destination altogether because of some unforeseen problem.

A generalized model of this problem-solving cycle is given in Fig. 5.2. Based on this overall framework, we can now continue to discuss in more detail the purpose of each of the individual steps and consider how they may be undertaken in practice.

5.2.2 Analysis of situation

Being a typical hard systems approach, the systems analysis technique follows a framework which breaks down the whole task into a set of broad steps, and demands certain fixed outputs from one stage before logically continuing with the next. As shown in Fig. 5.2, analysis of a problematic situation is the first stage of the systems approach. This stage perhaps plays the most important role in determining the nature of the analytic task as a whole.

The aim of this stage is to identify the problem to be solved by analysing the contents of the existing system. Attention here should be focused on the structure of the existing system, including its elements, relationships, boundaries, environment, functions, strengths and weaknesses. Scenarios for the development of the system and its environment should also be investigated. The successful completion of this stage should provide the correct answer to the key question 'Where are we now?'.

The 5W procedure for system identification and description (section 2.4.2) provides an obvious example of techniques which can be used for this purpose. The list of 'preconditions for effective operation of organization' (section 2.4.3) gives an example of a checklist which may also be used to assist problem identification at this stage.

5.2.3 Formulation of objectives

Having answered the question 'Where are we now?', it is clear that the next logical question to ask is 'Where should we be?'. The goal of this next stage of the systems approach is to define the required performance of the system, identify conflicting objectives and interests and recognize any constraints.

Within the systems analysis cycle of Fig. 5.2, each stage is dependent upon the successful completion of the previous stage. This requirement is clearly illustrated by the relationship between the first two stages: only with exact knowledge of the current location and state of the environment can one begin to plan the right direction to follow in order to reach the desired end result.

The combination of the first two stages of this problem-solving model has also been referred to as problem *formulation*. This is because, by establishing where we are and where we should be, these two stages together identify the gap between the present system state and what the wider system or the environment has set for the system. Dandy and Warner (1989), for example, provide the following checklist of possible activities which might be undertaken as part of this problem formulation function:

Identify the problem as part of a wider problem
Identify each system as part of a larger system
Determine the real underlying needs
Gather relevant background information
Search for possible side-effects of the project
Identify constraints
Specify objectives and identify possible conflicts among objectives
Specify any performance requirements and operating conditions
Devise measures of effectiveness.

At this stage, three points deserve particular attention. First, the importance of problem formulation must be fully recognized. The importance of the successful completion of this function should be quite clear by now – since its output represents the nature of the opportunity or challenge, the better formulated this is, the better we can hope to improve the situation. There is always a danger of concentrating too much on the analytical techniques and the resulting evaluation stages of the problem-solving process, while overlooking the importance of problem formulation in the first place. This represents a great pitfall in systems analysis, for there is a real danger that the project concerned may end up attempting to solve a problem that has little relevance to the real issue. Therefore, sufficient attention must be paid here to guarantee a sound definition of the systems project.

Second, problem boundaries and critical components must be accurately identified and precisely defined before a problem can be formulated. On the surface, problem formulation may appear to be a straightforward matter. A 'problem' is generally thought to be obvious when one is dissatisfied with the input to or output from a system at a certain level or state and wishes to improve the situation in the future. However, the 'gap' observed by someone analysing the situation does not always directly reveal the nature of the underlying real problem, only reflecting its symptoms.

For example, a systems analyst might initially be presented with a 'problem' concerning the need for improved planning policies to permit improved machine utilization and hence reduced production cost. After a detailed analysis of the production cost structure of the particular manufacturing operation concerned, however, it could be discovered that the level of machine utilization is unlikely to be the key factor which affects the production costs significantly. Instead, the time-weighted level of work-in-progress (WIP) might be identified to be the key issue, so that the problem would need to be redefined accordingly.

Finally, any difficulties must be acknowledged. The systems analysis model as presented in this section is a typical hard systems approach which is essentially based on an engineering perspective. It has been argued by some, therefore, that such an approach does not have universal application to all types of problem. To clarify the situation, Checkland (1981) has distinguished between *structured* (or *hard*) problems which can be well defined by an explicit language and hence solved effectively by the mathematically orientated hard systems approach, and *ill-structured* (or *soft*) problems which cannot be dealt with in the same manner without risking oversimplification of the situation (as quoted by Cleland and King, 1983):

> Problems . . . in the world are unbounded, in the sense that any factors assumed to be part of the problem will be inseparably linked to many other factors. Thus an apparently technical problem of transportation becomes a land-use problem which is seen to be part of a wider environment-conservation problem, which is itself a political problem. Do these then become part of the original problem? Can any boundary be drawn? How can the analyst justify the limits that practicality forces him to impose?

In contrast to the more 'scientific' elements of the systems analysis process (such as modelling, calculation and simulation), problem formulation is an artistic activity that requires skill and craftsmanship on the part of the decision-maker or systems analyst. This is illustrated by the fact that so far there are no universally accepted 'scientific' rules to solve the kinds of question posed above. To stand a better chance of success, therefore, the analyst should follow certain guidelines.

First, he or she must be aware of the limitations and pitfalls of the hard systems approach and its associated analytical techniques. One of the lessons learnt from those not-so-successful operations research projects carried out in the past is that, being trained in a particular analytical discipline (usually mathematical), researchers tend to favour certain modelling techniques and apply them to the wrong types of problem. What is observed tends to be a distortion of reality. Therefore, the systems analyst needs to be fully aware of the advantages and disadvantages of different approaches and techniques so as to adopt the right approach without the risk of distorting the real issue by focusing on inappropriate aspects of the problem.

Second, the analyst must apply systems principles to bound the problem in a realistic manner. When the problem situation gets 'messy', the analyst should attempt to tackle it by defining system boundaries encompassing the relevant factors that represent formulated problems which are more abstract and less complex. Frequently the level of problem abstraction is influenced by factors such as the time and resources available, and the level of risk of oversimplification.

The first obvious systems concept to direct the systems analyst here is that of systems perspective. The hierarchical nature of systems can also provide some useful guidance in this regard. When defining objectives, it is useful to distinguish between strategies and tactics. In a chess game, for example, a player may adopt an overall *strategy* such as that of 'king's side attack' and then use a variety of *tactics* by combining his pieces in such a way as to achieve this strategy and gain an advantage over his opponent. In systems terms strategies are exemplified by the objectives set up for the systems by their wider systems (the ends), whereas tactics are the ways in which the systems attempt to attain these objectives (the means). A strategy at one level within the hierarchy may be present as a tactic at another level and vice versa, depending on how the situation is considered.

In many cases, thinking of objectives in terms of a hierarchy of strategies and tactics (or ends and means) allows the complex task of problem formulation to be tackled in a more structured manner. In particular, the application of this concept in the design and analysis of manufacturing systems is extremely important. For example, the management of a manufacturing company may have decided on a manufacturing strategy which is 'to satisfy our customers' requirements by being flexible in our production organization in terms of . . .' As a result the following objective hierarchy may be drawn up:

satisfy customers' needs
|
achieve flexibility
|
organize production facilities into operation cells
|
design and implement individual cells

Although 'designing and implementing individual cells' is shown here as a means to achieve the higher-level objective of reorganizing production facilities, it is obvious that it can be seen as an objective in itself (indeed, it represents a typical project in the area of manufacturing systems design and analysis). The purpose of such a project, however, is now much clearer. By adopting such a structured way of thinking, it may often be possible to sort out a difficult situation so that the analyst or project team can set out to solve the right problem. It can be seen that the formulation of manufacturing strategy and the design of manufacturing systems are in fact coherent in nature and hence inseparable in practice.

Third, the analyst must consider the alternatives. This includes not only alternative systems approaches as indicated above, but also alternative objectives, constraints and measures of effectiveness. For example, the following are a few such alternatives which are frequently encountered in a manufacturing environment:

Profit or market position?
Sale price or sale volume?
Low production cost or high quality level?
Workload or customer requirement?
Reduce or expand size of operation?

New facility or reorganization?
Buy or make?

To sharp-eyed readers it may be obvious that again consideration of alternatives should be closely related to the hierarchical nature of objectives and constraints. That is, at any given level a set of alternative means should be explored to satisfy a particular end. Take the above manufacturing strategy example again: at the middle level of the hierarchy there may exist a number of alternative methods to achieve flexibility:

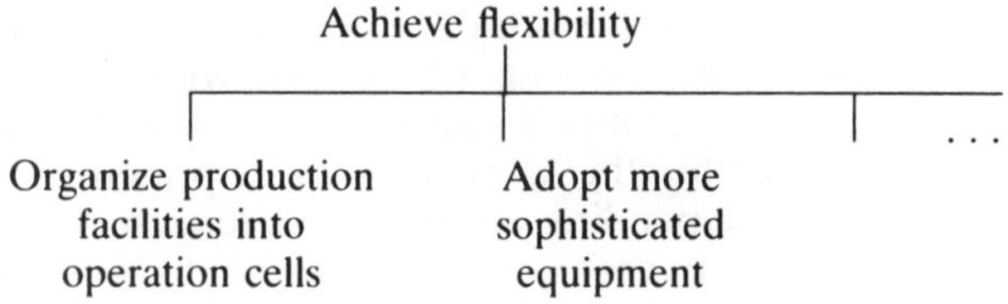

Similarly, alternatives will also exist at other levels within the structure. All of these possibilities should be looked into so as to draw as clear a picture of the problem situation as possible.

5.2.4 Synthesis and analysis of concepts

On completion of the problem-formulation phase, the project situation will have been thoroughly analysed, and the analyst should have committed to paper a statement of the problem proper which, as far as possible, clearly expresses the nature of the problem, the limitations placed upon the possible solution and the criterion of satisfaction to be worked to. Having done so, the next action is to explore the ways in which these objectives may be achieved. At this point, therefore, the model of the systems analysis process progresses to the next phase, which is that of decision-making and problem-solving. This will involve all the rest of the activities shown in Fig. 5.2.

The first task involved in problem solving is to generate a set of possible alternative routes to the objectives. This is what is meant by the 'synthesis of concepts' and 'analysis of concepts' in the conceptual model of Fig. 5.2. Synthesis of concepts requires the creation of a comprehensive set of alternative routes. It should be an exhaustive process, that is to say, the number of ideas generated should be as great as possible given the time and resource constraints of the project. The quantity of ideas at this early stage of problem solving is perhaps as important as their quality – the greater the number of choices, the greater the likelihood of including the best possible choices among them. This explains why idea-generating techniques such as brainstorming often prove useful. What is required here, therefore, is a divergent style of thinking for effective searching and synthesis of the initial concepts. However, the whole decision-making process requires this to be supplemented by convergent thinking. The general nature of the problem-solving phase may be described as a continous refining process, as depicted in Fig. 5.3. With this structured approach the original set of possibilities is gradually narrowed down, or converged, to eliminate the least attractive. In this way greater and greater precision is achieved until one or more solutions are left which satisfy the project objectives. Figure 5.3 shows that, as a first convergent step, the wide range of 'wild ideas' generated by the synthesis function should be refined next by the analysis function to prepare for the

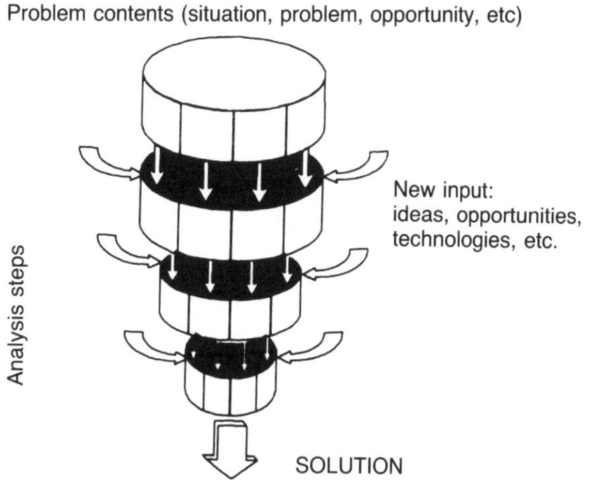

Fig. 5.3 Continuous refining process of systems analysis.

more rigorous analytic processes which are to take place at the evaluation and decision stages.

Judgement at an intuitive level is usually sufficient for 'analysis of concepts'. Unfortunately it is not possible to give specific guidelines as to how to judge the short-listed alternatives in the best way. Generally speaking, in addition to the specified measures of performance, other relevant criteria such as technical, financial, political and social considerations should also be used to expose those suitable candidates which at an intuitive level appear both to meet the objectives and not to violate constraints. The practical experience and professional judgement of both the systems analyst and the project client are of great value in this regard. Analysis of concepts is one of the functions which are likely to initiate iterative processes, as shown by the loops of Fig. 5.2. In fact, iteration plays an extremely important role in systems analysis. This will become clear later in this chapter.

5.2.5 Evaluation of concepts

The aim of this function is to identify from the alternatives the solutions that have the greatest outcome value, as measured by the performance criterion, for the least risk. This function perhaps involves the most 'scientific' elements in the cycle of systems analysis. The tasks involved here can be divided into two categories – model building and outcome evaluation.

Model building

Modelling is usually involved here to provide the analytical tool. The types of modelling technique used to evaluate the alternatives are highly diversified, and include mathematical, physical and simulation. Further, they can be either quantitative or descriptive. The choice of evaluation process depends not only on the contents of the problem itself (type of problem, nature of performance measure and objectives) but also on other factors such as the amount and type of quantitative information available, the amount of time and money at the analyst's disposal and the facilities available (for instance, computer hardware and software).

Choice of model

Within the scope of this book it is not possible to give a detailed account of the available modelling techniques. For further information on this aspect of systems analysis, a large number of books may be consulted (for example, Blanchard and Fabrychy, 1981). Analytic models, Wilson (1984) observed, could be described as, on the one hand, steady-state or dynamic and, on the other, as deterministic or non-deterministic. This led him to a four-way classification of analytic models.

Steady-state deterministic models are based on algebraic relationships. They are used when the mechanisms governing a behaviour are understood and thus the relationships among system variables and parameters are transparent to the model builder. For example, an accountant may calculate the pre-tax profit or loss made by a company over a certain period by using the following simple algebraic model:

$$\text{Profit} = \text{Total Sales} - \text{Total Costs}.$$

It is clear that some of the production planning models presented in this book – among them linear decision rules, linear and integer programming models, and fixed-cost models – belong to this category.

By contrast, *steady-state non-deterministic* models are used when the governing mechanisms are incompletely understood. As such the relationships among the system parameters, or sometimes the value of system parameters, can only be estimated through techniques provided by statistical analysis. A typical example is regression analysis which utilizes historical data to relate system variables in the following form:

$$Y = a_0 + \sum_{i}^{n} a_i X_i$$

The production models based on queuing theory are another example of this type of model.

Like steady-state deterministic models, *dynamic deterministic* models are used when the system relationships can be observed by the model builder. They differ from the former category, however, in that an extra dimension – that of time – must now be taken into consideration. A major effort has been devoted to the development of such models and their analytical solutions mainly as a result of their applicability to solving control problems. One of the most common modelling techniques of this type has already been discussed – the use of differential equations. Typical applications of this type include the modelling of mechanical and electronic control systems.

Dynamic non-deterministic models are used when the governing mechanisms are only partially transparent to the modeller. In addition, the value of systems parameters cannot be predetermined in a fixed fashion. As a result, the power of mathematical modelling techniques is rather limited when the problems concerned fall into this category. Consequently computer simulation models are frequently relied upon to carry out the evaluation process on a statistical basis.

In fact, among the wide variety of modelling techniques available, when it comes to a project concerned with design and analysis of manufacturing systems, computer simulation provides a flexible and powerful evaluation technique which frequently stands out clearly from the rest. This is perhaps due to the fact that the majority of problems in this subject area belong to the dynamic non-deterministic type. Taking this important observation into account, Chapter 6 is devoted to the topic of computer simulation.

Having carefully analysed the nature of the problem concerned and chosen the appropriate model type accordingly, the analyst can now set out to tackle the actual task of model construction.

Verification and validation

A very important issue that deserves particular attention here is data verification and model validation. Generally speaking, modelling implies simplification because real-world problems are usually too complex to be modelled totally realistically. A major part of modelling, therefore, are the decisions as to which aspects of the real world should be included in the model and which ignored. Consequently, the model builder must take great care to ensure that the decisions he or she makes are sound and valid, so that the resulting model is realistic at least for those aspects of the real world that he or she is interested in. A successful model should effectively simplify the problem situation while at the same time being capable of predicting the behaviour of the system concerned in such a way that it provides a solid basis for sensible decision-making. This important aspect of modelling will be discussed in more detail in Chapter 6.

Outcome evaluation

With the help of a properly constructed and validated analytic model, the performance of the system under each of the alternatives can be tested to produce either quantitative or qualitative results for the purpose of comparison. Although superficially it may appear a simple matter to tabulate the performances of the system under different operating conditions so that the decision-maker can choose the best combination from the set of alternatives, in fact the analysis of output data is not always straightforward. The analyst should pay particular attention to the following points at this stage of the analysis process.

Design of the evaluation operation

In the general context the alternatives may be viewed as controllable factors or actions which can be altered during an evaluation operation to predict their impact on the world. In addition, there may be uncontrollable factors – interest rates and raw material prices, for example (for the individual manufacturers at least) – whose effects on the outcome must also be taken into consideration. The situation may be generalized in a two-dimensional array in which each element indicates a specific factor in one of its specific states.

	S_1	S_2	S_3	S_4
F_1	A_{11}	A_{12}	A_{13}	A_{14}
F_2	A_{21}	A_{22}	A_{23}	A_{24}
F_3	A_{31}	A_{32}	A_{33}	A_{34}
F_4	A_{41}	A_{42}	A_{43}	A_{44}

In the above table, the factors (either controllable or uncontrollable) are labelled $F_1, \ldots, F_m$, their associated states are denoted as $S_1, \ldots, S_n$, and the alternative outcomes $A_{11}, \ldots, A_{mn}$. Listing the alternatives involved in a problem situation in this way clearly reveals how the experimental variables will be changed to predict alternative system outcomes. A simple illustration will serve to clarify this. Suppose that we are faced with the problem of evaluating the efficiency of a machining operation. We may have three factors to consider – the hardness of the work material, the cutting speed and the feed rate. Each of these factors may be set at a number of particular levels (say, two), as shown in the following table:

Factor	*State 1*	*State 2*
Material hardness	soft	hard
Cutting speed	slow	fast
Feed rate	low	high

Clearly the decision-makers should be provided with a complete picture of the predicted outcome under combinations of these factors so that the best alternative may be chosen. Whenever possible, it is also highly desirable to be able to identify the underlying causes of the effects. Since an evaluation operation can be an expensive exercise because of the time, effort and resources involved, it is important to design it and then analyse the output in such a way as to squeeze as much information as possible out of the investigation. We may use some of the available techniques to aid our evaluation process in this regard. One such technique is the so-called evolutional operation (EVOP). EVOP is a simple but useful technique initially introduced to improve the efficiency of laboratory investigations (Box and Draper, 1969). However, it may also be used in general to design and analyse experiments involving multiple system variables and as such has a wide range of possible applications. We will use the above example to illustrate the basic principles involved in this technique and in doing so highlight the issue of effective outcome evaluation in general.

According to EVOP, a factorial test scheme could be adopted for the above simple evaluation process which employs each of the eight possible combinations of alternative conditions, as shown in the following table:

Test no.	*Cutting condition*					
	Material		*Cutting speed*		*Feed rate*	
	Soft	*Hard*	*Slow*	*Fast*	*Low*	*High*
1	X		X		X	
2	X		X			X
3	X			X	X	
4	X			X		X
5		X	X		X	
6		X	X			X
7		X		X	X	
8		X		X		X

This is known as a 2^3 factorial design because we are interested here in investigating the effects of three alternatives, and have decided to try two different states for each alternative. That is, in using this particular arrangement we are actually studying the three alternatives in two states each. This 2^3 factorial design will enable us to study not only the effect of each of the

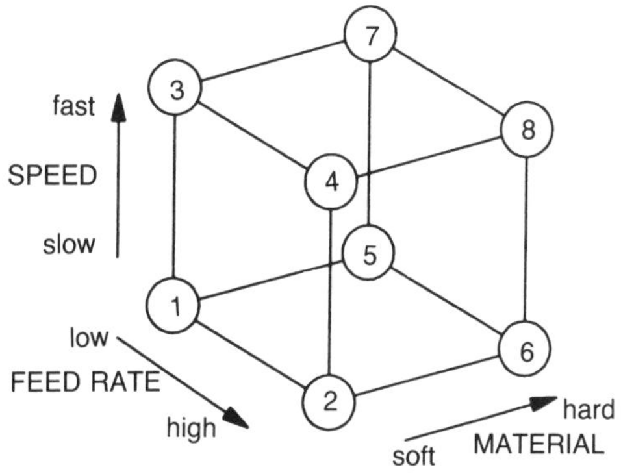

Fig. 5.4 Evaluation scheme for a machining operation.

three alternatives in turn but also their interactions. This evaluation scheme may be represented by the eight vertices of a cube, numbered 1 to 8 as shown in Fig. 5.4. Some of the features of such a design can be readily seen in this figure. For instance, a comparison of the results of cutting conditions 5 and 1 would provide one measure of the effect of work material on the machining efficiency at identical conditions for the other two alternatives (slow cutting speed and low feed rate). Similarly, the effect of work material may be measured by the comparison of the results of cutting conditions 6 and 2, 7 and 3, and 8 and 4. This complete set of four comparisons allows us to determine the effect of this particular factor under all four tested combinations of cutting speed and feed rate. Thus all the results can be used to assess the effect of work material. Likewise, comparison of conditions 3 and 1, 4 and 2, 7 and 5, and 8 and 6 will enable us to measure the effect of cutting speed – again in each pair the two remaining variables, feed rate and work material, are held constant. Finally, via comparison of conditions 2 and 1, 4 and 3, 6 and 5, and 8 and 7 allows us to measure the effect of changing the feed rate. As a result, in this design each piece of outcome data supplies information on the effect of each of the three alternatives. This feature is of particular value because it produces the maximum averaging out of evaluation error.

The following shows how the situation can be conceptualized to utilize the outcomes from this 2^3 factorial design to the maximum. First we look at estimates of the effects of each of the three alternatives individually. Then we extend this to examine how to derive estimates of the degree of interaction between any two alternatives.

Suppose a 2^3 factorial evaluation scheme has three alternatives A, B and C, as shown in Fig. 5.5. The Ys in this diagram denote the outcomes from the system under the relevant conditions. Then the effect of alternative A on the system can be measured by the difference between the results from the fifth and the first condition ($Y_5 - Y_1$) under which both B and C are held fixed at their lower levels. Similarly, three more estimates of the effect of changing A are given by ($Y_6 - Y_2$, $Y_7 - Y_3$ and $Y_8 - Y_4$). The main effect of A can then be estimated by averaging the values of these simple effects:

$$\begin{aligned} E_A &= \tfrac{1}{4}[(Y_5 - Y_1) + (Y_6 - Y_2) + (Y_7 - Y_3) + (Y_8 - Y_4)] \\ &= \tfrac{1}{4}(Y_5 + Y_6 + Y_7 + Y_8) - \tfrac{1}{4}(Y_1 + Y_2 + Y_3 + Y_4) \end{aligned}$$

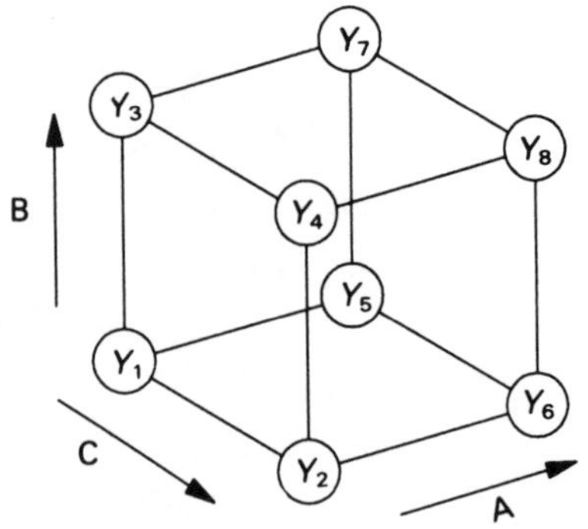

Fig. 5.5 Diagrammatic representation of a 2^3 factorial design.

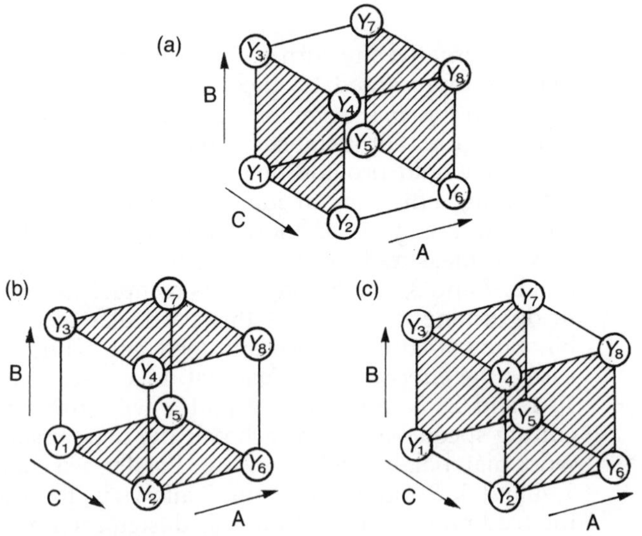

Fig. 5.6 Main effects of a 2^3 factorial design. (a) A effect, (b) B effect and (c) C effect.

Thus the main effect is a comparison of the average response on the high-level face of A with the average on the low level face of A (Fig. 5.6a). Similarly the main effects of B and C are given as follows (as shown in Figs 5.6b and 5.6c):

$$E_B = \tfrac{1}{4}[(Y_3 - Y_1) + (Y_4 - Y_2) + (Y_7 - Y_5) + (Y_8 - Y_6)]$$
$$= \tfrac{1}{4}(Y_3 + Y_4 + Y_7 + Y_8) - \tfrac{1}{4}(Y_1 + Y_2 + Y_5 + Y_6)$$
$$E_C = \tfrac{1}{4}[(Y_2 - Y_1) + (Y_4 - Y_3) + (Y_6 - Y_5) + (Y_8 - Y_7)]$$
$$= \tfrac{1}{4}(Y_2 + Y_4 + Y_6 + Y_8) - \tfrac{1}{4}(Y_1 + Y_3 + Y_6 + Y_7)$$

A similar approach is used to estimate the combined effect of any two alternatives, that is to say, the interaction between those two alternatives. For instance, the interaction between A and C is given by:

$$E_{AC} = \tfrac{1}{2}[(\text{average effect of A at the high level of C}) - (\text{average effect of A at the low level of C})]$$

Consider again the model shown in Fig. 5.5. The results of such an analysis may be summarized as follows:

A effect	*Level of B*	*Level of C*
$(Y_5 - Y_1)$	low	low
$(Y_6 - Y_2)$	low	high
$(Y_7 - Y_3)$	high	low
$(Y_8 - Y_4)$	high	high

Averaging over B, $[(Y_5 - Y_1) + (Y_7 - Y_3)]/2$ gives the average effect of A at the low level of C, and $[(Y_6 - Y_2) + (Y_8 - Y_4)]/2$ the average effect of A at the high level of C. The difference between these two values provides a measure of the extent to which the A effect is different at the different levels of C, that is, the interaction between A and C. Half this difference is called the two-factor interaction, E_{AC}, between factors A and C:

$$E_{AC} = \tfrac{1}{4}(Y_1 + Y_3 + Y_6 + Y_8) - \tfrac{1}{4}(Y_2 + Y_4 + Y_5 + Y_7)$$

It can be shown, and the reader should check using a similar argument, that $E_{AC} = E_{CA}$.

Figure 5.7b illustrates the two oblique planes intersecting on the cube's A and C faces. It can be seen that the difference between the averages of the two sets of four outcomes lying on these two planes provides the interaction E_{AC}. By the same argument it can be shown that the AB and BC interactions are the contrasts of averages of four outcomes on the appropriate diagonal planes of the cube (Figs 5.7a and 5.7c). The remaining two two-factor interactions are therefore given by:

$$E_{AB} = \tfrac{1}{4}(Y_1 + Y_2 + Y_7 + Y_8) - \tfrac{1}{4}(Y_3 + Y_4 + Y_5 + Y_6)$$

$$E_{BC} = \tfrac{1}{4}(Y_1 + Y_4 + Y_5 + Y_8) - \tfrac{1}{4}(Y_2 + Y_3 + Y_6 + Y_7)$$

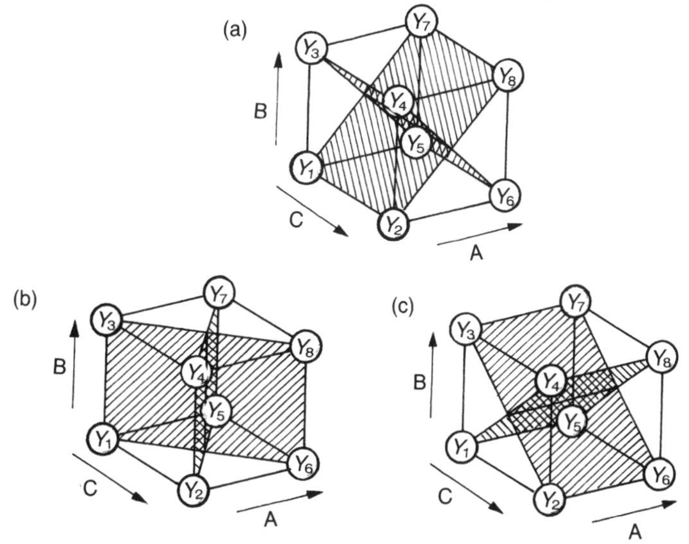

Fig. 5.7 Two-factor interaction. (a) Interaction AB, (b) interaction AC and (c) interaction BC.

It should be pointed out that the above example is based on a very simple problem. In practice not only are more alternatives and states often included in the problem situation, but also certain statistical techniques have to be applied to reduce the average outcome error. However, this example illustrates that the design of the evaluation operation is an important issue which deserves careful consideration. If properly planned, the evaluation process may be executed in such a way that effective use of both time and resources may be achieved. When the evaluation operation involves expensive and time-consuming exercises – involving expensive equipment or large-scale computer models, for example – effective planning and execution of an investigation scheme becomes a particularly important issue.

Accuracy of the evaluation outcome

Like the validation considerations required for data input and model construction, the outcome from the model should be treated with a great deal of care. It is always necessary to validate the outcome from this stage. When a non-deterministic simulation model is used for evaluation, for instance, it is essential to make certain that the estimated error of the result is reduced to a desirable level. Furthermore, the initial conditions must also be carefully chosen so that the corresponding system responses are valid for analytical purposes.

Presentation of results

The outcome from the evaluation stage must be presented clearly in a format that is easy to communicate and understand. This is necessary for the ensuing comparison and decision processes. Numerical output data listed in tables are the most common and straightfoward means of presenting results. However, whenever possible it is desirable to use graphic means such as line diagrams, bar and pie charts to increase the quality of presentation. Some computer tools provide very useful assistance in this normally tedious task. For example, a PC-based spreadsheet such as Lotus 1-2-3, is very helpful in this regard. It can be used not only to reduce considerably the time and effort needed for sorting the data, but also to increase significantly the quality of presentation.

Another point to remember at this stage of the project is that when presenting the results from the evaluation stage, the analyst should start to consider how the results are to be 'sold' to those decision makers who have to put together the strategy which they feel will best meet the demands of their problem or opportunity. The analyst must bear in mind that the best strategy may well involve a combination of alternatives, for it may have been revealed by the evaluation process that one such alternative does well on some measures of performance but not so well on others. The implications of this must be made clear through a coherent set of reports or plans which can be understood by the people who will have to make a choice and implement the chosen strategy.

5.2.6 Decision-making process

Having evaluated all of the alternatives using the chosen modelling technique and predicted their possible future yields against the measure of performance, the decision must now be made on the choice of options to implement. The objective of this particular stage is therefore quite clear: given the results of evaluation on the set of alternatives, as far as possible to choose the best strategy, taking into account the quantifiable and non-quantifiable objectives and constraints.

Analysis of weighted objectives

Very rarely do we find a problem situation in which a single alternative is found to stand out clearly from the rest. Suppose, in a particular decision situation, that a production manager has set objectives concerning the 'on-time delivery of products' and 'average level of WIP' achieved by alternative scheduling rules. The evaluation process might reveal the following two possible outcomes:

Outcome of scheduling policy 1:
on time deliver = 85%
average WIP level = 5000 (items)
Outcome of scheduling policy 2:
on time deliver = 90%
average WIP level = 8000 (items)

Obviously there is no clear basis for choice between these outcomes. Whereas scheduling policy 1 is likely to achieve lower production costs because of the lower average WIP level, it also yields an inferior delivery performance compared with policy 2.

Frequently, therefore, the various alternatives tested will be evaluated in terms of different measures of performance because different measures may be regarded as having different 'utility' in comparison with each other. In order to aid decision-makers in their choice, what is required at this stage is usually some means for the performances of the alternatives to be assessed and compared across the whole set of objectives. A key method for this purpose is to reduce the multiple outcomes to an overall single measure which will reflect their aggregate utility.

The simplest situation is when it is possible to formulate the problem in such a way that quantitative assessment can be made of the system performance achieved by the alternative means on all of the objectives and also in terms of similar units. We may use the scheduling policy example given above to illustrate this point. Consider a similar decision situation, but now involving two outcomes described in a single unit:

Outcome of scheduling policy 1:
cost of holding WIP = £4000
cost of machine idleness = £6000
Outcome of scheduling policy 2:
cost of holding WIP = £6400
cost of machine idleness = £5000

Two of the system outcomes from the problem situation, the average level of WIP and the average level of machine idleness, are now both described in terms of cost. It can be seen that the system outcome achieved by scheduling policy 1 is better in terms of WIP, while the outcome from scheduling policy 2 is better in terms of machine utilization. In spite of this, one would tend to conclude that the outcome associated with scheduling policy 1 is better because it has produced a total cost of £10 000 which is lower than that of £11 400 produced by scheduling policy 2.

This simple example shows that it will be a relatively simple task to develop an overall utility indicator if all the outcomes can be assessed in the same terms through arithmetical operations. Unfortunately, however, it is normal in practice to have multiple objectives including technical, economical as well as other factors, making it difficult, if not impossible, for this type of straightforward assessment to be carried out. For instance, given the following set of outcomes:

Outcome of scheduling policy 1:
on time delivery = 85%
cost of holding WIP = £4000
cost of machine idleness = £6000
Outcome of scheduling policy 2:
on time delivery = 95%
cost of holding WIP = £6400
cost of machine idleness = £5000

then at least one of the outcomes is not immediately suitable for quantitative assessment in terms of cost. In general, therefore, some kind of assessment method must be sought to aid the comparison process when this situation arises. The weighted objectives method is one such approach which may be of value. This particular analysis procedure involves the following steps.

First, list the project objectives. Having gone through the previous stages of the analysis cycle, this is now just a simple routine task of laying out all of the objectives in a simple order and labelling them appropriately.

Second, sort the objectives in order of importance. Among the objectives listed, some will be considered by the decision maker to be more important than others. This step aims to rank-order the objectives according to how important each is considered to be. When there are only a few objectives, it is usually possible to list them in rank-order of importance through simple intuitive judgement. However, this method may not work well when the number of objectives included in the list becomes large. When this is the case, a pairs-comparing method may be useful.

The procedure starts by taking the first objective and comparing it against each of the other objectives in turn and recording the comparison results s_{ij} in a chart like the one shown below. The importance indicators, s_{ij}, will take values 0, 1 or 2 depending, respectively, on whether objective i is considered to be less important than, as important as or more important than objective j (by convention $s_{ij} = 0$ if $i = j$). For example, if objective A is considered a more important objective to achieve than objective B, then s_{AB} should be given a value of 2. Note that it follows that $s_{BA} = 2 - s_{AB}$. This process continues until each objective has been compared with every other objective and the list is complete.

Objectives	*A*	*B*	*C*	*D*	*E*	*Total Score*
A	0	s_{AB}	s_{AC}	s_{AD}	s_{AE}	S_A
B	s_{BA}	0	s_{BC}	s_{BD}	s_{BE}	S_B
C	s_{CA}	s_{CB}	0	s_{CD}	s_{CE}	S_C
D	s_{DA}	s_{DB}	s_{DC}	0	s_{DE}	S_D
E	s_{EA}	s_{EB}	s_{EC}	s_{ED}	0	S_E

Having filled in all of the matrix cells in the chart, sum the row scores to give the overall rank-order of importance for each objective. Thus the complete list of objectives can be ordered according to the values in the last column of the chart. For example, if we have a chart as follows:

Objectives	*A*	*B*	*C*	*D*	*E*	*Total Score*
A	0	2	2	1	0	5
B	0	0	0	2	0	2
C	0	2	0	2	0	4
D	1	0	0	0	2	3
E	2	2	2	0	0	6

then obviously the objectives should be rank-ordered E, A, C, D, B. That is, E is ranked as the highest-priority objective and B the lowest-priority objective.

The third step is to weight the objectives according to their rank-order of importance. There are a number of ways to assign a numerical value to an objective which represents its weight relative to the other objectives. As long as the positions of relative importance are defined, any one of these methods can be used for the task. For example, they might simply be allocated along a horizontal axis as shown below:

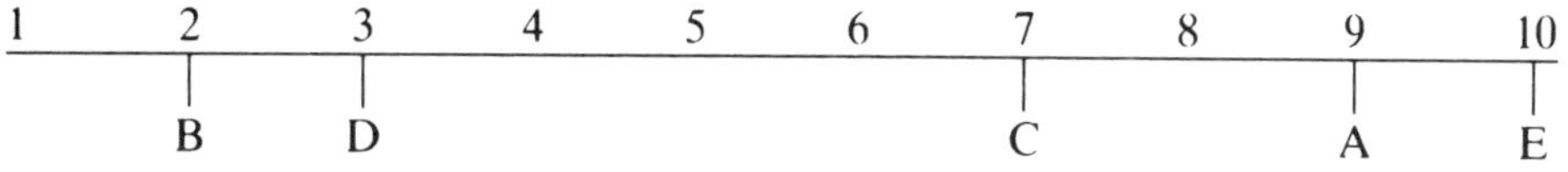

This clearly shows the importance of the objectives relative to each other. In the example above, objective D is seen roughly is half as important again as B, objective E is only 11% more important than A (we will use f_j to indicate such a weighting factor for objective j).

The fourth step is to estimate utility scores for each of the objectives. Much as in the previous step, in order to convert the problem contents into a form suitable for arithmetic operations, it is now necessary to assign utility scores to the system outcomes obtained by the alternatives for each of the objectives. This involves working out what a particular level of system outcome actually means when measured against an objective. For example, suppose that we have predicted a machine utilization level of 85%, then we must ask whether this is a good system outcome or a totally unacceptable one. Once the situation becomes clear in this particular regard, the possible system outcomes may be placed on a scale of utility like this:

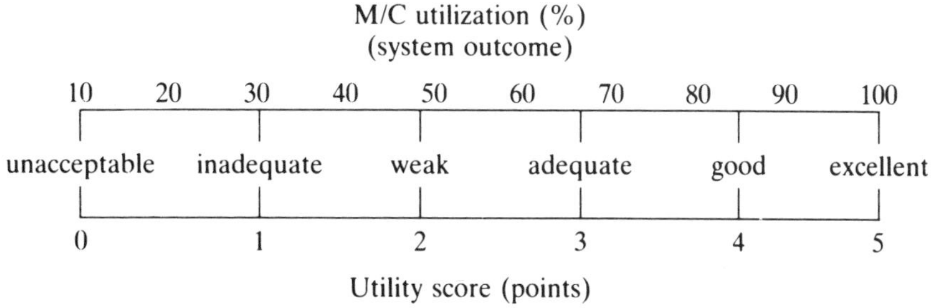

Then each of the system outcomes can be assigned a utility score according to such a scale. This conversion process is a useful technique because it allows both qualitative and quantitative performance measures to be compared on a similar ground. Here we will use u_{ij} to indicate a utility score for the system outcome given by alternative i, against objective j.

Finally, having converted the problem contents to an acceptable format, comparison between alternatives can now be carried out on the basis of their relative utility values. The relative utility value of an outcome given by a particular alternative is obtained simply by multiplying its utility score by its weighting factor. The utility value of the outcome given by alternative *i* against objective *j* is therefore given by:

$$v_{ij} = u_{ij} f_j$$

As a simple measure of comparison, these individual utility values can be summed up to give a single utility value which indicates the relative overall 'worthiness' of the alternative concerned. Thus the relative worthiness of alternative *i* is given by:

$$V_i = \sum_j v_{ij} = \sum_j u_{ij} f_j$$

The idea behind this weighted objective approach – reducing the problem contents to a single dimension – is of great importance in decision-making. Unfortunately it is not always easy to adopt such a seemingly simple approach effectively in practice. The major problem arises from the fact that there is as yet virtually no formal guidance as to how to measure a utility. It must be emphasized that such a procedure requires skill, experience and the participation of all parties in the systems project in order to succeed. If the necessary care is not taken, then there is a danger that the analyst could follow the procedure mechanically step by step and in the end arrive at the wrong conclusion by using some misleading overall utility values.

System implementation

The determination of the choice among the alternatives is not the end of the story. An extremely important and frequently difficult action is yet to follow – the implementation of the decision. It is clear that a decision or an intended system will not be of much use until properly and effectively implemented. For example, many manufacturing companies nowadays have come to the conclusion that some kind of rationalization action either on their organization or on its operation, such as the adoption of MRP II or JIT, should be undertaken in an attempt to achieve more effective production. The decision behind this move may be very sound and the working of the intended system proven. However, many cases have shown that it is the difficulties associated with the implementation stage that have remained as the major obstacle to full implementation of the choices made.

The entirety of the decision block, therefore, can be viewed as consisting of two actions – making a choice and then implementing the change associated with that choice. Implementing change is never an easy task. It is obvious that the level of difficulty associated with implementation depends on the amount of change required to the existing system, and how its structure and operation are to be effected. A carefully thought-out strategy will normally be required to carry out this last phase of a systems project. According to such a strategy, a coherent set of detailed plans and instructions must be prepared to guide through the necessary actions to be taken. Both the strategy and the instructions on action should be clearly understood by the people who will actually implement the changes envisaged. An implementation plan should, therefore, outline why change is needed, how it will come about, what tasks will need to be carried out, what is required of personnel (in terms of man-hours, skill levels, and so on), what resources are needed, as well as a timetable for the project.

Once started, the progress of implementation tasks will need to be continuously monitored to see that the changes are indeed taking place according to plan, and, if necessary, feedback actions should be taken to adjust the individual tasks or even the course of the whole plan.

The procedure outlined above may look familiar to those who have read Chapter 3 of this book. This is because the procedure conforms to the basic procedures involved in project management (section 3.3.4). Indeed, project management can be regarded as an intrinsic part of the overall framework of systems analysis: the analysis and evaluation procedures assist us in the choice of alternatives, whereas the methodology provided by project management helps make the chosen option operational (Cleland and King, 1983).

Finally, once implemented the results of the changes brought about by the chosen option should be quantified in some way. They can then be compared with the predicted outcome. If necessary, the cycle of actions can be initiated all over again – this time taking consideration of the results obtained from the previous cycle of analysis. The overall process of systems analysis therefore has the inherent characteristics of a feedback structure, as shown in Fig. 5.2.

5.2.7 Case study: development of a WIP cost-related measure of effectiveness

In this section we have introduced a conceptual framework which outlines the general procedures of the systems analysis approach. Being a conceptual model, it can only be expected to provide us with the general ideas of the methodology. Thus for those readers with no previous experience in this area, the best way to illustrate it is perhaps through case studies or examples of its actual applications. In this sub-section, therefore, we will follow the analytical processes of such a case. We will describe the design and development of a cost-related measure of effectiveness, which is based on a detailed analysis of the production-cost structure of a manufacturing company (Wu, 1990). This cost-related performance criterion allows simple and effective analysis of the overall production operation and so significantly simplifies the integration of a computer simulation model of the company's manufacturing operation with an intelligent back end (IBE) for the purpose of knowledge-based system evaluation (Chapter 6).

Analysis of manufacturing-cost structure

In many previous projects concerned with the performance of manufacturing systems, profit has been used as a composite indicator of company success. However, it should be realized that from the production management's point of view, the problem of maximization of profits is equivalent to the problem of minimization of production costs, since sales volume and price are beyond its control. As a result, a major management objective must be to minimize production costs. It is, therefore, necessary to identify the most important cost elements in the particular system under study, and use them to form the most appropriate system performance indicators. It is noted that the structure and operation of the company concerned are in good agreement with the conceptual model presented in Chapter 3. A detailed analysis of the cost structure of the company revealed that the main cost elements of the company could be associated with the relevant manufacturing functions as defined in the conceptual OHMS model. These are as follows:

1. costs of marketing, sales and distribution;
2. general administrative costs of production;
3. costs of product improvement;
4. costs of the design process;
5. costs of direct material/items;
6. costs of indirect materials;
7. direct labour costs;
8. indirect labour costs;
9. work-in-progress (WIP) costs.

Choice of performance indicator

The list above provides a fairly detailed account of all the cost elements incurred in the company. Among these costs the production management is less concerned with those associated with marketing and sales, development, and the design activities. The reason for this is quite obvious: although not totally unconnected with the production process itself, costs associated with these functions are unlikely to be significantly affected by production decisions in the medium term. For example, it would be feasible to quote the cost incurred at the design stage of the production as a constant rate per contract, because at this stage a similar number of activities and similar amount of effort are needed for each new contract, and the magnitude of costs associated with them would not be influenced by how the production programmes are made up, or how the components stores are controlled. These cost elements could, therefore, be excluded from the performance criteria to assess the effectiveness of production management decision policies. This leaves the cost elements 2, 5, 6, 7, 8 and 9 as the relevant factors.

A detailed analysis of the production costs was carried out based on one year's production of the company involving 22 contracts totalling approximately 4000 product units. A summary of these cost data is given in the table below, in which a cost unit, rather than the actual money unit, is used to show the relative cost weighting of each of the four major sub-assemblies which constitute the company product.

	Sub-assembly				
	No. 1	*No. 2*	*No. 3*	*No. 4*	*Total*
		Component cost			
Mean value	1761.62	642.65	351.70	71.70	2827.66
Standard deviation	340.11	164.08	102.98	23.15	557.93
		Estimation error (95% confidence)			
Absolute	124.79	60.20	37.79	8.49	204.72
Relative	0.071	0.094	0.107	0.118	0.072

The amounts listed in this table are the total costs for the associated items. They are the 'pure' production costs because the profits added values are not included. Also, the cost elements given here are all 'tangible' elements. WIP costs are not yet taken into consideration. The table shows that contract cost deviations are very small. This is justifiable by the observation that although the product units in each contract in such an OHMS operation are tailored to the needs of their particular user, their main structure and hence the costs of the key components are always similar, as shown by the statistical analysis given in the above table. This result implies that it would be feasible to quote constant cost rates for the components listed in the table to estimate the production cost for any of the contracts involved in a contract series, if so desired in the later stages of the investigation.

Further analysis of the cost information also revealed that the 'original' production materials – the raw materials and the semi-processed materials – account for up to 56% of the total production cost. The rest of the total is added gradually as the items progress through the various stages of the production process. These 'company-added' costs include a number of elements, shown in the table below.

At this point, a decision must be made as to how to use this production-cost information to enhance the system performance. A performance-measurement system can either be orientated to a disaggregated level, that is, to an operational level so that the performance of an individual function can be evaluated, or to an aggregated level, such that the throughput efficiency of the system can be assessed.

Breakdown of company-added cost elements

Element	*Percentage of total*	*Description*
machining	24	various machining operations
processing	24	cleaning welding plating polishing, etc.
sub-assembly	25	sub-assembly of various components
assembly	27	assembly of No. 1 assembly of No. 3

At the disaggregated level of assessment, the individual rates of items such as those listed in the table, together with their associated variables of system response, can be used to carry out detailed cost analysis for each of the 'cost centres'. For instance, the direct material cost rate, together with the amount of the material in a store, may be used to indicate the value of material inventory in that store; a particular WIP rate, together with the average level of WIP involved, may be used to measure the significance of minimizing the WIP level at a given stage of production; and the labour and machining rates, together with the under-utilization time, may be used to calculate the relative importance of maximizing the labour and equipment utilization at a workstation. A refined analysis of this kind can result in accurate indications of departmental performance, and hence results in optimal production control at local levels. However, evaluation of the overall performance of the total system cannot be readily carried out from the results, and consequently general conclusions about the system at the company level are difficult to draw.

On the other hand, at the aggregated level the rates of all the relevant cost elements, together with the associated system variables, can be put together to form an overall measure of the system on the plant-wide scale, as in the examples quoted previously. Since with this approach a single value is all that is considered necessary to provide a satisfactory measure of the system performance, the calculation and evaluation processes at the company level are made much easier than with the disaggregated approach. However, the disadvantage of this integrated approach is that it is usually less accurate than the detailed cost analysis, because important information could be destroyed by the 'blending' process. Furthermore, the underlying causes of any result that may have been yielded by the experiment are hidden from the researcher, and this limits the inside understanding of the system under study.

Having examined the composition of production costs and the measurement options, it was concluded that the best combination was to take advantage of the situation by utilizing both approaches. That is, an aggregated indicator should be developed to evaluate the company-level performance, and in the meantime detailed information about the performance of each cost centre should also be made available for detailed analysis where necessary. This combination is possible in practice because of the overriding importance of the WIP problem in this particular type of manufacturing system, as will be discussed later. The cost indicator used for the system concerned (Equation (5.1)) is a WIP-related weighted average:

$$C = \sum_{i=1}^{m} \alpha_i Q_i \tag{5.1}$$

where m is the total number of cost centres in the system; Q_i is the time-related average level of in-process inventory at cost centre i; and $\alpha_1, \alpha_2, \ldots, \alpha_m$ are the weighting factors. In this formula, the Q_i are the system variables whose values will be dependent on how the system responds to different operating environments and to different management decision policies; and α_i are the system-inherent parameters. The values of these weighting factors are determined by the relative value of the WIP items at their associated cost centres, taking into account the material and operational cost rates at the centres and the values added to the items by the previous operations. Therefore, this indicator recognizes the fact that in a manufacturing system some WIP items take more values with them than the others, either because they have higher rates of initial cost or because of the company-added values at the later stages of production. The meaning of this WIP-related indicator, the definition of Q_i and the values for α_i, together with the justification of the suitability of this indicator for this particular system, are given below.

Justification and development of WIP cost-related measure

The cost elements listed previously may be classified into four types: *material-related* costs, including costs of materials, bought-out items, consumables, and the like; *resource-related* costs, such as machine rates; *workforce-related* costs, including wages and salaries; and *decision-related* costs, which are WIP costs. The utilization of a WIP cost-related indicator rather than the more usual 'objective function' type of indicator, which tries to cover all the cost elements of a system, is justified on the following grounds.

The first and most important consideration is based on the special characteristics of an OHMS company; that is to say, in this type of system WIP cost is the 'key' cost element of concern to the production management, because of the high value of the materials and bought-out items involved, the complexity of the production, and hence the long lead times and the high value of the products. It is also the cost element in the system that is directly and solely influenced by the immediate production decisions.

Second, the prices of materials, tools, production consumables, and other bought-out items are, to a large extent, exogenous parameters to the production management since they are beyond its control. Therefore, they may be excluded from the objective criteria for comparative experiment.

Third, the operations costs, including regular payroll payments to all personnel (direct and indirect), costs of machining and overheads, depend mainly on the size of the workforce employed in the company and the production structure of the organization. Therefore, it is considered feasible to treat the workforce as fixed, distributing the associated labour costs, whether direct or indirect, over various production operations involved to reflect their significance in the production system.

Fourth, the reason for treating the attributions of machine rates and other items as constant is based on a similar analysis of that used above.

The last two reasons are both based on one dilemma. When these capacity-resource-related costs are treated as variables in an objective function, they are usually used as indications of the expected significance of resource utilizations, as reflected in terms such as 'costs of machine idleness' or 'costs of operator idleness'. While it is true that in general the management of a manufacturing organization will always try to achieve a higher utilization of the production facilities in its plant, it should be realized that high utilization of facilities is not in itself an end, merely a means to the end. High utilization of resources itself does not directly save production costs, because the capacities of the resource will have been paid for in any case, whether they were utilized or not. Therefore, the significance of high utilization of production resources, and its related 'lost opportunity costs', varies from system to system, dependent on the nature of the problems to be tackled. Here, the management is more concerned with the end purpose of improving its system performances (such as the delivery) while at the same time keeping unit production costs as low as possible, under the assumption that the workforce and basic operation structure remain relatively stable. As far as the production management is concerned, the opportunity costs paid to keep the necessary capacities are constant under these circumstances. Therefore, besides being a criterion that measures the delivery performance of the company, the cost measurement of the system should be related to the management decision-dependent factors. That is, it should be WIP-related. This also illustrates the major reason why the underlying philosophy of successful manufacturing methodologies such as JIT production regards the WIP level as a much more important factor than the conventional parameters such as resource and workforce utilization.

Within the manufacturing structure of the company there are 20 major buffers where entities such as material, components and sub-assemblies are stored, as shown in Fig. 5.8. These may be regarded as 'WIP pools'. In order accurately to evaluate the WIP-related production costs in the whole system, it is necessary to measure the WIP value associated with each of the above WIP pools. This can be established by assigning a unit cost rate, α_i, to WIP pool i, and multiplying it by the average WIP level of the pool to give the WIP-related production cost of this pool:

$$C_i = \alpha_i Q_i \tag{5.2}$$

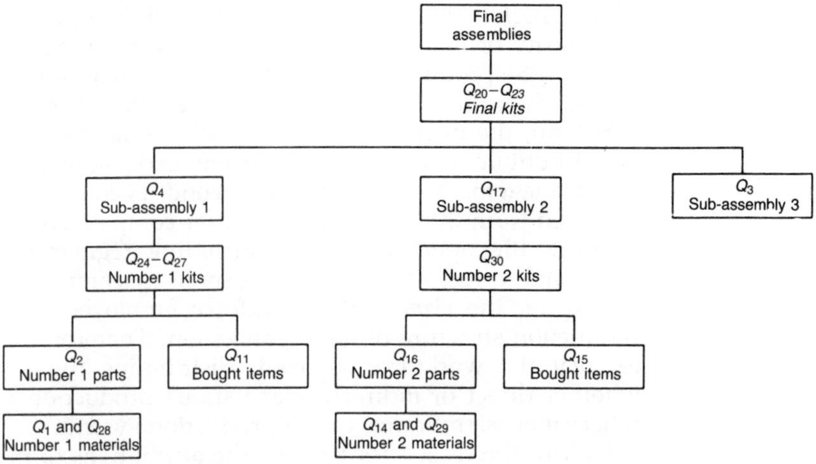

Fig. 5.8 Company WIP pool structure.

The company cost information summarized in the table on page 199 was used to determined the value of the WIP cost rate α_i. By following its progress through production, the manufacturing process of sub-assembly no. 1 can be used to illustrate how these WIP cost rates were calculated (Fig. 5.8). This production process involves five WIP rates (material, parts, bought-out components, kit and assembly) at nine WIP pools (i.e. Q_1, Q_{28}, Q_2, Q_{11}, Q_{24-27} and Q_4). These cost rates are given as follows:

1. WIP pool: Q_1 and Q_{28}
   Entity : materials
   WIP cost rate (cost/unit):

$$\begin{aligned}\alpha_1 &= \alpha_{28}\\ &= \text{total cost of assembly} \times (\%\ \text{cost of material})\\ &= 2776.88 \times 56.0\% = 1555.05\end{aligned}$$

2. WIP pool: Q_2
   Entity : part
   WIP cost rate (cost/unit):

$$\begin{aligned}\alpha_2 &= \text{cost of material} + \text{cost of processing} + \text{cost of machining}\\ &= \text{cost of material} + \text{total cost of assembly}\\ &\quad \times (\%\ \text{cost of machining} + \%\ \text{cost of processing})\\ &= 1555.05 + 2776.88 \times (23.0\% \times 24.0\% + 23.0\% \times 24.0\%)\\ &= 1861.62\end{aligned}$$

3. WIP pool: Q_{11}
   Entity : bought-out components
   WIP cost rate (cost/unit):

$$\begin{aligned}\alpha_{11} &= \text{total cost of assembly} \times (\%\ \text{cost of bought items})\\ &= 2776.88 \times 12.0\% = 333.23\end{aligned}$$

4. WIP pool: Q_{24}, Q_{25}, Q_{26}, Q_{27}
   Entity : kits
   WIP cost rate (cost/unit):

$$\begin{aligned}\alpha_{24} &= \alpha_{25} = \alpha_{26} = \alpha_{27}\\ &= \text{cost of part} + \text{cost of component} + \text{sub-contract added}\\ &= 1861.62 + 333.23 + (2776.88 \times 9.0\%) = 2444.77\end{aligned}$$

5. WIP pool: Q_4
   Entity : assembly
   WIP cost rate (cost/unit):

$$\begin{aligned}\alpha_4 &= \text{cost of kit} + \text{assembly cost}\\ &= 2444.77 + 2776.88 \times 23.0\% \times (27.0\% + 25.0\%)\\ &= 2776.88\end{aligned}$$

The rest of the weighting factors were obtained through similar calculation procedures. Since the WIP cost is related to how long the entities stay in a pool, the measurement of the average WIP level, Q_i, must be obtained as a time-weighted average. Therefore, Q_i here is defined as a time-weighted average queue length given by

$$\begin{aligned}Q_i &= \sum_{j=1}^{2N-1}\left(\frac{T_{i,j+1} - T_{i,j}}{T_{i,2N} - T_{i,1}}\right)Q_{i,j}\\ &= \frac{1}{T_{i,2N} - T_{i,1}}\sum_{j=1}^{2N-1}(T_{i,j+1} - T_{i,j})Q_{i,j}\end{aligned} \tag{5.3}$$

where N is the total number of contracts to go through the system; $T_{i,j}$ is the time when the jth event occurs at pool i (the jth event at pool i will be either the arrival or the departure of an entity at the pool), $j = 1, 2, \ldots, 2N$; and $Q_{i,j}$ is the status of WIP pool i after the jth event (that is, the queue length after the jth event).

Similarly, the standard deviation of the time-weighted WIP level at pool i, S_i, is defined as:

$$\begin{aligned}S_i &= \left\{\sum_{j=1}^{2N-1}\left(\frac{T_{i,j+1} - T_{i,j}}{T_{i,2N} - T_{i,1}}\right)Q_{i,j}^2 - Q_i^2\right\}^{1/2}\\ &= \left\{\left[\frac{1}{T_{i,2N} - T_{i,1}}\sum_{j=1}^{2N-1}(T_{i,j+1} - T_{i,j})Q_{i,j}^2\right] - Q_i^2\right\}^{1/2}\end{aligned} \tag{5.4}$$

It should be pointed out that C, as defined by equation (5.1), is only a WIP-related cost indicator, and so its value does not represent the actual WIP cost of the system.

Applications of the performance indicator

Supported by a wide spectrum of experimental opportunities provided by a simulation model of the manufacturing system of the company, the WIP cost-related performance indicator has been utilized to evaluate the company's production situation in relation to its operational policies. For example, it was shown that significant differences in production costs were associated with different production practices. On average a maximum difference of approximately 25 per cent was observed in a number of cases. Bearing in mind the very high production costs involved, these differences represented considerable possible savings on working capital (in money-time units). It was also revealed that within the existing operational structure of the company, although the average physical WIP level of certain items was modest, due to their high WIP weighting factors their WIP-related costs contributed significantly to the total. This situation clearly calls for a more accurate control of ordering and delivery times of these particular items, since evidence strongly indicates that the concept of JIT and an appropriate ordering procedure are more important for these costly items than any other parts under this particular operational setting.

Therefore, together with an appropriate analytical tool, the WIP cost-related performance indicator developed in this case study was shown to be capable of producing a clear, dynamic picture of the production-cost situation of a manufacturing company. Of particular interest is the fact that the result matrix produced will be in a format which is ideally suited for a knowledge-based analysis process, that is, it not only provides a relatively comprehensive picture of the production situation at both aggregate and detailed operational levels, but also is sufficiently limited in variety for effective expert system applications.

5.3 Overview of the 'soft' systems approach

The 'hard' systems approach, as exemplified by the conceptual model of a systems analysis cycle, has been applied extensively in many situations where the basic objectives can be clearly defined and where an analytical model for qualitative analysis of alternative means to achieve the objectives can be built. However, when applied to a problematic situation where the problems are ill-structured, the hard system approach appears not to be equipped well enough to guarantee fruitful results. As a result, the hard systems approach has attracted a certain amount of criticism, some of which is probably justified.

Why should a methodology so powerful when applied in some cases appear so weak when applied to others? The key issue here appears to be associated with the fact that, in order to operate, the hard systems approach insists on a 'problem' to work with, which must be clearly identified and well defined. The trouble is that in many real-life situations it can be difficult to describe the problem in such a way that it can readily be tackled by the method. Even worse, in some cases we do not even know exactly what the problem is. Depending on their perceptions, individuals may differ on what the trouble is with the situation – and this is exactly why the situation concerned is problematic in the first place! As a result of this failure of the hard systems approach when applied to an ill-structured problem situation, a new methodology known as the 'soft' systems approach has emerged.

There is only space here to outline the basic issues of the soft systems approach, including the general framework of the methodology and the differences between the hard and the soft approaches. A detailed treatment of the soft systems approach will be found in Checkland (1981) or Wilson (1984).

5.3.1 The hard and soft systems approaches compared

First of all, it should be emphasized that the soft systems approach should not be viewed as a candidate for replacement of the traditional hard approach, which has been highly successful in many of its applications and to which there is as yet no real alternative. Although at the moment there is a debate at the 'intellectual' level between the adherents of the two approaches about what is the right way to view the world and hence what is the right way to approach its problems, from a practical point of view the development of the soft systems approach, for what it is worth, perhaps can only be considered as an attempt to compensate for the weaknesses shown by the hard approach when applied in certain situations. These two approaches, then, are complementary. Bearing this in mind, let us now examine the similarities and differences between them and try to identify their respective strengths and areas of appropriate applications.

Similarities

The most important similarity between the two approaches is that they are both based on the view that systems ideas are essential to decision-making in complex situations. No matter what exactly the situation concerned involves (system design, system analysis or problem solving), the systems concepts such as those of organized interconnectedness, emergent properties, control and communications should be applied so that the situation is approached in a holistic manner, rather than in the traditional reductionist way with its emphasis on the 'proper' functioning of individual elements.

Second, both of the approaches are typical 'methodologies' which consist of a series of stages designed to guide the analyst in a step-by-step and/or iterative manner (see the beginning of this section). They are both methodologies in the sense that neither is intended merely to solve a single type of problem. Instead, they are aimed at a wide range of applications.

Third, in their actual application both methodologies rely upon the use of models. Whereas the hard approach puts great emphasis on the use of simulation and quantitative models, the soft approach utilizes a 'conceptual model' for the purpose of evaluation.

Finally, and perhaps most importantly from an applications point of view, some aspects of the two approaches are complementary. Since each of these two methodologies has a different emphasis in its approach to a problem situation, we may place the two approaches in the same overall framework of application, so that the outputs from a 'soft' analysis of the problem situation provide the kind of information required by a subsequent 'hard' project for problem formulation.

Differences

The most significant difference between the two approaches is due to the fact that the 'hard' method relies on simulation and quantitative modelling techniques, and requires a very clear picture of the problem contents so that the problem can be approached in a precise and rigorous way, whereas the 'soft' method uses an abstract activity (verbal) model which is intended to handle the 'soft' factors such as ideas and feelings which cannot be easily described in a quantitative manner. Consequently, while the hard systems approach always aims to 'solve' the 'problems' with precision whenever it is possible to do so, the soft approach emphasizes 'appreciating' and 'improving' problems situations when such precision is not achievable but progress must be made. This difference, however, may in practice be utilized to one's advantage, as outlined above.

5.3.2 The general structure of soft systems approach

Although supposedly founded on highly intellectual and complicated philosophical thinking, the underlying logical structure of the soft systems approach is really quite easy to understand. When faced with an ill-structured problem situation, where goals and objectives cannot be easily specified and hence where the relevant data for any particular problem cannot be clearly qualified, instead of concentrating on any particular elements of the situation, the analyst should approach the situation with a broad mind, survey its contents as thoroughly as possible and resist the temptation to jump to conclusions about what 'type' of situation it is or what type of 'problem' it involves.

The analyst is then required to postulate a system considered likely to be of assistance in appreciating and improving the situation. This system will be further developed and refined to produce a 'conceptual model' which is detached from the problem situation and represents the idealized setting. That is, this model will indicate what the system would logically have to do if the system were to be effective in the role intended for it. By comparing this abstract conceptual model with the perceived reality of the real-world situation, the possibilities for future changes are detected from the discrepancies between the two. Debate on the feasibility and desirability of the possible changes will then identify the course of action that should be taken so as to improve the situation.

An overview of the general structure of the methodology is given in Fig. 5.9. As can be seen, like the hard systems approach, this methodology also consists of a series of discrete stages, with the outcome of each stage constituting the input to the next stage. Again, progress through the stages is generally iterative rather than purely sequential.

Stage 1. The problem situation: unstructured

The first stage of this methodology is concerned with finding out about the problem situation. It aims to produce some statements and basic facts about what makes the situation problematic. At this stage the problem situation will appear to be 'unstructured'. That is, the analyst should approach the situation with as broad a mind as possible and collect every possible view from different individuals or groups of individuals involved in the problem situation. The analyst should pay equal attention to all of these views at this stage, even if some may appear to be in conflict with others. That is, he or

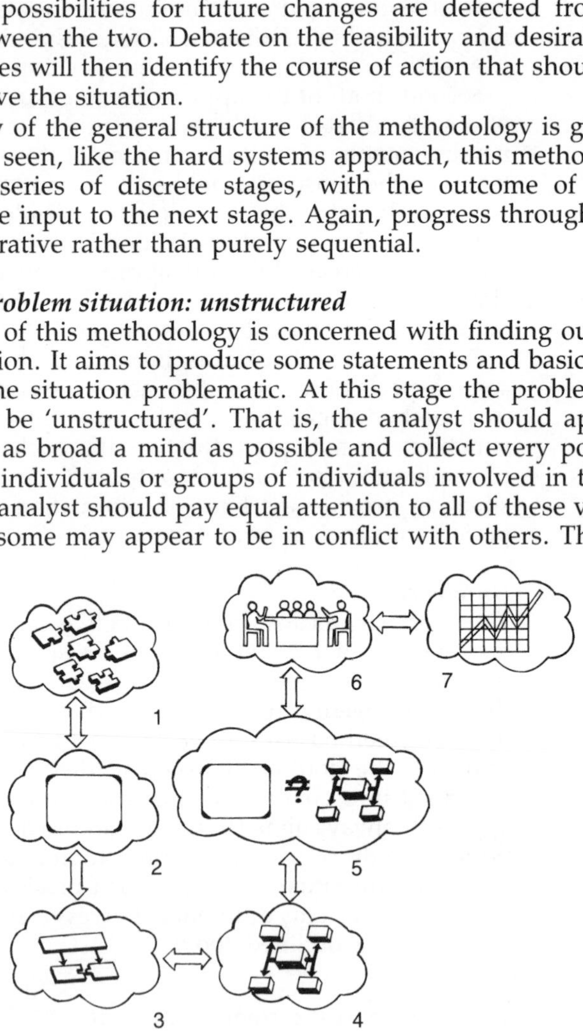

Fig. 5.9 Overview of the soft systems approach. 1. The problems situation: unstructured, 2. The problem situation: expressed, 3. Relevant systems: root definition, 4. Conceptual modelling, 5. Comparison, 6. Debate and 7. Implementation (action to improve situation).

she must try to stand firmly on neutral ground and resist all attempts to impose a particular structure on the situation by anybody – including themselves. For example, in their own mind they should avoid conclusive statements such as 'This is obviously a production scheduling problem' or 'Apparently a new manufacturing facility should be put into operation in this department'. It should also be emphasized that at this stage the analyst must not attempt to provide any answers or to offer advice. Their job is merely to listen and observe. With such an open-ended and non-judgemental exercise of fact-finding, the analyst enters the situation in which the one certainty is that one or more aspects of the situation are considered problematic.

Stage 2. The problem situation: expressed

Although still concerned with finding out about the problem situation, at this stage some systems thinking about the situation is required. This is the first step of the methodology to get from finding out about the situation to taking action to improve it, by expressing or summarizing the situation through the use of a so-called *rich picture*.

A rich picture is a summary of everything the analyst knows about the situation which can be depicted by means of graphs, tables and charts, words, or any combination of these. This portrait of the situation will usually convey organizational data (departmental structure, company physical layout); operational data (transformation processes, production factors, management practice, production procedures, historical statistics); environmental data (market situation, financial situation, any other environmental constraints and considerations); personnel data (managers, operators, other individuals); and judgemental data (feelings, perceptions, observations, opinions, or even gossip).

At this stage it may be helpful to use a more structured approach to picture the unstructured mess of the problem situation. For example, the following general rules may be followed:

1. Search for the permanent entities in the situation which remain stationary or change slowly over time (i.e. the organizational information).
2. Next search for the temporary entities whose attributes (states) change frequently over time (i.e. the operational information such as processes, activities, materials, parts, products, information, financial factors).
3. Then determine how these entities are interrelated (relational, operational, personnel and environmental information).
4. Finally, look for any relevant subjective information and include these in the rich picture (i.e. the judgemental information).
5. While carrying out the above steps, keep the following points in mind:
   (a) A rich picture is not a systems description of the situation, for the elements of the mess which it is trying to describe may in general not be logically interrelated in the way in which the elements of a system must be linked together.
   (b) Try not to overlook any relevant elements of the situation – use meaningful footnotes and remarks where appropriate so that the resultant picture is as complete as possible.

In summary, a rich picture is a representation of the general patterns or aspects within the situation in which there is perceived to be a problem. This picture will allow one to visualize the content of the situation concerned

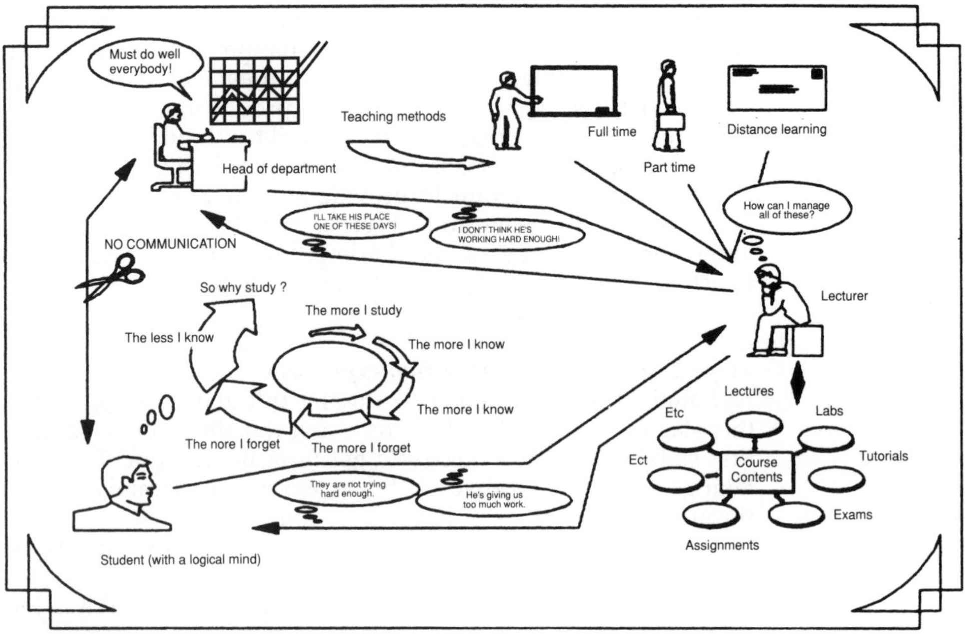

Fig. 5.10 Possible rich picture of a university department.

and help communicate one's impression of the situation to others. Figure 5.10 gives an example of what a rich picture of a situation may look like in practice.

Stage 3. Relevant systems: root definition

The name chosen for this particular systems approach certainly does the methodology justice – for there are some seemingly quite 'soft' or vague ideas associated with it. The concept of *relevant systems* is one of these.

According to this approach, having gathered sufficient information about the situation and built up the richest possible picture of it, the next task for the analyst is to 'choose to view the problem situation in ways that they believe will produce insight' (Wilson, 1984) in terms of relevant systems. In general a relevant system is not a description of a system of the problem situation that is already in existence, nor is it a system designed to solve the problems inherent in the situation that ought to be implemented in the real world. It is a hypothetical, or notional, system which is somehow 'relevant' to the problem situation in the sense that it might bear on significant aspects of it and hopefully provide some potentially constructive way of comprehending the situation. (Sounds pretty baffling, doesn't it? Now perhaps we can see why the soft systems approach has been criticized as fuzzy, lacking in rigour and precision, lacking solid content and over-possessed of subjective judgement.)

However, the procedure involved in the development of a relevant system is not entirely unlike that of a general systems description process. First, as a 'system', a relevant system must possess all the usual features that a normal system should have; and second, based on the same rich picture a number of different relevant systems could be conjured up dependent upon one's perspective (or what, in the jargon of the approach, is called the *Weltanschauung* – a German word meaning 'world view'). For example, looking back at the pub example quoted at the beginning of Chapter 2, it can be seen that based on its contents as reflected in a rich picture developed for it, the following relevant systems might be named:

Pub: an enjoyment-generating system
Pub: a profit-generating system
Pub: a wage-generating system

A relevant system must be precisely described in terms of what is known as a *root definition*. A root definition should be a concise and precise verbal description of the nature of the relevant system; its completeness can be checked against the following checklist (note the similarities between this checklist and the 5W procedure for systems identification and description discussed in Chapter 2):

Customer: client of the activities who is affected (e.g. beneficiary or victim, affected sub-systems).
Actor(s): those who are responsible for carrying out the activities of the envisaged relevant system (e.g. types of agents or groups of individuals).
Transformation: this refers to the core process of the system that transfers its input(s) into output(s). Definition of the transformation process of the relevant system is the part of a root definition which reflects the nature and main object of the system.
World view: the viewpoint from which the system defined becomes relevant and meaningful. This viewpoint is not often explicitly expressed in the root definition, but is always implied.
Ownership: the wider system that exercises control over the system, and has enough power to determine the existence or non-existence of it.
Environmental constraints: the given environmental factors which affect the system activity or activities.

As an example, Wilson (1984) quoted a root definition that was drawn up for the Engineering Department of British Airways which was responsible for aircraft maintenance:

> A British Airways owned system concerned with the continuously effective and efficient planned maintenance of aircraft belonging to BA and their contracted airlines, with a performance acceptable to operating departments but within statutory, local, and BA-applied constraints.

In this root definition the 'customers' are clearly specified as 'BA and their contracted airlines'. The 'owner' is British Airways and the 'transformation' is to provide 'continuously effective and efficient planned maintenance'. Also explicitly specified are some of the 'environmental constraints' including a performance standard which is 'acceptable to operating departments but within statutory, local, and BA-applied constraints'. It is also implied that the 'actors' here would involve a maintenance unit which is responsible for carrying out the actual maintenance activities, a planning unit for efficient aircraft maintenance scheduling and a technical support unit that is necessary in order continuously to adopt new technologies so as to satisfy the requirement of 'continuously effective . . . maintenance' of the aircraft.

It should be pointed out once again that for any given transformation process, there is no single root definition that can be regarded as *the* definition for the situation, because it is always possible to envisage different root definitions for exactly the same process but based upon different viewpoints.

Stage 4. Conceptual modelling

Having stated, as precisely and concisely as possible, what the system is in terms of a root definition, the next step is to expand the definition so as to specify what activities should be logically necessary for the system to perform its intended tasks. The result from this stage is called a *conceptual model*; each root definition must have a conceptual model developed for it.

Once again, a great deal of subjective judgement is involved in the development of a conceptual model in terms of both the characteristics of the model and the degree of detail it contains. This explains why this conceptual modelling has been described by the developers of this methodology as an 'intellectual activity' which may be carried out in isolation from the real world. Like relevant systems, a conceptual model is neither a portrait of some real-world system nor a design for a system that is to be implemented in the future. It is just a picture of a system that is developed through deductive logic from the root definition, which indicates what *should* be in existence, or what activities *should* be carried out, in order for a system to correspond to the system specified by the root definition. For this purpose, some of the systems concepts outlined in Chapter 2 can be used to aid the modelling activity involved. These include:

- *System hierarchy*. The concept of system hierarchy and system decomposition is of great value here. To ease the task, the development of a conceptual model may follow a top-down approach. That is, the conceptual model can first be defined at the top level in accordance with the root definition which, it is suggested, should contain no more than 12 activities for the purpose of clarification (this is based on the observation that the human mind is only capable of engaging itself with a limited number of entities at any one time). The activities at this level may then be decomposed into the next resolution level in a similar manner.
- *Formal system model and checklist*. Since the main purpose of a conceptual model is to indicate what should exist in the system, and what activities should be carried out by the system, it is obvious that the outline of the formal system model and its associated checklist, as given in Chapter 2, will provide useful guidance in checking the basic 'shape' and completeness of the model developed.

Being a tentative exercise, the choice of the actual modelling language used here is again dependent very much upon the modeller's preference. Methods such as systems dynamics or IDEF, for example, are both equally suited for this purpose. However, a conceptual model is usually presented in graphical form so that it can be conveniently compared in the next stage with the rich picture of the situation.

It should again be emphasized that conceptual modelling is *not* a form of systems description – it involves the considerations of an ideal and logical model which should not contain any real-world elements. This is intended to serve the purpose of keeping the conceptual model detached from the real problem situation, so that the gap between the ideal situation and the reality can be identified later. It is also important to bear in mind that the relationship between root definition and conceptual modelling is iterative. Somewhere in the development process of a conceptual model it may become desirable or necessary to go back one step to modify the root definition. The final version of the conceptual model may only emerge after a number of such cycles.

Stage 5. Comparison

This stage is concerned with comparison between the idealized situation as represented by the conceptual model and what is actually in existence in the real world, so as to highlight the discrepancies between them for further debate.

The basic way to carry out this comparison is by looking at the conceptual model and the rich picture together, and asking questions such as: what is involved in the model and how *should* it operate, and what is involved in the real world and how *does* it operate? However, there are other structured techniques for carrying out this comparison. Wilson (1984) outlined four methods of comparison that are most frequently used for this purpose: general discussion, question generation, historical reconstruction, and model overlay.

General discussion about the nature of the two situations, as outlined above, should ideally involve the participation of all parties to the problem situation. This type of comparison tends to raise strategic issues rather than issues at a detailed procedural level.

Question generation is concerned with a more detailed level of discussion. In this case the contents of the conceptual model (structure, activities, and operations, and so on) are used as a checklist to generate a set of questions. The answers to these questions are then sought in relation to the real world. The following is a typical sequence of such questions:

For any activity, A_i, in the conceptual model,
↓
does A_i exist in the real world?
↓
If it does, how is it carried out and by whom?
Has A_i been satisfactorily conducted?
How has it been measured?
Is there anything that can be done to improve its performance?
↓
If not, what change should be proposed?

This method of comparison is the most commonly used in practice.

Historical reconstruction involves looking at what has happened in the real world, and then comparing this with what should have happened according to the operational sequence of the conceptual model. Suppose the problem situation concerned involved the process of design and implementation of a certain system, and as a result the conceptual model developed was similar to the model of systems analysis cycle as shown in Fig. 5.2. With the method of historical reconstruction, the comparison stage would be carried out by looking at how the design tasks had been accomplished in practice, and then comparing this with how the model would have caused this to be done if its framework had been followed. This particular method of comparison is not as widely used in practice as the preceding two because of some of the technical difficulties associated with it (Checkland, 1981).

The model overlay is the most direct type of comparison of the four. This involves first drawing a picture each for both of the rich picture and the conceptual model on transparent paper, and then literally placing one picture on top of the other. Any differences between the two will then be highlighted through this direct comparison process. For this method to be effective, it is necessary to present both the rich picture and conceptual model in the same format and to construct them in such a manner that one reflects the other as closely as possible.

Stage 6. Debate

The differences highlighted by the comparison stage typically suggest certain omissions in the real-world situation, that is to say, that some of the activities which the conceptual model considers essential do not actually occur in reality. Having directed our attention to these activities, the obvious question to ask is 'Should they happen?'. In this methodology this question is not to be answered by the analyst. Instead, the answer is to be sought through debate with the people who are involved in the problem situation, including the client and the problem owners.

The purpose of this stage is to have an orderly discussion on the possibilities for change, so as to identify those possibilities which are both desirable and feasible. For this purpose, the agenda which the analyst puts to the other participants for discussion should always be constructed in terms of 'what' (what is missing or what is possible, for example) but never in terms of 'how' (specific statements about how things should be done). For instance, if the comparison stage highlights certain questionable issues in the production control area of a manufacturing system, then the topic for discussion should be announced in terms like 'A systematic way of effectively planning and controlling the production process seems to be necessary' – and never in a statement like 'An MRP II system must be implemented'. The reason for this again lies in the hierarchical structure of objectives. A 'what' here is analogous to an 'end', while a 'how' to a 'means' (section 5.2.3). Since within the hierarchy of objectives there usually exist a number of possible means to achieve a particular end, then for the purposes of debate all of these possibilities should be explored. In the example given above, 'implementing an MRP II system' is but one of the possible ways to achieve more effective production control. It should not, therefore, be considered at this stage as the end itself.

There are three possible outcomes of this debating stage:

1. It is agreed by all that certain changes are both desirable and feasible, in which case the changes concerned are implemented and the project progresses to the next stage.
2. None of the proposed changes is considered by the actors as both desirable and feasible, because they fail to agree that the relevant system concerned is really 'relevant'. In this case the problem must go 'back to the drawing board' again – another relevant system must be examined.
3. After an exhaustive debating process, possibly involving extensive discussion of a number of relevant systems, still none of the ideas for change survives this filtering process. This indicates that there is really no need for change within the existing situation. In this case both the analyst and problem owner should be prepared to accept the fact that, under the particular circumstances, it may be wise to change nothing at all.

Stage 7. Implementation

Having assembled the set of changes which are regarded as both desirable and feasible, it is now time to consider how to implement these changes.

The nature of the changes to be implemented differs from case to case. Experience has taught that some of the most common changes are changes in organization, in operation, in strategies, and in attitude. In general, the issues and problems associated with implementation, as discussed in section 5.2.6, apply equally well here for the soft systems approach.

5.4 Conclusion

In this chapter the general systems approaches to real-world problem solving have been reviewed. In particular, two different but complementary systems techniques have been discussed, together with some of the technical issues associated with them – one being the well-established hard systems approach, with its wide range of applications; the other being the soft approach, a relatively young cousin of the hard approach which attempts to overcome certain weaknesses inherent in the hard approach.

Both approaches assume wide applicability, and as such are not restricted in any way to the problems specific to manufacturing systems design and analysis. This is illustrated by the fact that the issues involved in both the hard and soft systems approaches to problem solving have been discussed in this chapter at the conceptual level and in general terms. That is, they have been presented in such a manner that the approaches may be adapted to a variety of different situations.

However, many of the problems involved in the area of manufacturing systems are well suited to the application of systems approaches such as the two discussed here. Having discussed the general principles involved, it is necessary to show how the general approaches can be adapted to the specific tasks of systems design and evaluation involved in manufacturing organizations. The development of these specific methodologies is the subject of Chapter 7. While not specifically aiming to explain the detailed workings of each methodology, it is intended to show the trends, commonality and, where possible, to look at the effectiveness of the design methodologies.

Further reading

Blanchard, B.S. and Fabrychy, W.J. (1981) *Systems Engineering and Analysis*, Prentice-Hall.

Cleland, D.I. and William, R.K. (1983) *Systems Analysis and Project Management*, McGraw-Hill.

Wilson, B. (1984) *Systems: Concepts, Methodologies, and Applications*, John Wiley.

Questions

1. Identify at least two disadvantages of comparing system outcomes by using the overall utility approach described in section 5.2.5.
2. Revise the hard systems approach by rereading section 5.2. Pay particular attention to both the overall characteristics of the approach and the particular issues associated with the specific steps. Repeat this process if necessary until you are confident that you understand the approach well

enough to try applying it yourself. Then apply the methodology to a suitable, real-world problem situation in relation to your job or within your organization. Prepare a report of this project (2000 words).

3. Revise the soft systems approach by reading section 5.3, and then apply the methodology to a suitable, real-world problem situation in relation to your job or within your organization. Prepare a report of this project (2000 words).
4. Having carried out Questions 2 and 3, compare the hard and soft systems approaches. Highlight the fundamental differences and similarities through your own experiences of their applications and discuss why these two different approaches should be regarded as complementary.

6 Computer simulation in manufacturing systems analysis

6.1 Introduction

Computer simulation can be considered as an experimental approach for studying certain functional properties of an organization by experimenting with an appropriate computer model rather than with the real system itself. It is basically an experimental methodology using the power of a computer to process and analyse the large amount of data involved in a problem which otherwise would be extremely difficult to handle (Shannon, 1975; Zeigler, 1976; Poole and Szymankiewicz, 1977; Pritsker, 1979; Pritsker and Pegden, 1979; Law and Kelton 1982; Ellison and Wilson, 1984; Gottfried, 1984; Pidd, 1984; Carrie, 1988). It provides an efficient and economical – and sometimes even the only possible – way to analyse a system. Compared with direct real experimentation, the computer simulation approach has the advantages of lower cost, shorter time, greater flexibility and much smaller risk. As a result, this methodology has been extensively used in the area of manufacturing systems studies by both academic researchers and practical engineers.

Engineers or managers use simulation to improve the performance of existing manufacturing organizations, as well as to plan and design new systems. For example, it is very unlikely that anyone today would spend a substantial amount of money on building new plant without having thoroughly checked its future performance by some sort of simulation study. On the other hand, academic researchers have been using simulation techniques to seek a better understanding of the characteristics of various manufacturing systems in an attempt to develop more rational organization structures, more effective manufacturing strategies and more efficient production control policies.

This chapter is concerned with computer simulation techniques and systems that are currently used in the manufacturing environment. It first provides an overview of applications, including a discussion of the desirable features, as well as the limitations, of this experimental method as compared to analytical (or mathematical) approaches. Technical information is then provided to illustrate how different modelling techniques can be used to solve manufacturing engineering problems in various application areas.

Some theoretical techniques, such as those of probability and statistical analysis and some results from queuing theory, are also discussed, with particular attention paid to their applications in relation to computer simulation.

Having discussed the structure and techniques of computer simulation in general, attention is then directed to carrying out simulation modelling with special-purpose simulation languages. In particular, one of the commercially available, graphics-based simulation systems is used to demonstrate the functions, features and facilities provided by these systems. The discussion of this chapter is supported by a case study, which is based on an investigation of the operations of a manufacturing organization.

Finally, one of the future trends of computer simulation – the integration of simulation and expert systems – is reviewed.

6.2 Overview of the applications of computer simulation

It has already been pointed out that problem solving in industry is growing increasingly complex, so that managers require increasingly effective guidance in their decision making. Since the 1950s the computer has been a key factor in improving management and engineering decisions. However, computer usage here has been limited primarily to providing quick and accurate access to status information. Only in a few specific applications has the computer been used in conjunction with managerial expertise to improve the actual process of decision making. Being a method for predicting the dynamic characteristics of a situation and thus improving the basis of the decision process, computer simulation is one of these applications. With a computer simulation model, a systems analyst or manager is able to observe the behaviour of a process without the necessity of experimenting with the actual system. They may try out, for example, different manufacturing runs, new operational conditions, new equipment layouts or different cycle times so as to evaluate the system's performance given various disturbances, or to identify the bottlenecks.

It is possible to separate two basic types of simulation application in the area of MSE. In *systems evaluation* an existing process is simulated so that it may be adapted or expanded according to changing operating conditions. In *systems design* a new process is investigated to avoid expensive pitfalls and to observe the reaction to extreme operating conditions.

6.2.1 Design of advanced manufacturing systems

The implementation of AMS creates major problems for the system designers and engineers who are responsible for the successful design, implementation and operation of the systems concerned. The design, planning and scheduling of FMS, for instance, has received substantial interest in recent years due to the high initial investment costs involved, as well as the mixed success of their adoption. Much capital and effort is now invested in an attempt to make their implementation more successful. Generally speaking, it has become necessary to perform thorough analyses of the possible effects of an AMS implementation before it is actually introduced. Only in this way can a satisfactory financial justification for the large capital sums involved be made before the commitment is undertaken.

As a result, computer simulation has found a wide range of applications in the design of AMS, due to the fact that computer capabilities, combined with the versatility of simulation techniques, form a very powerful and flexible tool to assist the tasks involved. A computer simulation model can cope with the complexity of the systems and, in addition, provides an effective communication tool to explain their operation clearly to prospective investors. By using simulation the risks of introducing AMS can be considerably reduced.

Two levels of system definition can be distinguished for evaluating alternative AMS performances using computer simulation. In the first, an assessment of the gross operating characteristics is made – for example, the overall means and variances of work station utilization, production rate, part waiting time and transporter utilization. For this type of assessment a relatively coarse model will suffice and a typical simulation run corresponds to a production program of six to twelve months. More detailed simulation models can then be used for the determination of equipment requirements – for example, the number of transporters, number of pallets, bottleneck problems and effects of equipment interference. This type of information will be required by the design team to formulate the operating strategies needed to optimize the performance of the system. Detailed models demand much more computation time than coarse models and the simulation runs have to be restricted to a few days or a few weeks of actual production. In fact the problems involved in the design and operation of an AMS may be described by the following three-level hierarchy:

Level 1: basic components of the AMS
Level 2: order processing and planning
Level 3: scheduling and machine control

The techniques of computer simulation may be used effectively at each of these problem levels. A number of surveys have been carried out to investigate the uses of simulation in the design of manufacturing cells (Kiran *et al.*, 1989). It is not surprising that it has been concluded that simulation has been proved a powerful tool in the exploration of the ramifications of the detailed designs. Consequently, many companies now require that simulation studies of proposed manufacturing facilities be performed before a final decision on the implementation of a new design is made.

However, it should be pointed out that, when applied to AMS, design simulation is not without its drawbacks, the major one being that it can be very costly and use expensive resources. While cost has been identified as the major factor in dissuading companies from using simulation, it is suggested that expenditure of between 2% and 4% of the total investment would be an acceptable yardstick.

6.2.2 Other applications

Recent studies (such as that by Mellichamp and Wahab, 1987) have concluded that graphical input capabilities, microprocessor-based languages, expert system integration and advanced information processing techniques all tend to increase the use of computer simulation. It has been shown that although recently fewer firms are using simulation in some areas, those firms which do use simulation are doing so to a greater degree. Simulation has been used at all levels of management in the 500 largest corporations in the United States, most frequently at the product or process level. Within functional areas of business, simulation has been most heavily used in the production area. Although applications in system layout, investment justification, scheduling and inventory analysis and control are most common, frequent use in other manufacturing areas has also been reported, including quality control, maintenance studies and system reliability.

However, it should be pointed out that there seems to have been little increase in the use of simulation modelling since 1975. On the surface this is surprising considering the emphasis on quantitative analysis in business and engineering schools at both undergraduate and postgraduate level, and the increased availability of mini and personal computers and specialized simulation software. Reasons for this may be that simulation modelling is too sophisticated and time-consuming to accommodate quick decisions in a dynamic environment. Also the surveying methods used may have caused the results to understate the actual use of simulation throughout the organization.

It is felt that as far as the planning and design of AMS is concerned, the primary group of potential users – including process designers, practical engineers and managers – does not have the necessary expertise to take full advantage of the benefits of the methodology. This is the main reason why computer simulation is regarded as an important subject area in this book.

6.3 Characteristics of computer simulation

As an aid to decision-making, the technique of computer simulation has many desirable features and, unfortunately, some disadvantages when compared to other approaches.

6.3.1 Benefits of computer simulation

Flexibility
One of the most significant advantages of computer simulation is that it is a very flexible approach. Once a model is developed, with a reasonably flexible structure, it can be quickly and cheaply modified to include new features. Frequently only input data need to be changed for the same model to be able to evaluate additional alternatives; remodelling a system using an analytical approach would take much longer.

Study of transient behaviour
Most systems are very complex and since analytical models can only be created using simplifying assumptions, often they can at best predict only steady-state situations. In reality, however, a manufacturing process involves various parts being continuously released into the system at short irregular intervals. In such a situation it is sometimes necessary to analyse the transient behaviour of the system. Computer simulation in many cases is the only viable tool available to perform this task.

Communication
One of the major advantages of computer simulation is its ability to animate the process under investigation. Animation results in benefits for the model builder, beneficial communication between the model builder and the model user, and benefits in presentation to users and management (Smith and Platt, 1988). Because of the increased ease of communication that animation allows, the model user can be actively involved with a simulation model throughout the model-development cycle. This leads to significant improvements in the model and in the benefits the user derives from it. A graphics display can give a very good idea of the logical behaviour of the simulated process. The programmer therefore benefits by being able to follow the logical consequences of state changes in the system. Well-designed graphics can aid effective experimentation with the model, for it is easy to spot whether a particular experiment is fruitful by watching a dynamic graphics display.

Table 6.1 A summary of model types

	Model type			
	Descriptive	*Physical*	*Analytical*	*Procedural*
Method of prediction	judgement	manipulation	mathematical	simulation
Method of optimization	?	experiment	mathematical	experiment
Cost	low	high	medium	high
Ease of communication	poor	good	poor	excellent
Limitation	not repeatable	cannot represent information process	only able to cope with simplified cases	optimal solution not guaranteed

Furthermore, animation with appropriate notation allows information to be conveyed in a very short amount of presentation time that would otherwise require several pages of written documentation. As a result, this tends to make the users feel that they are being presented with something they can believe in. With the use of animation as a tool to enhance the accessibility and credibility of a simulation study, simulation is likely to grow more rapidly as a valuable and accepted technique for manufacturing systems engineers.

Limitations of other types of approach

Although computer simulation is no panacea and, realistically, models may be large and complex and as a result take a long time to build, they can still have many advantages over other analytical methods. In comparison to direct experimentation, where a real system may have to be physically modelled, computer simulation has the advantages of lower cost, time savings, precise replication and safety. A computer simulation model also has advantages over mathematical modelling, which is usually unable to cope with dynamic or transient effects. Queuing theory models provide an example of such approaches which can be used only to analyse the steady-state operation of simple systems and are only able to manage certain probability distributions. This limits the variety of problems that can be analysed. Other operational research methods are also frequently insufficient in aiding management decisions as conclusions may only be reached with too many constraints. Scheduling jobs through a workshop by analytical methods is one typical example. Box 6.1 provides an example which clearly illustrates the weakness of certain analytical approaches in this regard. As a summary, the features of different approaches are listed in Table 6.1.

Box 6.1 Limitations of the analytical approach when applied to production planning and control

It has been pointed out that the power of analytical approaches, even where computer-assisted, seems to be very limited when applied to the complex environment of actual manufacturing systems (Chapter 3). This is illustrated by the following discussion.

It is interesting to note an early observation by Conway *et al.* (1967) concerning the general $N/M/A/B$ scheduling problem (for the meaning of this and other notation used below, see section A.2.2):

> The general job-shop problem is a fascinating challenge. Although it is easy to state, and to visualize what is required, it is extremely difficult to make any progress whatever toward a solution. Many proficient people have considered the problem, and all have come away essentially empty-handed. Since this frustration is not reported in the literature, the problem continues to attract investigators, who just cannot believe that a problem so simply structured can be so difficult, until they have tried it.

At the time of this statement, it was not possible to explain why this should be. Recent work in the theory of complexity when applied to scheduling problems has offered an explanation in terms of a mathematical conjecture: that most $N/M/A/B$ problems are 'NP-complete', and that therefore no polynomial bounded constructive algorithms will ever be found for the majority of such scheduling problems (that is, $P \neq NP$). This implies that, even with the ever-increasing power of computers, optimal schedules for this class of problems are not likely to be obtainable with a practical computational time.

The theory of complexity divides combinatorial problems into two classes: the P-time class (polynomial bounded) and the NP-time class (non-polynomial bounded). P-time complexity problems can be tackled, because for them polynomial bounded algorithms can be found. On the other hand, it is very difficult to obtain optimal solutions for NP-time complexity problems, for which only algorithms with exponential behaviour have been found (Garey and Johnson, 1979). The differences between a P-time complexity problem, and a NP-time complexity problem can be illustrated quite easily by the following example.

The $N/1//F_m$ scheduling problem is a P-time complexity problem. It has been shown analytically that the optimal solution of this problem is given by the SI scheduling rule, and the number of computations required for an optimal solution using this rule can easily be estimated. For example, a straightforward method for sorting the jobs into the required order according to this rule is to list, compare and exchange the adjacent jobs and to repeat this comparison and exchange processes until the jobs are in listed order of increasing operation times. The number of comparison and exchange processes required to yield the optimal sequence for N jobs will be bounded by the following function:

$$2[(N-1)+(N-2)+\ldots+2+1]=N(N-1)$$

Approximately the same number of operations will be needed to carry out the executive controls during the actual computation. Therefore the total bounding function for this problem will be:

$$2N(N-1)$$

The $N/1//T_m$ scheduling problem is of NP-time complexity. There is no simple constructive algorithm such as that for the $N/1//F_m$ problem, but an optimal solution may be found by dynamic programming (dynamic programming algorithms have exponential time complexity). Following a similar analysis to that above, it can be shown that the time required to generate an optimal schedule is bounded by (French, 1982):

$$6N \times 2^{N-1} + 3(2^N - 1)$$

Assuming that a computer takes 1 microsecond to perform one of the operations needed to evaluate the solutions, then the above bounding expressions would give the computation times shown in the table for the two problems.

As the number of jobs increases, the computation time for a P-time complexity algorithm remains manageable, while the computation time for an NP-time complexity algorithm increases rapidly as N increases, to the point where practical solutions of the problems are impossible. This implies that if algorithms with polynomial time complexity have been found for a problem, then, with the increasing power of computers, solutions of the problem with much larger sizes will be readily obtainable. On the other hand, if for the problem only solutions with exponential time complexity are available, then it is unlikely that the problem will as yet be completely solvable.

Number of jobs (N)	$N/1//F_m$ (P-Time Complexity)	$N/1//T_m$ (NP-Time Complexity)
4	0.000024 sec	0.000237 sec
10	0.000180 sec	0.033789 sec
20	0.000760 sec	66.060285 sec
40	0.003120 sec	4.29 yr
60	0.007080 sec	6.69 $\times 10^6$ yr

Of those problems in the NP-time class, the most difficult of all are those problems in the NP-complete sub-class. (An NP problem Ω is called NP-complete if every other problem in the NP problem set can be polynomially reduced to problem Ω. Therefore, if one polynomial time algorithm could be found for Ω, then it would be also possible to solve all the other NP problems in polynomial bounded computational times, because they could all be polynomially reduced to problem Ω, and then solved polynomially using the algorithm developed for Ω.) One of the best-known conjectures of modern mathematics is that such NP-complete problems will never be solved polynomially. In other words, it is believed that an NP-complete problem can never be reduced to a P-time complexity problem, that is, $P \neq NP$. Unfortunately, it has been shown that most scheduling problems are NP-complete (Ullman, 1976; Garey and Johnson, 1979; Graham *et al.*, 1979). This is why many of the problems in this area have been studied extensively, but no algorithm with polynomial time complexity has yet been devised. As is evident from the current literature, traditional analytical approaches have not been very successful in finding optimal solutions to real industrial problems. Although many methodologies are available and a great number of various models have been reported, very few are successfully implemented for actual application.

Although a rather pessimistic picture has been drawn here, the applications of complexity theory have offered valuable guidance to the researchers in this field. That is, it is clear now that for the proven NP-complete problems it is more sensible to direct efforts towards identifying and finding efficient and practical techniques which will generate feasible plans rather than to waste time and effort trying to develop constructive algorithms. Not surprisingly, therefore, computer simulation is extensively used in the study of manufacturing operations, largely due to the complexity of the problems involved.

6.3.2 Limitations of computer simulation

As an aid to systems design and analysis, simulation also has certain shortcomings. Perhaps the most crucial of these are the requirement of expertise and the amount of time required to construct a simulation model. Generally, engineering personnel are rarely skilled in simulation methodology, and so a team development approach is often employed. Even given the best circumstances, however, simulation projects are always time-consuming – frequently needing many months before useful results are realized.

There are also other difficulties that limit the use of simulation (Kiran *et al.*, 1989). The first problem is the number of design alternatives that can be simulated in a reasonable time. The statistical properties of simulation outputs mean that a high number of simulation runs are required for each system configuration, preferably under every conceivable scenario. This limits the designer's freedom to generate more alternative designs. There is also always a danger that the effort required for a quick implementation of the simulation model is underestimated. Despite the fact that many standard simulation tools are available, every application has a unique set of decision rules and a structure which requires its own solution. The second issue is the data availability for simulation models. Simulation models mimic a real system and so must be provided with 'pseudo-data' that detail the conceived system, such as processing and demand information, and operation rules that are not readily available in the design phase. Another problem is the starting phase of the simulation model. As there are many undefined problems at the planning stage, an arbitrary state has to be set at which the simulation model is assumed to represent the process. Because of these problems a simulation project always contains a risk factor which neither the designer nor the user of the model can fully anticipate in the early stages of the design project.

In addition, it should always be remembered that simulation experiments are of a statistical nature, and the results must be treated as such. Optimal solution of a problem is not always guaranteed with such an approach. Furthermore, simulation cannot be the sole substitute for the planning operation – it always requires a preceding design phase.

6.4 Construction of simulation models

All simulation projects follow the following steps:

data collection
problem analysis
simulation model specification
model programming
model verification
simulation experimentation
evaluation and interpretation of simulation results
report generation.

This list may look rather familiar, for some of the steps have already been discussed in Chapter 5. Because of this, what follows will concentrate on the design and software realization of simulation models. This particular section examines the technical aspects of computer simulation – the nature of computer simulation, types of simulation model used in MSE, simulation languages, and programming approaches. Some of the important issues associated with simulation modelling, such as model validation and initial conditions, are also discussed.

6.4.1 Simple examples of computer simulation models

The nature of computer simulation may be revealed by examining the procedures involved in the following examples.

Monte Carlo simulation

The Monte Carlo technique is used to solve problems by means of sampling experiments. A dice game provides a simple illustration. The game involves throwing a pair of dice, adding together the scores from each die and recording the result. The probability of occurrence of a particular total score can then be estimated from the results obtained from a large number of trials. This direct approach is very time-consuming, however, and we can replace the dice by two independent series of random numbers – one from each die, since the individual die scores are independent of each other. Each series of random numbers must follow the same discrete uniform probability distribution (section 6.5.1), reflecting the fact that a fair die, when thrown, will give a 1, 2, 3, 4, 5 or 6 with equal probability $\frac{1}{6}$. If a large number of total scores are generated in this way, the probability of occurrence of a particular total score can now be estimated without throwing any actual dice. This technique is used in many electronic game machines.

The statistical process of computer simulation depends on a similar method, using probability assumptions to estimate the actual statistical features involved in a real-life situation. The following example shows how this works in practice.

A queuing situation

An example of the procedure of a typical computer simulation is provided by the following model of a queuing situation. For the sake of simplicity the simulation program of this model is given in BASIC.

When we wait to be served at the bank counter, or are being delayed at a computer terminal trying to use a multi-user CPU, we are in a *queue*. From the operational point of view a manufacturing system can usually be represented as a network of queues, in which raw materials wait to be processed and parts wait to be assembled. Therefore, it is important for us to be able to make predictions about the behaviour of a queuing situation so that we are in a position to manage the situation better, or perhaps even to eliminate the undesirable effects of queues. For this purpose, a computer is well suited to help us, and the following example illustrates how it can do this.

Parts manufactured on a machine system arrive in random order. They wait in a buffer if necessary, and are processed according to the first-in-first-served rule. After processing they are released from the system. The inter-arrival time of parts is uniformly distributed over the range (11.25, 18.75) minutes, and the processing time of the system has a negative exponential distribution with a mean value of 12.5 minutes. We wish to evaluate the

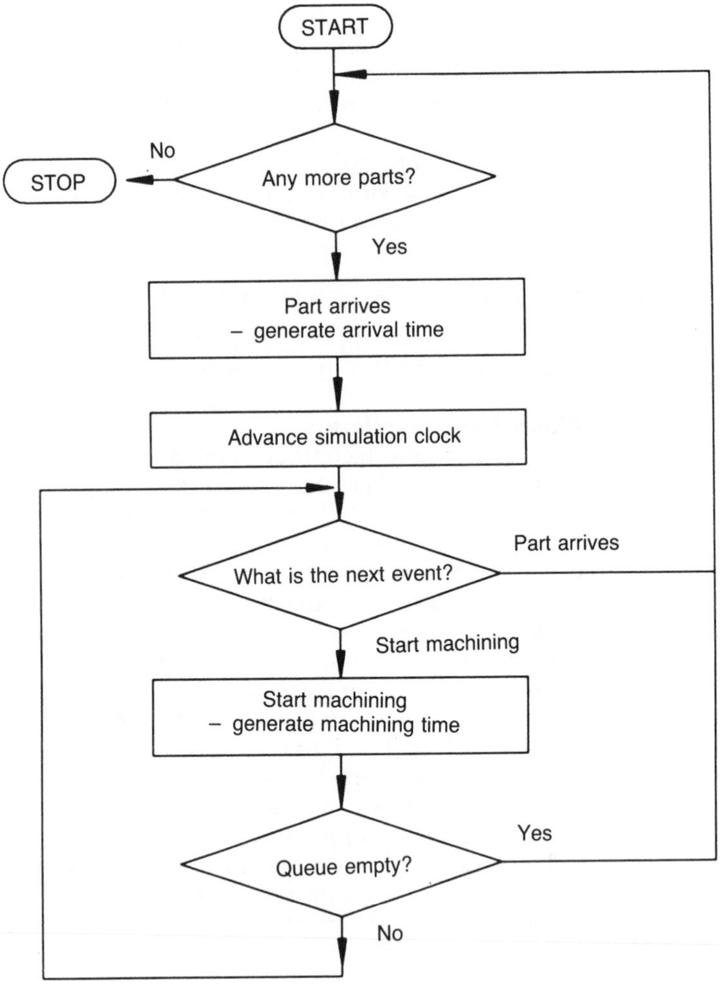

Fig. 6.1 Flow chart of a machining model.

performance of this machine system by simulating its operation over a period involving the processing of 1000 parts. The overall modelling logic of the program is shown in Fig. 6.1. The program itself is given in Box 6.2. With the help of the commentary lines, this program is self-explanatory.

Box 6.2 Program listing of a simple queuing model

```
90   '____________________________________________________________
95   '
100  'SIMULATION OF A SIMPLE MACHINING OPERATION
105  '
106  '____________________________________________________________
107  '
108  '
110  'PARAMETERS AND ARRAYS
```

```
115  '
120  ' TIME UNITS
125  '       CLOCK  - SIMULATION CLOCK
130  '       ARRI   - TIME OF NEXT PART ARRIVING
135  '       TNXT   - TIME WHEN MACHINE WILL BECOME READY
140  '
145  ' TIME RECORDS
150  '       TQUE   - TOTAL WAITING TIME OF PARTS
165  '       AWIT   - AVERAGE WAITING TIME OF PARTS
170  '       TIDL   - TOTAL IDLE TIME OF MACHINE
175  '       UTIL   - UTILISATION OF MACHINE
180  '       TARR( )- ARRIVING TIME FOR EACH PART
185  '       TPRO( ) - PROCESSING TIME FOR EACH PART
190  '
195  ' NUMBER RECORDS
200  '       IA     - NO. OF PARTS ARRIVED
205  '       IP     - NO. OF PARTS PROCESSED
210  '       IQ     - TOTAL NO. IN QUEUE
215  '
220  '
225  '------------------------------------------------------------------
230  '
235  '
240  DIM TARR(1000),TPRO(1000)
250  SEED=85:                          'RANDOM SEED
255  RANDOMIZE(SEED)
260  MARR=15:DARR=3.75:                'MEAN & DEVI. OF ARR. TIME
265  MPRO=12.5:                        'MEAN PROCESSING TIME
270  PARTS=1000:                       'SIMULATE 1000 PARTS
275  ARRI=0:TNXT=0:CLOCK=0:            'INITIALIZATION
280  IA=0:IP=0:IQ=0
285  '
290  '
360  IF IP>=PARTS GOTO 590:            'SIMULATION OVER ?
370  IF IA>=PARTS GOTO 450
380  IA=IA+1:GOSUB 2010:               '1. PART ARRIVES
390  TARR(IA)=UNIFM+ARRI
400  ARRI=TARR(IA):                    '  SET PART'S ARR. TIME
410  IF ARRI<TNXT GOTO 360:            '  & CHECK FOR NEXT EVENT
420  '
430  IF CLOCK<ARRI THEN CLOCK=ARRI     'ADVANCE SIMULATION CLOCK
440  '
450  IP=IP+1:GOSUB 1010:               '2. PROCESS PART
460  TPRO(IP)=EXPONT
470  IF TNXT>TARR(IP) GOTO 500
480  TIDL=TIDL+TARR(IP)-TNXT:          'SUM UP MACHINE IDLE TIME
490  TNXT=TARR(IP):GOTO 520
500  TQUE=TQUE+TNXT-TARR(IP):          'SUM UP WAITING TIME
510  IQ=IQ+1:                          'INCREASE QUEUE NUMBER
520  TNXT=TNXT+TPRO(IP):               'SET MACHINE'S NEXT FREE TIME
530  IF CLOCK<TNXT THEN CLOCK=TNXT     'ADVANCE SIMULATION CLOCK
540  '
550  IF IA>IP GOTO 410:                'QUEUE EMPTY ?
560  GOTO 360
570  '
580  '
590  UTIL=(CLOCK-TIDL)/CLOCK           'CALCULATE M/C UTILISATION
600  IF IQ>0 GOTO 620
610  AWIT=0:GOTO 660
620  AWIT=TQUE/IQ:                     'CALCULATE AVG. WAITING TIME
630  PRINT "SYSTEM:     NO. OF PARTS                SIMULATION TIME"
640  PRING TAB(15)PARTS;TAB(39):PRINT USING "######.##";CLOCK
650  PRING:PRINT "MACHINE:    UTILISATION         TOTAL IDLE TIME"
660  PRINT TAB(16):PRINT USING ".##";UTIL;
670  PRINT TAB(39):PRINT USING "######.##";TIDL
```

```
680  PRINT:PRINT "PART:      TOTAL NO. IN QUE.    AVEG. WAITING TIME"
690  PRINT TB(15)IQ;TAB(39):PRINT USING "######.##";AWIT
700  END
710  '
1000 '
1010 '________________________________________________________________
1015 '
1020 ' SUBROUTINE TO GENERATE EXP. DISTRIBUTION TIME
1025 '
1030 '________________________________________________________________
1035 '
1040 RAN=RND(IP)
1050 EXPONT=-MPRO*LOG(RAN)
1060 RETURN
1070 '
2000 '
2010 '________________________________________________________________
2020 '
2030 ' SUBROUTINE TO GENERATE UNIFORMAL DISTRIBUTION TIME
2040 '
2050 '________________________________________________________________
2060 '
2070 RAN=RND(IA)
2080 UNIFM=MARR-DARR+2*DARR*RAN
2090 RETURN
```

6.4.2 Design of simulation model: choice of model type

In choosing a modelling approach, one should consider the characteristics of the system in question and the nature of the problems to be tackled. In this sub-section, therefore, the various simulation approaches, together with their relevant applications, will be outlined. In general, the simulation models used in the area of MSE can be classified according to the following scheme:

Static/dynamic
Continuous/discrete
Stochastic/deterministic
Aggregated/disaggregated.

Static and dynamic simulation models

In accordance with the types of system outlined in Chapter 2, a static simulation model is a representation of a system at a particular point in time. Models of this kind are defined as having structure without activity, that is to say, the simulation process will not evolve over time. Typical examples of this kind involve those Monte Carlo simulation models and many spreadsheet-based analytical models.

In contrast to static systems, dynamic simulation models combine structural components with activity, that is to say, a dynamic simulation model is a representation of a system as it evolves over time. A typical example of this type is given by the simulation model of a production operation in which workpieces move into and out of the system as time passes. A dynamic simulation model of this kind would be utilized to find the appropriate values of the decision variables – for example, when and how much to produce such that the total operations cost is minimized.

Continuous and discrete simulation models
Continuous simulation utilizes a model in which the state variables change continuously over time. Typically, continuous modelling is based on abstraction, decomposition and aggregation, and involves one or more differential and/or algebraic equations which describe the relationships existing inside the system. The basic characteristics of continuous analysis (both analytical and by simulation) are presented in Chapter 2.

In contrast to continuous simulation models, which are usually based on mathematical equations, discrete-event simulation is concerned with the modelling of a system by a representation in which the state variables change at sudden distinct events. As pointed out in Chapter 2, there is no general standard way of describing discrete-event modelling. Nevertheless, many applications of discrete-event simulation involve queuing structures of one kind or another and therefore the following terminology is often employed:

The structural elements:

1. *Entities* (or *customers*). These are the basic temporary elements in the simulation model of a system, which can be individually identified and processed. Jobs waiting in a shop to be processed are typical entities in a manufacturing system.
2. *Attributes*. Each entity in the system may possess one or more attributes. Attibutes are relevant information concerning an entity. Job number, customer information, type of material, the colour and number of doors of a car body are some of the typical attributes encountered in a manufacturing environment. These attributes are useful in many ways during a simulation process. For instance, they can be used to identify each of the entities in the system or to keep track of the current status of the entities.
3. *Resources* (or *servers*). These are the basic permanent elements of the model structure. They are responsible for providing services for the temporary elements, the entities. These typically include the machines and other work-stations in a manufacturing system.
4. *Queues*. A queue is formed in the system whenever there are entities waiting for the service of a resource but the resource is not free for whatever reason. In general, resources and queues can be interconnected to form the network of a simulation model with entities passing through it.

The dynamic elements:

5. *Events*. These are the points in time at which changes of system state occur, as when a resource engages an entity and becomes busy or when it finishes its service for the entity and becomes free again.
6. *Activities*. The operations or procedures which are initiated at each event are called activities. Activities are responsible for the transformation of the states of system entities. For example, the activity of machining would transform a 'raw material' into a 'part'. Certain conditions may need to be satisfied before an activity can be initiated and, once initiated, an activity usually takes some time to complete.
7. *Processes*. A process is simply a group of sequential events. For instance, a raw material arrives at a work-station, is machined or processed, and then goes to the next work-station. In general, a process describes an entity's complete experience, or part of that experience, as it flows through

a system. Sometimes it is useful to use such processes to facilitate a simulation process.

8. *Simulation clock*. This is a simulation model variable used to keep track of the current value of simulated time as the simulation proceeds. According to the value of this variable, appropriate activities are initiated in the simulation process.

The general form of a discrete-event model may be outlined according to the above definitions. It can be assumed in general that a system will consist of I resources S_i ($i = 1, 2, \ldots, I$), each responsible for a queue Q_i. These resources and queues can be interconnected in an arbitrary way so that a particular system under study can be presented. During the dynamic simulation process, the temporary entities E_j ($j = 1, 2, \ldots, N$) arrive at resources; after one or more activities have taken place, the entities will normally proceed to a new queue Q_k, and hence to resource S_k. According to the state of Q_k (empty or non-empty), the status of S_k, and possibly the priority attribute of E_j, a waiting delay w_{jk} will in general be experienced by E_j at this stage (w_{jk} may take the value zero). When it is E_j's turn to be served by S_k, a processing delay P_{jk}, which is one of the primary system parameters, will again be experienced by E_j. When E_j leaves S_k, and if S_k still remains functional, the next entity may then proceed to S_k and the process continues. If there is no next entity, resource S_k stays idle until the next entity arrives. Additional delays may occur when any resource breaks down. This dynamic process will continue until a certain predetermined termination condition is met – for example, when the simulation clock has reached the required point of simulated time; when the whole set of N entities has passed through the system; or when some other criteria have been satisfied. Quite a variety of systems can be regarded as having a queuing structure and thus lend themselves well to discrete-event simulation.

The operation of a machining station is a good example. When a workpiece arrives for machining, it will wait in a buffer store if the required machine is not free. Once the machine is ready to take a job and there is a job waiting in the buffer, the machining process can start. Therefore, the state of the system changes discretely from one state to another with time. From a system analyst's point of view, the only concern with the machine is whether the machine is busy, idle, or broken down, rather than how the machine actually processes the job. Viewed from the perspective of discrete state change, there are a number of obvious system events in this machining station example, among them the arrival of materials; the loading of material to a free machine; the unloading of the finished workpiece from the machine; the machine becoming ready for the next entity; and machine breakdowns. The time taken for the machine to process a workpiece can either be sampled from some appropriate distribution (random simulation), or set to a known constant (deterministic simulation). This is also true for other activities in the system. Therefore, the next change in state of the system can be simulated by referring to these known activity times. In a discrete-event simulation the interest is in the state change of the system and the interaction of entities and resources due to such changes as the time moves on.

It can be seen, therefore, that the general problem of discrete-event simulation modelling is how to embody all of the interactions of the above-mentioned structural and dynamic elements. Almost all simulation languages use one of a few basic approaches to accomplish this process. There are at least four widely employed approaches to discrete-event simulation modelling, with each embodying a distinctive view.

In order to describe the functions of and relationships between the activities of resources and entities within a discrete system, we need a technique which is capable of embodying all of the elements involved. The *activity cycle diagram* technique is a means which serves this purpose well. It is a useful technique (particularly in simulation modelling) for describing the classes of resources, outlining the activities in which they engage, and linking these activities together.

Box 6.3 Describing a discrete system: Activity Cycle Diagrams

Activity cycle diagrams make use of only a few symbols: a circle to represent the 'idle state' of a resource, and a rectangle to indicate the 'active state'. In addition, a queue symbol is used to represent a buffer in which entities wait for something to happen. An active state involves cooperation between different resources and entities, whereas an idle state involves no cooperation and is generally a state in which the entities wait for something to happen. In a discrete system such as that of a manufacturing operation, each type of resource and entity may be considered to have a life cycle which consists of a series of states, and the entities move from one state to another as their life proceeds. Activity cycle diagrams are one way of modelling the interactions of these states. To illustrate the use of this technique, Fig. 6.2 gives an example of an activity cycle diagram which describes the activities of a simple machining

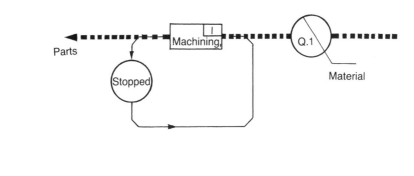

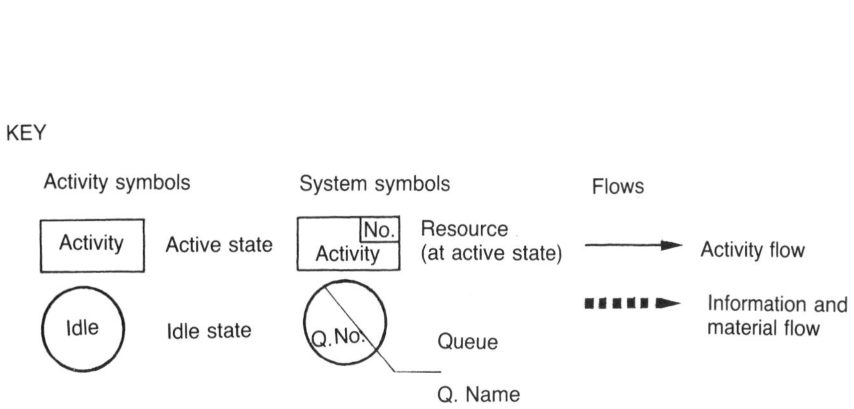

Fig. 6.2 A simple activity cycle diagram.

operation. This diagram shows that this simple discrete system involves only two types of item – a machine (resource) and a group of workpieces (entities), linked together through their interaction during the 'machining' activity that establishes a certain transformation process. During this machining operation, a workpiece enters the system and waits in a queue to be processed by the machine. When the workpiece is at the head of queue and the machine is ready to take on other job, then it is taken from the queue and loaded on to the machine so that the machining operation can take place. Therefore, the activity of machining can commence only if these two conditions are both satisfied, that is, if the machine itself is ready and if the queue is not empty. This is a common condition for any resource to become active in a discrete system. That is, an activity will always have some kind of a 'test head' attached to it, and can only take place if the conditions associated with this test head are satisfied.

When a machining operation is completed, the material will have been processed and transformed into a part. It will then be released from the system and perhaps sent for another operation. This is another of the common features of a discrete system: having been engaged in certain activity (usually a transformation process), the status of the entity will be

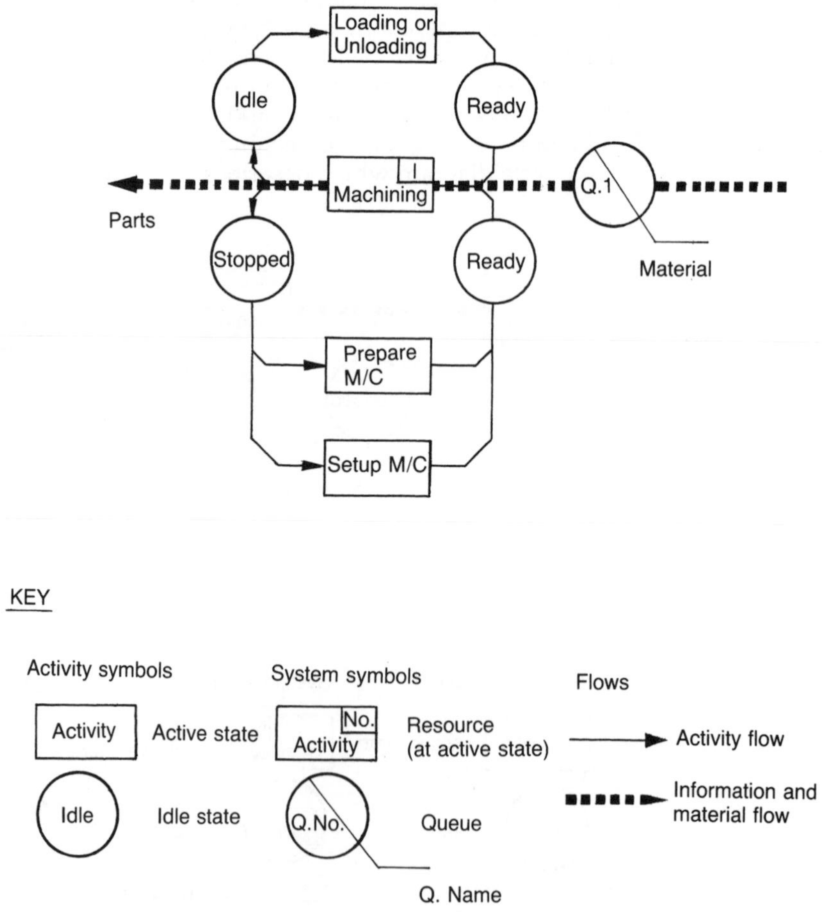

Fig. 6.3 Activity cycle diagram of a machining operation.

changed (in this case from a raw material to a part). During a simulation process, this is accomplished by updating the appropriate attribute(s) of the entities concerned after an activity is performed.

The activity cycle of the machine in Fig. 6.3 is a little more complicated. In this particular case the machine could engage itself at any time in one of the five possible active states – loading, unloading, machining, repairing machine, and setting up machine. It is obvious that each of the activities will have a time parameter associated with it because in practice an activity will normally take a certain time to complete. The way in which the value of these time parameters is determined during a simulation process decides the nature of the model – that is, it will be deterministic if it always takes the same value, or stochastic if takes a random value by sampling from a certain distribution each time an activity takes place. Once a resource or an entity is engaged in an activity, it becomes 'busy' and unavailable for any other tasks until it is released from that activity, and hence becomes 'idle' again.

Some activities may require more than one resource. For instance, if the systems boundary of the study is now drawn in such a way that a machine setter is also considered as a system resource, then in order to initiate the activity of 'setting up machine' both the machine and the setter must first be available.

An activity cycle diagram, therefore, integrates the entire system through two types of link. That is, it links resources and entities through the flow paths of the entities; and links the activity cycles of different resources through their common activities.

At least one of the simulation packages, HOCUS (developed by P.E. Inbucon), uses activity cycle diagrams as its modelling method and converts such diagrams directly into simulation models.

Stochastic and deterministic models

In accordance with the distinction between stochastic and deterministic systems, if a simulation model contains no random variable as its system parameters, it is said to be deterministic. For instance, the machining operation system referred to in Box 6.3 would be deterministic if the workpiece arrival interval and the machining time of a workpiece were entirely predictable. Given a set of inputs, a deterministic simulation model would produce a unique set of system responses. On the other hand, if one or more of the system parameters cannot be entirely predicted, but can only be sampled from a certain probability distribution, then the simulation model is considered stochastic. For a given set of inputs, a stochastic simulation model will yield only estimates of the true system response, because the outputs themselves are now random variables. In many cases of stochastic simulation, a number of repeated simulation runs are needed to estimate the true system response, in order to reduce the estimated system variance. The techniques of probability and statistics that are used in association with stochastic modelling and simulation will be discussed in section 6.5.

Aggregate and disaggregate models

The level of aggregation of information is one of the most important system variables to be decided in the design of a simulation model. This is because the proper aggregation of information is crucial to the cost of developing and maintaining a simulation model, and, more importantly, to the usefulness of the model (this is true for both simulation and analytical models). This critical question of aggregation level can be resolved by analysing the product structure and the operational structure of the manufacturing system under study with respect to the purposes of the simulation model to be built. Although not clearly defined in operations management studies, there is usually a link between the level of aggregation of information and the level of the hierarchy of production planning and control. That is, the level of aggregation of the simulation model often coincides with the production planning horizon as one goes down the hierarchy.

Aggregate models are usually employed to study the policies of long-term planning at the corporate level (Forrester, 1975). Studies utilizing this kind of simulation model are concerned with the setting of corporate objectives, the making of strategic decisions and the developing of plans to achieve these objectives. The major aspects here are, for example, the market position of the corporation, the development of new products, the buying or selling of plant, and the adjustment of workforce levels. Therefore, products, the workforce, equipment and other resources in the system may all be represented by aggregated entities in the model. Detailed production processes and production outputs are not tackled. Consequently, this kind of model cannot generally provide a useful tool of analysis at the production management level of a manufacturing organization.

At the other extreme, there are the very detailed models for short-term planning problems down to the workshop level. The entities of such models may conform to a detailed list of raw materials, parts and sub-assemblies required for production. Machines and operators will also be treated as individual items. The simulation process employed in an MRP system is an example of this category, in which the simulator takes full account of shop orders which specify item, quantity and requirement date together with detailed information about the capacity available. The movement of work-in-progress is simulated over the next few days or weeks in order to plan the completion dates of individual order as well as to determine capacity requirements. In contrast to online applications, this type of model may also be used offline for research purposes – for example, as an experimental tool for analysing production processes and evaluating alternative shop loading algorithms.

Some models fall in between these two extremes. The manufacturing simulation models of Hax and Meal (1975), Browne and Davies (1984) and Adshead and Price (1986) are examples of this type. There are various combinations in the construction of this type of model. However, two basic approaches are more often used to design such a model. The first approach uses a similar data structure to that used in an aggregate model, but with the information disaggregated to a lower level. The hierarchical model of Hax and Meal (1975) provides a good example of this approach. On the other hand, the data structure used by the second approach is closer to a detailed model, in which individual items are used as entities in the system. However, the difference is that instead of the full list of shop information, only the key items are included in the model. This approach has been employed, for instance, in the make-for-stock shop model of Adshead and Price (1986), in which only approximately 20% of the total production items are presented because Pareto analysis indicated that this 20% of items accounted for 80% of the total turnover of the particular production system. The argument for the use of this approach is that typically, in practice, production management will regularly carry out a Pareto analysis in order to

Table 6.2 Summary of model types

Uses of model	*Example of applications*	*Time horizon*	*Type of model*	*Data needed*
strategic research at corporate level	market study and forecast, total capacity planning, financial evaluation, conceptual design and specification of manufacturing system	long term	both static and dynamic; mainly continuous but can be discrete; both stochastic and deterministic; always aggregated	highly aggregate at corporate level
production management decision evaluation at company level	inventory and capacity planning; master production scheduling; evaluation of other production management decisions; conceptual design and specification of manufacturing facilities	medium to long term	often dynamic; continuous or discrete; deterministic or stochastic at medium aggregation level	aggregate data at company level but more detailed than above; or reduced set of data at shop level
production scheduling at workshop level	planning of detailed work loading; study of scheduling and job issuing policies; detailed planning and specification of manufacturing systems	short to medium term	dynamic, discrete, deterministic models for online planning; deterministic or stochastic for offline research	highly detailed data at shop level

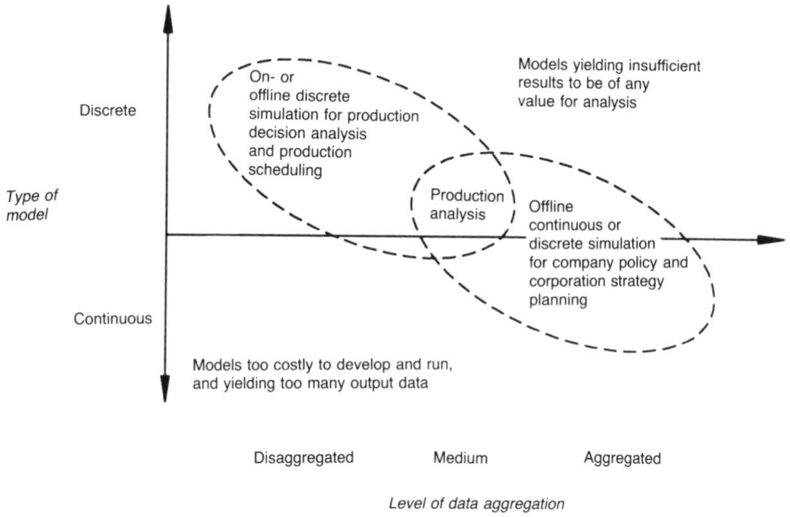

Fig. 6.4 Application of different simulation models.

identify the items which are most vital to the success of the company, so as to concentrate its managerial efforts upon them. Since these identified items to a large extent determine the overall performance of the company, it is sensible for the analyst to concentrate his modelling effort upon the same set of items. This second approach may be quite adequately termed the *reduced set of data* approach. As can be expected, these models are usually constructed to aid the production decision-making processes at the production management level in the medium term – for instance, to assess the overall effects of various production management policies and strategies on the operation and performance of the manufacturing company, such as decisions on stock control, leadtime assignment, batch size, production scheduling, load on shop and capacity employment.

The various approaches used to model manufacturing systems and their applications as discussed above may now be summarized in Table 6.2. The situation is also presented graphically in Fig. 6.4.

6.4.3 Simulation modelling

Choice of programming language

To develop a computer simulation model, there are two choices of programming approach available to the analyst: the model can be written either in one of the special-purpose simulation languages or in one of the general-purpose programming languages.

Simulation languages

From the vast number of simulation models developed since the early 1960s, some general principles of computer simulation have emerged. Certain general mechanisms are required in most simulation approaches, such as execution-control procedures (for all types of simulation), queue-handling procedures (for structured queue models) and random process procedures (for stochastic simulation). Consequently, special-purpose simulation languages have been developed which generalize these simulation procedures and make the programming of simulation models a relatively easy task, because, first, no special knowledge of computer programming is required to write a simulation model, so that this type of computer-aided system analysis can be carried out by analysts who are not computer-programming experts; and second, these specially written languages usually provide some debugging facilities to help modellers reduce the complexity of setting up a simulation model and thus shorten the simulation analysis time. The following are some of the common features which are provided by these simulation languages: event and simulation time handling; model logic for entities and states; probability distribution generation; model-initialization facilities; interactive simulation runs; graphic animation; results-analysis facilities; report-generation facilities; and other control facilities and error messages.

Box 6.4 gives two examples of simulation models for the same machining situation as discussed previously in section 6.4.1, but this time written in two well-known simulation languages – GPSS and SIMAN. As can be seen, the programming task involved in modelling this system is now significantly reduced.

Box 6.4 A simulation model of the machining operation written in GPSS

```
***********

GENERATE  300,100        ;create next part
QUEUE     BUFFER         ;part waits in queue
SEIZE     MACHINE        ;attempt to grab the m/c
DEPART    BUFFER         ;engage m/c and end of
                           waiting
```

```
ADVANCE     400,200                  ;the operation takes certain
                                       time
RELEASE     MACHINE                  ;operation done, free m/c
SAVEVALUE AVE_Q,QT@MACHINE           ;collect queue data
TERMINATE 1                          ;part leaves system
```

****

A simulation model of the machining operation written in SIMAN

****

Model file:

```
BEGIN;
  CREATE:RN(1,1):MARK (2);   create next part
  QUEUE,1;                   wait for processing
  SEIZE:MACHINE;             attempt to grab m/c
  DELAY:UN(3,1);             engage m/c for certain time
  RELEASE:MACHINE;           operation done, free m/c
  TALLY:1,INT(2);            collect resulting time
END;
```

Experiment frame:

```
BEGIN;
  PROJECT,MACHINE,MCH,28/2/1990;                title
  DISCRETE,80,3,1;                              system
                                                  organisation
  RESOURCES:1,MACHINE,1;                        define capacity of
                                                  machine
  TALLIES:1,PART STAY;                          define time to be
                                                  collected
  PARAMETERS:1,300,100:2,0:3,400,200;           distribution
                                                  parameters
  DSTAT:1,NQ(1),Q_LENGTH:2,NR(1),M_UTIL;        output data
  REPLICATE,1,0,20000;                          simulation run
                                                  time
END;
```

****

Apart from the simulation languages already mentioned, some of the commercially available packages specialize in modelling manufacturing systems, such as MAP/1, Speed, MAST and Witness. Also, since the FMS and CIM concepts have been widely accepted as the trend for the factory of the future, the need to simulate such systems has grown dramatically and therefore led to the development of more dedicated simulation software with special features suited for this kind of simulation – the provision of special functions for modelling the system's material-handling components and of graphic animation capabilities, for instance. Examples in this category include See-Why, Modelmaster and Simfactory. (Most of the simulation languages are themselves written in a general-purpose computer language. For example, GPSS, Simscript, Slam, ECSL, and the more recent SIMAN are all FORTRAN-based simulation languages.) To summarize, therefore, we may classify this wide range of simulation languages into two basic categories: *general application systems* and *special application systems*.

General application systems are those which can be used for simulating processes in various areas. They have basic simulation functions and statistic facilities. The advantage of this class is that they are suitable for most modelling problems. However, expert modellers will be required when a practical model is to be constructed. Examples of this class are: GPSS (General Purpose Simulation System), developed by IBM; HOCUS (Hand Or Computer Universal Simulator), developed P-E Incubucon; SIMAN/Cinema, developed by Systems Modelling Co.; and PCModel (PC Modelling system), developed by SimSoft Ltd.

Special application systems have been developed for certain classes of problems. They reduce the modelling and programming time and effort. Some of the advanced packages even eliminate the necessity of programming altogether. This is usually achieved at the expense of their applicability – the resulting model may not be very flexible and will always be strictly limited to the type of situation for which it was designed. Examples of this type of simulation system include: MAST (MAnufacturing System design Tool) developed by CMS Research Inc., which is particularly suitable for the simulation of computerized manufacturing systems; SIMFACTORY, which is used specifically for simulating job-shop systems; and WITNESS II, which may be used for both job-shop and computerized manufacturing systems simulation. These special-purpose simulation languages are all commercially available. Information about currently available computer simulation systems, including applications summaries, hardware and systems requirements, as well as current price, can be found in the appropriate journals (such as *Simulation*). Sufficient information is also available in these sources for the detailed evaluation of the financial commitment needed for both system and training requirements.

One of the most attractive features of the modern simulation language is perhaps the facilities for graphic animation that most of them provide. Graphic animation is used extensively in MSE to animate simulation processes. Systems layout, elements and movement of entities are represented either by icons or characters in an animation simulation model. In a character-based system, an entity will be represented by a character – for example a part by a moving letter *P* – whereas an icon-based animation system uses a programming technique known as the *bit-mapped* approach in which an entity is represented by a lifelike icon – a part can be shown by a figure resembling the real subject. Most of the animation systems restrict the movement of entities in a two-dimensional plane, although a few systems allow entities to move in three-dimensional space.

Programming languages

Although numerous simulation languages have been developed which are readily available in the market, using a general-purpose programming language (such as FORTRAN, BASIC or C) to write a simulation model still has advantages, the main one being the flexibility that such a language provides. The decision to choose such an approach may in practice be due to any of the following reasons:

First, it is generally true in the computing world that the higher the level of a programming language, the more restricted its applications. This is certainly the case for special-purpose simulation languages. Most of the special-purpose simulation languages that are commercially available are suited to only certain types of system. The format of output reports produced by these packages is rigidly dictated by the language used. For example, hardly any of the simulation languages allow time-dependent graphical output. The output reports from these packages are usually statistical summaries of the variable of a system, such as means and standard deviations. Because of the nature of this type of report – a summary of statistical data gathered over a considerable time-span – a good deal of information can be lost in presentation, in particular how the quantities involved vary with time. In some cases, apart from statistical reports, it is also important for the model to be able to produce detailed reports on those time-dependent system variables (this is especially important if the model is to be used for online production planning). Such customized reports may only be accomplished by using a general-purpose programming language. Furthermore, it may also be felt that some of the special features of the system under investigation could be better modelled by using a general-purpose programming language because such a language would provide much more flexibility than a special-purpose package.

Secondly, if the systems analyst has had previous experiences both of computer programming and of modelling with general programming languages, then this will have eliminated one of the most frequently quoted reasons for using a special-purpose simulation language, as outlined above. Also, in practice a company may already have a significant investment and expertise in a general-purpose language – it may be one of the major programming languages used by all the system engineers in various departments of the company. Therefore, a simulation model written in this language would be easily understood and maintained by the user. From an academic point of view, this would also increase the portability of the model if further research were required on the model in the future.

Third, it could be more expensive for a company to use a special purpose simulation package. Since in general a manufacturing company is not a dedicated organization in simulation studies, it may not be considered necessary for the company to invest money and time to exploit the full benefit of such packages.

Even if the model is to be developed with one of the special-purpose languages, knowledge of the structure and logic behind the programming approach will still help the model builder in the modelling process of simulation.

The modelling logic of discrete simulation

Since discrete simulation is perhaps the most important type of computer simulation used in the design and analysis of modern manufacturing systems, the rest of this chapter will be devoted to this particular type of modelling.

As many manufacturing processes are not continuous in nature, discrete-event simulation is often used to look ahead to each occurrence of a change of state in the system and then evaluate the situation. Discrete-event computer simulation skips from one event at one time to the simulated time for the next event. In this way there is no waste of computer capacity to scan situations that have not changed. There are at least four programming approaches in which a discrete-event simulation model can be written. These are: the *event* approach, the *activity* approach, the *process interaction* approach and the *three-phase* approach. A brief description of these different approaches is presented below. In what follows, the simple machining operation quoted in Box 6.3 will be used as an example to illustrate the

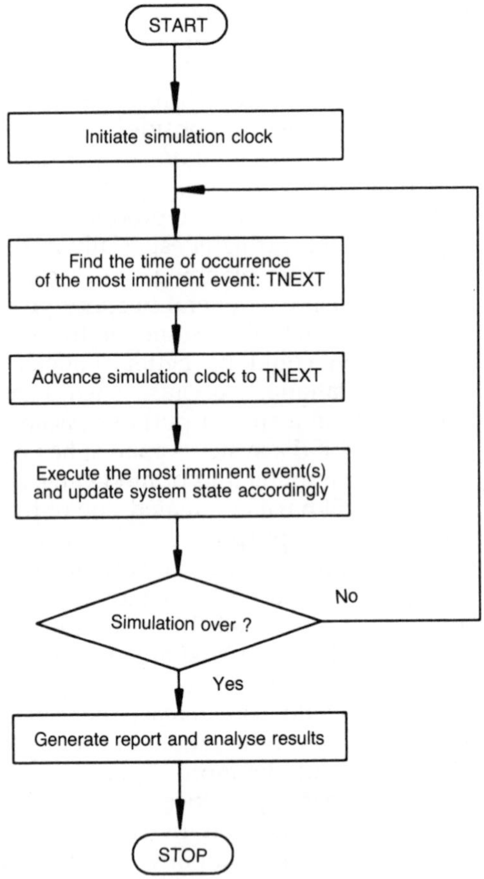

Fig. 6.5 Executive control of event-based simulation model.

techniques used by the different programming approaches for discrete simulation. The terminology used here is as defined before.

The event approach

With the event approach, the simulation clock of a model is first initialized at the start of a simulation run, and the points in time at which future events will occur are determined. The simulation clock is then advanced to the point where the most imminent future event occurs, and the state of the system, together with the knowledge of the times of occurrence of future events, updated accordingly. This cycle is repeated until the end of the simulation, as shown in Fig. 6.5.

A discrete simulation model constructed with this approach will consist of a set of event routines, each of which defines the actions to be initiated when an event occurs to change the system state. A system will be modelled first by identifying its characteristic events and then the event routines can be written accordingly. In the simple machining operation, for instance, the basic state changes may be described by two event routines – the arrival of a workpiece, and the end of machining. The flow charts of these routines are presented in Fig. 6.6.

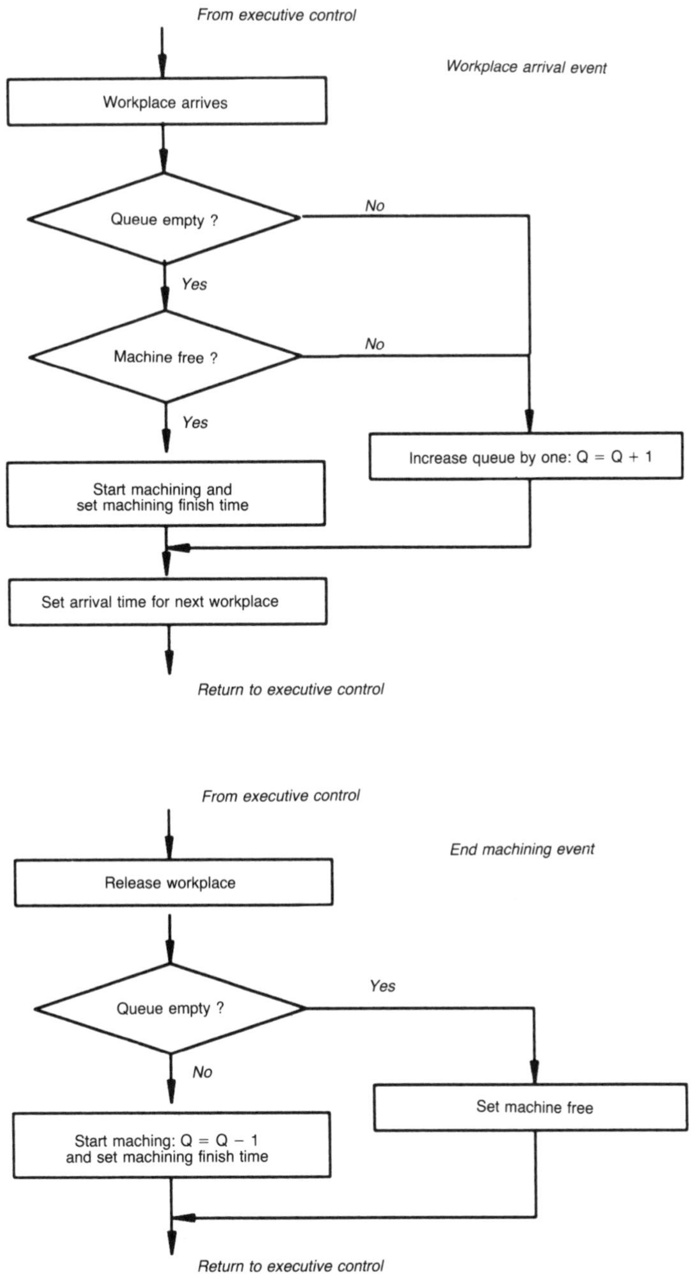

Fig. 6.6 Flow charts of event routines for simple machining operation.

The activity approach

Like the event approach, a simulation model based on the activity approach moves from point to point where changes of system state occur. Unlike the event approach, however, the basic system routines here are based on activities rather than events. An activity routine describes only the action

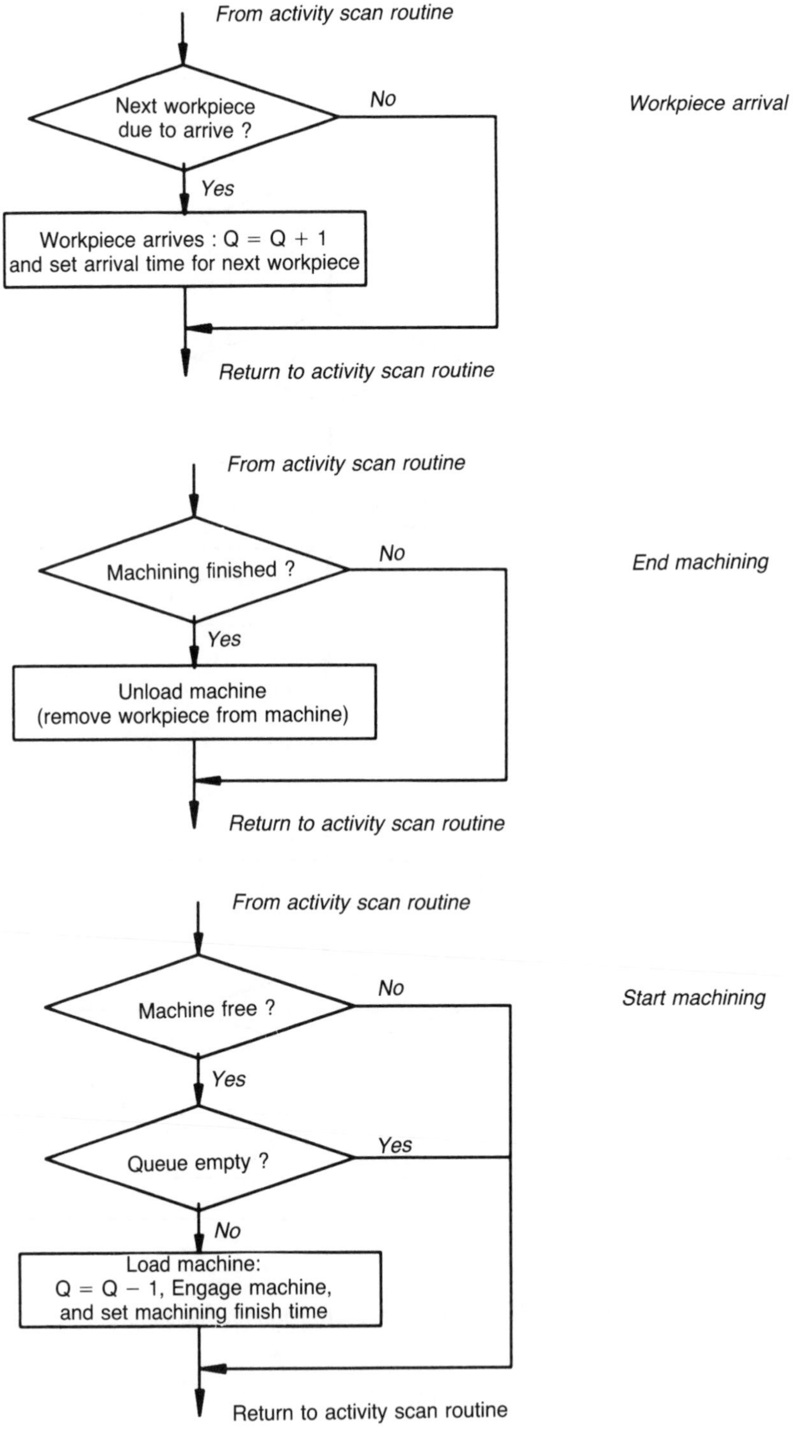

Fig. 6.7 Flow charts of activity routines for simple machining operation.

that will be immediately triggered by a state change. With this approach, the activities required to carry out the machining operation will include the arrival of a workpiece to the machine, loading and starting the machining, and finishing machining and unloading the machine. Compared with the event approach, each action will now be modelled independently. When the simulation clock is advanced to the next point where the system is to change its state, the control routine of the simulation flow scans the whole set of activity routines and initiates in turn all the actions that should be taken. Obviously, an action here can only be carried out if certain conditions are satisfied, as illustrated by the flow charts for the three activity routines shown in Fig. 6.7.

The process approach

With the process approach, a model is constructed with a set of process routines. A process routine outlines the history of an entity as it moves through the system with the explicit passage of simulated time. It can be said that while the event and activity approaches model a dynamic system from the viewpoint of the system structure based on resources and queues, the process approach models a system from the viewpoint of the entities as they move through this structure.

To illustrate how to model the sequence of operations through which an entity passes during its life within a system, Fig. 6.8 shows a flow chart for a workpiece process routine in the example of the machining operation. A flow chart as shown in Fig. 6.8 can be considered to exist for each of the entities in the system. At any given time, there may be many workpiece processes interacting with each other while competing for a system resource – in this case, the machine. During its life within the system, a workpiece will experience two possible delays: *conditional* delay (waiting in the queue for the machine to be available for it) and *unconditional* delay (time spent on the machine being processed). These delays make it necessary for a process routine to have possibly more than one entry point. For example, the workpiece process routine shown in Fig. 6.8 has three entry points corresponding to the events of workpiece arrival, workpiece removal from queue and workpiece removal from machine. When a new workpiece arrives, it is guided through the system following the flow chart of the process routine, until a delay, conditional or unconditional, is encountered (in this example when the workpiece reaches blocks 3 or 5). At these points control is returned to the flow-control routine. An event list is used to store the status of all delayed entities. The flow-control routine determines from this list which workpiece and what process's event is the most imminent, and the appropriate workpiece process will then be activated at the corresponding entry point. The simulation evolves over time by repeating the above process until eventually some prespecified stopping condition is satisfied.

The three-phase approach

Each of the above three approaches has advantages and disadvantages with respect to ease of programming, model debugging and the efficient running of the model developed; these are listed in Table 6.3. This table shows that from a programming point of view the activity-based approach is superior to the other two approaches. Because the resultant model structure is based on small, independent building units, this approach makes model modification and enhancement very easy. However, the most undesirable feature of this approach is its inefficient use of computer time when running the model. A

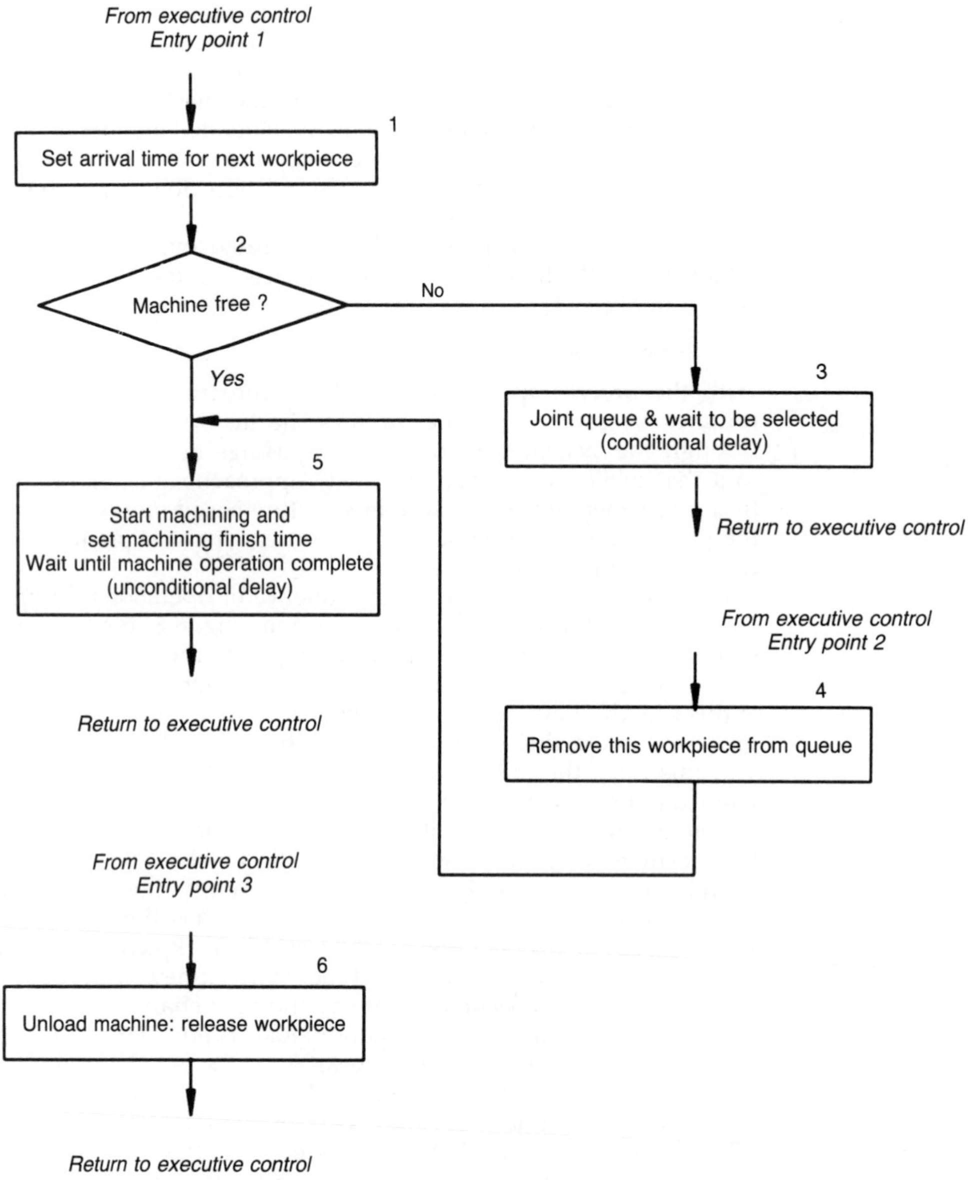

Fig. 6.8 Flow charts of workpiece process routine.

significant amount of computer time is wasted in unnecessary scanning of the whole set of activity routines each time the simulation clock is advanced. On the other hand, although the event and process approaches make much more efficient use of computer time, the difficulties involved when modifying and enhancing a developed model make them less attractive for large and complex systems.

A fourth approach, the three-phase approach, is available which retains the simple model structure of the activity approach, yet allows the efficient execution of the event-based approach. The most important feature of this

Table 6.3 Comparison of programming approaches

Approach	*Advantages*	*Disadvantages*
Activity	Uses small independent activity blocks, so it is easy to write, debug, update and modify the model – a very useful feature for developing large and complex models.	Very inefficient in computer running time.
Event	Simulation runs faster than activity-based model, so is more efficient in terms of computer running time.	Bigger program segments, which are dependent on the activity sequence; models therefore not easily modifiable.
Process	More natural framework for modelling, fewer lines of code needed for programming a model; efficient in terms of computer running time.	Requires complex control routine; can be difficult to modify.

approach is the combination of the efficient execution of the event-based approach with the programming simplicity of the activity-based approach. As with the activity approach, a three-phase model consists of a set of small, discrete and mutually independent activity routines. However, there are two types of activities defined in this approach: B activities (bound or unconditional activities), which should be initiated by the control routine whenever their scheduled time is reached; and C activities (conditional activities), whose execution depends on either the cooperation of entities and resources, or the satisfaction of certain specific conditions within the simulation. For the machining operation example, therefore, there will be two B activities and one C activitity:

arrival of a new workpiece	(B activity): a workpiece will arrive according to the preset schedule regardless of the state of the system
finish machining and unload	(B activity): a workpiece will be taken off the machine as soon as the operation is completed regardless of the state of the system
load workpiece and start	(C activity): dependent on whether the queue is empty or if the machine is available.

Based on the classification of system activities according to these two types, a three-phase execution carries out the simulation process in three steps. Phase A, the *time scan*, finds the most imminent event, decides which B activity is due to occur at that time, and advances the simulation clock to that time. Phase B, the *B call*, initiates the B activity identified in Phase A, and updates the system state accordingly. Phase C, the *C scan*, scans the whole set of C activities, initiates all possible activities whose conditions are now satisfied, updates the system state accordingly and then goes back to Phase A to start the cycle all over again. A flow chart of the three-phase execution process is presented in Fig. 6.9. The advantages of the three-phase approach are obvious. First, like the activity approach, it will lead to simulation programs with well-structured independent subroutines of a man-

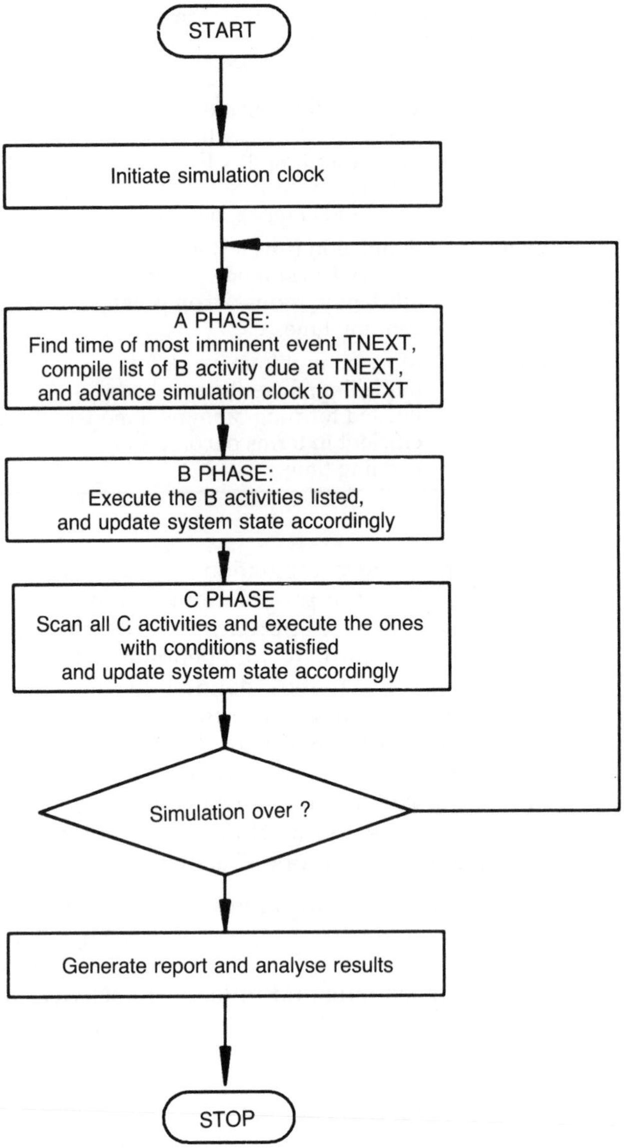

Fig. 6.9 Executive control of three-phase simulation approach.

ageable size. This not only allows the easy initial modelling of a system but also makes later modification much easier. Second, as with the event approach, the time-wasting process of scanning each activity in turn is avoided, making the simulation run more efficiently.

6.4.4 Simulation validity

Qualification, verification and validation

An extremely important issue that deserves particular attention here is data verification and model validation. Past experience of model building and

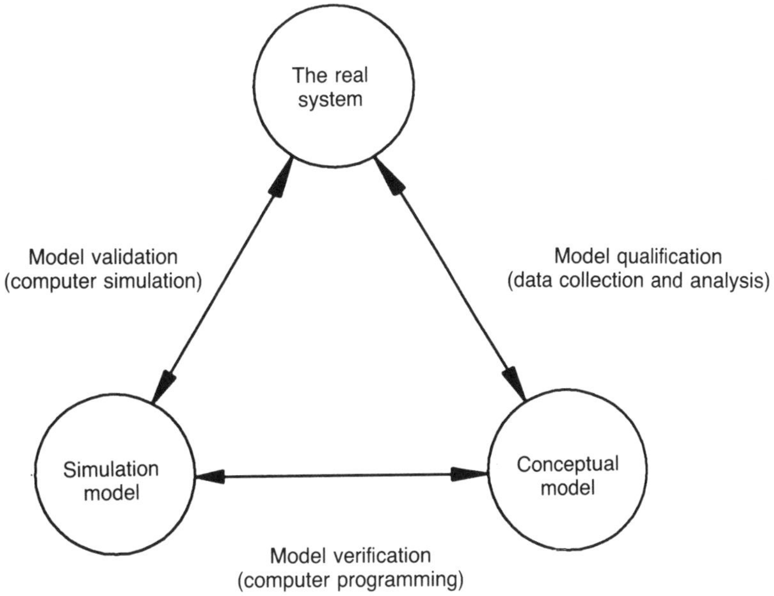

Fig. 6.10 Simulation model development process.

validation has shown that the model development process undertaken in a systems analysis project should follow a framework which may be graphically presented as in Fig. 6.10. This framework consists of three basic elements. Model *qualification* is the procedure of determining the structure, elements and data of the model which are required to provide an acceptable level of agreement with the actual system. Model *verification* involves ensuring the proper functioning of the model developed for the system. This is mainly concerned with the correctness of the model itself. For example, if the model concerned involves computer programming tasks, then it must be made certain that the routines are programmed correctly, and that they function exactly according to the logic as designed. Finally, model *validation* is concerned with determining whether the model developed is indeed an accurate representation of the real system of interest.

The qualification stage of the model involves the activities of data collection, data clarification and parameter calculation, and the analysis of system structure and functions. Because of the initial lack of adequate information on which to base the model, the first step in the majority of model-building tasks involves extensive data collection. The data-collecting process is usually very laborious, and considered by many as the largest and most costly task in the construction of a model. The necessary information, both descriptive and quantitative, may in practice be gathered through various channels – for example, through an extensive range of interviews and meetings with the key personnel who are involved in various areas of the system under study; or by accessing an existing formal information system, such as the management information system and/or engineering data bases of the company. Normally a large amount of data need to be collected through these channels, and must then be analysed to identify the important information. For example, they may have to be clarified, grouped, aggregated, or derived by calculation, to describe the specifications of the products, the configuration and capacity of the shops, and the relevant operational policies of the management of the manufacturing system. This gathering and analysis of data can be a continuous process throughout the modelling stage of the investigation, as shown in Fig. 6.10.

As the first stage of model construction, the qualification stage of the modelling process plays a vital role in guaranteeing the quality of the final model. Great care must be taken to make certain of the validity of the input data to the model, in order that the output data from the model can be valid in turn. There are data-analysis techniques available to help this process. For example, if the model we are dealing with is of the dynamic non-deterministic type, then the data available are often in the form of a time series based on previous observation. We may want to use a fitted distribution of some kind – the normal or Poisson distribution, for example – to describe these data in order that they may be reproduced statistically at a later stage during the investigation – perhaps for the purpose of experimentation. In this case we must examine whether the intended fitted distribution is in agreement with the observed data series before we can use such a distribution in the model with confidence. A number of statistical techniques known as *goodness-of-fit* tests can be used to carry out such an examination. These tests include informal visual assessment, the chi-squared test, Kolmogorov–Smirnov tests, and the Poisson process test (Law and Kelton, 1982). Other goodness-of-fit tests have been designed for a number of specific distributions. The basic principles of these techniques will be discussed later in this chapter. The theoretical details of these and other similar data-validation techniques can be found in standard textbooks on probability and statistics.

Model verification and validation are both concerned with the problem of trying to determine whether the model is accurate. Model verification is basically concerned with the proper construction of the model itself. In computer terms this can involve repeated processes of program writing and debugging. Numerous repeated test runs may have to be conducted to check that all the activity routines in the model function exactly according to the logic defined by their flow charts and that these elements of the model have been properly embodied within the overall modelling frame through the executive control routines, so that they affect the system states in the manner expected.

Although also concerned with the correctness of the model, validation is a process distinct from model building and debugging. It is concerned with whether the model's 'behaviour' is acceptable as compared with the real system. There has been considerable effort to provide validation techniques which are practical. The overall validation process may include many possible actions. When it comes to computer simulation models, certain actions should be taken to increase model validity: a model with high face validity should be developed; use should be made of existing research, experience, observation, or other available knowledge to supplement the model; simple empirical tests of means, variances and distributions should be conducted using available data; Turing-type tests should be run; complex statistical tests should be run on available data; queuing theory results should be applied; and special data collection should be carried out (Van Horn, 1971).

Although all these techniques may be applied to a model to enhance its validity, it has been recognized that the involvement of the decision-makers in the modelled system has the most significant impact in model validation. For example, Gass (1977) argued that in most cases the pressures of the real world (as well as other factors) do not permit the interrelated steps to be completed as described. Furthermore, a model will not be used purely on the strength of the modeller's confidence in it, but only to the extent that the decision-makers believe that it is useful and usable. It seems, therefore, that an involvement with the user throughout the development of the model is crucial to its successful implementation and validation. Therefore, during model development it is important to attempt to answer the following questions:

1. In the experts' opinion, is the initial impression of the model's stated hypotheses credible?
2. Do sensitivity analysis studies (varying input parameters) show output patterns which compare with their assumed counterparts in the real world?
3. Does the model 'act' like the real world?

Thus there are a number of ways of achieving the face validity of a model. The first is to let the user access the model when it is still in the early stages of development. The user can then gain increasing confidence in the model's validity as it develops in size and complexity. The second is through the use of more effective communication means such as graphics capabilities built into the model, so that the user can 'watch the model work' and hence judge its behaviour. The effect of these is that close collaboration can be maintained between the analyst and the client (typically the management at various corporate levels) so that their professional opinion can be consistently invited about the model's validation and then modifications of the model made accordingly.

The initial condition

Frequently during a simulation study it is important to eliminate the effects of the transient period when the simulation is started with no activity occurring and with all the queues empty. This is because the system outcomes during this period will not reflect the true steady-state behaviours of the system – which is in general the real condition encountered in the day-to-day operation of a system that we are interested in – thus introducing biased observations to the outcomes of the investigation. When analysing the output data, therefore, it is desirable to use only the outcomes of the system when the model has reached its steady state.

One of the important issues of this type of steady-state simulation study is to determine when during the simulation process the model has reached its steady state, and this has always been a difficult task. There are two commonly used approaches to reduce the effects of the initial bias on the simulation (Pidd, 1984). One is to utilize a *setting period*. With this approach, the simulation is run first for a setting period during which no output from the simulation is recorded. Output data collection only starts at the end of this setting period when the transient phase is over. The other is to utilize typical *initial conditions*. Here the simulation process begins with non-zero conditions so that the system might be in a steady state right at the start of the run.

If the run-in period were long enough then in many cases the setting-period approach would be preferable to the initial-conditions approach, because with the former approach it could be made sure that a steady state had indeed been reached before starting the output recording process, while with the latter approach a certain amount of bias is possible depending on the adequacy of the initial conditions chosen. However, the problem with using a long setting period is that a considerable amount of computer time may be wasted. With large-sized models, in particular, efficient use of computer time must be considered as one of the most important factors.

It might be possible to use a compromise approach which achieves a steady state for the experiment, while maintaining relatively high efficiency in terms of computer time. Trial simulation runs could first be carried out for long enough for the system to attain steady state; the system states would then be recorded and used as the initial system conditions for the subsequent proper simulations. In this way, the effect would be as if a relatively long run-in period proceeded each simulation run, but computer time would not be wasted on an actual run-in period for the simulation. It should be noted that with this approach some bias might still result because of the non-zero initial conditions. That is, using the initial condition obtained in this way would be like linking two time series together at the point where the actual simulation starts (the first one being the trial series generated to derive the initial conditions, and the second one the real series used for the actual simulation), with the result that at this point a certain degree of discontinuity might be introduced. If this were the case, then this discontinuity would inevitably cause a certain degree of transience in the dynamic process of the system, resulting in bias in the simulation outcomes. The analyst must determine to what level this undesirable effect could be tolerated.

6.5 Applications of probability and statistics in stochastic simulation

Most of the simulation work in manufacturing systems design and analysis involves stochastic simulation models. This is illustrated by the simple machining model of section 6.4.2. As a result, the use of probability and statistics is usually an integral part of a manufacturing systems simulation study. Generally speaking, during a simulation study the techniques of probability and statistics are needed on various occasions:

- During model development, to choose the correct probability distribution for those system parameters which are stochastic.
- Before experimentation, to design test schemes so that the purpose of the study will be properly served.
- During the simulation run, to generate random variables according to a given distribution so that the process concerned can be statistically represented.
- During the evaluation process, to analyse the output data in an appropriate manner so as to reduce estimation errors.

Therefore, it is necessary for us to review the basics of these techniques. The particular applications of these techniques in relation to simulation will also be pointed out where appropriate.

6.5.1 Random variables and probability distribution

This section reviews some of the concepts and techniques of probability which are relevant to simulation model construction.

Random variables, probability density functions and distribution functions

If we know that a machining operation takes on average 12 minutes, so that a production planner can schedule the production plan to allow five jobs per machine per hour to come into the shop, no job will have to wait for a machine. Right? Wrong. This will never be the case in real life because individual machining operations can take more time than the expected average, and jobs do not always arrive exactly on time. This is a typical example of a process in which the variables are random and can only be described in terms of probability and statistics.

The basic concept of probability is closely related to everyday life. For example, we use words such as 'likely' or 'possibly' to describe the probability that some specific event will occur in a particular situation. Probability and statistics are a collection of mathematical methods which numerically analyse the behaviour of such a situation and estimate the characteristics of an entire population by taking a finite set of samples from it. In the theory of probability, a random variable X is a variable which can take on a value x with probability

$$P\{X = x\} = p(x)$$

where $p(x)$ is known as the *probability density function* (or *probability distribution*) and has the following properties:

1. If $I = [a, b]$, with $a < b$, then the probability of X taking a value between a and b is given by:

$$P(X \in I) = \int_a^b p(x)\mathrm{d}x$$

2. Similarly, the probability of X taking on a value less than or equal to x is given by:

$$P(X \leqslant x) = \int_{-\infty}^{x} p(z)\mathrm{d}z$$

3. Therefore, the fact that X is *bound* to take a value in the range $[-\infty, +\infty]$ gives:

$$\int_{-\infty}^{+\infty} p(x)\mathrm{d}x = 1$$

The distribution function, also known as the *cumulative distribution function*, of X is defined as:

$$F(x) = P(X \leqslant x) = \int_{-\infty}^{x} p(z)\mathrm{d}z$$

It is easy to see that for I defined as above, the probability that X is contained in I is given by:

$$P(X \in I) = \int_a^b p(x)\mathrm{d}x = F(b) - F(a)$$

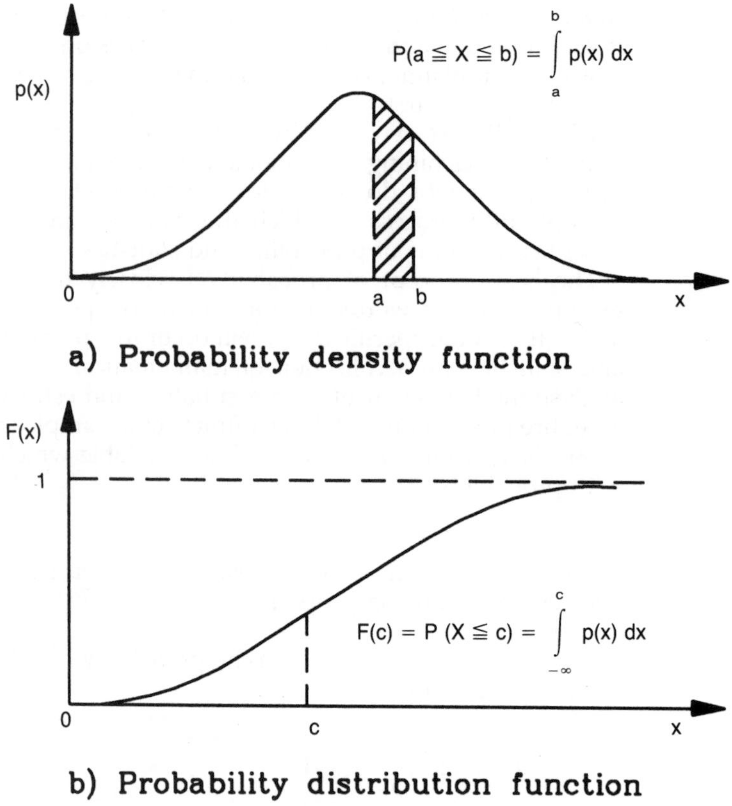

Fig. 6.11 Probability density (a) and probability distribution (b) functions.

Note also that

$$F(-\infty) = 0$$

and that

$$F(+\infty) = 1$$

Figure 6.11 illustrates these two functions.

Examples of useful probability distributions and their distribution functions
For a stochastic modelling process we need to specify the probability distributions of those system parameters which are random variables. Some of the frequently encountered probability distributions for simulation application are outlined below. The characteristics of these probability distributions are also shown graphically in Fig. 6.12.

Uniform distribution: U(a, b)

This is the simplest type of probability distribution in which there is an equal probability of the random variable X taking any value over the given range $[a,b]$, as shown in Fig. 6.12a. In simulation studies this type of distribution has a number of applications because of its simplicity. The probability density function and distribution function of $U(a, b)$ are given respectively as follows:

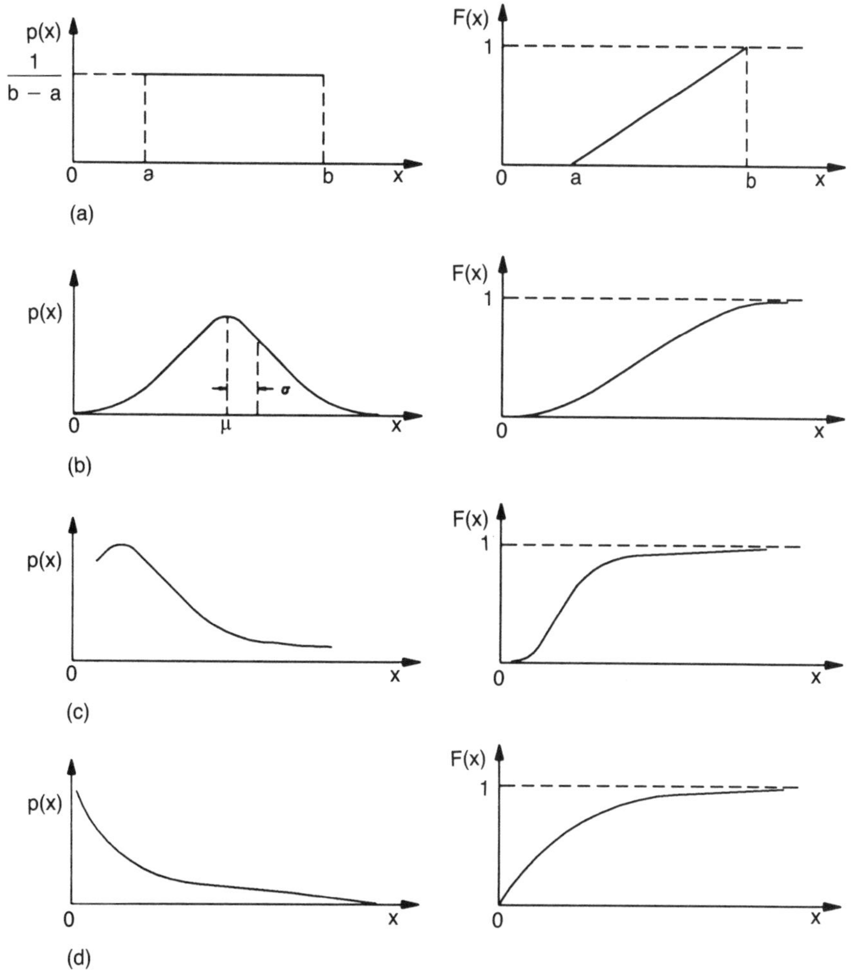

Fig. 6.12 Common probability distributions. (a) Uniform distribution, (b) Normal distribution, (c) Poisson distribution and (d) Exponential distribution.

$$p(x) = \begin{cases} \dfrac{1}{b-a} & \text{for } a \leqslant x \leqslant b \\ 0 & \text{otherwise} \end{cases}$$

$$F(x) = \begin{cases} 0 & \text{for } x < a \\ \dfrac{x-a}{b-a} & \text{for } a \leqslant x \leqslant b \\ 1 & \text{for } x > b \end{cases}$$

The special case of the uniform distribution where a = 0 and $b = 1$, $U(0,1)$, is of particular importance. This is the probability density function of what are known as *random numbers*, which are essential in generating random variables from all other types of distribution. It is easy to verify that $U(0,1)$ has the following probability density and distribution functions:

$$p(x) = \begin{cases} 1 & \text{for } 0 \leqslant x \leqslant 1 \\ 0 & \text{otherwise} \end{cases}$$

$$F(x) = \begin{cases} 0 & \text{for } x < 0 \\ x & \text{for } 0 \leqslant x \leqslant 1 \\ 1 & \text{for } x > 1 \end{cases}$$

Because of the importance of this particular distribution, a random number generator is actually included in BASIC as a standard library function which generates random values uniformly distributed over the range (0,1). The applications of this kind of random number generator in simulation will be shown later.

Normal distribution: $N(\mu, \sigma^2)$

This is one of the most commonly used distributions in modelling and simulation. Many quantities associated with physical activity are known to be normally distributed including, for example, the resultant diameters of a cutting operation and the measurement errors of experiments. This is due to a well-known property of random situations, described by the *central limit theorem* which states that the sample means obtained from any kind of distribution are approximately normally distributed provided that the sample size is sufficiently large.

The normally distributed random variable is characterized by a symmetric and bell-shaped probability density function with values spread around a *mean* denoted by μ, as shown in Fig. 6.12b. Its probability density function is given by:

$$p(x) = \frac{1}{\sigma(2\pi)^{1/2}} \exp\left[-\frac{1}{2}\left(\frac{x-\mu}{\sigma}\right)^2\right]$$

where σ is known as the *standard deviation*. There is no simple definition of the normal distribution function because the probability function of this distribution as given above cannot be analytically integrated.

One of the most important features of the normal distribution is the relationship between the area bounded by the probability curve and the parameter σ. This is used particularly in forming confidence intervals. The area bounded by the probability curve and the x-axis ordinates $\mu \pm j\sigma$, for different values of j, is given in Table 6.4.

Poisson distribution: $Psn(\lambda)$

The Poisson distribution is a discrete distribution (a random variable is discrete if it can only take on a finite number of values $x_1, x_2, \ldots$), whose

Table 6.4 Area under the normal probability curve bounded by x-axis ordinates $\mu \pm j\sigma$, for different values of j

j	*Area under curve*
1	0.6827
1.9600	0.9500
2	0.9545
2.5758	0.9900
3	0.9974

importance in modelling and simulation is demonstrated by the fact that it is one of the two fundamental distributions used to describe the stochastic process of a queuing situation – the other one being the negative exponential distribution (see below). In a manufacturing environment, variables which are typically represented by the Poisson distribution include those concerned with sequences of arrivals and departures, and with the size of random batches (for instance, the number of items involved in a customer order or the number of parts taken from the inventory store by a withdraw order). The probability density function of the Poisson distribution is given by:

$$p(x) = \frac{\lambda^x}{x!}\exp(-\lambda)$$

and its distribution function by:

$$F(x) = \sum_{i=0}^{x} \frac{\lambda^i}{i!}\exp(-\lambda)$$

where $x = 0, 1, 2, \ldots$ and λ is the mean (and also the *variance*, which is the square of the standard deviation) of the distribution.

Negative exponential distribution: Expn(α)

As pointed out above, the negative exponential distribution is the other kind of distribution which has a wide range of applications in simulation problems. Basically it describes the distribution of the inter-arrival times of a random process in which the arrivals occur at an average rate, α. A typical example of this kind of random variable is the inter-arrival time of the workpieces in a machining operation which on average processes a constant number of workpieces over a certain time (say, 10 000/month). The probability density function and distribution function of Expn(α) are given respectively by:

$$p(x) = \alpha \exp(-\alpha x)$$
$$F(x) = 1 - \exp(-\alpha x)$$

where the mean is given by $1/\alpha$.

The Poisson distribution and the negative exponential distribution are closely related. If the likelihood of an event, such as the arrival of a customer order, occurring at any instant of time remains unchanged with time (in queuing theory, this is known as a *completely random process*), then the probability of this event having a frequency of occurrence of x (number of occurrences per unit time) is given by Psn(λ), where λ indicates the mean frequency of occurrence (or mean arrival rate). Further, the probability of any two successive occurrences of this process having an inter-arrival time of t is given by the negative exponential distribution, where the mean inter-arrival time is given by $T = 1/\lambda$. Thus both Poisson and negative exponential distributions in fact describe the same random process, but in different ways. A process that can be described by these two distributions is known as a *Poisson process*. The basic results obtained from the analysis of queuing systems based on the Poisson process will be summarized in section 6.5.3.

Other distributions

Only a few examples of the more commonly used probability distributions have been given above. Many other types of distribution function may be used in various types of simulation study. Some of these are listed in Table 6.5 together with their possible applications. If required, the reader should consult textbooks on probability and statistics for detailed information.

Empirical distribution

In a manufacturing environment some control is always exercised over an operation, so that the process concerned is not likely to be completely random. For example, the machining rate of an NC machine centre can be determined quite accurately by the length of the part programme that is being used to control the process (though it may still be regarded as a random process because of the small variations in processing time and the likelihood of machine breakdowns). Under these circumstances not only do the analytical results based on the Poisson process become invalid, but also it may be impossible to use any of the known theoretical distributions to describe the system variables. This is why sometimes it is necessary to employ a user-defined distribution in order faithfully to model the system under study and simulate and predict its behaviour.

If the data available consist of a finite number of points to which no theoretical distribution can be adequately fitted, so that they have to be used directly in the simulation model, then an empirical distribution must be constructed. The first step then is to collect data. Carry out a study of the process under investigation and collect a number of samples of the variable of interest. Obviously the bigger the size of the sample, the more accurate the resultant distribution will be – though this will have to be balanced

Table 6.5 Some of the less common probability distributions and their possible applications

Distribution	*Applications*
Beta	Non-conforming proportion of items in a batch of delivery, time to complete a task, etc., thus often used in project management.
Gamma	Time to complete a task, e.g. repair of machine, used mainly for maintenance planning and reliability design, etc.
Weibull	Similar to gamma distribution with a relatively wide range of applications in the study of stochastic processes.
Log-normal	Applications similar to those of the normal distribution, i.e. to represent quantities that result from a large number of other quantities, such as measurement errors.
Binomial	A discrete distribution representing the probability of n occurrences of a certain event out of a possible N; used for discrete variables such as the number of defective items in a batch.
Erlang	Represents the sum of S exponential variables each with a mean of $1/S$, used for queuing and inventory control applications.
Geometric	Represents the number of trials before a certain event occurs, such as the number of items inspected before the detection of a non-confirming one.
Triangular	Used as a rough approximation when the necessary data are not available.

against the time and resources available. Now the probability distribution can be compiled. A simple way of doing this is to display the data in a number of intervals, for example N, of equal width. The interval width w can then be determined by dividing the difference between the largest sample value, S_M, and the smallest, S_m, by N:

$$w = (S_M - S_m)/N$$

Having determined the interval size, count the number of values which fall into the first interval, $[S_m, (S_m + w)]$ (including S_m and $S_m + w$), then the half-open interval, $[S_m + w, S_m + 2w]$ (greater than $S_m + w$, but less than or equal to the upper value of $S_m + 2w$) and so on, until finally all of the samples are accounted for. This will result in a probability distribution of the samples which can then be tabled or plotted. As an example, suppose the following processing times of a machining operation have been observed:

37, 46, 25, 50, 43, 85, 28, 58, 71, 56, 45, 29, 70,
45, 80, 46, 14, 25, 48, 75, 91, 68, 47, 51, 52, 58,
70, 78, 30, 62, 82, 35, 57, 31, 50, 28, 65, 35, 46,
55, 19, 70, 53, 54, 39, 35, 93, 20, 37, 51, 70, 55.

In this sample set, the largest sample value is 93, the smallest value is 14. If the value of N is chosen to be 10, then the interval size is given by:

$$(93 - 14)/10 = 7.9$$

If, for convenience, we take this to be 8, then the upper values of the intervals will be 22, 30, 38, 46, 54, 62, 70, 78, 86, and 94. Now it is possible to construct a table showing the number of observations which fall into the intervals [14, 22], [22, 30], [30, 38], . . . , [86, 94], as shown below:

Interval no. (i)	*Time interval* (x_i)	*No. of observations (frequency of occurrence)* (f_i)
1	14–22	3
2	23–30	6
3	31–38	6
4	39–46	7
5	47–54	9
6	55–62	7
7	63–70	6
8	71–78	3
9	79–86	3
10	87–94	2

This is shown graphically in Fig. 6.13.

Choice of probability distributions

Selecting the appropriate probability distribution for a system variable involves three tasks: determining the type of distribution to be used, estimating the parameter value(s) for the chosen distribution, and checking the goodness-of-fit of the proposed distribution.

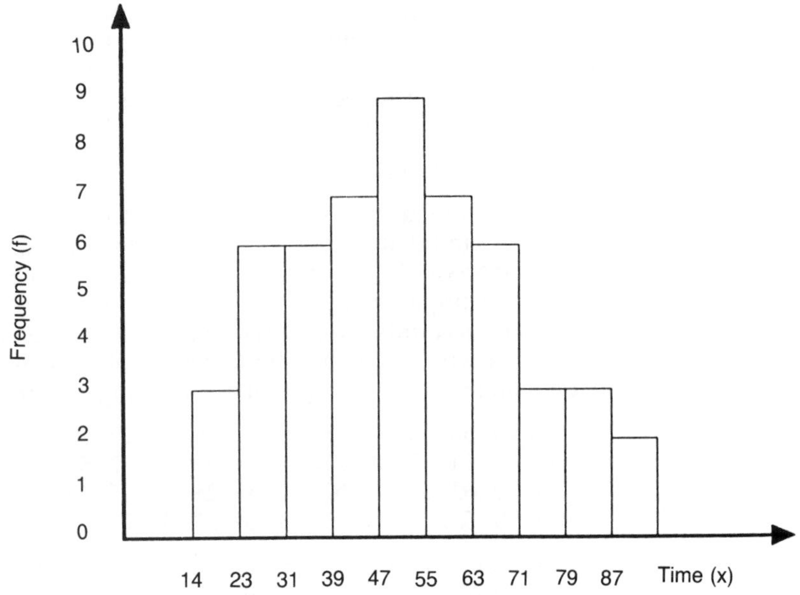

Fig. 6.13 An empirical distribution.

Type of distribution

There are a number of heuristic methods which can be used to help select an appropriate type of distribution (Law and Kelton, 1982). Some of these are described below.

Point statistics is a method that makes use of the distinct features which are characteristics of certain types of distribution. For example, the exponential distribution is known to have a coefficient of variation (the standard deviation divided by the mean) of 1. To take advantage of this fact, we can estimate the value of this particular coefficient from the original data. If the resultant value is near 1, then this might suggest a variable that is exponentially distributed.

Probability plots are used to assess whether a distribution constructed from empirical data has the same shape as one of the theoretical distributions by graphically comparing the two. The approach involves some analytical procedures aimed at easing the comparison processes by assessing the extent of coincidence of the two along a straight line.

These two methods both involve analytical procedures and numerical evaluation. The method of *histogram plots*, however, is a purely intuitive and visual approach. A histogram is essentially a graphical representation of a user-defined distribution, which is constructed directly from the original data collected. If the true distribution of the variable concerned does belong to one of the theoretical types, this will often be revealed by the shape of the histogram plot, provided the amount of data gathered is sufficiently large. This is illustrated by Fig. 6.14(a) and (b), which respectively reveal the recognizable shapes of a uniform and a normal distribution.

Parameter estimation

Having specified the type of distribution, its appropriate parameter(s) must then be estimated. One useful method for this purpose is provided by the approach known as maximum-likelihood estimation. The idea behind this method is very straightforward. Having specified the type of distribution, we know that its probability density function is given by:

$$P(X = x) = p(x)$$

but the parameter(s) of $p(x)$ are not yet known to us. However, from the basic characteristics of probability, we know that the probability of our having observed our data, $x_1, x_2, \ldots, x_n$, can be calculated from the probability density function of this variable, that is:

$$p(x_1)p(x_2)\ldots p(x_n)$$

Obviously the above expression is a function of the unknown parameter(s) of the distribution, say β, whose value is to be estimated. The above can then be written as:

$$L(\beta) = p(x_1)p(x_2)\ldots p(x_n)$$

The maximum-likelihood estimator of β, here denoted as β_m, is the value of β which maximizes $L(\beta)$. This means that if β_m is used as the parameter value for the chosen probability density function $p(x)$, then it is most likely that we would have obtained the data, $x_1, x_2, \ldots, x_n$, in the way we have obtained them. To put this another way, the particular distribution $p(x)$ with a parameter value of β_m is most likely the distribution that we are looking for.

Take the exponential distribution for example. Given that $p(x) = \beta\exp(-\beta x)$, then:

```
    **** DISTRIBUTION TESTING ****
     RANGE        FREQ. PERCENT|           HISTOGRAM DISPLAY
------------------------------+----+----+----+----+----+----+----+----+----+
    0 -    4     206     4.1   |*****************************************
    4 -    8     195     3.9   |**************************************
    8 -   12     202     4.0   |****************************************
   12 -   16     191     3.8   |**************************************
   16 -   20     209     4.2   |*****************************************
   20 -   24     197     3.9   |***************************************
   24 -   28     200     4.0   |***************************************
   28 -   32     196     3.9   |***************************************
   32 -   36     203     4.1   |****************************************
   36 -   40     198     4.0   |***************************************
   40 -   44     204     4.1   |****************************************
   44 -   48     190     3.8   |*************************************
   48 -   52     204     4.1   |****************************************
   52 -   56     208     4.2   |*****************************************
   56 -   60     204     4.1   |****************************************
   60 -   64     183     3.7   |************************************
   64 -   68     201     4.0   |****************************************
   68 -   72     218     4.4   |*******************************************
   72 -   76     199     4.0   |***************************************
   76 -   80     195     3.9   |**************************************
   80 -   84     198     4.0   |***************************************
   84 -   88     207     4.1   |*****************************************
   88 -   92     199     4.0   |***************************************
   92 -   96     198     4.0   |***************************************
   96 -  100     195     3.9   |**************************************
------------------------------+----+----+----+----+----+----+----+----+----+
                              0         50        100       150       200

            *****     DISTRIBUTION TESTING *****
          TOTAL NO. OF OBSERVATION :       5000
          MINIMUM VALUE            :       0.02
          MAXIMUM VALUE            :      99.95
          MEAN VALUE               :      49.97
          STANDARD DEVIATION       :      28.85
```

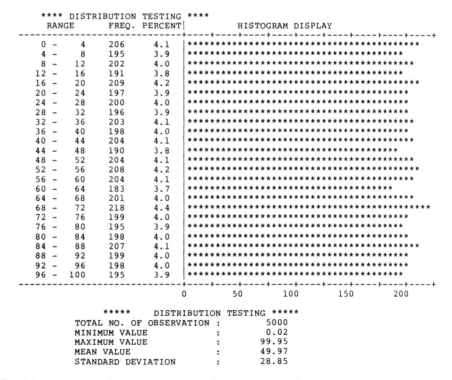

Fig. 6.14a Distribution of random numbers from a uniform distribution.

```
**** DISTRIBUTION TESTING ****
  RANGE        FREQ. PERCENT|          HISTOGRAM DISPLAY
----------------------------+----+----+----+----+----+----+----+
-600 - -550       1     0.0 |*
-550 - -500       3     0.1 |*
-500 - -450      10     0.2 |*
-450 - -400      19     0.4 |*
-400 - -350      35     0.7 |**
-350 - -300      47     0.9 |***
-300 - -250      84     1.7 |*****
-250 - -200     119     2.4 |******
-200 - -150     192     3.8 |**********
-150 - -100     248     5.0 |*************
-100 -  -50     343     6.9 |******************
 -50 -    0     425     8.5 |**********************
   0 -   50     483     9.7 |*************************
  50 -  100     502    10.0 |**************************
 100 -  150     501    10.0 |**************************
 150 -  200     471     9.4 |************************
 200 -  250     398     8.0 |********************
 250 -  300     360     7.2 |******************
 300 -  350     253     5.1 |*************
 350 -  400     166     3.3 |*********
 400 -  450     133     2.7 |*******
 450 -  500      95     1.9 |*****
 500 -  550      49     1.0 |***
 550 -  600      25     0.5 |**
 600 -  650      18     0.4 |*
 650 -  700      10     0.2 |*
 700 -  750       6     0.1 |*
 750 -  800       3     0.1 |*
 800 -  850       1     0.0 |*
----------------------------+----+----+----+----+----+----+----+
                            0        200       400       600       800

          *****    DISTRIBUTION TESTING *****
      TOTAL NO. OF OBSERVATION :          5000
      MINIMUM VALUE            :       -562.29
      MAXIMUM VALUE            :        816.02
      MEAN VALUE               :         99.91
      STANDARD DEVIATION       :        198.71
```

Fig. 6.14b Distribution of random numbers from a normal distribution.

$$
\begin{aligned}
L(\beta) &= p(x_1)p(x_2)\ldots p(x_n)\\
&= [\beta\exp(-\beta x_1)][\beta\exp(-\beta x_2)]\ldots[\beta\exp(-\beta x_n)]\\
&= \beta^n \exp\left(-\beta\sum_{i=1}^{n} x_i\right)
\end{aligned}
$$

The standard method of differential calculus can be used to find the value of β_m, by first differentiating $L(\beta)$ and then letting the resultant expression equal zero, and then solving this equation for β. In this particular case, however, it is much easier to apply this analytical procedure on the logarithm of $L(\beta)$ rather than on $L(\beta)$ itself (this will not affect the value of β_m since $\ln[L(\beta)]$ is strictly increasing). That is:

$$\ln[L(\beta)] = n\ln(\beta) - \beta\left(\sum_{i=1}^{n} x_i\right)$$

Then let

$$\frac{d\{\ln[L(\beta)]\}}{d\beta} = \frac{n}{\beta} - \sum_{i=1}^{n} x_i = 0$$

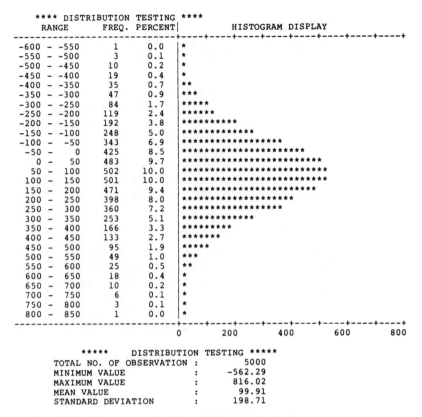

which gives

Table 6.6 Maximum likelihood estimators of commonly used distributions

Distribution	*Maximum likelihood estimator*
Negative exponential	$1/x_a$
Poisson	x_a
Normal	$\mu_m = x_a \quad \sigma_m = \left[\sum_{i-1}^{n}(x_i - x_a)^2/n\right]^{1/2}$

$$\beta_m = 1\bigg/\left[\sum_{i=0}^{n} x_i \bigg/ n\right] = 1/x_a$$

where x_a indicates the average value of $x_1, x_2, \ldots, x_n$.

The maximum-likelihood estimators of the three commonly used distributions are given in Table 6.6.

Goodness-of-fit

The final task involved here is to check whether the resulting distribution as specified by the last two steps is indeed in agreement with the original data, $x_1, x_2, \ldots, x_n$. There are a variety of tests specially designed for this purpose. Here we will examine one of these methods – the *chi-squared test*.

The idea behind this method this simple and straightforward – to estimate the deviation of the fitted distribution from the observed data and, if this is found to be too large, to reject the hypothesis that these data are random variables with the specified distribution. Provided the size of the observed data set, n, is sufficiently large, this method of goodness-of-fit test can in practice be carried out as follows.

From the original data, $x_1, x_2, \ldots, x_n$, compile a user-defined distribution in the manner described earlier in this section. This will consist of n observations split into m intervals (z_i, z_{i+1}), i = 1, 2, . . . , m, each containing f_i observations.

For each interval (z_i, z_{i+1}) the expected frequency of occurrence is given by:

$$e_i = n \int_{zi}^{z_{i+1}} p(z)dz = 1$$

The observed and expected frequencies of occurrence are now compared using the following formula:

$$x^2 = \sum_{i=1}^{m} \frac{(f_i - e_i)^2}{e_i}$$

Finally the decision as to whether or not to reject the specified distribution can be made by comparing χ^2 with the appropriate value found from the chi-squared distribution. Due to certain considerations on both the usefulness of the method and the possible decision errors, it is recommended that, provided the resolution of the user-defined distribution is relatively high (that is, the number of intervals, m, is relatively large), with a significance level of β, the proposed distribution should be rejected if:

$$\chi^2 > \chi^2_{(m-1,\beta)}$$

Table 6.7 Critical values of the chi-squared distribution

DOF	β			
	75.0%	*25.0%*	*5.0%*	*1.0%*
9	5.899	11.389	16.919	21.666
10	6.737	12.549	18.307	23.209
19	14.562	22.718	30.144	36.191
20	15.452	23.828	31.410	37.566
29	23.567	33.711	42.557	49.588
30	24.478	34.800	43.773	50.892
50	42.942	56.334	67.505	76.154
100	90.133	109.141	124.342	135.807

where $\chi^2_{(m-1,\beta)}$ is the upper β critical point for the chi-squared distribution with $(m - 1)$ degrees of freedom (DOF), whose value can be easily found from standard statistical tables. β is referred to as the *rejection probability*, that is, the probability of the fitted distribution being wrongly rejected. Some of the useful critical points of $\chi^2_{(m-1,\beta)}$ are given in Table 6.7.

Example 6.1

Suppose that, on the basis of 300 observations, a distribution is specified which has probability function $P(x)$. By arranging the original data according to a 30-interval distribution and from the theoretical function $p(x)$, χ^2 is found to be 32.50. Can we confidently assume that it is indeed from this fitted distribution that the observed data were sampled?

Solution

In practice some reasonably small rejection probability is usually used, say β = 5%. From the values given in the table we note that:

$$\chi^2_{(29.5\%)} = 42.557$$

Since χ^2 does not exceed $\chi^2_{(m-1,\beta)}$ at the 5% significance level, the fitted distribution should not be rejected.

The chi-squared test is one of the best established methods for checking the goodness-of-fit for an assumed distribution and has found a wide range of applications. However, this approach does have some drawbacks. First, it demands that the number of samples, n, must be very large if the result of the test is to be valid. This is because the whole approach is based on the fact that, if the hypothesis about the assumed distribution is true, then χ^2 tends to be distributed in a chi-squared form as n approaches infinity. Second, there is no simple guidance on how to choose the right level of resolution that should be used in the distribution, that is, the number of intervals, though 10–40 intervals are usually used in practice.

6.5.2 Generation of random variables

The purpose of identifying and describing the appropriate distribution of a stochastic systems variable is to be able adequately to represent that variable during an actual simulation process. Having defined the distribution, therefore, the next important issue is to be able to generate a series of random values for the systems variable concerned which follow the proper distribution. To simulate the activity of a machining operation, for example, we need to schedule for the processing time after a workpiece is loaded on the machine. In practice, this can be achieved by taking a random sample from the probability distribution of the machine's processing time – or, in other words, by generating a random value from an appropriate distribution. In this section, therefore, we will discuss how such a random value can be generated according to a given distribution during a simulation run, so that the stochastic process concerned can be simulated as faithfully as possible.

A general approach: the inverse transformation method

The inverse transformation method provides a simple approach to the problem of generating a random value that has a probability density function of the correct type. A number of types of random variable, whose cumulative distribution can be analytically inverted, can be generated by this approach. The basic principle of this method is also utilized when generating random values from user-supplied distributions. The approach makes use of the observation that the value given by the cumulative distribution function:

$$y = F(x) = P(X \leqslant x) = \int_{-\infty}^{x} p(z)\mathrm{d}z$$

has a value range of (0,1) – that is to say, that the probability of a random variable X taking a value less than $-\infty$ must be 0, and the probability of X taking a value less than ∞ must be 1. If $F(x)$ is solved for x, we obtain its inverse transformation as follows:

$$x = F^{-1}(y)$$

which gives the particular value of x for a given y, as shown in Fig. 6.15. Now it is obvious that if the random variable X takes a value less or equal to x_i, its cumulative distribution function will give a corresponding value which is less than or equal to y_i. That is, the probability of Y taking a value less than or equal to y is given by the probability of X taking a value less or equal to x:

$$P(Y \leqslant y) = P(X \leqslant x) = F(x)$$

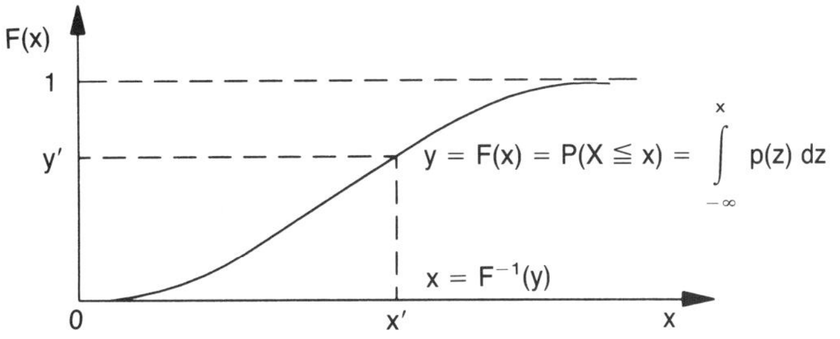

Fig. 6.15 The inverse transformation method.

Therefore

$$P(Y \leqslant y) = y$$

Since the above is exactly the cumulative distribution function of random numbers with distribution $U(0,1)$, we can see that the values of $Y = F(x)$ are in fact the random numbers which are uniformly distributed within the range (0,1). Therefore, a straightforward way of generating a value for X is simply first to generate a uniformly distributed random number, r, and then find its corresponding value for X using the inverse transformation:

$$X = F^{-1}(r)$$

Examples of inverse transformation functions

The inverse transformation method of generating random variables can be used directly in a computer simulation model if the probability function is known and can be mathematically described and if this function can be analytically inverted. Here we give two examples of distributions which satisfy both these conditions.

To generate a random value from the negative exponential distribution we proceed as follows. Since the cumulative distribution function of this distribution is given by:

$$F(x) = 1 - \exp(-\alpha x)$$

we can write:

$$r = 1 - \exp(-\alpha x)$$

where r is our uniformly distributed random number. We can now solve this equation for x:

$$\exp(-\alpha x) = (1 - r)$$

$$-\alpha x = \ln(1 - r)$$

$$x = -(1/\alpha)\ln(1 - r)$$

Consequently, implementation of an exponential random generator in a computer program is just a simple matter of first generating a random number r in the range (0,1), and then using the above equation to calculate the corresponding value of x. The procedure is illustrated by lines 1000–1060 of the simple BASIC simulation model (section 6.4.1).

To generate a random value from the uniform distribution $U(a,b)$, we proceed in an analogous manner. Given

$$F(x) = \frac{x - a}{b - a} = r$$

then

$$x = a + r(b - a)$$

An example of the application of this equation is given again by lines 2000–2090 of the BASIC model.

Generation of empirical random variable

The basic principle of the inverse transformation method is frequently used to generate random variables from user-supplied distributions. Using the same sample data as given above, the following shows how this can be done in practice (this is know as the table look-up method).

First, the probability distribution is converted into a cumulative distribution. The construction of a cumulative distribution involves summing up the number of observations less than or equal to the upper value of each interval and expressing this as a proportion of the total sample size, as shown below:

Interval no. i	*Time Interval* x_i	*No. of observations* f_i	*Cumulative frequency* $F_i = \sum_{j=1}^{i} f_i$	*Cumulative distribution* $\left[F_i \Big/ \sum_{j=1}^{N} f_i\right]$
1	14–22	3	3	0.057
2	23–30	6	9	0.173
3	31–38	6	15	0.288
4	39–46	7	22	0.423
5	47–54	9	31	0.596
6	55–62	7	38	0.731
7	63–70	6	44	0.846
8	71–78	3	47	0.904
9	79–86	3	50	0.962
10	87–94	2	52	1.000

A graph of this cumulative distribution can now be plotted against time, as shown in Fig. 6.16.

Now, a random number generator is used to produce a number between 0 and 1. For each random number on the vertical axis of the cumulative distribution graph, read across to the curve and then down to the horizontal axis to give the sampled value of the random variable, x. Thus a random number of r, for example, will give an activity time of t, as shown in Fig. 6.16. In a computer program, the value of x corresponding to r can be obtained by the technique of linear interpolation. That is, if the value of r happens to fall in between (F_i, F_{i+1}), then the corresponding value of x is given by:

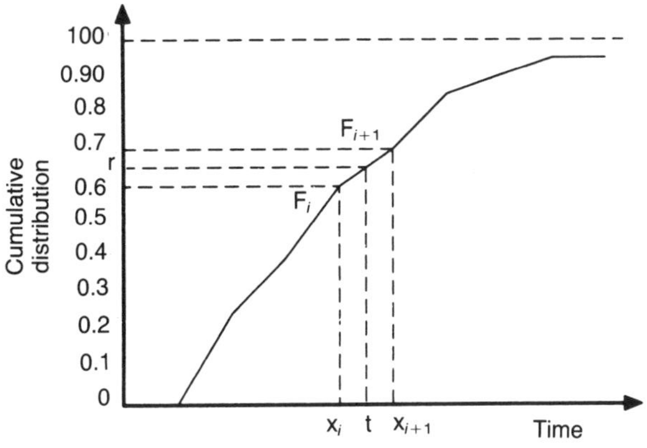

Fig. 6.16 Linear interpolation.

$$x = x_i + \left[\frac{r - F_i}{F_{i+1} - F_i}\right](x_{i+1} - x_i)$$

Other random variables

Unfortunately the straightforward method of inverse transformation cannot be applied to all distributions, and alternative ways have to be used. These include, for example, the rejection method and the composition method. It is not the intention of this chapter to go into the theoretical background of these methods; an adequate treatment of this subject is given by Law and Kelton (1982). Our concern here is to outline the methods that can be used in a computer program so that a particular type of random variable can be easily obtained. A number of such methods for the generation of the more common distributions are given below.

Random number

The importance of this particular type of random variable is now apparent. Many different random number generators exist, of which the *linear congruential* procedure is perhaps the most frequently employed:

1. To generate a series of random values in the range 0, 1, . . . , $m - 1$, first define the following parameters (constants) for the generator:
   a: a multiplier
   c: the increment
   x_0: the random seed (or initial value).
2. The next integer random value is then given by:

   $$x_i = (ax_{i-1} + c) \bmod m$$

   where mod stands for *modulus*. This means that $(\alpha x_{i-1} + c)$ is divided by m and the remainder used as x_i.
3. Obtain random number r_i in (0,1) by letting

   $$x_i = x_i/m$$

Normal distribution

There are a number of ways in which a normally distributed random variable can be generated. The *polar* method proceeds as follows:

1. Generate two random numbers r_1 and r_2 ($0 \leqslant r_i \leqslant 1$, $i = 1, 2$)
2. Let $q_1 = 2r_1 - 1$, and $q_2 = 2r_2 - 1$
3. Let $Q = q_1^2 + q_2^2$
4. If $Q > 1$, then go back to step 1
5. Otherwise, let $W = [(-2\ln Q)/Q]^{1/2}$
6. Finally, a pair of random variables with normal distribution $N(0,1)$ are given by $x_1 = q_1W$ and $x_2 = q_2W$.

Binomial distribution

If p is the probability of success and n is the size of the sample, then a quite intuitive method can be used for this distribution:

1. Generate n random numbers $r_1, r_2, \ldots, r_n$
2. Count the number of times, x, that $r_i \leqslant p$. Then x is the resulting value of the binomially distributed random variable.

Poisson distribution

Random values from the Poisson distribution with mean μ can be generated as follows:

1. Let $a = 1$, and $n = 0$
2. Generate random number r from (0,1)
3. Let $a = ar$
4. If $a < \exp(-\mu)$, then $x = n$ is the resulting random value
5. Otherwise let $n = n + 1$, and go back to step 2.

The above procedures will generate random variables from some of the most frequently used types of probability distribution. In general, since the majority of manufacturing systems are stochastic in nature, the generation of random values is an operation that is frequently required in the modelling and simulation of manufacturing systems. It should be pointed out that since random sampling is such an important feature of stochastic simulation, to be qualified as a simulation language a package must provide a facility for random variable generation which is capable of producing random variables from, for example, the uniform, normal, binomial, Poisson and exponential distributions, as well as from customer-supplied data. Box 6.5 gives an example which shows how this is done by a particular simulation package – PCModel.

Box 6.5 PCModel's instructions for random variable generation

PCModel uses a random value [RV( )] instruction to generate random values from a number of different types of distribution. The generated random value can be assigned to a systems parameter in order to simulate the stochastic process concerned. Instructions have the following formats:

RV(x,low,high)
RV(U,x,mean,variance)
RV(N,x,mean,deviation)
RV(E,x,mean)
RV(C,x,disname)
RV(D,x,disname)

where

x	is a variable which receives the random value
low	is lower random value limit
high	is higher random value limit
U	means to sample from uniform distribution
N	means to sample from normal distribution
E	means to sample from exponential distribution
C	means to sample, in a continuous manner, from a user-defined distribution
D	means to sample, in a discrete manner, from a user-defined distribution
mean	is the mean value of the distribution

variance	is the variance limits of the uniform distribution
deviation	is the standard deviation of the normal distribution
disname	is the name of a user-defined distribution (a two-dimensional array within the paired values of the cumulative distribution are stored)

For example:

```
''''''''''''''''''
RV(@RANDOM,1,100)          assign a random value from 1 to 100 to
                           variable @RANDOM
''''''''''''''''''
''''''''''''''''''
RV(N,@MCING,60,30)
WT(@MCING)                 sample from a normal distribution with
''''''''''''''''''         a mean value of 60 and standard devia-
''''''''''''''''''         tion of 30 and assign this to variable
                           @MCING, then make the entity con-
                           cerned wait over a period of @MCING
                           time units
RV(C,@MCING,@@MCTIME)      this time sample from a user-defined
WT(@MCING)                 distribution named @@MCTIME. This
''''''''''''''''''         instruction tells PCModel to extract
                           values from the user-created array of
                           values defining the distribution. The
                           array @@MCTIME should first be filled
                           with pairs of numbers representing the
                           value to be extracted along with its rep-
                           resentative accumulative distribution.
                           The function will then return a random
                           value interpolated from the distribution
                           using straightline interpolation between
                           entry points in the array. If the data
                           given in our previous example of sam-
                           ples of machining time were used to
                           define @@MCTIME, then its contents
                           would be as follows (in a PCModel dis-
                           tribution the probabilities are scaled to
                           parts/10000).
```

```
Cumulative probability          Time value
@@MCTIME(00,00) = 0000;         @@MCTIME(01,00) = 14;
@@MCTIME(00,01) = 0570;         @@MCTIME(01,01) = 22;
@@MCTIME(00,02) = 1730;         @@MCTIME(01,02) = 30;
@@MCTIME(00,03) = 2880;         @@MCTIME(01,03) = 38;
@@MCTIME(00,04) = 4230;         @@MCTIME(01,04) = 46;
@@MCTIME(00,05) = 5960;         @@MCTIME(01,05) = 54;
@@MCTIME(00,06) = 7310;         @@MCTIME(01,06) = 62;
@@MCTIME(00,07) = 8460;         @@MCTIME(01,07) = 70;
@@MCTIME(00,08) = 9040;         @@MCTIME(01,08) = 78;
@@MCTIME(00,09) = 9620;         @@MCTIME(01,09) = 86;
@@MCTIME(00,10) = 10000;        @@MCTIME(01,10) = 94.
```

6.5.3 Evaluation of manufacturing operations using the basic results from queuing theory

As a natural development of the previous discussion on probability distribution and random variables, this sub-section aims to provide a practical and useful guide to the applications of queuing theory in the analysis of manufacturing systems. As far as simulation modelling is concerned, this sub-section is also concerned with some of the analytical techniques which may be used for checking the validation of a simulation model, for the theoretical results to be presented here are well suited for this purpose.

Manufacturing operations and Poisson process

We can use the terminology introduced in section 6.4.2 to describe the main parties involved in a simple manufacturing operation and the basic ways in which they can be arranged to form different manufacturing systems, as illustrated in Fig. 6.17. Since queuing theory is concerned with the probability analysis of queue systems in order to predict their performance, its results can be used to evaluate the operations of these manufacturing systems.

One of the most important conditions which must hold for the results presented in this sub-section to be valid is the assumption that both the arrival time and the service time in the queuing system have Poisson distributions. That is, the processes are based on an underlying complete randomness. Fortunately, this has been shown to be the case in many real-life situations. Although, as pointed out earlier in section 6.5.1, sometimes it might not be reasonable for the controlled manufacturing operations to be considered as completely random processes, the approach is still a valuable tool, at least in terms of long-term average predictions, to provide rough estimates of the systems parameters such as capacity requirement, production rate, facility utilization, throughput time, and overall WIP situation.

The basic results of queuing theory provide a convenient method of estimating system performance, at least at the preliminary stage of system specification, provided basic statistics of the likely average workload and mean machine capacities are available. For example, the initial specification of a manufacturing operation can be quickly checked to see whether it is appropriate by employing the results to calculate the machine capacities required to satisfy the intended production load, and to evaluate the buffer-storage requirement before a work-station for holding parts waiting for service.

Also, for simulation applications these results may be employed to check the validity of a simulation model. This can be done by comparing the data outputs from a simulation with the theoretical predictions, so as to check the degree to which the two sets of data match each other. If for the same situation there exists an obvious mismatch, then this is a clear sign that there is something wrong with the simulation model, provided the method used to perform the theoretical analysis is sound.

A number of conditions must hold if the basic results from queuing theory which are presented below are to remain valid. These conditions are as follows:

- Both the arrival rates of the workpieces and the processing rates of the work-stations follow Poisson distributions, with a mean arrival rate of λ and a mean processing rate of μ, respectively. Further, if the processing rate is distributed as Psn(μ), then the processing times are negative exponentially distributed with an average processing time of $T_s = 1/\mu$.

- The parts arrive randomly from a source of infinite size.
- The operation is allowed to run continuously without disruption.
- There is no limit on queue size.
- The parts are processed according to first-in-first-served (FIFS) queuing discipline.
- The overall capacity of the system must exceed the overall workload, e.g. in a single work-station system the value of the mean processing rate μ must be greater than that of λ. This is necessary if the system is to reach a steady state of operation, for otherwise the queue in the system would grow unchecked.

The system variables involved in this type of analysis are summarized as follows:

λ is mean arrival rate of parts
μ is mean processing rate of workstation
T_a is mean inter-arrival time of parts $= 1/\lambda$
T_s is mean work-station processing time $= 1/\mu$

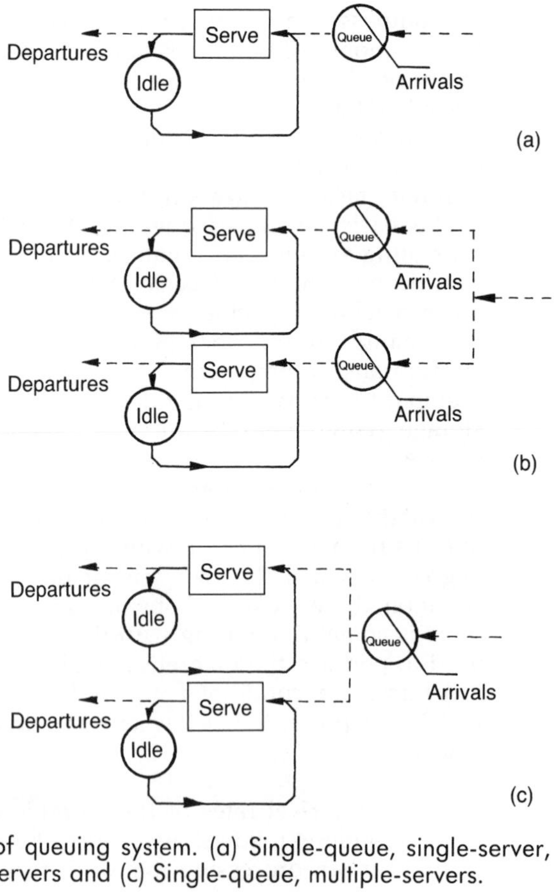

Fig. 6.17 Types of queuing system. (a) Single-queue, single-server, (b) Multiple-queues, multiple-servers and (c) Single-queue, multiple-servers.

S is the number of work-stations in system
u is the utilization factor = λ/μ
L is the average number of parts in system
L_q is the average queue length
W is the average time spent in system
W_q is the average time spent queuing
$P(n)$ is the probability of n parts being in the system
$Q(n)$ is the probability of n parts being in a queue

Single-work-station system (Fig. 6.17a)
This is the simplest case of a queuing situation. Provided the conditions listed above are met, the following results hold:

1. Work-station utilization is given by u, where $u < 1$.
2. $L = \dfrac{u}{1 - u}$
3. $L_q = \dfrac{u^2}{1 - u} = (L - u)$
4. $W = \dfrac{1}{\mu - \lambda}$
5. $W_q = \dfrac{u}{\mu - \lambda} = (W - T_s)$
6. $P(n) = u^n(1 - u)$
7. $Q(n) = \begin{cases} 1 - u^2 & \text{for } n = 0 \\ (1 - u)u^{n+1} & \text{for } n \geqslant 1 \end{cases}$

In addition, the probability of a part spending a period of time t in the system is given by:

$$(\mu - \lambda)\exp[-t(\mu - \lambda)]$$

Hence the probability of a part spending a period of time less than t in the system is given by:

$$\int_0^t \{(\mu - \lambda)\exp[-x(\mu - \lambda)]\}dx = 1 - \exp[-t(\mu - \lambda)]$$

and the probability of a part spending a period of time longer than t in the system is given by:

$$\int_t^{+\infty} \{(\mu - \lambda)\exp[-x(\mu - \lambda)]\}dx = \exp[-t(\mu - \lambda)]$$

Example 6.2

Workpieces arrive randomly in a one-machine work-station at an average rate of 18 every hour. The processing time of the work-station was found to be negative exponentially distributed with a mean of 3 minutes. Find:

(a) The machine utilization.
(b) The average throughput time of the system.

(c) The average queue length.
(d) The probability that a workpiece will go through the system in less than 15 minutes.
(e) The machine capacity that would reduce the average throughput time to 15 minutes.

Solution

(a) For this manufacturing system:

mean arrival rate $\lambda = 18/60 = 0.30000$ parts/min
mean processing rate $\mu = 1/3 = 0.33333$ parts/min

Therefore the machine utilization is given by:

$$u = \lambda/\mu = 0.90 = 90\%$$

(b) The average throughput time of the system is calculated as follows:

$$W = \frac{1}{\mu - \lambda} = 1/(0.33333 - 0.30000) = 30 \text{ minutes}$$

(c) Average queue length:

$$L_q = \frac{u^2}{1 - u} = 0.9^2/(1 - 0.9) = 8.1 \text{ parts}$$

(d) The probability that a workpiece will go through the system in less than 15 minutes can be evaluated as follows:

$$\begin{aligned} 1 - \exp[-t(\mu - \lambda)] &= 1 - \exp(-15.000 \times 0.03333) \\ &= 0.3935 \end{aligned}$$

(e) Here we need to find the mean processing rate, μ, of the machine that would reduce the average throughput time of the workpieces to the level required. Since

$$W = \frac{1}{\mu - \lambda} = 1/(\mu - 0.3) = 15$$

then

$$\mu = 0.3 + 0.066667 = 0.36667 \text{ parts/min}$$

That is, we would need a machine which is capable of processing on average one workpiece in every $1/0.36667 = 2.7273$ minutes. As a result, the machine utilization would be reduced to $0.3/0.36667 = 0.8182 = 81.82\%$.

Example 6.3

An FMS manufactures a mix of parts, within which a number of AGVs are used for the purpose of material handling. One of the tasks that the AGVs must perform is to take a finished part to an inspection station, where the part is checked for quality. The part being checked by this station will remain on the pallet of the AGV during the inspection operation. A time study over a long period of operation has established that the AGVs visit the inspection station randomly at an average rate of 2 per hour.

It is undesirable to keep the AGV at the inspection station for more than 15 minutes each time it visits the station, since during this period it will not be able to perform any other tasks. How fast, then, must the inspection operation be carried out if this limit is not to be exceeded for at least 90% of the time? Assume that the inspection time is negative exponentially distributed.

Solution

In this case the inspection station can be regarded as the server and the AGVs as customers. The objective here is to find the working rate of the inspection station, μ, so that the probability of an AGV spending less than 10 minutes in the system is 0.9, i.e.:

$$\int_0^{15} \{(\mu - \lambda)\exp[-x(\mu - \lambda)]\}dx = 1 - \exp\{-15(\mu - \lambda)\} = 0.9$$

Rearranging the above equation gives:

$$\exp[-15(\mu - \lambda)] = 0.1$$

$$\mu = \lambda - [\ln(0.1)/15]$$

Given that $\lambda = 2/60 = 1/30$, the required average rate of inspection is therefore:

$$\mu = 0.03333 + 0.15351 = 0.18684$$

This tells us that to satisfy the operational requirements, the inspection station must have the ability at least to complete on average one inspection operation every $1/0.18684 \approx 5.35$ minutes.

Multi-work-station single-phase system (Fig. 6.17c)
The multi-work-station single-phase system represents the more complicated situation in which, although the workpieces again arrive randomly in the system at an average rate λ, this time they are served by one of a number of k possible work-stations. These work-stations are assumed to be identical in the sense that they all have a processing time that follows the same negative exponential distribution with mean 1/μ. The workpieces wait in a central buffer and then, again following FIFS queuing discipline, are taken to the first work-station that becomes free.

The analytical process has now become much more complicated – as shown below, the results from queuing theory when applied to this situation appear to be rather messy:

1. Workstation utilization is given by $\lambda/(k\mu) = u/k$ (<1).
2. The probability that there are no parts in the system is given by:

$$P(0) = \left(\frac{u^k}{(k-1)!(k-u)} + \sum_{n=0}^{k-1}\frac{u^n}{n!}\right)^{-1}$$

3. The probability that there are N parts in the system is given by:

$$P(N) = \begin{cases} \dfrac{1}{N!}u^N P(0) & \text{for } N < k \\ (k!k^{N-k})^{-1}\, u^N\, P(0) & \text{for } N \geqslant k \end{cases}$$

4. The average number of parts in the queue is given by:

$$L_q = \frac{u^{k+1}}{(k-1)!(k-u)^2} P(0)$$

5. The average time spent in the queue is given by $W_q = L_q/\lambda$, that is:

$$W_q = \frac{u^k}{(k-1)!(k-u)^2} \frac{P(0)}{\mu}$$

6. The average number of parts in the system, $L = L_q + u$.
7. The average time spent in the system, $W = L/\lambda$.
8. The probability of a part spending a period longer than time t in the system is given by:

$$P(W > t) = \exp(-\mu t) + \frac{u^k P(0)}{(k-1)!(k-u)(k-u-1)} \times \{\exp(-\mu t) - \exp[-(k\mu - \lambda)t]\}$$

Example 6.4

Jobs randomly arrive at a machining cell which consists of a number of identical machining centres. The average arrival rate of the jobs is 10 per hour. Also, the due dates of the parts are determined by the constant decision rule:

$$d_i = r_i + 70$$

where d_i the due date of part i, r_i is the arrival time of part i, and 70 represents the throughput allowance (in minutes). Assuming that the machining time of the machine centres is negative exponentially distributed with a mean of 11.5 minutes per part, calculate the average throughput time of the parts on the basis of there being 1, 2 and 3 machining centres.

The hourly cost of running the machining centres is M (in pounds per hour per machine). There is a fixed penalty for late deliveries from this cell, which is C (pounds per late part). What is the right way to make the choice between employing 2 or 3 machining centres in the cell?

Solution

For one machining centre ($k = 1$):

$$\lambda = 10/60 = 0.1666667 \text{ parts/minute}$$

$$\mu = 1/11.5 = 0.0869565 \text{ parts/minute}$$

therefore

$$\lambda/k\mu > 1$$

This means that the system is unstable, and so one machining centre on its own is not able to cope with the workload.

For two machining centres ($k = 2$), and given that $u = \lambda/\mu = \frac{1}{6} \times 11.5 = 1.9166667$:

$$\begin{aligned}
P(0) &= \left(\frac{u^k}{(k-1)!(k-u)} + \sum_{n=0}^{k-1}\frac{u^n}{n!}\right)^{-1} \\
&= \left(\frac{u^2}{1!(2-u)} + \sum_{n=0}^{1}\frac{u^n}{n!}\right)^{-1} \\
&= \left(\frac{u^2}{(2-u)} + \frac{u^0}{0!} + \frac{u^1}{1!}\right)^{-1} \\
&= \left(\frac{u^2}{(2-u)} + 1 + u\right)^{-1} \\
&= (44.083333 + 1 + 1.916667)^{-1}
\end{aligned}$$

$$P(0) = 0.0212766$$

The average number of parts in queue is then given by:

$$\begin{aligned}
L_q &= \frac{u^{k+1}}{(k-1)!(k-u)^2}P(0) \\
&= \frac{u^3}{1!(2-u)^2}P(0) \\
&= 21.572695 \text{ parts}
\end{aligned}$$

Finally, the average throughput time of parts is as follows:

$$\begin{aligned}
W &= L/\lambda \\
&= (L_q + u)/\lambda \\
&= (21.572695 + 1.916667) \times 6 \\
&\approx 140.94 \text{ minutes}
\end{aligned}$$

For three machining centres (k = 3):

$$\begin{aligned}
P(0) &= \left(\frac{u^k}{(k-1)!(k-u)} + \sum_{n=0}^{k-1}\frac{u^n}{n!}\right)^{-1} \\
&= \left(\frac{u^3}{2!(3-u)} + \sum_{n=0}^{2}\frac{u^n}{n!}\right)^{-1} \\
&= \left(\frac{u^3}{2!(3-u)} + \frac{u^0}{0!} + \frac{u^1}{1!} + \frac{u^2}{2!}\right)^{-1} \\
&= \left(\frac{u^3}{2(3-\mathrm{u})} + 1 + \mathrm{u} + \frac{u^2}{2}\right)^{-1} \\
&= (3.2497329 + 1 + 1.916667 + 1.8368056)^{-1} \\
&= 0.125
\end{aligned}$$

Again the average number of parts in queue is given by

$$\begin{aligned}
L_q &= \frac{u^{k+1}}{(k-1)!(k-\mathrm{u})^2}P(0) \\
&= \frac{u^4}{2!(3-u)^2}P(0) \\
&= 0.718691 \text{ parts}
\end{aligned}$$

Therefore, where three machining centres are being used, the average throughput time of the parts is given by:

$$W = L/\lambda = (L_q + u)/\lambda$$
$$= (0.718691 + 1.916667) \times 6$$
$$\approx 15.81 \text{ minutes}$$

Finally, as to whether we employ two or three machining centres, the probability of a part being late is given in either case by $P(W > 70)$, calculated respectively as follows. For two machining centres:

$$P(W > 70) = \exp(-70\mu) + \frac{u^k P(0)}{(k-1)!(k-u)(k-u-1)} \times \{\exp(-70\mu) - \exp[-70(k\mu - \lambda)]\}$$
$$= \exp(-70\mu) + \frac{u^2 P(0)}{1!(2-u)(1-u)} \times \{\exp(-70\mu) - \exp[-70(2\mu - \lambda)]\}$$
$$= 0.0022723 + (-1.02321) \times (0.0022723 - 0.6021504)$$
$$= 0.6160736$$

For three machining centres:

$$P(W > 70) = \exp(-70\mu) + \frac{u^k P(0)}{(k-1)!(k-u)(k-u-1)} \times \{\exp(-70\mu) - \exp[-(k\mu - \lambda)70]\}$$
$$= \exp(-70\mu) + \frac{u^3 P(0)}{2!(3-u)(2-u)} \times \{\exp(-70\mu) - \exp[-(3\mu - \lambda)70]\}$$
$$= 0.0022723 + 4.8745946 \times (0.0022723 - 0.0013683)$$
$$= 0.0066789$$

That is, while the number of late jobs from the three-machine case is very small, on average approximately 62% of the jobs from the two-machine case would fail to meet the pre-planned delivery date, and as a result cause loss due to their lateness. The average saving made by employing the third machining centre to reduce the number of late jobs is given by:

$$(0.6160736 - 0.0066789)\lambda C$$
$$= (0.60939347)10C$$
$$\approx 6.09C$$

Therefore, as long as the cost of running the third machining centre, M, is less than $6.09C$, the machining cell should employ three machining centres.

6.5.4 Statistical analysis of simulation outputs

So far we have reviewed the probability and statistical techniques associated with model construction, simulation execution and model validation. We now discuss the analysis of output data from simulation experimentation.

Estimation of the means of response

The estimation of the mean and variance of the response of a model under a particular set of inputs and operating conditions is of particular importance to simulation study, because the mean value of individual observations is often used as the system performance criterion in many manufacturing problems – for example, mean system throughput time, mean machine utilization, mean order tardiness, and mean work-in-progress level. Generally speaking, therefore, the primary purpose of experimenting with a dynamic non-deterministic simulation model is to obtain estimates of the mean value, μ, and, to a lesser extent, the standard deviation, σ, of certain response variables from a finite number of observations obtained from the simulation results.

For the non-terminating simulation experiment (that is, the simulation terminated not by a time limit, but by the event when all of the entities have passed the system), the simulation objective may be defined as (Law and Kelton, 1982):

$$U(X \mid t_0 = t_s) = E\left[\frac{\sum_{i=1}^{n} x_i}{n} \Big|_{t_0 = t_s}\right]$$

where $U(\ )$ is the simulation objective; the vertical bar (|) is read as 'given that'; n is the total number of entities to go through system; $t_0 = t_s$ defines the initial condition, that is, that the simulation commences at time t_s; $\Sigma_{i=1}^{n} x_i/n$ is a random variable (also denoted X) that defines the average of a system variable x_i based on the n entities; and $E(X) = \mu$ is the mean value of X. The above equation indicates that the experiments here are concerned with estimating a mean value $E(X)$ given the initial system condition at time $t_0 = t_s$.

As discussed above, among the simulation expriments carried out to evaluate system responses some will need the model to run in deterministic mode – for example, when comparing the relative effectiveness of alternative production decision rules – whereas others are concerned with the experimental results obtained from stochastic simulation runs – for example, when assessing the effects of uncertainties in operation lead times. When running in deterministic mode, the system outcome from a simulation run when subjected to a particular condition (i.e. one of the alternative routines to objectives, such as a production decision rule) may be treated as the 'typical' system responses under that particular operating condition. Therefore, with everything else remaining unchanged, the relative effectiveness of that particular condition can be assessed under these controlled, though rather artificial, experimental conditions.

On the other hand, when the model is used in stochastic mode it will no longer be sufficient to make a single simulation run and then treat the resulting simulation estimate as being the true system response. It is important to bear in mind that under stochastic experimental conditions the estimate X as defined above becomes a random variable itself, and therefore the resulting estimates of system response from any particular simulation run can only be regarded as one sample from the outcome distribution. This sample can differ from the corresponding true response.

Now we know that, as the average of a number of other quantities, X should be a normally distributed random variable. We also know that the maximum likelihood estimator of the mean of a normal distribution is simply given by the average value of the samples. As a result, therefore, we can conclude that the average value of $X_1, X_2, \ldots, X_m$, denoted μ', may be used as an effective estimator of the true system response $\mu = E(X)$, where μ' is given by:

$$\mu' = \frac{\sum_{j=1}^{m} X_j}{m}$$

where m is the number of observations obtained from different simulation runs (mutually independent, see below), which are to be used for the purpose of estimating.

Confidence interval of estimation

As an estimate, the average of observation obtained in the above manner should be treated with caution. If misleading conclusions are not to be drawn from the study, the systems analyst must be aware of the extent to which this type of estimate approximates the true systems response. If we define μ' as above and

$$\sigma'(m) = \left\{ \sum_{j=1}^{m} [X_j - \mu'(m)]^2 \Big/ (m-1) \right\}^{1/2}$$

then, based on the central limit theorem, we can predict that at a confidence level of $\Gamma\%$, the true systems response of the simulation model μ lies within the interval given by:

$$\mu'(m) \pm C[(\sigma'(m)/m^{1/2})]$$

where m is the number of independent observations used to be obtain the estimate, and $C = t_{(m-1,\Gamma)}$ is the upper Γ critical point for the t-distribution with $m - 1$ degrees of freedom (DOF). $[\sigma'(m)/m^{1/2}]$ in the above formula is often referred to as the *standard error* (that is, when m is sufficiently large), and $C[\sigma'(m)/m^{1/2}]$ as the *absolute precision* of the estimation. Some of the more useful values of the upper Γ critical point from the t-distribution are given in Table 6.8.

Example 6.5 Replications of simulation runs

As can be seen from Table 6.8, the values of the upper Γ critical point for the t-distribution decrease as the DOF increase. This implies that the accuracy of

Table 6.8 Critical values of C from the t-distribution

DOF	Γ				
	80.0%	*90.0%*	*95.0%*	*97.5%*	*99.0%*
1	1.376	3.078	6.314	12.706	31.821
2	1.061	1.886	2.920	4.303	6.965
4	0.941	1.533	2.132	2.776	3.747
6	0.906	1.440	1.943	2.447	3.143
8	0.889	1.397	1.860	2.306	2.896
10	0.879	1.372	1.812	2.228	2.764
20	0.860	1.325	1.725	2.086	2.528
40	0.851	1.303	1.684	2.021	2.423
75	0.846	1.293	1.665	1.992	2.377
100	0.845	1.290	1.660	1.984	2.364

the estimate increases as the number of observations increases.

To increase the confidence of the results, therefore, the expected average value $E(X)$ may be estimated by averaging the values of X resulting from making as large a number of independent replications of the simulation as practically possible, with the lengths of each replication determined by the same prespecified event (the completion of the nth job), and each replication subjected to the same initial condition. For example, since it has been shown above that statistically the error range of the estimates can be expressed as:

$$\pm[CS/m^{1/2}]$$

where S is the standard deviation, then at a confidence level of 90%, averaging the outcome of five independent simulation runs will reduce this standard error by approximately 69%, as compared to the estimate based on only two simulation runs. This is because (refer to the Γ-values given before) for $m = 2$

$$e(2) = [CS/m^{1/2}]_{m=2} = t_{(1.90\%)}(S/\sqrt{2}) = 3.078(S/1.414) = 2.176S$$

whereas for $m = 5$

$$e(5) = [CS/m^{1/2}]_{m=5} = t_{(4.90\%)}(S/\sqrt{5}) = 1.533(S/2.236) = 0.686S$$

where e(m) is known as the absolute error of estimation (the relative error of estimation is given by e(m)/(m)). Therefore

$$e(5)/e(2) = 0.686/2.176 \approx 0.315$$

That is, the error of estimation is reduced by approximately 69 per cent.

In practice, there are a number of ways in which independent replications of a simulation run can be obtained. Perhaps the most popular method is to use different random numbers for each replication. This method is based on the fact that a particular random seed will produce a particular stream of random numbers. If, therefore, a number of different random number seeds are used for the replication simulation runs under each condition, then this will result in a different and unique stream of random variables being used in each run. The simulation results from these replication runs can, therefore, be averaged to estimate the expected system response under that particular condition.

Another approach is to use a single long simulation run, and then divide it into separate portions in such a way that there are equal numbers of individual observations contained in each of the portions. If the division is done properly, it is possible to obtain mutually independent observations from these portions, which can then be used to reduce the error of estimation.

Example 6.6 Confidence interval of estimation

Suppose that the replication of five simulation runs produced the following average throughput times:

29.60 20.85 32.02 27.71 23.33

Under the particular operating conditions concerned, what is the mean response value of the system under study?

Solution

From the data given:

$$\mu'(5) = \frac{\sum_{j=1}^{5} X_j}{5}$$
$$= \tfrac{1}{5}(29.60 + 20.85 + 32.02 + 27.71 + 23.33)$$
$$= 26.702$$

and

$$\sigma'(5) = \left[\sum_{j=1}^{5} [X_j - \mu'(5)]^2 \Big/ (5-1)\right]^{1/2}$$
$$= \tfrac{1}{2}(8.3984 + 34.2460 + 28.2811 + 1.0161 + 11.3704)^{1/2}$$
$$= 4.5638$$

From Table 6.8 we obtain:

$$t_{(4,80\%)} = 0.941$$
$$t_{(4,90\%)} = 1.533$$
$$t_{(4,95\%)} = 2.132$$
$$t_{(4,975\%)} = 2.766$$
$$t_{(4,99\%)} = 3.747$$

At a confidence level of 80%, the true system response can be predicted to be within the confidence interval given by:

$$\mu'(5) \pm t_{(4,80\%)}[\sigma'(5)/5^{1/2}]$$
$$= 26.702 \pm 0.941[4.5638/5^{1/2}]$$
$$= 26.702 \pm 1.921$$
$$= \{24.781, 28.623\}$$

The confidence intervals at other confidence levels can be evaluated in a similar manner. These are summarized below.

Confidence level (%)	*Confidence interval*
80.0	{24.781, 28.623}
90.0	{23.573, 29.831}
95.0	{22.351, 31.053}
97.5	{21.057, 32.347}
99.0	{19.054, 34.350}

Finally, it should be pointed out that the above method of evaluating the confidence interval of an estimation is valid only if the X_js are themselves the averages of individual samples (e.g. those of x_is). For example X_j might be the average of 1000 throughput times given by the entities involved in the jth simulation run. If the individual samples – that is, x_is instead of X_js – are to be used to estimate the mean systems response, as in the case when a single simulation run is used, the method as described above will not apply.

A similar approach can be used to evaluate the corresponding confidence interval when only a single simulation run is used, provided that the number of individual observations n is very large. This is necessary because the 'strong law of large numbers' in probability theory states that the average

$$X = \frac{\sum_{i=1}^{n} x_i}{n}$$

will tend to approach the true value of μ only if n approaches infinity.

6.6 PCModel

Having discussed the structure and techniques associated with computer simulation in general, in this section we examine the working of a particular simulation package – PCModel – for the purpose of demonstrating the common features of today's simulation packages. PCModel is a PC-based simulation system intended for use by non-computer specialists to develop and experiment with discrete computer simulation. The relative simplicity and flexibility of the way in which this system builds a simulation model, together with the extensive graphic animation facilities provided, make this system an ideal tool for both education and practical simulation studies involving small-scale models.

A PCModel simulation model is essentially a model which is built around an 'overlay', that is, a graphic representation, of a system being simulated. The model uses a set of instructions to guide the system entities through the overlay along various routes as required in the real system and to collect statistical data where and when appropriate. The latest versions of PCModel are high-resolution graphics systems which use lifelike icons (created through the PC-based CAD system AutoCAD) for the purpose of animation. In this section, however, we will base our description and demonstration on an earlier, character-based version of the system, because this version is not as hardware-demanding as later ones.

Section 6.6.1 provides an overview of the basic concepts of PCModel. Using a simple example of a machine shop, section 6.6.2 looks at how a model can be developed in practice with this simulation language. Having worked through these two sub-sections the reader should be able to develop his or her own simulation models to solve the problems set in the exercises.

A brief guide to PCModel's instruction set is presented in Appendix C. The programs, exercise files and demonstration models of this section can all be obtained by contacting the author. This will enable you to run all the supplied demo applications, experiment with the run-time interactive controls, and use the overlay files supplied to create and run your own simulation models for the exercises included.

6.6.1 Overview of PCModel

PCModel is an easy-to-use and flexible simulation system which uses graphic animation to model the movement of entities through system processes. These entities are referred to as *objects*, which during a simulation are

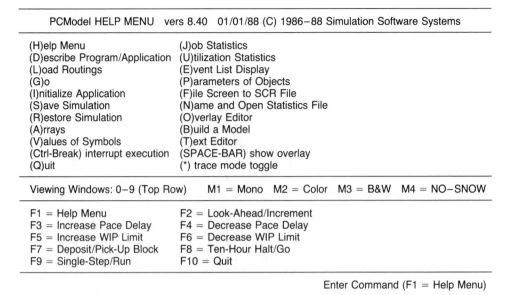

PCModel HELP MENU vers 8.40 01/01/88 (C) 1986–88 Simulation Software Systems

(H)elp Menu	(J)ob Statistics
(D)escribe Program/Application	(U)tilization Statistics
(L)oad Routings	(E)vent List Display
(G)o	(P)arameters of Objects
(I)nitialize Application	(F)ile Screen to SCR File
(S)ave Simulation	(N)ame and Open Statistics File
(R)estore Simulation	(O)verlay Editor
(A)rrays	(B)uild a Model
(V)alues of Symbols	(T)ext Editor
(Ctrl-Break) interrupt execution	(SPACE-BAR) show overlay
(Q)uit	(*) trace mode toggle

Viewing Windows: 0–9 (Top Row) M1 = Mono M2 = Color M3 = B&W M4 = NO–SNOW

F1 = Help Menu	F2 = Look-Ahead/Increment
F3 = Increase Pace Delay	F4 = Decrease Pace Delay
F5 = Increase WIP Limit	F6 = Decrease WIP Limit
F7 = Deposit/Pick-Up Block	F8 = Ten-Hour Halt/Go
F9 = Single-Step/Run	F10 = Quit

Enter Command (F1 = Help Menu)

Fig. 6.18 PCModel main system menu.

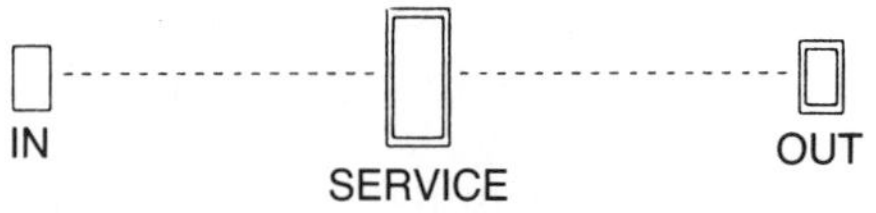

Fig. 6.19 PCModel overlay.

processed for certain specified lengths of time at one or more *stations* located along a number of user-defined *routines*.

A complete simulation model created through PCModel will always consist of two files: an *overlay* file (filename.OLY), which is a character-graphic overlay, or a graphic background of the system being simulated over which the objects will be cruising; and a *model* file (filename.MDL) which describes system variables, provides job information (for example, types of objects), specifies patterns of object movement, and establishes all of the logical interactions within the model.

At the systems level, PCModel provides an overlay editor and a text editor which are used to create and modify the overlay and model files, respectively. These editors can be evoked by choosing the relevant options from the main systems menu, which is shown in Fig. 6.18.

An overlay is actually a picture of the field of the simulated system. Figure 6.19 shows an example of such an overlay depicting a single-channel, single-phase queuing situation.

A model file contains the program of the model. A model program in PCModel consists of two sections – *system definitions* and *simulation routines*. System definitions is the section where the modelling environment is configured by using the so-called *directives*. Every item that is to be used in the subsequent routines must be specified in this section. A directive is a one-letter command, which together with its parameters carries out a particular data-declaration or system-specification task. Five types of directive are used in this section to initialize and direct the modelling frame: storage allocation;

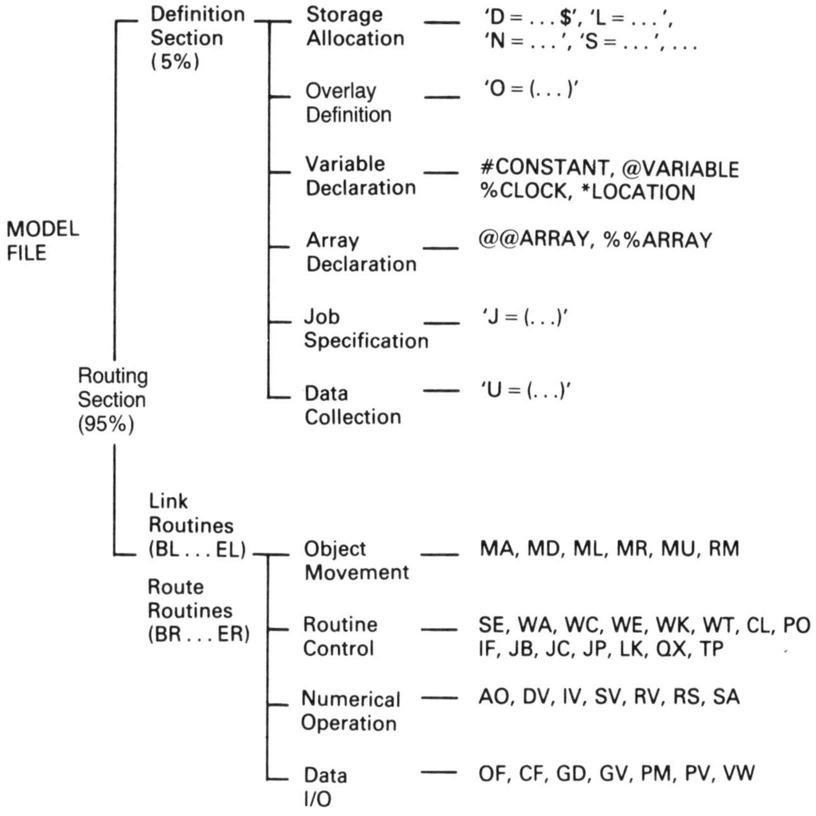

Fig. 6.20 Model structure, directives and instructions.

overlay definition; variable and array declaration; job specification; and requirement of data (utilization) collection.

Simulation routines constitute the major part of the program. Routeing instructions are two-letter words with or without parameters, which define the flow of objectives through the simulation. Five kinds of instruction are used to construct routines: object movement; routeing control; arithmetic/logical operation; data input/output; and execution control. Two different types of routine can be defined using these instructions – *routes* and *links*. The use of these two different types of routine is described in Appendix C. The structure of a model program, together with the associated directives and instructions, are summarized in Fig. 6.20.

6.6.2 Simple PCModel simulation program: a machine shop

To illustrate the structure and working of this simulation package, the following simple queuing situation will be modelled to provide an overview of the basic concepts.

The problem

Parts manufactured on a machine system arrive in random order according to the distribution given below:

Inter-arrival times

No. of parts	time (sec)
99	1
159	4
119	7
119	10
89	12
79	15
79	17
69	21
69	27
59	35
49	45
29	60

They may have to wait in a buffer if necessary. The machining times of the system are also randomly distributed as shown below:

Machining times

No. of observations	time (sec)
259	6
119	9
119	12
89	14
79	18
49	21
69	25
69	32
59	42
49	55
29	72

As a manufacturing systems analyst, you have been asked by the management to evaluate the mean throughput time, and the mean machine utilization of this machining operation. In addition, you are to investigate the effects of increasing machining capacity by adding an identical machine to the system, and indicate how in practice this simulation approach can be used to justify equipment investment.

Simulation modelling: single machine

Based on the information given, the following steps are required to build a simulation model using PCModel.

First, a little preparatory work is needed. As discussed, the data should be converted into a format suitable for the program, that is, as a table of cumulative probabilities. The cumulative distribution of the inter-arrival times is as follows:

Inter-arrival times

No. of parts	*time (sec)*	*Fraction of total*	*Cumulative value*
99	1	0.0972	0.0972
159	4	0.1562	0.2534
119	7	0.1169	0.3703
119	10	0.1169	0.4872
89	12	0.0874	0.5747
79	15	0.0776	0.6523
79	17	0.0776	0.7299
69	21	0.0678	0.7976
69	27	0.0678	0.8654
59	35	0.0580	0.9234
49	45	0.0481	0.9715
29	60	0.0285	1.0000

and that of the machining times as follows:

Machining times

No. of observations	*time (sec)*	*Fraction*	*Cumulative*
259	6	0.2620	0.2620
119	9	0.1203	0.3822
119	12	0.1203	0.5025
89	14	0.0900	0.5925
79	18	0.0799	0.6724
49	21	0.0495	0.7219
69	25	0.0698	0.7917
69	32	0.0698	0.8615
59	42	0.0596	0.9211
49	55	0.0495	0.9707
29	72	0.0293	1.0000

These data can then be entered into the model by using a dummy job and a route to set their values (SV – Set Value) at the appropriate place in the appropriate table, as shown in Box 6.6.

The second task is to create an overlay of the machine shop which should define the entry point into the system, the route line and the machine

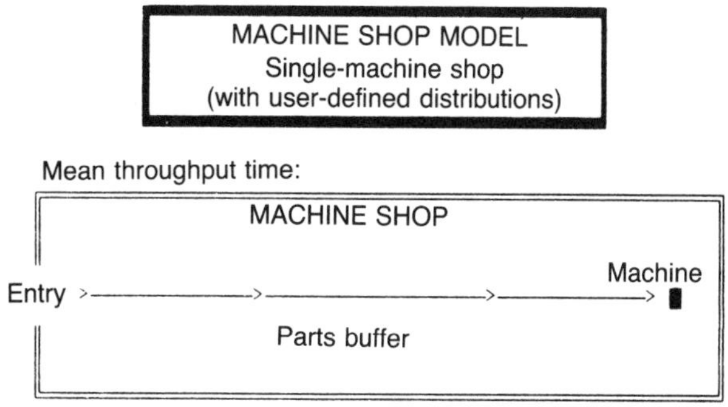

Fig. 6.21 Single-machine shop model.

location. A possible overlay for the single-machine case is shown in Fig. 6.21.

Finally, we can turn our attention to the program design. According to the structural requirements already set out, this model is structured as follows.

1. Define and initialize variables, define arrays for the distribution tables, define entry, machine and message points, as listed in Box 6.6.
2. Define job descriptions, overlay file and data collection point, etc., as shown. The J = directive specifies the make-up of the job:

   Job number = 1
   Symbol used to represent part on screen = P
   Route to take = No. 1
   Controlling parameters = the parts in this job are to be released only after job 2 is completed, i.e. it is necessary in this case first to initialize the distribution tables
   Number of parts in the job = 10 000

3. Define routes. Route 2 enters the cumulative distribution, whereas Route 1 defines the main part of the program where the machining process actually takes place:

   Enter parts into the shop
   Move around buffer area
   Move to machine
   Perform machining operation
   Calculate mean throughput time
   Exit system

   The coding of this route is as shown in Box 6.6. The BR instruction specifies the beginning of the route, and shows that a part should enter the system at screen location *ENTRY, and with a delay corresponding to %ARRIVAL. The functions of the rest of the instructions in this routine are very straightfoward and self-explanatory.

Box 6.6 A Sample PCModel Simulation Program

```
;************************************************************************
;
;  SIMPLE MACHINING MODEL
;
;  SIMULATION OF MACHINING OPERATION WITH USER
;  DEFINED ARRIVAL AND MACHINING TIME DISTRIBUTIONS
;
;************************************************************************

D=

                    SIMPLE MACHINING MODEL
                   SINGLE MACHINE SITUATION
$
;─────────────── VARIABLE INITIALIZATION ───────────────

  %ARRIVAL=(00:00)              ; Random time of arrivals
  %MCING=(00:00)                ; Random time machining
  @BF_SIZE=(50)                 ; Maximum number allowed in buffer
```

```
    *ENTRY=(XY(7,15))                 ; Entry location on the screen
    *MACHINE=(XY(65,15))              ; Machine position
    *MSG_FLW=(XY(35,9))               ; Message location on screen
    @NO_DONE=(0)                      ; Number of objects through system
    %T_TOT=(00:00)                    ; Total flow time
    %T_MN=(00:00)                     ; Mean flow time

    %%D_ARRIV=(1,12)                  ; Table of arrival time distribution
    %%D_MCING=(1,11)                  ; Table of machining time distribution

;
;———————— DIRECTIVE INITIALIZATION ————————

    O=(=)                             ; Overlay File with the same time
    L=(50)                            ; Wip Limit to 50 objects
    U=(1,Machine,*MACHINE)            ; Collect the utilization of machine

    J=(1,P,1,2,2,50,10000)            ; Job #1 - parts
    J=(2,C,2,0,0,0,1)                 ; Job #2 - a dummy object to initiate
                                      ;           distribution tables
;
;———————— MAIN ROUTE DEFINITION ————————

BR(1,*ENTRY,%ARRIVAL)                 ; Begin Route 1 at entry point

    RV(C,%ARRIVAL,%%D_ARRIV)          ; Get arrival time from distribution
    MR(@BF_SIZE,0:00.00)              ; Move object right @BF_SIZE steps
    MA(*MACHINE,0:00.00)              ; Load machine
    RV(C,%MCING,%%D_MCING)            ; Get machining time from distribution
    WT(%MCING)                        ; Carrying out machining operation
    MR(2,0:00.00)                     ; Unload and then leave shop
        LK(!CAL_FLW)                  ; Call !CAL_FLW to calculate mean
ER                                    ; End of Route 1

;
;———— DEFINITION OF USER SUPPLIED DISTRIBUTIONS ————

BR(2,XY(1,1),0)                       ; Route to set distribution tables

 RS(1234)
;                   TABLE 1
;Probabilities                              Interarrival times

 SV(%%D_ARRIV(00,00),0:00.00)           SV(%%D_ARRIV(01,00),0:00:00)
 SV(%%D_ARRIV(00,01),0:10.00)           SV(%%D_ARRIV(01,01),0:00:01)
 SV(%%D_ARRIV(00,02),0:25.00)           SV(%%D_ARRIV(01,02),0:00:04)
 SV(%%D_ARRIV(00,03),0:37.00)           SV(%%D_ARRIV(01,03),0:00:07)
 SV(%%D_ARRIV(00,04),0:49.00)           SV(%%D_ARRIV(01,04),0:00:10)
 SV(%%D_ARRIV(00,05),0:57.00)           SV(%%D_ARRIV(01,05),0:00:12)
 SV(%%D_ARRIV(00,06),0:65.00)           SV(%%D_ARRIV(01,06),0:00:15)
 SV(%%D_ARRIV(00,07),0:73.00)           SV(%%D_ARRIV(01,07),0:00:17)
 SV(%%D_ARRIV(00,08),0:80.00)           SV(%%D_ARRIV(01,08),0:00:21)
 SV(%%D_ARRIV(00,09),0:87.00)           SV(%%D_ARRIV(01,09),0:00:27)
 SV(%%D_ARRIV(00,10),0:92.00)           SV(%%D_ARRIV(01,10),0:00:35)
 SV(%%D_ARRIV(00,11),0:97.00)           SV(%%D_ARRIV(01,11),0:00:45)
 SV(%%D_ARRIV(00,12),0:100.00)          SV(%%D_ARRIV(01,12),0:01:00)
;
;                   TABLE 2
;Probabilities                              Machining times

 SV(%%D_MCING(00,00),0:00.00)           SV(%%D_MCING(01,00),0:00:00)
 SV(%%D_MCING(00,01),0:26.00)           SV(%%D_MCING(01,01),0:00:06)
 SV(%%D_MCING(00,02),0:38.00)           SV(%%D_MCING(01,02),0:00:09)
 SV(%%D_MCING(00,03),0:50.00)           SV(%%D_MCING(01,03),0:00:12)
```

```
SV(%%D_MCING(00,04),0:59.00)        SV(%%D_MCING(01,04),0:00:14)
SV(%%D_MCING(00,05),0:67.00)        SV(%%D_MCING(01,05),0:00:18)
SV(%%D_MCING(00,06),0:72.00)        SV(%%D_MCING(01,06),0:00:21)
SV(%%D_MCING(00,07),0:79.00)        SV(%%D_MCING(01,07),0:00:25)
SV(%%D_MCING(00,08),0:86.00)        SV(%%D_MCING(01,08),0:00:32)
SV(%%D_MCING(00,09),0:92.00)        SV(%%D_MCING(01,09),0:00:42)
SV(%%D_MCING(00,10),0:97.00)        SV(%%D_MCING(01,10),0:00:55)
SV(%%D_MCING(00,11),0:100.00)       SV(%%D_MCING(01,11),0:01:12)
ER
;———————— CALCULATE MEAN THROUGHPUT TIME ————————
BL(!CAL_FLW)
    IV(@NO_DONE)
    SV(OBJ%1,CLOCK)
    AO(OBJ%1,-,OBJ%ST)
    AO(%T_TOT,+,OBJ%1)
    SV(%T_MN,%T_TOT)
    AO(%T_MN,/,@NO_DONE)
    PV(*MSG_FLW,%T_MN)
EL
; ***************************************************************************
```

Simulation modelling: two machines

The overlay for this modified model is as shown in Fig. 6.22. The modelling process is very similar to that of the single machine case – in fact, only some minor modifications are required to turn model MC1 into MC2. This illustrates one of the most important features of computer simulation – its flexibility in modelling and evaluating different system configurations. The only difference between the two cases is that in the two-machine situation, once a part is moved to the head of the queue, it is necessary now to check the availability of either of the two machines. In the program this task is done by the five statements starting from label: TRY_MC.

Results of simulation

The two models were each run for approximately 8000 parts. The data collected are as follows:

One-machine system

Mean throughput time = 12 mins 36 sec
Mean machine utilization = 99.95%

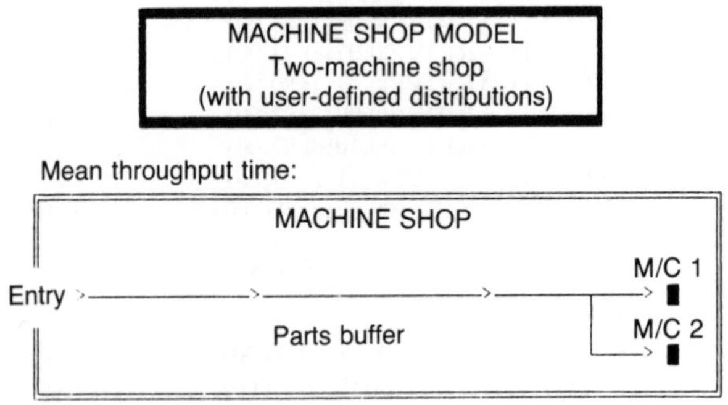

Fig. 6.22 Two-machine shop model.

```
;
;------------------------ MAIN ROUTE DEFINITION ------------------------
BR(1,*ENTRY,%ARRIVAL)              ; Begin Route 1 at entry point

   RV(C,%ARRIVAL,%%D_ARRIV)        ; Get arrival time from distribution
   MR(@BF_SIZE,0:00.00)            ; Move object right @BF_SIZE steps
   RV(C,%MCING,%%D_MCING)          ; Get machining time from distribution

 :TRY_MC
     JC(A,*MACHN1,:MACHN1)         ; Jump to label: MACHN1 if not busy
     JC(A,*MACHN2,:MACHN2)         ; Jump to label: MACHN2 if not busy
     WE                            ; Wait till next event, (some one moves)
     JP(:TRY_MC)                   ; Then try the machines again

 :MACHN1
     MA(*MACHN1,0:00.00)           ; Load machine 1
     WT(%MCING)                    ; Carrying out machining operation
     MR(2,0:00.00)                 ; Unload machine 1 and
     LK(!CAL_FLW)                  ; Call !CAL_FLW to calculate mean
     JP(:END_MC)                   ; Then Jump to :END_MC
 :MACHN2
     MA(*MACHN2,0:00.00)           ; Load machine 2
     WT(%MCING)                    ; Carrying out machining operation
     MR(2,0:00.00)                 ; Unload machine 2 and
     LK(!CAL_FLW)                  ; Call !CAL_FLW to calculate mean
     JP(:END_MC)                   ; Then Jump to :END_MC

 :END_MC                           ; Machining done, leave the shop

ER                                 ; End of Route 1
;
```

Two-machine system

Mean throughput time	= 39 sec
Utilization of machine 1	= 67.3%
Utilization of machine 2	= 53.8%

The difference in performance of the two systems is clearly revealed. In the single-machine case, the machine is overloaded. The utilization of the machine was 100% nearly all the time, and parts had to wait for a long time in the buffer. The situation is clearly illustrated by Fig. 6.23. It is quite obvious that a single machine would not have sufficient capacity to cope with the intended workload. The system was not stable, as illustrated by the fact that the mean throughput time was still slowly increasing even after such a relatively long simulation time.

As one might expect, with the addition of an identical machine the queuing situation was significantly improved, with many fewer WIP items. As far as machine capacity and workload are concerned, this represents a much more adequate situation than before. However, this was achieved at the expense of machine utilization. The average utilization levels of the two machines are now perhaps both slightly lower than the level that one would usually consider satisfactory.

In practice, information such as this is obviously useful to a manager or systems designer, in that they can vary system parameters to strike a desired balance between them. In this particular case, for example, they can deter-

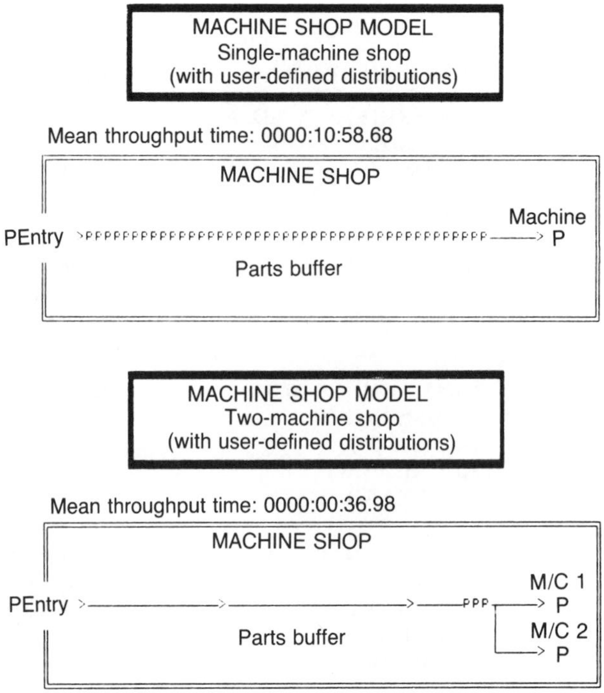

Fig. 6.23 Simulation display.

mine from the results whether or not a second machine is justified, depending on the relative importance assigned to machine utilization, throughput time and WIP situation.

The above description has provided some simple models to illustrate the basic concepts and techniques of a simulation language. It has been illustrated that a simulation package, such as PCModel, provides a simple enough tool to model easily the less complicated problems frequently encountered in manufacturing. However, much larger and more complicated simulation models can also be created with ease.

6.7 Computer simulation and artificial intelligence

Right from the start, artificial intelligence (AI) has been a somewhat controversial subject area. The fear of computer-controlled automatic devices taking over jobs and causing national disasters has resulted in many disputes about the application of AI techniques. Despite these fears, the first practical application of AI, known as *expert systems*, appeared in the early 1970s. By the 1980s governments were encouraging the development of AI through initiatives such as the Esprit programme in Europe, the Fifth Generation programme in Japan, and the Alvey programme in the UK. This recent surge of interest in AI and, in particular, expert systems has lead to many simulation researchers and practitioners discovering a similarity between AI and simulation techniques. In this section we examine the approaches and techniques of the recent trend to integrate the two.

6.7.1 Expert systems versus computer simulation

The nature and applications of expert systems

Expert systems have found practical application in a number of areas, among them diagnosis, monitoring, interpretation, design and planning, and control. An expert system is basically a software program that contains knowledge and process rules with the purpose of emulating the decision-making process of an expert in a particular area. It therefore uses a knowledge base consisting of high-quality knowledge about a specific problem domain, which is organized into rules or some other representation.

Expert systems differ from other computer software in their architecture: an expert system does not work with a fixed logic pattern where all the steps of a problem are predetermined, as in a simulation model. Instead, it uses a knowledge base which is composed of a series of general rules and an inference system or, in expert systems terminology, an *inference engine*. The inference engine governs how the facts provided by a user should be dealt with by the rules, and decides what inferences should be drawn based on changing facts. Particular emphasis is also placed on the user interface. The user interface presents questions, prompts, explanations and conclusions to the user, all of which must be totally free of ambiguity and open to interrogation.

Practical expert systems are usually built by using either an AI-type programming language such as LISP or PROLOG, or with an *expert shell*, which is a computer program specially designed for the easy construction of expert system applications. The idea of using an expert shell to develop expert system applications is exactly like that of simulation packages – it is a tool to allow non-specialists to build working applications without having to become too involved with the underlying logic of the control mechanism involved. In fact, even the tasks of building expert system applications are extremely similar to those involved in simulation model construction, including:

1. *Identification*: this is concerned with how to characterize the key points in the process.
2. *Conceptualization*: this is equivalent to the conceptual modelling stage of a simulation study. The concepts, relations, logic, and level of details to be represented, etc., are considered.
3. *Formalization and implementation*: during this stage the conceptual model of the system is to be transferred into a working system using an application building tool.
4. *Validation*: the resulting system should be tested for its validity. Refinements, modifications, or even redesign of the system may be necessary before the system can be put to practical use.

When deciding whether expert system technology should be applied to a problem many factors must be considered. The practicality of the proposed system must be judged. A good guideline is to find a well-documented area where a skilled person is having to carry out what is considered to be a mundane job. Then the possibility of improving efficiency (in terms of cost savings, for example) by applying cheaper, faster, better or more thorough analysis must be investigated. If successful, an expert system implementation can automatically perform tasks for which specially trained people have conventionally been employed. The advantages and failings of expert systems over human experts are summarized in Table 6.9.

Table 6.9 Expert systems and human compared

Expert systems	*Human experts*
consistent	inconsistent
multiple locations	one location
cheap	expensive
easily monitored	difficult to monitor
permanent	temporary
easily modified	difficult to modify
easily documented	difficult to document
easily integrated with other systems	difficult to integrate
inflexible	adaptive
unimaginative	creative
limited knowledge	wide range of knowledge
limited (or no) real intelligence	with real reasoning capacity
needs updating	self-improving
only symbolic input	many input types

Expert systems and simulation compared

Simulation is a means of studying the behaviour of complex problems that are too difficult to solve mathematically. Knowledge-based expert systems are very similar in that they are also a means of making decisions in complex situations in which mathematical techniques are difficult or impractical to apply.

However, there is a marked difference between the approaches in terms of the solution they provide. Expert systems usually give *prescription*, whereas simulation models provide *prediction*. That is, unlike a simulation model, an expert system suggests a course of action and provides a rationale or explanation for the suggested course of action. An expert system is also able to absorb more knowledge and hence become more knowledgeable (relatively speaking, of course) and has the potential of making itself a more useful decision-support tool. Consequently, for certain applications the traditional simulation tools alone do not compare with the benefits of knowledge-based expert systems in terms of prescription and explanation. (This point is well illustrated by the previous examples of simulation experimentation, section 6.6.2, in which a better solution might only be found through trial and error by experimenting with the models.) On the other hand, expert systems are not constructed to deal with the complex dynamic behaviour of large systems.

Simulation and expert systems can be used as complementary technologies in order to provide a more powerful decision technique than either on its own. Traditionally, for example, the model builder would construct a simulation model using a general-purpose programming language or a special simulation language. The model would then be validated, executed and then the results analysed and interpreted in the manner already described. Many aspects of this process could be automated through the use of AI techniques and, more specifically, expert systems. The ideal compromise for the use of a management and design decision tool, therefore, would be to have an integration of the two tools.

6.7.2 Knowledge-based simulation models

Investigations into the potential use of AI techniques in simulation modelling began as early as 1977 when AI techniques were proposed for the purpose of assisting computer model building. When simulation problems become complex the expertise required to solve them becomes very varied. Also, the number of models and the number of experiments carried out with them can become very large so that their efficient management becomes difficult. This is one of the main underlying reasons why much attention has so far been given to the possibility of creating systems that organize and process knowledge about simulation models.

There are two main applications of expert systems in simulation modelling – to help the user build and run a model, and to help the user interpret the results of a simulation run and develop the model further. An ideal knowledge-based simulation system, therefore, should be able to: accept a problem description and synthesize a simulation model by consulting an appropriate knowledge base; accept a goal in the form of expectations and constraints and produce a scenario that satisfies the stated goal; explain the rationale behind any decisions made; learn from past experience; present both the resulting model and the results with a high degree of accuracy and readability so as to increase user confidence.

Model construction

Given that expert systems can emulate human experts in certain fields, it appears logical that they should have an application to programming. Much programming in general-purpose languages is too diversified for expert system application. However, programming for simulation, whether in a special-purpose or a general-purpose language, is sufficiently limited in variety for this purpose.

For simulation and expert systems to be 'integrated' in this regard, in addition to the traditional need for a simulation model to express logically both the static structure and the dynamic behaviour of a system being modelled, there is a fresh need to link the model structure into an environment which is suitable for a knowledge-based program generator to work in. Consequently, for the automatic construction of simulation models two main bases are required: a model base which contains the executable numerical sub-models and a knowledge base that contains the semantic information necessary for the use of these models. It is not very difficult to satisfy these requirements in practice. One way to realize this, for example, is to build a simulation model by selecting and then arranging system entities and procedures from a standard library. An expert system can then be built to generate simulation models using this library according to the user's requirements (a bottom-up approach). One scheme for the knowledge representation of simulation modelling, therefore, is the so-called *object-orientated representation* in which every element of a model may be thought of as an object that is described by its *properties*, as discussed in Chapter 4.

Based on the above reasoning, among the current trends, model generators are one of the most favourable environments for the implementation of an expert system in relation to simulation. These systems are referred to as intelligent front ends (IFEs), which have been described as expert systems that interface the user with the package and generate the program code of a simulation model following the dialogue with the user. (Most of the existing simulation program generators, however, perform only some of the functions of IFEs.) A knowledge-based approach in this regard should allow the combined automation of both the model-specification and the model-construction processes. Basically there are three alternative approaches to automating these traditional information-acquisition and model-programming tasks of simulation:

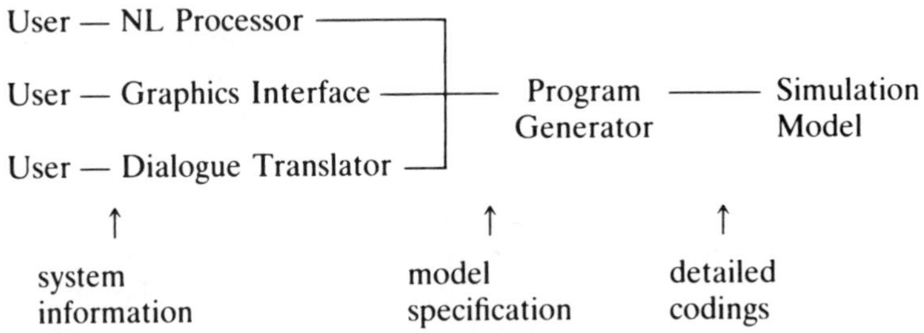

As shown, a detailed definition of a simulation model may be obtained from a natural language description of the system to be simulated by a user through a natural language (NL) processor. This specification may also be produced through a graphics interface, which allows the structure of the layout of components to be described. Similarly, this can also be achieved through a dialogue translator, which obtains the necessary information through question/answer communication. Once the detailed information has been obtained by the system through these processes, the model itself can then be generated by an automatic programming system. Each of these approaches interactively produces the detailed model specification to satisfy the user's requirements.

An NL approach

An example of the NL-based interface is provided by the expert simulation system for an electronics manufacturing plant developed by Ford and Shroer (1987). This system includes a transformer and understander, a simulation writer, a simulation analyser, and a simulation language. It transforms normal English sentences into an internal representation through the transformer, and makes the necessary inferences in the understander. The output from this NL interfacing function is then in a format suitable as the input for the simulation model generator.

A dialogue translator

An example of model generators using a dialogue translator is given by a system known as Knowledge-Based Model Construction (KBMC) (Murray and Sheppard, 1988). In this system, automatic programming is used to construct simulation models of queuing systems, utilizing the simulation functions provided by SIMAN.

KBMC requires the user to input the pertinent information about the system being modelled through a structured and interactive dialogue. The information is then retained within the system. Next, the system uses its knowledge, through a combination of program-transformation and goal-directed search, to guide the dialogue for model specification and to construct the simulation model automatically. The model generator uses a knowledge base of models and model parts which has been developed previously. Such a data base of sub-models provides the basis for the expert system that allows the user to search the data base for an existing suitable model or to create a new one from the existing models.

Graphic-based modelling systems

In the last section we examined one particular simulation system. Although adequate here for the purpose of revealing the structure and working of typical simulation processes, compared with some of the more recent packages the modelling approach discussed is not the easiest to use in practice, because it is still very much based on a programming exercise. To overcome this problem, many of the latest simulation packages now provide a graphic interface through which the modeller enters a complete description of the system under study.

Generally these systems are menu driven, and only require the user to perform the following simple tasks:

1. Provide configuration information. A standard graphic library of workstations and other items is provided. A graphic presentation of the model can be constructed simply by selecting the required elements from this library and then placing them at the desired location on the screen, through a point-and-click process. For instance, SIMFACTORY provides a set of five such building blocks: Receivers, Stations, Queues, Transfers and Conveyors (Fig. 6.24). These building blocks are closely related to the typical structural and the dynamic element of a discrete-event system, as identified earlier in this chapter. That is, Receivers model the entry point where materials arrive for processing; Stations represent the resources which are responsible for the processing tasks; Queues are the buffer areas where work waits for the service of a Station; Conveyors model various material-handling equipment; and Transfers, with their built-in control rules, specify how a system keeps the work flow logically through its various processes. Figure 6.25 shows an example of a graphic background thus constructed.

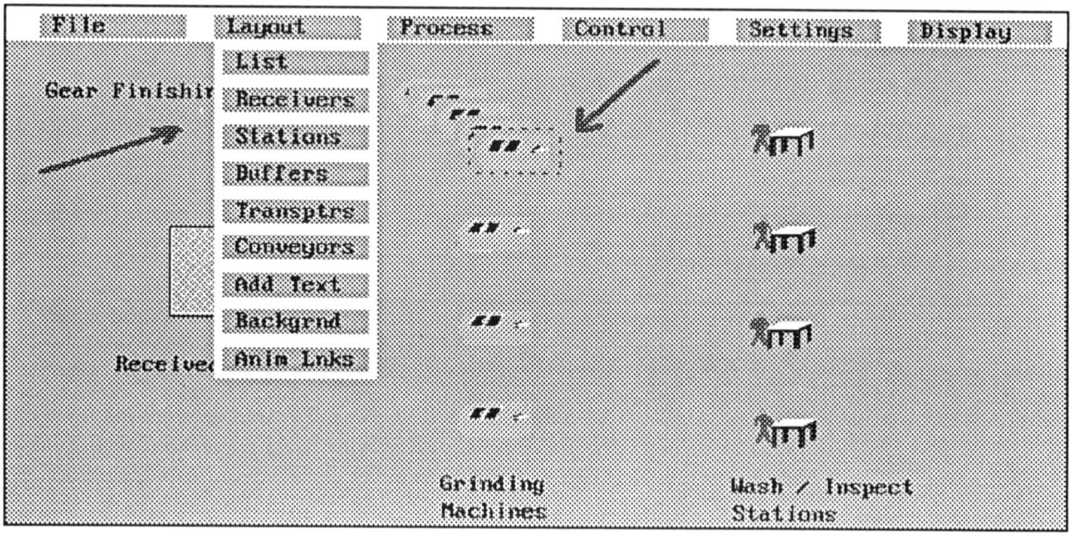

Fig. 6.24 SIMFACTORY's graphic menu.

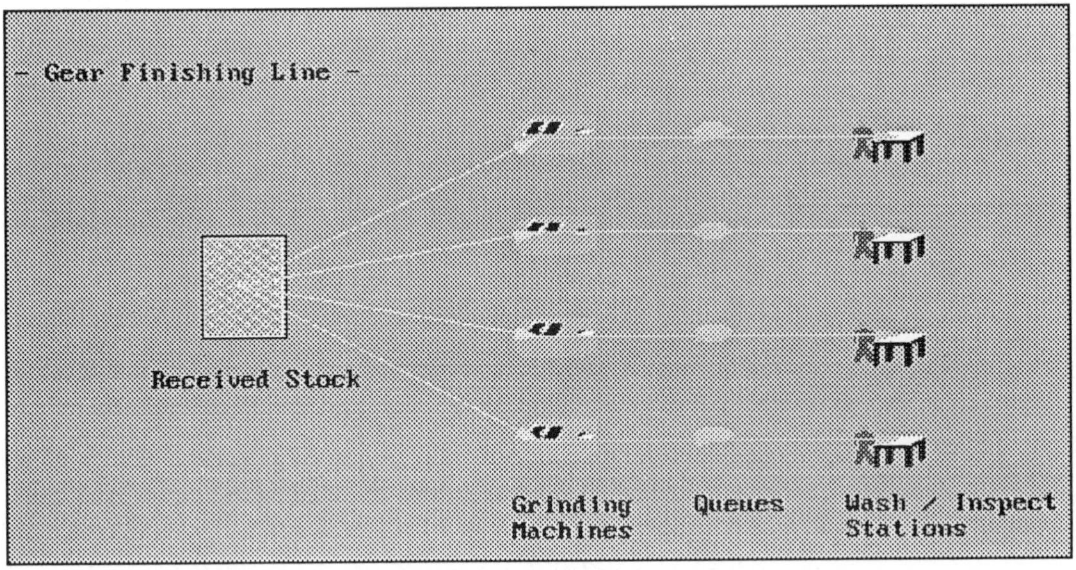

Fig. 6.25 Example of a graphic background.

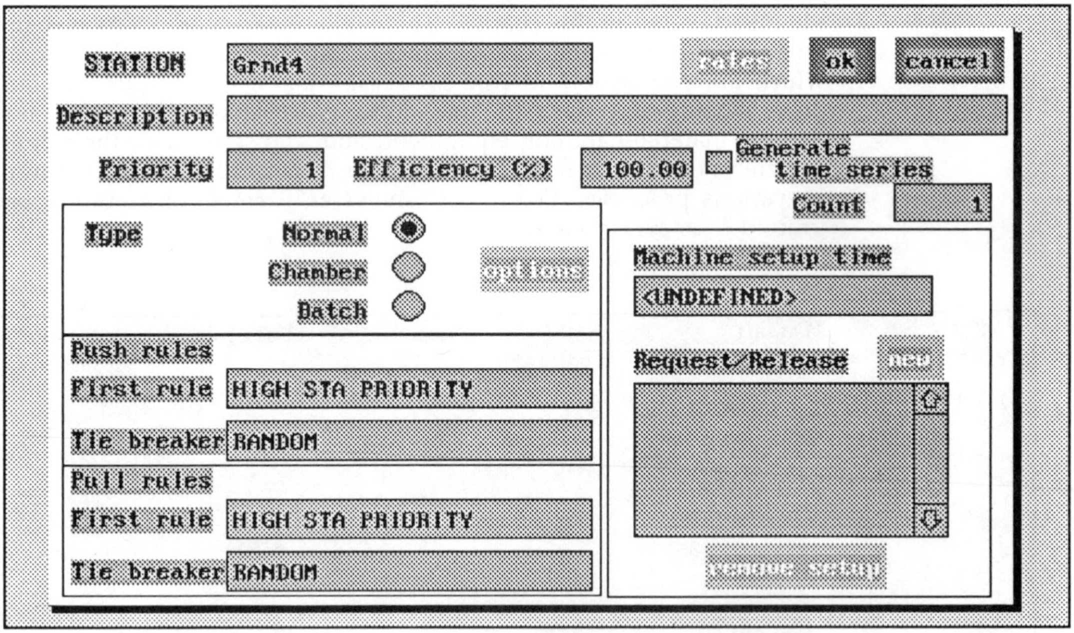

Fig. 6.26 SIMFACTORY's input menu for work-station description.

2. Provide work-station information. The standard work-station information such as processing times and the type of distribution can then be defined through a predetermined format, as shown by the examples given in Figs 6.26 and 6.27. In addition, other relevant information such as queuing rules and flow controls can be specified explicitly.

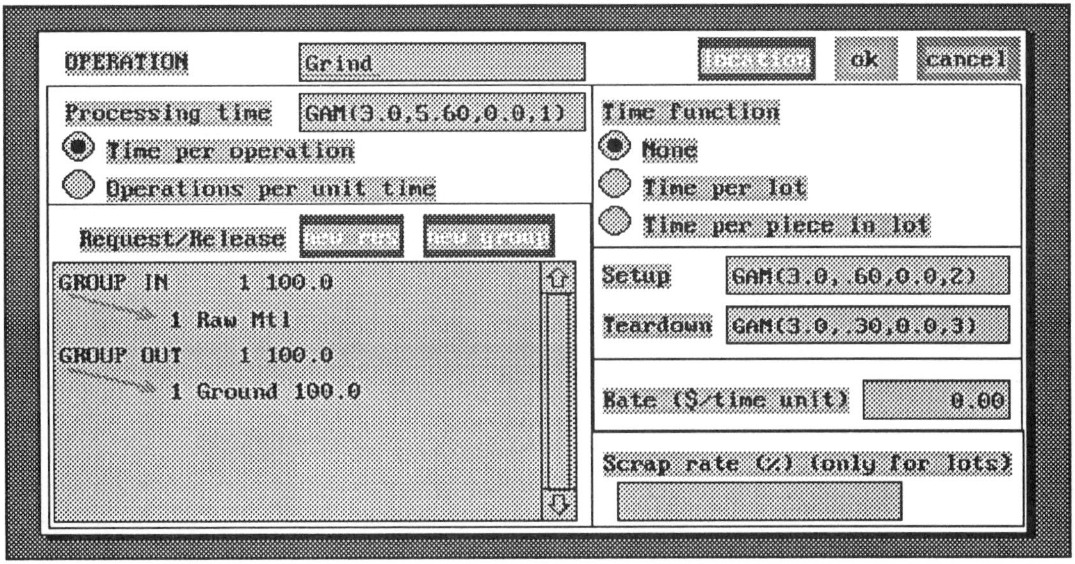

Fig. 6.27 Description of individual operation.

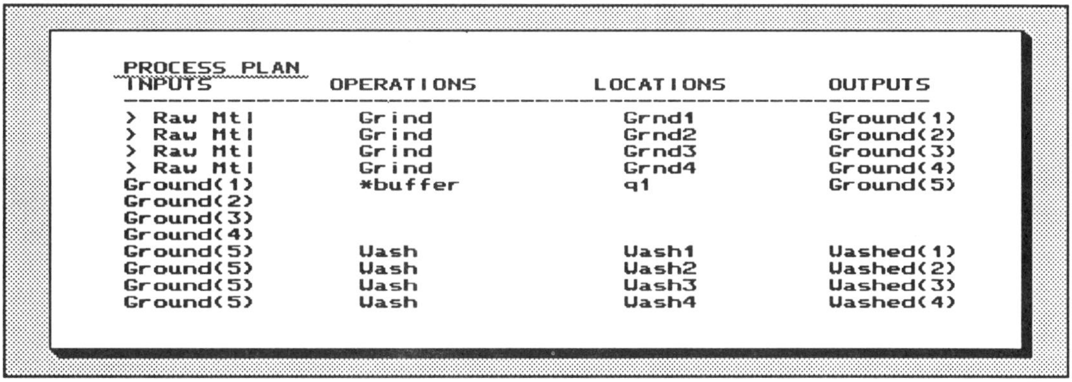

Fig. 6.28 Route specification.

3. Provide job information. This is concerned with the jobs to be tackled, describing the materials, parts and sub-assemblies involved. In addition, the technical routes that these jobs are to follow through the system (the arrival, load, work, and unload sequences) should be specified. In relation to the factory layout and other technical information described so far through (1) and (2), such information can again be provided to the system through input screens, as shown in Fig. 6.28.
4. Provide system information. Finally, some of the common system parameters need to be specified. These are mainly concerned with run-time control of the simulation process specifying, for example, the length and number of simulation runs needed and/or the confidence level to be achieved by the simulation results.

SFII.5 (v3.10) RUN7 Processing Station Report
(Data collected for 4800.0 MINUTES in 5 replications)
Degree of Confidence - 95.0%

Statistic	Status (%)								Parts.
	Work	Idle	Reqst	Setup	Blckd	Prty	Pass	Trdwn	
				--- Grnd4 ---					
Mean	68.8	0.0	0.0	8.6	18.6	0.0	0.0	4.0	40.2
Std. Dev.	5.3	0.0	0.0	0.8	5.4	0.0	0.0	0.4	0.3
Lower C.I.	63.8	0.0	0.0	7.9	13.4	0.0	0.0	3.7	39.9
Upper C.I.	73.9	0.0	0.0	9.3	23.7	0.0	0.0	4.3	40.5
				--- Wash4 ---					
Mean	73.7	26.3	0.0	0.0	0.0	0.0	0.0	0.0	40.4
Std. Dev.	2.6	2.6	0.0	0.0	0.0	0.0	0.0	0.0	0.3
Lower C.I.	71.2	23.8	0.0	0.0	0.0	0.0	0.0	0.0	40.1
Upper C.I.	76.2	28.8	0.0	0.0	0.0	0.0	0.0	0.0	40.7

Fig. 6.29 Example of simulation output.

Once the detailed information has been obtained in such a manner, the simulation systems then automatically generate the coding of the simulation model, and during simulation keep statistics of the relevant system parameters and generate detailed reports in a number of standard formats (such as pie charts, histograms, time series and tables; Fig. 6.29). Therefore, the use of a graphic interface by these systems for the purpose of system modelling has totally eliminated any need for programming. Examples of this type of package include, among others, ProModel, MicroSaint, WITNESS and SIMFACTORY.

Simulation evaluation

The interpretation of simulation results is another major area where expert systems and simulation have been used together as complementary techniques. The value of this becomes clear when it is realized that while simulation is normally used to evaluate suggestions, expert systems make suggestions based on evaluation. Operating together, skilled manual intervention can be reduced and a solution may be arrived at more reliably and more rapidly.

One major disadvantage identified for the simulation approach to problem-solving is that only one set of circumstances can be tested by an individual simulation run. An expert system designed and integrated into the modelling language could overcome this by selecting which variables are likely to give a desired effect, and how they should be altered and controlled. Such a system could then adjust the variables and repeat the simulation, iterating until an optimum is found. This could be applied, for instance, to change the assumed load to utilize the maximum capability, or to experiment with different routines or line-balancing arrangements. Expert system applications used for this purpose has been referred to as *intelligent back ends* (IBEs).

One obvious back end application of expert systems, for example, is concerned with the statistical analysis aspects of simulation design and result evaluation. As outlined in section 6.5, the theory and techniques used in this regard are well developed and have been already clearly defined in a format ideally suited for expert system applications. It should be a straightforward process to transfer this body of knowledge into an expert system for the purpose of, for instance, choosing probability distributions or evaluating the accuracy of estimation.

Knowledge-based simulation of FMS

To illustrate the use of expert systems as an IBE of simulation, let us examine one piece of research work that has been carried out into the development and application of an expert system that could analyse the output from an FMS simulation model, determine whether operational and financial objectives were met, identify design deficiencies and then propose designs which could overcome the identified deficiencies (Mellichamp and Wahab, 1987). The work was based on the observation that the design decision for FMSs appeared to be an excellent application area for expert system techniques, because it was a well-defined and narrowly bound area from which design experts could make good design decisions. The decisions also required heuristic problem-solving as opposed to algorithmic or optimization approaches.

In developing the expert system the design approach of human FMS designers was analysed, and then the knowledge used by these real experts to perform FMS design was documented and represented in a framework that could facilitate the development of an expert system. The various financial and operational objectives of the model were also outlined.

The primary goal of the operational design was output. During such a design project, when the production goals were not met, the designers had to discover ways of modifying the system to make it more proficient. This was usually carried out by looking for production bottlenecks in the system, and by eliminating these bottlenecks using more efficient equipment or procedures. Bottlenecks can usually be spotted by considering equipment utilization and queue lengths. Very high or low utilizations and excessive queue lengths with respect to desired values or to utilizations and queue lengths of other equipment in the system usually indicate an imbalance among the production facilities. Upon the discovery of a problem such as a bottleneck, the expert system developed would recommend design changes and incorporate them into the simulation model. The process would be repeated until an acceptable design resulted.

6.7.3 Conclusion

It is evident from literature that the integration of computer simulation and artificial intelligence has been regarded by many as the next natural step in this particular application area of computer technology. The use of artificial intelligence with traditional simulation techniques can provide a very useful analytical tool. Simulation generators can assist system analysts or managers in model creation and model updating, as well as in the analysis of alternative scenarios. As a result, some believe that the future uses of expert simulation systems within manufacturing will be extensive, because they have the potential to allow manufacturing systems to be evaluated and analysed under various design changes, alternative control actions, different procedures and new policies, with a minimal need for expert intervention. It has even been suggested that expert systems will ultimately change the way in which information is managed in manufacturing, and that the future of computer simulation will be determined by the future of AI!

However, there is still much work to be done in the area of general goal processing and explanation of recommendations. Although these approaches do open up new possibilities in the analysis and design of manufacturing systems, the use of the AI programming approach does not automatically make such a tool so powerful that very complex problems can be solved without intervention of a user. The developers of the system quoted in section 6.7.2, for instance, concluded from their research that the validity of the statement that expert systems can be developed which perform at the level of human experts is questionable. They indicated that the particular system was only a step towards the ultimate goal as it had a limited knowledge base in terms of heuristics for analysis. A true expert may use a dozen or so guidelines or indicators to locate a bottleneck in a system, whereas the FMS design expert system used only a few primary rules. In addition, a human expert may also backtrack over previously tested or discarded strategies, whereas the expert system is more restrained in generating design modifications.

One major disadvantage of using expert systems in a manufacturing environment is the diversity of manufacture. This implies that there may be few situations where the same expert system will apply to different companies, or even different sites of the same company (the use of PCModel's model generator, for example, is limited strictly to a narrow range of manufacturing systems). Therefore, in most cases when a specialized expert system is to be used in simulation, it must first be established and tested before being incorporated, adding significantly to the time and cost of the project.

It may be, however, that there is a set of problems and situations which have broadly similar properties, and which, while requiring different simulation models, may be able to use a common expert system to suggest qualitative improvements. A good guideline, therefore, when looking for suitable applications is to find the key process in the manufacturing system where an improvement of, say, 1% in efficiency can bring large cost savings, as well as the processes which share certain common key features between them.

Finally, it should be pointed out that not everyone is in favour of this new technical trend. Hill and Roberts (1987), for example, argue that the potential of AI techniques in simulation modelling design must be challenged because such a process is creative and unique in nature. Issues of problem objectives, level of model details, solution perspectives, use of models, and so on, all require the full use of human intelligence and discretion. What can be accomplished in term of the integration of AI and simulation, therefore, remains open to debate.

6.8 Investigation into the operation of OHMS: a case study

This case study describes a project which utilized a large-scale simulation model to investigate the operation of an OHMS in its original settings (Wu, 1990). The investigation involved the analysis of the manufacturing systems of two major UK companies. The simulation model integrated their key features, and concentrated on the information and material flow patterns within the manufacturing system. The model was then applied to company data on contracts, procurement and operational lead times, and work-in-progress. Some experimental results from the model, which reveal certain overall systems behaviours of the manufacturing operation, are presented.

The project was concerned with various aspects which influence the efficient operation of an OHMS, including the effects of the inherent system characteristics and the decision policies used by management to plan and control production within such a system. Two major UK capital equipment manufacturers, both in the same market, both with characteristics typical of an OHMS organization, were involved in the study. It was desirable to see how the managements of different companies operated within a similar business environment, and how they tackled a similar set of problems. From an academic point of view, it is interesting to note that due to their make-to-order nature, both companies have adopted similar manufacturing practices. They carry out comparable design, prototype making and testing, parts manufacturing, and final assembly activities. Due to the similarities in their structures of organization and operations, it was decided that modelling efforts would be concentrated on one of the companies. It is generally true that the conclusions drawn from the model of a particular system are applicable to that system only. In this special case, however, the possibility of obtaining conclusions having a wider applicability could be anticipated. This makes the particular investigation a more general case of system evaluation rather than a particular case study, which is what the majority of the simulation studies usually accomplish.

6.8.1 Company data and the simulation model

A large-scale model of the company was developed, validated and installed at the engineering department of the company which is a dynamic, discrete computer simulation model capable of simulating the manufacturing activities of its OHMS operation for either online deterministic planning or offline deterministic/stochastic research operations, at both medium and more detailed levels of data aggregation. A large quantity of company data was involved in the development and operation of the model, which may be classified into three relevant data groups with respect to the structure of the model constructed: production-process data; production-contract and production-capacity data; production-cost and performance-assessment data. These three groups of information were interrelated to form the experimental frame of the model (presented graphically in Fig. 6.30). As can be seen, the frame consists of three sub-frames – the *model* sub-frame, the *input and control* sub-frame, and the *output and analysis* sub-frame.

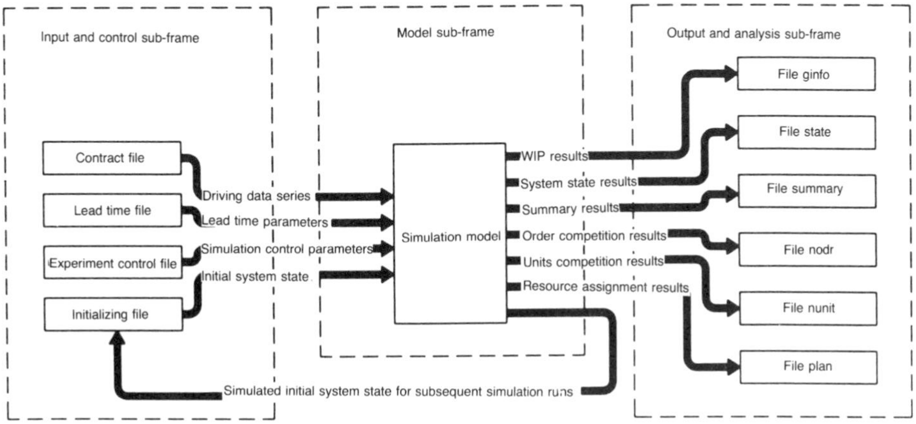

Fig. 6.30 Overview of the modelling frame.

The first group of information, concerned with the manufacturing process itself, is used to develop the model sub-frame, which can be viewed as the core of the total experimental frame and performs the task of simulating the activities of the organization.

The second group of information is used in the input and control sub-frame. The data of this information group are stored in several separate data files of the sub-frame – the real contract series file, the real lead-times file (including material and other bought-out item supply lead times, and operation processing lead times) and the control file. The first two files are used to provide the model with input data and system parameters, and the control file is used to control the execution of the simulation runs. While real company information is stored in the first two files, the control file stores simulation control variables for a set of simulation runs. These data can be provided to each single simulation run in the set in a predefined manner, allowing a large number of simulations to run on the computer without the attention of the analyst. These control variables are defined by the user according to the requirements of the simulation runs.

The third group of data is used in the output and analysis sub-frame. As can be seen from Fig. 6.30, this sub-frame consists of a number of output files and a set of analysing routines. In the output files are stored the model system responses to the inputs – the time series of contract-completion information together with other system-performance information. The analysing routines not only format and display the data contained in these output files, but also carry out assessments on system performance according to certain performance criteria. The company information of the third data group is utilized to define these criteria.

6.8.2 Simulation applications

The model has provided a wide spectrum of experimental opportunities concerning OHMS operation. To carry out a systematic investigation into the operation of the company under study, a series of offline experiments with the model were designed, which covered several problem areas that had been identified as particularly significant in such a system:

Effects of system uncertainties

1. Internal lead-time uncertainties.
2. External lead-time uncertainties.
3. Customer-demand uncertainties.

Effectiveness of production decision rules

4. Due-date assignment decision rules.
5. Production-scheduling rules.

WIP cost analysis of the system

6. Cost analysis at system level.
7. Cost analysis at detailed WIP pool level.

The first three areas are concerned with the effects of changing system parameters, while the last four are concerned with employing alternative production decision rules. These can both be readily achieved by using the built-in features in the input and control sub-frame. For example, difficulties associated with the production planning and control of this type of manufacturing can be related to the following sources of uncertainty:

1. External lead-time uncertainties: concerned with the lead time(s) of function block A_{322}, that is, the function of material and bought-out items acquisition.
2. Internal lead-time uncertainties: involve the lead-time uncertainties occurring in function blocks A_{222}, A_{421}, A_{422}, A_{423}, and A_{424}.
3. Environmental (contract due-date) uncertainties: for the OHMS operation, the due-date assignment process is always a balancing act between the needs of customers and the ability of the manufacturer to satisfy these. From the manufacturers' point of view, the due-date situation is deterministic when the market allows the manufacturers to have the control of the due-date. On the other hand, if market power is with customers, then the process becomes stochastic. Therefore, uncertainty exists within function block A_{11} concerning how due-dates are assigned to the contracts.

With the model set in stochastic mode, it can be used to estimate the system's response to these uncertainties. For instance, Fig. 6.31 illustrates

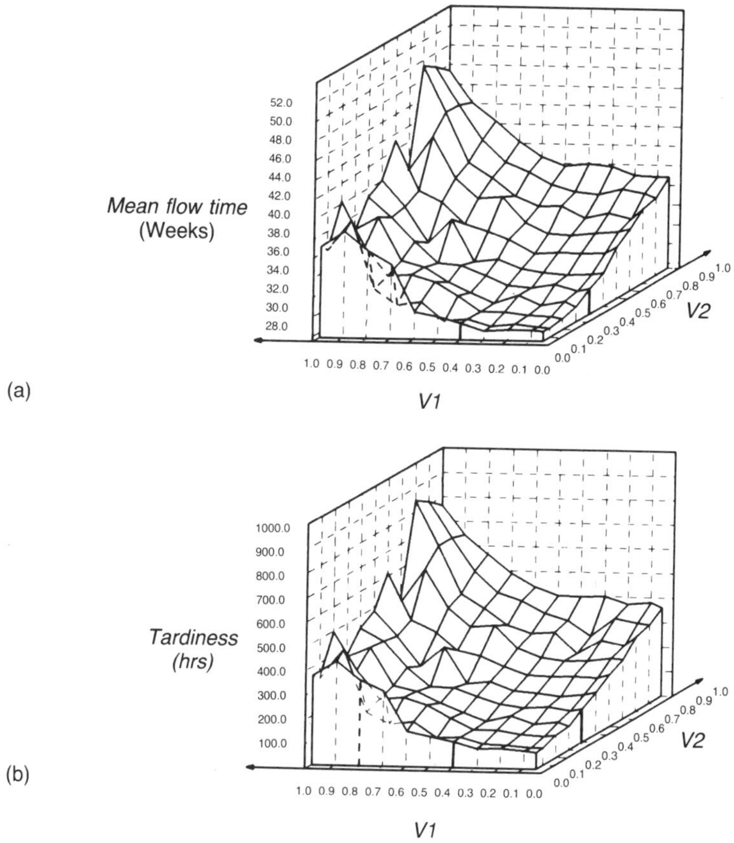

Fig. 6.31 Effects of lead-time uncertainties on (a) flow time; (b) order tardiness.

the simulation results revealing the system responses (mean contract flow time and mean contract tardiness) to a set of different uncertainty conditions. It was established from the experimental evidence that the system was much more sensitive to the external supply uncertainties than to the internal operational uncertainties, confirming management intuition that the availability of key bought-out items is one of the dominant factors that determine the overall efficiency of the organization. It was also concluded that variation of internal operation times could possibly be tolerated to a certain extent without resulting in any seriously undesirable effects on the overall system performance.

However, beyond a certain limit the system performance would deteriorate rapidly. It is of vital importance, therefore, for the company management to restrict the relevant system variation to within such a limit. Information on how these variations affect the overall company performance is very valuable to management in the planning and control of manufacturing programmes. Such information is also useful when developing or installing formal control systems for such firms, for the problem of uncertainty has long been recognized as of vital importance for these control systems. Therefore, these experiments not only provide practical guidelines for the management, but also have significant implications for the development of corresponding computer-integrated manufacturing systems which are to function in this kind of operation mode.

6.8.3 Conclusion

Advanced manufacturing systems involve a total systems concept which embraces a wide range of manufacturing aspects from customer enquiry to product delivery, involving all of the operational functions. As a result, one of the key issues for the successful design and implementation of AMS is a thorough understanding of the nature of the organization concerned within the entire systems context. In order to develop the strategies and policies regarding the efficient operation of AMS, it is no longer sufficient to study only artificially decoupled hypothetical and low-dimensional systems. Instead, analysis of manufacturing systems in their entirety to reveal their characteristics and behaviours should be regarded as a prerequisite. This will require an analytic approach which is more realistic and problem-specific. This case study has shown that in this situation a relatively large-scale computer model which is capable of simulating the entire operation of a proposed manufacturing system is likely to be a valuable decision aid for the design and analysis tasks involved. With such a model it would be possible to simulate and monitor a vast number of operational activities simultaneously, so as to provide a clear view of the total manufacturing process. Consequently, various hypotheses put forward during the design stage of a system may be tested in an exhaustive and realistic setting.

In the past (particularly since the early 1970s) there have been many, and frequently justified, criticisms of the real values of large-scale simulation models (or *macro* models, as they are sometimes called). For example, large complex models have often been criticized for requiring a large amount of both human and computing resources to develop, making them too expensive and time-consuming to be of any real use. It is not difficult to defend the soundness of this kind of argument: it is clear that there is not much point in spending £X on developing and implementing a simulation model of a system which could in the end yield a saving of £Y, unless £Y is significantly more than the initial £X. Nevertheless, when it comes to the design and evaluation of AMS, the use of large-scale simulation models would seem to be readily justifiable for two reasons (Wu and Price, 1989): if used properly, these large-scale simulation models are well suited to reflecting the complex interactions which are inherent in an AMS environment; and AMS usually require long-term and expensive investments, thus making the time and effort spent on modelling worthwhile.

Further reading

Browne, J. and Davies, B.J. (1984) The Design and Validation of a Digital Simulation Model for Job Shop Control Decision Making, *International Journal of Production Research*, **22** (2) pp. 335–57.

Carrie, A. (1988) *Simulation of Manufacturing Systems*, John Wiley.

Ellison, D. and Wilson, J.C.T. (1984) *How to Write Simulations Using Microcomputers*, McGraw-Hill.

Ford, D. and Schroer, B. (1987) An Expert Manufacturing Simulation System, *Simulation*, May.

Ketcham, M. *et al.* (1989) Information Structures for Simulation Modelling of Manufacturing Systems, *Simulation*, February.

Kiran, A. *et al.* (1989) An Integrated Simulation Approach to Design of Flexible Manufacturing Systems, *Simulation*, February.

Law, A.M. and Kelton, W.D. (1982) *Simulation Modelling and Analysis*, McGraw-Hill.

Pitt, M. (1984) *Computer Simulation in Management Science*, John Wiley.

Wildberger, M. (1989) AI and Simulation, *Simulation*, July.

Questions

1. Based on the characteristics of computer simulation reviewed in section 6.3, give some examples of manufacturing problems which are best

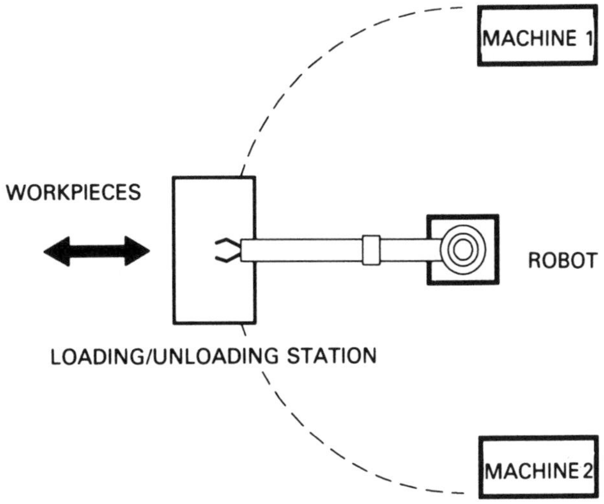

Fig. 6.32 Cell consisting of two machines and a loading robot.

solved analytically and of problems which are best tackled by simulation. Outline the reasons for your choice.

2. Read and try to understand the underlying logic of the BASIC program presented in Box 6.2. Run this program on your computer and then attempt the following questions by modifying the relevant parameter values in the model:
   (a) What are the values of performance data given by the system under the existing operation conditions?
   (b) What is the throughput time of 500 parts instead of 1000?
   (c) If the machine capacity is reduced by 30%, how does this affect the operation of the system?
   (d) What is the effect if the machine capacity is increased by 30%?
   (e) What is the effect if the workload to the system is increased by half? What if it is reduced by half?
   (f) What is the effect if the parts arrive in the system in a more orderly manner (that is, if their distribution has a smaller variance)? How about if they arrive more randomly?
3. Figure 6.32 illustrates a machining cell which consists of two identical machining centres and a robot. The robot serves the two machining centres by loading and unloading them at the start and the end of a machining operation respectively. Draw an activity cycle diagram to describe this discrete manufacturing system. (Hint: first draw the activity cycle of the machines, then draw the activity cycle of the robot, and finally integrate these through their common activities.)
4. Choose a real-life activity (such as a machining operation) and collect a number of samples of the time the activity takes (40–50 will be sufficient). Compile a cumulative distribution for this set of samples according to the steps outlined in this chapter.

   Use the resulting graph and a table of random numbers to produce a series of values, and then construct a frequency distribution graph for these values. How well does this graph resemble that of the original data? (Remember that the number of random values generated should be sufficiently large.)
5. Modify the simulation program that you have developed in Question 2 so that it now allows the use of user-defined distributions. Then use this to simulate the process involved in Question 4. Compare your results with the real situation.
6. Jobs randomly arrive at a single-station machining cell at an average rate of 8 per hour. The processing time of the work-station is negative exponentially distributed with a mean of 7 minutes per job. Find:
   (a) The machine utilization.
   (b) the minimum size of buffer store to cope with average queue length.
   (c) The average time for a job to go through the system.

   If the due dates of these jobs are to be assigned by the following formula:

   $$d_i = r_i + 100$$

   where d_i is the due date of job i,
   r_i is the arrival time of job i, and
   100 represents the throughput allowance in minutes:

   (d) What is the probability of a job being late for delivery?
   (e) At what speed must the machine operate so that on average the proportion of late jobs is restricted to less than 20%?

(f) If the hourly cost of running the machining centres is £M per hour per machine and the fixed penalty for late deliveries is £C per late part, for what values of M and C is it worth employing another identical work-station?

7. The production operation of a product in a small factory involves a particular process which is performed on a single facility. During an eight hour working day, on average 45 products require to be processed by this facility, which completes the production process at an average speed of 8.5 minutes per product at a cost of £50. The management feels that in general the products are being delayed for too long by this operation, and hence wishes to increase the capacity of the facility concerned.

   If the situation can be described by a Poisson process, how much will the process cost if the average queue length before this facility is reduced to half its present value, assuming that it costs an additional £5 for every minute of reduction in the processing time?
8. Discuss the possible ways in which the theoretical results of queuing theory as listed in this chapter may be used to check the validity of a computer simulation model.

   Modify the BASIC simulation model given in section 6.4.1 so that both the arrival and service time are now represented by Poisson processes. Run the simulation model and then compare the various system statistics with the results from the analytical approach. Do you think your simulation model is correctly predicting the system behaviour?
9. Having discussed the modelling approaches of discrete simulation models using a general-purpose programming language, and reviewed the necessary techniques of probability and statistics such as probability distribution and random variable generation, the reader is now in a position to write his or her own discrete simulation models.

   As a summary of the concepts and techniques learned in this chapter, use BASIC to write a three-phase simulation model to solve the problems composed in Example 6.4. Compare your results with the theoretical solution provided. (Hint: The program should consist of a main program block which controls the overall execution of the simulation by initializing the model, advancing the simulation clock to the time of the next event, checking the simulation time so as to terminate the simulation process when appropriate. It should call three-phase sub-routines which respectively determine the time of the next event, call the immediate activity routine(s) concerned, and then scan and initiate any other possible activities; and a set of task-specific activity routines which, when their condition(s) are satisfied, simulate the activities of the system being modelled and update the system state accordingly. Use a two-dimensional array as an activity list which lists the next due times of the B-type activities. The 'next due' time can then be found easily by scanning this list.)
10. If you have obtained the demonstration diskettes, run simulation model MC1. Use the interactive simulation facilities provided by PCModel to look at the simulation parameters and machine utilization, etc. Note the queuing situation and the average throughput time of the parts.
11. Following on from Question 10 run simulation model MC2. Note the queuing situation and the average throughput time of the parts and compare these to those given by MC1.
12. Following on from Question 11, modify model MC2 so that the machin-

ing time in the two-machine model is now exponentially distributed, then experiment with the model to find the machine capacity (i.e. mean machining time) to achieve:

(a) mean machine utilization around 90% between the two machines;
(b) mean WIP level of around 20 items;
(c) mean throughput time of around 1.5 minutes.

13. Figure 6.33 is the overlay of a machining cell consisting of four identical machining centres which are loaded and unloaded by a robot (supplied as CELL.OLY on Demonstration Diskette No. 1). Each loading operation will take 10 seconds to complete, and each unloading 8 seconds. Parts arrive at random according to the distribution given below.

Inter-arrival times

No. of observations	*Time (min)*
220	1
110	2
110	3
70	4
40	5
60	6
60	8
50	9
40	14
20	18

The machine times are distributed as follows:

Machining times

No. of parts	*Time (min)*
90	1
150	4
110	8
110	12
80	14
70	17
70	19
60	25
60	27
50	35
40	45
20	60

Write a PCModel program to simulate the cell, and use the model to answer the following questions:

(a) How long on average will this cell take to put through a batch of 2000 parts? What about 5000 parts? Your estimate should be based on the average of a number of independent samples (use different random seeds to obtain independent samples by using the instruction RS( )). Calculate the confidence interval of the estimates.
(b) Estimate the mean throughput time. Discuss the WIP situation in the cell.

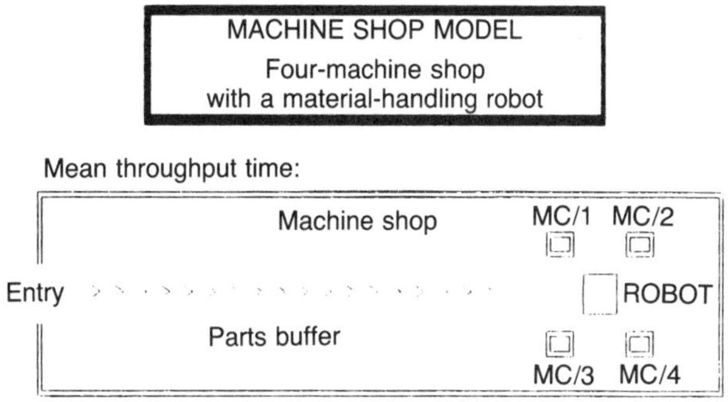

Fig. 6.33 Four-machine shop model.

(c) Estimate the mean utilization of the robot. Is it playing a satisfactory role in the system?

Hint: The program should have a similar structure to that of MC2.MDL. These is, however, an additional requirement to test the availability of the robot each time a part is to be loaded or unloaded. This can be easily accomplished by using the following instructions:

TP (Test Position): Test the screen location(s) specified for occupancy. Continue execution with the next instruction only if the location is clear.
PO (POst position): Post the specified location(s) as occupied.
CL (CLear position): Clear posted location(s).

Once a free machine is found, the program could try to grab the robot by, for example, calling a link such as:

```
;
BL(!LOAD)                       ; Begin link

   TP(*ROBOT)                   ; Test location *ROBOT for occupancy, and
   PO(*ROBOT)                   ; When free, grab the robot and set busy
      WT(%LOADING)              ; Load part for a time %LOADING
   CL(*ROBOT)                   ; Release robot
EL                              ; End of loading and return
;
```

The unloading operation can be simulated in a similar manner. The useful X- and Y-coordinates on this overlay are as follows:

Entry point	:	(6,15)
Message point	:	(35,9)
Robot	:	(58,15)
Machine 1	:	(55,12)
Machine 2	:	(62,12)
Machine 3	:	(55,18)
Machine 4	:	(62,18)

Part Four

Manufacturing Systems

7 A design and evaluation methodology of manufacturing systems

7.1 Introduction

The aim of this chapter is to present a framework of tasks which are involved in the design and evaluation of modern manufacturing systems. As discussed in Part Two, advanced manufacturing systems are considered to be those which include high levels of programmable automation – they may include robotic devices for the handling of materials or computer-controlled machines which need relatively little operator input. However, the set of manufacturing operations which fits the category of modern manufacturing systems is much wider, and includes all systems which are designed specifically to deal with the environment in which it must operate.

When today's most successful manufacturing organizations are considered, it is clear that there is more to their success than simply running all their machines at maximum capacity, regardless of the performance of other parts of the organization or of the environment. When used in the past, the local maximizing approach resulted in huge inventories, and poor throughput times and response. Therefore, optimization of part of an organization has not generally resulted in optimum performance of the overall system. In summary, manufacturing operations exhibit many system characteristics, and should be considered as such. This is the major reason why it is felt nowadays that a systems approach should be viewed as a more adequate framework for the analysis of problems which are generated by modern manufacturing operations. The most effective manufacturing operations always have clearly defined objectives. While the setting of these objectives can be a difficult and unstructured problem, the actual process of systems design which must create a system capable of fulfilling these objectives is often a structured problem to which a relatively hard systems approach is suited.

7.1.1 General problems faced

In general, manufacturing systems must be kept as simple as possible. The main attribute of systems theory in the area of manufacturing systems design and analysis is therefore *simplicity*. This is maintained by dividing the system as a whole into sub-systems; by clearly identifying the input, process and output elements of each sub-system; by determining the appropriate controls for monitoring achievement against predetermined standards; and by formulating and initiating the necessary corrective action.

The division of the system into sub-systems during a design or analysis process breaks up the complexity inherent in manufacturing systems into controllable units which can then be managed through a coordinated approach. However, the purpose of such an approach is not to create a collection of separate departments, but a system in which the effective interaction and flows of information, materials, personnel, equipment, money, and so on, guarantee a smoother and more effective overall operation in support of relevant business goals. Such an approach is necessary because of the following problems which are generally faced by modern manufacturing industries.

The need for frequent restructuring
As stated at the beginning of this book, industry is becoming more competitive, resources increasingly scarce, consumers are demanding more variety and higher quality, and, in spite of all this, international markets are causing prices to fall. Firms must increasingly react to initiatives from their competitors, or must use new technology to give themselves an operations advantage. This means that firms are having to restructure more frequently than ever before.

The need for accurate initial specifications
Manufacturing technology has increased in complexity and functional ability with the advent of the microprocessor. This has allowed a higher degree of automation than previously possible, combined with the ability rapidly to reprogram the equipment and thus improve flexibility. Significant improvements have also been seen in production management systems as computer power allows more information to be stored and processed quickly. However, this has resulted in problems which are usually encountered within the computer software industry being met also in manufacturing industry.

One of the valuable lessons learned from previous projects concerned with large-scale computer systems is the unacceptably high cost of work carried out after delivery of the systems. Analysis revealed that most of the problems which caused the increased expense stemmed from errors made during the early analysis and design of the systems. These errors were up to 100 times more expensive to correct once a system was installed than in the early phases of development (Marca and McGowan, 1988). This led to the realization that better methods had to be used in the early phases of the development process. The level of technology being implemented in today's manufacturing systems exhibits the same need for careful planning during the early design phases. The system components require high levels of interaction with each other to bring about the desired transformations. As there is less and less human intuition in their operation to deal with ambiguities, careful planning of such relationships is becoming increasingly critical.

The need for more regular operations decisions
The levels of flexibility required today in order to satisfy customer demands force decisions to be made more regularly. To make the correct decisions, it is necessary to have the correct information available. Having made the decisions, it is then necessary to ensure that the results are sent to all areas which need to be aware of them. As computer power is enhanced, allowing higher levels of automatic decision-making and at the same time demanding increasing levels of clarity of information, systems designers not only have to deal with the traditional manufacturing problems such as materials flow and production sequencing, but are also faced with increasingly tight con-

straints placed upon them – that is, they will also have to face the problems usually experienced by the computer software industry.

The need for faster new-product-to-market cycles
The costs of failing to implement a system correctly are also much greater today than ever before, because of the higher cost of capital equipment and the need for a more rapid response to the market. Again this requires a set of guidelines or a methodology to reduce the likelihood of mistakes in systems design and enhance systems effectiveness.

7.1.2 Specific aims of a design methodology

In relation to the requirements outlined in section 7.1.1, therefore, there exists a number of main areas of concern.

Improved quality of specification
Many of the problems outlined in section 7.1.1 concern the quality of the design, in terms of how well a designer can translate his vision into a system design which can be implemented in practice without major difficulties. A useful methodology should therefore ensure that the resultant design is of high quality so that the design is 'right' before implementation, thus reducing costs; so that the system can be rapidly commissioned to allow rapid repayment of the capital costs which are invested in it; and so that new products can be brought to market promptly.

Reduced systems design costs and cycle times
The expertise currently required for this type of project is relatively scarce and expensive. Therefore, it is desirable to reduce the time for which these resources must be employed. A design methodology will provide guidance, thus allowing more time to be spent considering the *design* and less time considering the design *process*. Another benefit in this regard is gained where a methodology is able to 'routinize' parts of the process. This will allow them to be carried out by less skilled people, so that only the more demanding parts of the process require the expensive services of experts. Finally, a routinized methodology will increase the speed with which the process can be carried out. Since system design occurs increasingly often, this is a very important consideration.

Fitness for purpose
So far we have dealt primarily with the efficiency of the design process. However, the actual quality of a design, in terms of its fitness for purpose, should also be considered. Thus, the methodology should offer some means of assessing whether the correct decisions are being made during the design process.

Summary of objectives
The methodology must serve as a guide to the designer, reducing the amount of effort expended in deciding what to do. In guiding the process, it should also ensure that all aspects which require consideration for the design of a successful system are given that consideration so that a better quality solution results. Although the methodology cannot offer solutions for specific problems, it can suggest tools which can be used in certain situations. However, it should not in any way restrict the design team in

terms of creativity. It must help ensure that innovative ideas are supported so that the effectiveness of the design operation can be improved.

In summary, therefore, the methodology should help decide which aspects of AMT are required and how they should be integrated, and provide a means of assessing the likelihood of success in order to allow an informed decision to be made as to whether the required investment can be justified.

Section 7.2 will review the models and current techniques that are already in use, while sections 7.3–7.8 describe in detail a design and evaluation approach which will bring together all the previously discussed concepts and techniques into a complete framework.

7.2 Review of current design methodologies

Having looked at the need for methodologies and the basic features they should offer, it is now necessary to identify the techniques which have already been developed and to evaluate these so as to establish how they may be adapted and expanded to fulfil the need for a structured approach specifically for the design and evaluation of AMS. This section, therefore, investigates current work in the development and application of systems design and evaluation methodologies.

As will be seen, most of the methodologies created to date break up the total design process into a set of broad phases. There is usually no set of rules to define what these phases are: they are arbitrary and reflect the observations of designers as they go through the process. Some methodologies require fixed outputs from one stage before continuing to the next, but most demand iteration. Thus the flow represents main areas of interest during the process, although other areas may be considered at the same time. These project phases are then broken down into smaller, more manageable steps (see, for example, the problem-solving cycle discussed in Chapter 5).

In order to review current work, it is first necessary at this stage for us to distinguish between methodologies and tools. Since the word *methodology* is often used rather loosely, Checkland (1981) has attempted to give it a precise definition by suggesting that a methodology is a set of principles of method, which in any particular situation guide the designer or analyst towards a method uniquely suitable to that particular situation. On the other hand, a *tool* offers no guidance but is simply a mechanism to allow the generation and clarification of ideas or thoughts. Such a tool could be a diagram on a piece of paper, a table of figures, a mathematical procedure or a particular modelling technique. Many of today's tools have computer support. Computer-aided design and analysis tools are extremely useful in many ways. For example, in addition to increasing effectiveness, the application of computer techniques may force a designer to abide by logical rules, thus reducing the possibility of error. The tools are as essential to successful system design and analysis as the overall framework of the guiding methodologies themselves. One of the most important tools used in the design and evaluation of manufacturing systems – computer modelling and simulation – was discussed in Chapter 6.

7.2.1 AMS design methodology developed by the GRAI Institute

Chapter 2 revealed that there must exist three essential sub-systems within a manufacturing organization – an auditorial sub-system; an operational sub-system; and a managerial sub-system. The operational sub-system consists of the physical resources, including humans and machines, used to transform raw materials into end products. This sub-system is controlled by the managerial sub-system which makes the periodical decisions necessary to reach the objectives set by the wider system. These decisions must be made based on accurate information provided by the auditorial sub-system.

We will now begin our review of system design methodologies by looking at an approach specially developed for the design of managerial systems. This approach is known as the GRAI (Graphe à Résultats et Activités Interliés) (Doumeingts *et al.*, 1987). Our overview of this methodology will begin by considering the tools developed for the implementation of this approach. The structure of the methodology itself will then be reviewed.

The tools

When applying the GRAI methodology, two graphic tools are used.

GRAIgrid is a tool intended to aid in the analysis of the production management systems within a manufacturing operation, and as such is concerned with the control systems in the manufacturing organization. The structure of a GRAIgrid is shown in Fig. 7.1. This tool uses a top-down approach to identify what *decision centres* (DCs) are required to achieve a coordinated system.

As will be seen, a standard set of functions is represented in the columns of the grid. The desired decision time-scales are represented in the rows. This provides a network of squares. Each square is considered to be a potential DC in the related function and time-scale. These are then examined in the context of the desired system, so as to establish whether there is indeed a set of decisions to be made in this response period (that is, whether a DC is required). Significant flows of information are shown on this grid, as are transmissions of decision frames (these occur when a decision comes under the control of two or more decision centres, so that the partly processed result from one decision centre must be sent to another decision centre for completion). Figure 7.2 gives an example of a GRAIgrid representing the structure of DCs of an OHMS operation (see Chapter 3). Only

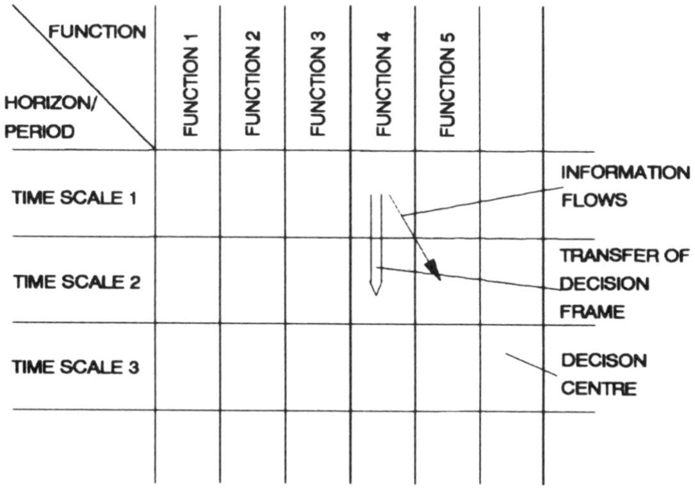

Fig. 7.1 Structure of a GRAIgrid.

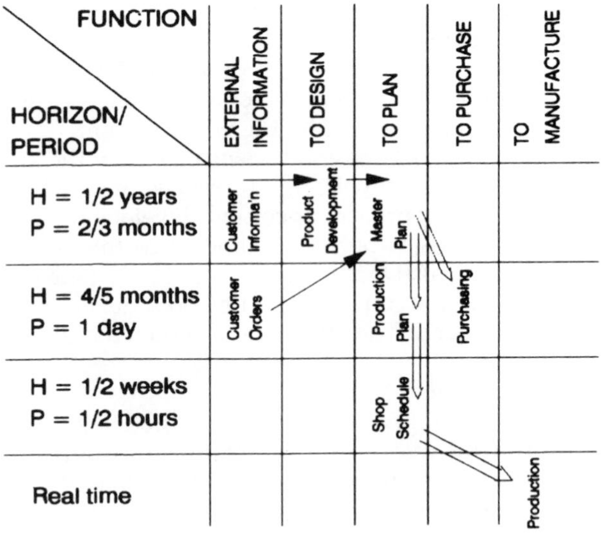

Fig. 7.2 OHMS decision centres.

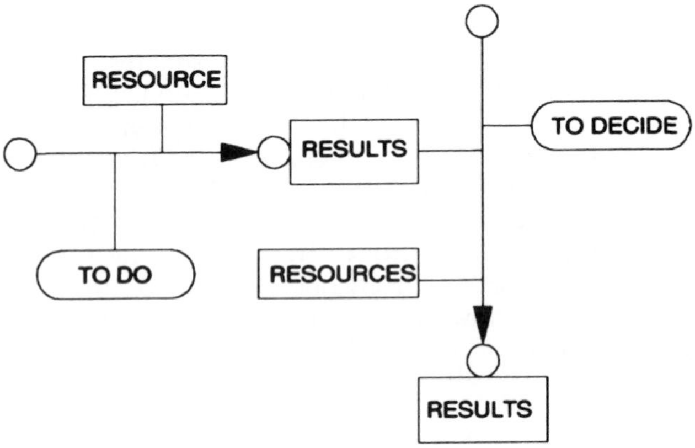

Fig. 7.3 GRAInet showing activities and decisions (source: Doumeingts *et al.*, 1987).

after the above information has been acquired are the detailed decision processes considered.

GRAInet, on the other hand, is a bottom-up approach used to describe the actual activities involved in each DC. An example of this is given in Fig. 7.3. The arrows together with their associated circles are used to represent the activities to be carried out. These can be either decisional activities (vertical) or process activities (horizontal). The resources and information requirements of the activities are specified in the rectangles. A GRAInet is developed for each DC which is identified as being necessary through the GRAIgrid analysis.

The methodology

The development of GRAI methodology is again based on the argument that using formalizing rules in the early phases will reduce errors in design and

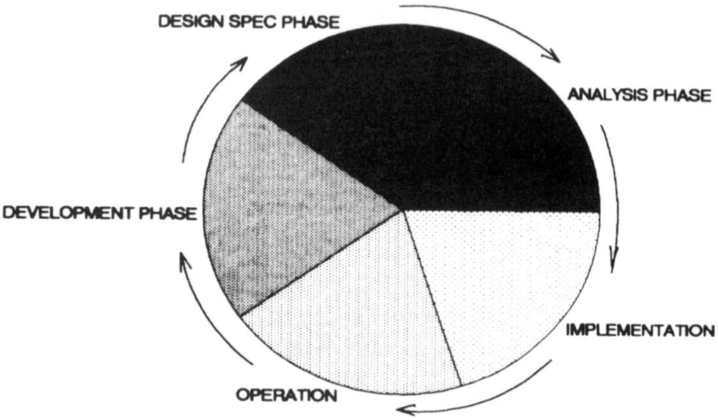

Fig. 7.4 GRAI interpretation of an AMS life cycle.

therefore improve overall system performance. The GRAI methodology identifies the existence of an AMS life cycle, whose basic structure is shown in Fig. 7.4. Following this cycle, the methodology has the following major phases:

1. *Analysis phase*. This is a study of the current structure and behaviour of the system, disclosing constraints, goals and possible inconsistencies. This analysis uses the GRAInets and GRAIgrids of the current system to identify faults such as missing information flows, lack of updated information, poor synchronization, wrong procedures, inadequate criteria, non-defined decision frames, missing vital DCs or incoherent coordination between DCs.
2. *Design specification phase*. This phase determines the functional specification, the basic framework and general behaviour of the desired AMS. This stage of the GRAI AMS life cycle can be further decomposed into three stages:
   (a) *Conceptual modelling*. A conceptual model of the manufacturing system under study will provide direction for the design team's work. A GRAI conceptual model of a manufacturing system consists of three sub-systems, as illustrated in Fig. 7.5. The physical system transforms materials into products, and is coordinated by a hierarchy of control or decisional systems. The information system carries out all data transfers between and within these systems.
   (b) *Structural modelling and functional specifications*. Here the designer attempts to create a system to fulfil certain criteria, using the conceptual model as an outline structure. (However, limitations such as technology or finance may prevent the system design from being an exact copy of the conceptual model.) The methodology utilizes the same GRAI tools to validate structural models, as using these will ensure that all control aspects are considered and that the processing of information required to support the decision making is well understood. Once a system design has been agreed the functional specification can be created using the model for guidance.
   (c) *Operational-level specifications*. This stage is concerned with the detailed specification of system components.

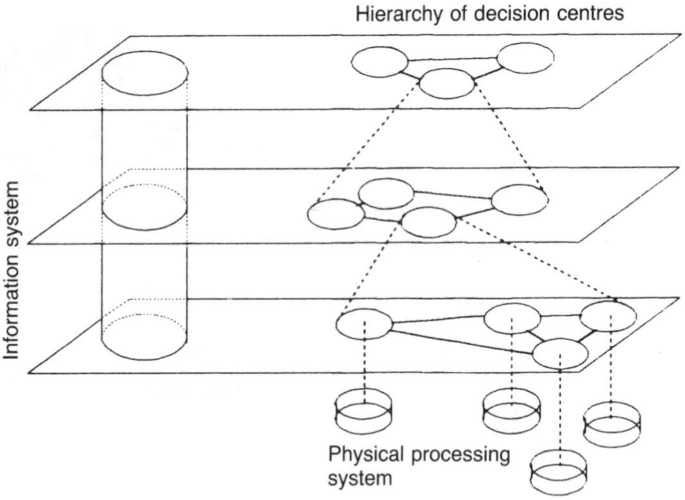

Fig. 7.5 GRAI conceptual model (Doumeingts *et al.*, 1987).

Comments

The GRAI methodology supplies techniques and tools for dealing with the first two phases within the design frame, which are the areas of prime concern in reducing total system costs (the later phases are adequately dealt with using computer support).

This approach, however, does not encourage the designer to check for technological, economical or financial feasibility during the design process. Also there is no reference to company strategy, or the multitude of other 'business' questions which should be studied. For instance, just how well the standard set of functions represented in the columns of a GRAIgrid reflects on every type of manufacturing operation is not clear. To fulfil the requirements of a typical industrial user, it may be necessary to give these areas further consideration.

In addition, the GRAIgrid approach to the construction of a decision or control system would appear to follow the ideas of the past two decades – of relying on large, centralized computer systems. This is reflected in the identification of single DCs to deal with work with a particular time-scale. It is noted that in many cases it is increasingly desirable to have distributed systems (Barekat, 1990). This particular methodology, therefore, may have certain weaknesses in dealing with the design of this type of systems.

7.2.2 SSADM

The Structure Systems Analysis and Design Method (SSADM) (for example, Ashworth, 1988) is a standard structured method used in computer system development projects undertaken by or in UK government departments. Although this method does not relate specifically to AMS design, we can learn general lessons from its structure and approach. As before, therefore, the tools which have been developed for SSADM will first be reviewed, and their use will then be considered in relation to the methodology itself.

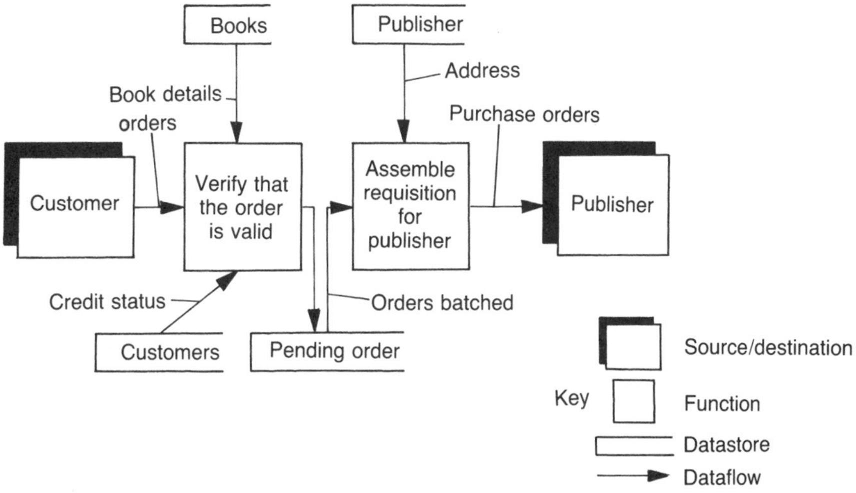

Fig. 7.6 Example of a data flow diagram (Gaine, 1979).

The tools

Four techniques are involved in this systems design method.

A data flow diagram (DFD) is a functional picture of the flows of data through a system. An example is given in Fig. 7.6. There are detailed rules for its application. For example, functions which are not descriptive enough for the purposes of design should be further decomposed. It is also advised that there should be no more than three levels of decomposition. Although this tool has been designed specifically for a computer environment, it can also be used to describe the control functions in a manufacturing context. Similarly, the flows could be used to represent either material or information and the store symbols could represent data or materials stores. Compared to the $IDEF_0$ and SADT (section 7.2.3) approaches, the graphic symbol for data stores is a welcome addition. However, $IDEF_0$ and SADT limit the number of functions on any level to ensure ease of understanding and manageability, while DFDs allow any level of complexity. Later developments of the method (for example, Downs *et al.*, 1988) also use different symbols, including two different types of arrows, one for physical flows and one for data flows. This makes the method more appropriate for the description of manufacturing functions.

Logical data structuring (LDS) is an entity modelling technique. An entity is considered to be anything about which the company wishes to store information. The first stage of LDS design is to identify these entities, either from information flows shown on the DFDs or from expert knowledge of the system. Following this a two-dimensional matrix is created, listing all the identified entities on both axes. An asterisk is used to indicate direct logical relationships between two entities. From this information an LDS diagram representing the logical data structures is created. Figure 7.7 shows some of the principles of the diagrams. The access paths for each data type are traced to ensure that all users have access to their required information. LDS is a useful tool for the structuring of data bases. However, like $IDEF_0$ models, it provides only a static representation.

Entity life histories (ELHs) aid the validation of DFDs. They explore the

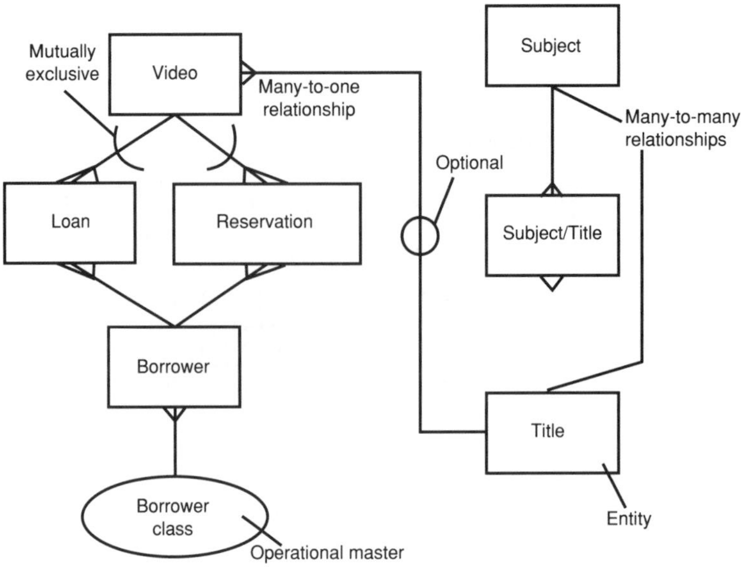

Fig. 7.7 Example of a logical data structure diagram (Gaine, 1979).

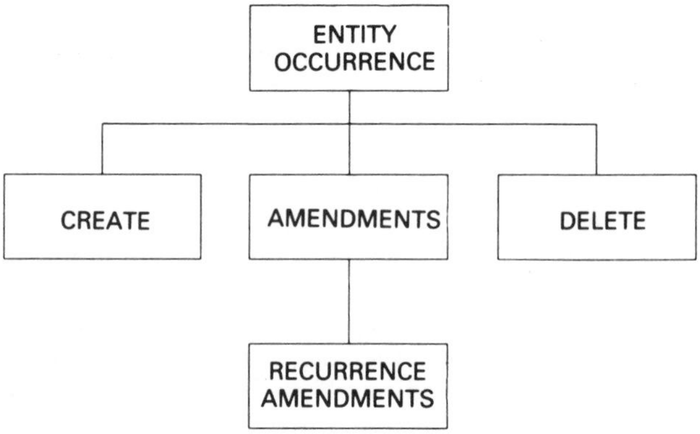

Fig. 7.8 Diagram of an entity life history (Downs *et al.*, 1988).

processing requirements of entities, identify state change transactions, record time sequences of changes and clarify event transactions. A two-dimensional entity/event matrix is created and the events which change the state of entities are marked with an asterisk (Fig. 7.8). Although ELHs are a tool intended for investigating data dynamics, they can become rather messy in practice. It is, therefore, difficult to imagine an entire set of system characteristics being studied using it unless significant computer support can be utilized.

Relational data analysis (RDA) is used in the structuring of data. Data are initially sorted into data tables. The structure of an address book gives a typical example of a data table. For a given table there will be a *key* which is a piece of data that uniquely defines a set of data in the table (for example, the name of a person in an address book). The first task in RDA is to sort out repeating data, and to select a key such that only one unique set of data

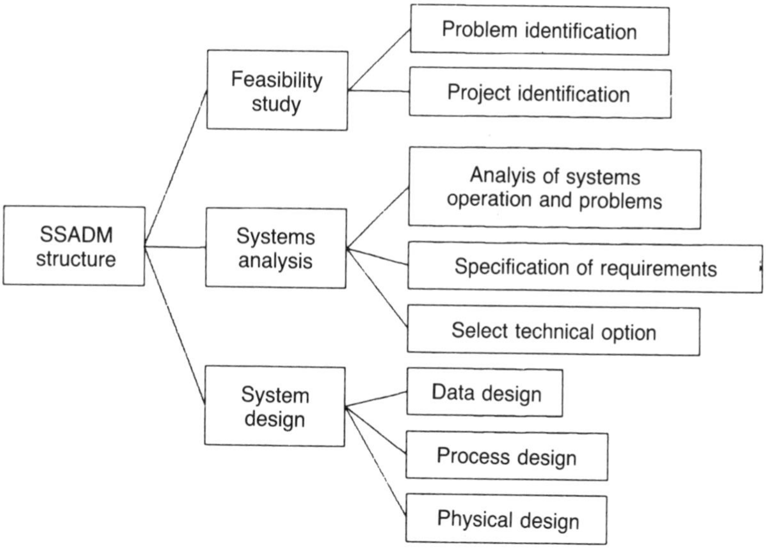

Fig. 7.9 SSADM structure (Downs *et al.*, 1988).

is identified. Any data fields not fully dependent on the whole compound key are then removed. Then optimization will take place, which requires merging of relations that have identical keys.

The methodology

An outline of the SSADM structure is given in Fig. 7.9. The individual parts of this methodology are discussed below.

1. *Feasibility study*. For a project to have been initiated there must be a certain amount of knowledge available about the system under consideration. From this an initial list of problems and requirements can be generated. An initial level 1 (the highest level) current physical DFD is created of the present system, including a description of each function. An overview LDS is also created, and these serve to help in the planning of the project by identifying which areas need to be investigated. After this the DFD is decomposed and the next level is validated through interviews with the end user. The DFD is then developed into an overview of the current logical system by merging duplicated functions and any functions which relate to the current environment on which the system runs. In the light of the work carried out, the problem/requirement list is updated.

   Several outline project options are formulated for level 1 which are reviewed with the user to eliminate any 'non-starters'. Resources required for the development and implementation of each option are estimated and the impact of changes and the disadvantages of each option are analysed on a cost–benefit basis. An option is then selected from these potential projects and a feasibility study report is produced.
2. *Analysis*. Here current physical system level 1 DFDs are generated. These are then decomposed to level 2 and level 3, with narrative descriptions of each process attached. Utilizing information gathered for the DFD, an LDS is created. The DFD and LDS are used to validate one another. A more

detailed problem/requirement list can now be generated which will reveal present organizational problems, the need for provision of new facilities, the obvious constraints on solutions, and so on.

3. *Specification of requirements*. The system is next 'logicalized' by redrawing to show what is to be achieved. All user requirements are formally included and functions are considered for their relevance to the system being designed. These are recorded as level 1 DFDs, called Business System Options (BSOs). Based upon the selected option, a new system design specification can be created. The LDS will have to be updated to ensure that all the required data are available. ELHs are then created to give a dynamic view of the data.
4. *Selection of technical options*. SSADM considers that at this stage there will be enough knowledge for the designer to select the hardware needed. (This may be true for information systems, but for an AMS and its mixture of hardware involved – for example, computer networks, CNC machine-tools and robots – the selection of hardware will be more difficult. Therefore, to allow SSADM to be useful in AMS design, it may be necessary to carry out this phase later in the design cycle. It may be desirable to carry out dynamic analysis through, for example, Petri net analysis (section 7.2.4) or computer simulation of workshop performance. These tools are not specifically mentioned in SSADM, but will obviously be useful for AMS design tasks.)
5. *Logical data design*. This stage builds up the logical data design, so that all the required data will be included. It applies the relational data analysis technique to groups of data items. Sources within the required system are analysed to produce data-grouping, and to show their relationships. This completes the LDS. The LDS is combined with the RDA to give an overall logical data design.
6. *Logical process design*. The specification of requirements is expanded to give detail necessary to build the system. Dialogue design is used to design the interactions between the system and the operator.
7. *Physical design*. The logical data and processing designs are converted into a design which will run on the selected environment and then detailed program design and coding take place (it is generally necessary to create the physical data design by combining the rules of the selected data base management system and required data structure).

Comments

It is clear that the SSADM technique is comprehensive in taking the designer through the development of a software, data-processing and data-storage type of system and as such is suitable for use with many aspects of an AMS design. Due to the interaction of data processing and physical processing in an AMS, the later phases of the design process will be of extreme importance to the success of the whole project. However, for industrial use there may be a requirement for further structuring of the SSADM approach to the later phases.

For further details of the SSADM approach, see Downs *et al*. (1988).

7.2.3 The SADT methodology

The system development process defined by SADT (Structured Analysis and Design Technique) is divided into several phases – analysis, design, implementation, integration, testing, installation and operation (Marca and

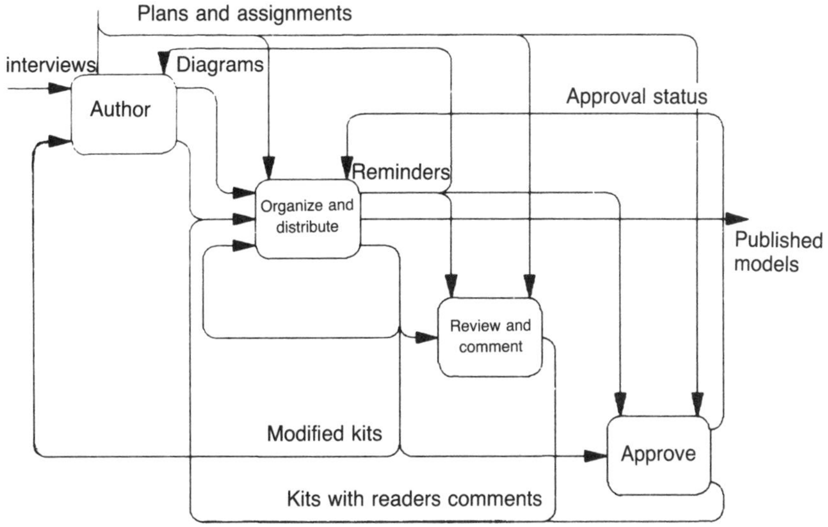

Fig. 7.10 'Creation of models': example of an activity diagram (Marca and McGowan, 1988).

McGowan, 1988). The SADT technique was developed for use mainly in the analysis and design phases (like the GRAI approach, it was felt that the later stages had already been dealt with sufficiently elsewhere).

The tools
A complete SADT model represents the structure of a system through the use of activity diagrams, data diagrams, node lists and data dictionary, of which the first two are graphical tools. Like an $IDEF_0$ block, a SADT activity diagram (Fig. 7.10) shows the functions which occur, together with their resources, inputs, controls and outputs. The method is top-down in nature, starting with a single function and decomposing this in order to achieve the necessary level of detail. Activities at the next level should describe precisely what occurs in this sub-system, with the same inputs and outputs. Data diagrams are formed in a similar way to activity diagrams. Instead of an activity being represented with data lines flowing into and out of a box, however, the data are represented by the box with activities flowing into it. For further and much more detailed explanation of these tools, see Marca and McGowan (1988).

The methodology
1. *Analysis*. For SADT to be successful, a number of steps must be taken early in the process:
   (a) *Bounding the subject*. To ensure that the analyst knows where the system ends and the environment begins, this is done by generating the top-level diagram.
   (b) *Determining the purpose of the study*. The models are created to allow the user to answer questions. These questions should be laid down before any modelling occurs, so that the objectives are known and the model can be developed to tackle them.
   (c) *Determining the perspective required*. If the perspective of the model is

not defined then the model will cease to be relevant to the purpose (Chapter 2).

2. *Creation of model.* The steps of analysis and design are intricately related in the SADT methodology. The activity diagram in Fig. 7.10 is an actual description of this process (Marca and McGowan, 1988). Various interviews are to be undertaken, through which knowledge can be obtained and experience from previous designs utilized. Thus where SSADM explicitly sets out to build on previous knowledge. SADT uses the current system in a much less formal way. When the analyst has decomposed the sytem to a level which answers the questions initially set, then the model is sufficient for his purposes. There are some further essential points which a user should be aware of before attempting to apply the methodology. Further details of these can be found in Marca and McGowan (1988).
3. *Design.* The design process refers to the actual creation of a specification. When the analyst is satisfied with the model, a detailed specification can then be constructed. This is done by succinctly describing each activity, including its non-functional aspects. These descriptions are ordered according to the numbering of the activities. When all the activities have been described, a further description is written of how these functions relate to one another. The process is repeated for all the diagrams; the result is then ordered so as to give a complete specification of the system.

Comments

The activity diagram used by SADT has a number of drawbacks. For instance, it does not help the designer understand the system in terms of storage space and timing (other methodologies, such as $IDEF_2$, have specific ways of overcoming these problems). The SADT methodology does, however, warn the user of the limitations of the activity diagram, and data diagrams could prove to be useful for at least some of these aspects.

The SADT methodology is, however, a well-proven approach. Because of its simplicity it has attracted a wide range of industrial users since the early 1970s. There are many items of literature which the user can refer to. It has computer support in the form of packages like $AUTOIDEF_0$ and SPECIF_X, which allow the user to be more confident of a logically correct model before any decisions are made on the basis of that model, thus improving the quality of the system which will be developed, and also reducing commissioning time.

7.2.4 Other methodologies

Reviewed below are a number of other methodologies which are of relevance to our purpose.

The IDEF methodology

As discussed in Chapter 3, IDEF has received a lot of attention, but does not appear to be a methodology in the same sense as GRAI, SSADM or SADT. Rather, it appears to be a collection of tools. It is true that the tools would lead the designer through the design process if there were no guidelines, but they would be much more effectively used as part of an overall framework. In most treatments of the applications of IDEF, $IDEF_0$ is by far the easiest tool to apply practically.

The DRAMA decision process model

The decision process model used by the Design Routine for Adopting Modular Assembly (DRAMA) (Bennett *et al.*, 1989) makes use of a number of tools, including 'narratives, flowcharts, decision trees and checklists', to lead a design and development process through the stages of decision-making in a concise and comprehensive fashion.

This particular methodology appears to take much more of a systems view than the other methodologies. This is based on the view that in modern manufacturing situations there is now a greater tendency for operational processes to be dictated by factors in the strategic and organizational domains. DRAMA acknowledges that decisions taken at any stage in the design process will affect the significance and appropriateness of decisions taken at other stages in the design process. Therefore three significant domains are to be considered:

1. *Strategic*. Here the effects of decision processes are considered in relation to wider boundary-spanning activities, and in particular to the organizational and operational implications of environmental influences relating to cooperative, marketing and manufacturing strategies.
2. *Organizational*. Factors which interrelate vertically are analysed in order to identify the variables which determine or are dependent on the different options within the strategic domain.
3. *Operational*. Here factors that interrelate laterally within manufacturing are identified. This domain covers the physical design, implementation and operation of new systems.

Petri nets

Petri nets were originally designed for modelling computer systems, but are being used increasingly in AMS modelling. This is because they are a method especially suited to systems with interdependent components in which it is possible for some events to occur concurrently but there are constraints on the concurrencs, precedence or frequency of these occurrences – a typical situation obtaining in manufacturing systems (Peterson, 1981; Favrel and Kwang, 1985). They may therefore be used to model the flow of material, information or other resources in a manufacturing operation. Like computer simulation or $IDEF_0$, however, they are more of a tool than a complete methodology.

7.2.5 Conclusion

The approaches reviewed in this section are not an exhaustive selection. However, they do illustrate that there are currently several methods available for the investigation of all aspects of a manufacturing system, as identified in the GRAI conceptual model. These methodologies all follow a similar pattern of problem-solving and they provide a range of specific tools for carrying out the necessary design tasks of a system.

There does not seem to be a single approach, however, which offers the designer a complete framework for the development of a modern manufacturing system. The approaches are at best fragmented – there seems to be something lacking in all the approaches we have considered. One of these is the absence of any formal method of evaluating the resulting designs. A common attitude appears to be that, because the methodology has been applied, the system will be of an acceptable standard, when in fact only the quality of the system itself should be regarded as the acceptable indication of the worthiness of any methodology. Since the actual ability of a system to fulfil tasks and objectives desired of it is still very much dependent upon the creative ability of the system designers and analysts, some form of evaluation is needed to assess this (among the methods reviewed in this section, only SSADM makes any form of evaluation of the benefits which will be achieved and the costs of disruption).

Another important observation about design methodologies is that the situation should always be approached from the viewpoint of a responsible manager who is actually involved in the operation of the system concerned, rather than from that of a technical specialist. The failure of many AMS projects (for example, Ingersoll Engineers, 1982) illustrates this clearly.

It is clear, then, that there is still a need for certain enhanced methodologies which will draw together more of the considerations important to the designer, and which will be useful in real-life business situations rather than for the purposes of theoretical research units alone.

7.3 Overview of a general design framework

The remainder of this book will be devoted to a general design methodology of modern manufacturing systems, which is an attempt to bring together the previously discussed concepts and techniques into a complete framework.

7.3.1 The overall structure

It is observed in the literature that two distinct approaches can be taken to systems design. One starts with a set of objectives and then creates a model which fits the intended purpose, with little consideration of the current system. Consequently, this approach may result in a design which requires total replacement of the current system. This, clearly, is less likely to be acceptable on financial grounds. The other approach is to consider the existing system, analyse its workings and then try to modify it to fulfil future requirements. This approach allows the incorporation of experience already gained, but since it is unlikely that the control systems within a previously established manufacturing unit have been designed using a systems approach, the options available could be constrained and ideas severely limited. The approach to be presented in this book is a hybrid of these two approaches. By analysing the current and desired positions before setting objectives, the current position will first be given consideration. This in turn will constrain the objectives set. A new system model is then designed which will fulfil these objectives. The creative thought for the new system is not therefore confined by the workings of the old system – only the objectives are constrained by the realistic starting position adopted.

The structure of the design process will be based upon the general problem-solving cycle model discussed in Chapter 5 (Fig. 5.2). The actual approach itself is shown in Fig. 7.11. The first two stages of the general model are the *analysis of the situation* and the *formulation of objectives*. When applied to manufacturing systems design, both fall into a field of study which could be loosely defined as 'manufacturing strategy'. This first requires an analysis of the current state of operation of the manufacturing organization concerned. Following this, an analysis of the current markets, their future prospects and the prospects of markets which could be entered

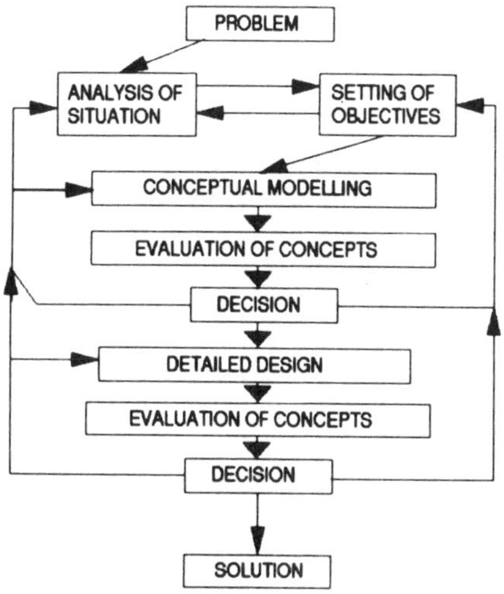

Fig. 7.11 Structure of the design approach.

is carried out. In setting objectives, a view is formed of where the firm should be in the future.

The next two phases identified in Fig. 5.2 are the design phases – *synthesis of concepts* and *analysis of concepts*. These can be considered to be the development of a plan which will transform the operation from the current to the future desired state. In the design approach shown in Fig. 7.11 these have been labelled *conceptual modelling* and *detailed design*. These will first form a framework which will achieve the desired results, and then fill in the detail to ensure that the output is a complete design.

The final two stages are also closely related, as the *evaluation of concepts* forms the basis for the *decision*. In designing a manufacturing system, there are two sets of criteria to be assessed. The first question is whether the system developed fulfils the requirements which were initially set out in the objectives. The second criterion in any real setting will be that of financial performance – whether the system generates a return sufficient to justify the investment when compared to the option of leaving things unchanged. On the basis of the evaluation it is possible to make a rational decision on whether to implement the system, to consider further development of the design, or to terminate the project.

As can be seen, therefore, the major emphasis here will be on design and evaluation. Each of the stages involved will be described in detail in the following sub-sections. Before we go into the details of the design methodology itself, however, it is necessary for us first to discuss one of the most important issues associated with this particular subject area – that of investment in AMT.

7.3.2 Advanced manufacturing investment

An extremely important issue here is concerned with the financial appraisal of advanced manufacturing technology. The implications of this subject are so significant that it must be regarded as an area of prerequisite knowledge for any systems designer involved in today's manufacturing industry.

Although AMTs are being increasingly adopted throughout the industry, they are becoming more complicated and expensive, both for stand-alone machines and for systems. This has made acquisition of the necessary investment capital difficult. This is because the traditional methods of investment appraisal (such as payback or discounting cash flows) demand rapid repayment, while the new technologies are increasingly infrastructural, and provide long-term rather than short-term benefits. Further, the suitability of the technology and the likelihood of success in employing it in any individual case is not yet entirely understood.

In the applications of earlier and simpler technologies, the improvements associated with investment were usually quite obvious – for example, in terms of productivity, which is easily measurable. As far as AMTs are concerned, however, benefits other than these tangible gains can also be achieved. An example is the improved quality and consistency achieved when numerical or computer-numerical control replaces manual control of machines. These intangible gains can in turn lead to improved sales. While it is clear that the company gains from this situation through improved cash flow, with the traditional methods it is unlikely that these improvements would be identified during the assessment of investment opportunities, and what their implications for the firm's position would be. One important feature in this regard is, therefore, that while today's manufacturing technologies are becoming more costly, the benefits which they bring about are less tangible.

It should be realized, therefore, that although there have been cases of firms undertaking major AMT investment programmes without first assessing their suitability, with quite disastrous effects, it is very clear that by turning down an investment opportunity simply because the tangible aspects do not seem to provide the returns will be likely to put a firm at a disadvantage compared to those of its competitors which do invest. As a result, it has been suggested that the intangible benefits brought about by AMT must be taken into consideration, as financial considerations are only a part of the evaluation of a system. In addition, to provide direct returns, a firm must also make investments to improve its competitive position if it is not to lose out in the marketplace. It has been pointed out, for example, that when it comes to advanced manufacturing investment, the successful Japanese companies place greater emphasis on logic and common sense than on sophisticated discounted cash-flow calculations (Swann, 1988).

If systems projects are to be used to enhance a company's position, therefore, there must be a method of monitoring the financial justification of AMT, and this will have to be a more complete method than those traditionally used. For this purpose there is a need to gain a better understanding of the implications of the intangible gains, so that the viability of advanced manufacturing investment can be properly assessed.

Current financial appraisal

The two main methods currently used are based on *payback* period, and on *internal rate of return* (IRR). These methods both take account of the time value of money, that is to say, that close cash flows are more valuable than distant cash flows. IRR requires a specified level of return, or a *hurdle rate*, to be achieved by a prospective investment. One problem with this approach is that return rates can often be set too high – in a survey 75 per cent of the cases studied used a hurdle rate of over 14% and 33 per cent used a rate of over 18%, when the true cost of capital was probably less than 10% (Hendricks, 1989). The likely result is that potentially profitable projects are rejected. The pay-back method appears to be even worse in this regard, as payback periods of two or three years (commonly used in industry) actually demand an IRR of 49% and 31%, respectively. Under this situation investment on AMT would be difficult to justify, because the high levels of investment required rarely provide the rapid returns demanded. However, the problem is compounded by the fact that many of the advantages claimed for AMT by managers do not improve productivity – for example, FMS achieves little advance in productivity over stand-alone CNC technology, yet it is significantly more expensive.

Intangible benefits

An intangible benefit is one which causes improvement in some facet of the business and its operations, the result of which is not clearly seen in financial terms. Some items on the following list of intangible improvements, for example, could be brought about by the implementation of AMT:

- ability to respond to customer consistently and predictably;
- rapid response to market changes with respect to product volume, product mix and product changes;
- shorter product lead time;
- reduced inventory;
- improved manufacturing controls;
- real-time control of components;
- better quality;
- high utilization of key equipment;
- reduced tooling;
- simplified fixture design;
- reduced direct labour content;
- reduced fitting and assembly requirements;
- reduced overhead costs;
- addition of new disciplines to the planning process;
- the possibility of integrated manufacturing.

These improvements can be transformed to improvements in profits through better figures of sale, reduced inventory costs, and reduced operating costs. While the list provided covers many intangible benefits, it is by no means exhaustive. Although when included in an evaluation process they may improve the image of the project, they will still not provide an ideal representation.

Dealing with risk and uncertainty caused by intangibles

To improve the situation, several approaches can be used to consider the problems of the vagueness and risk associated with intangibles. It must be remembered that there is a premium to be paid above the 'risk-free' price of capital for projects which involve risk (Primrose and Leonard, 1985). The analyst should first identify the difference between the cost of capital for the level of risk (inclusive of the premium) and the return which the investment will generate. The projects selected should be those with the highest differential. Some methods deal with this more explicitly than the others. Several approaches are reviewed below.

Analytical hierarchy process (AHP)

The AHP (Ajderian, 1989) employs a strategic approach. It is claimed to be a theoretically sound alternative to traditional financial analysis. An objective is initially set for the required technology. This objective is then split into sub-objectives which must be fulfilled to achieve the higher objective. The objectives are established and agreed by wide-ranging discussion. Intangible benefits can be considered in relation to the set objectives. A team of experts then runs through the list, assigning priorities. Finally a figure is set for the level of fulfilment for each of the objectives. These can then be manipulated to allow a 'utility vector' to be calculated for alternative projects (Chapter 5).

While this approach is not as rigid and formalized as are financial techniques, it has the major advantage that it builds upon skills and experience, and thus is more likely to include within it those intangible benefits which are not even explicitly recognized by the team. The treatment of risk must be dealt with intuitively by the panel of experts, that is, by setting figures which reflect high risk by a low score against the criteria.

Board intervention

It is also argued (Hendricks, 1989) that if benefits can be identified and qualified, they should then be taken into account in the analysis. This will make the discounted cash flow (DCF) model more accurate. While claiming this, it is accepted that in some cases the information is difficult to quantify or has low significance. In order to save time and effort, only the major items should be included. Only if the project fails to fulfil the required selection criteria when the main intangible and tangible benefits are considered, should the analysis be taken further. This requires the analysis to be presented to the board, along with the size of shortfall, and the list of intangible benefits which have not been included in the analysis. The board should then use its experience to decide whether the non-quantified items are valuable enough to the firm to compensate for the shortfall. This method combines the rigour of numerical techniques with the strategic flexibility which, it may be argued, is needed if the firm is to survive in a competitive market. It also allows the wide experience of the board to be used in some risky areas.

Probability

Purely numerical approaches deal with the uncertainty of estimated values in a clear and numerical way. In some cases the analysis uses only pessimistic figures for all the situations which arise. Uncertainty and risk can be expressed in terms of probability values, and thus the estimated values for key items used in the analysis can have a related probability value. It is then possible to calculate a single figure which will incorporate money values and risk figures.

It has been stated that intangible benefits from AMT investments will involve higher levels of uncertainty. These combined value/probability figures provide one way of dealing with such a situation. It should be pointed out that although, if adopted conservatively, this can effectively create a relatively risk-free situation, a firm should always try properly to manage risks rather than to avoid them.

Sensitivity analysis

With this approach, the uncertain factors are varied one at a time, to bring the project to the break-even value. This effectively shows how sensitive the overall outcome of a project is to each variable. This approach has the advantage of being able to identify the variables which management should concentrate on if the project is actually undertaken. However, although this approach shows how much a figure can vary, it gives no guidance on the likelihood of the variance occurring.

Conclusions
In the case of AMT investment, the systems involved can be very complicated and expensive, and this in turn can lead to difficulties in evaluating both costs and returns. One of the main difficulties is that the benefits which are achieved by AMT are not always easily measurable. As a result, AMT investments can be withheld not because the potential returns from investment opportunities are poor, but because of the possibility that the techniques of appraisal are not properly applied.

However, the gains from AMT do increase system revenues and improve its performance in certain ways. Therefore, if financial approaches are used to judge the feasibility of AMT investment, it is essential to remember their limitations, that is to say, that some items will not have been identified. Also, it should be realized that such an approach rarely considers the full consequences of rejecting an investment opportunity. Although a project may look very marginal, by not investing, the company's position may become considerably worse. To summarize, the more specific, numerical approaches should be applied as an input to a wider decision-making process. The probability method, for example, can be an effective approach provided it is combined with board intervention to allow decisions of strategic importance to be assessed. This will prevent a piecemeal approach being taken to what must be an integrated solution.

7.4 Analysis of situation

The rest of this chapter is taken up with a detailed discussion of the stages of the design methodology shown in Fig. 7.11.

7.4.1 Initiation

The signs of impending change
In a manufacturing environment there exists a number of possible causes for change. The first reason for change comes from the need to meet objectives issued from the wider system. The second reason for change comes from a perceived change in the environment which will allow the firm an opportunity for innovation, but which will require modification of the system in order to take full advantage of it. It is likely to be triggered by a marketing exercise or by the product-development function, which will highlight future requirements. The third and probably the most acute reason is the unsatisfactory performance of an existing system – for example, slower throughput times or poor profit figures. These are illustrated as inputs in Fig. 7.12. The reasons which trigger the change process are not necessarily the sources of problems themselves. For example, increasing inventory will not be a problem in itself, but is likely to be a symptom of control-system weakness. Similarly, a wider system objective to reduce inventory costs will not in itself be a problem, but is again an indication that the system must be modified to allow the objective to be achieved (Chapter 5).

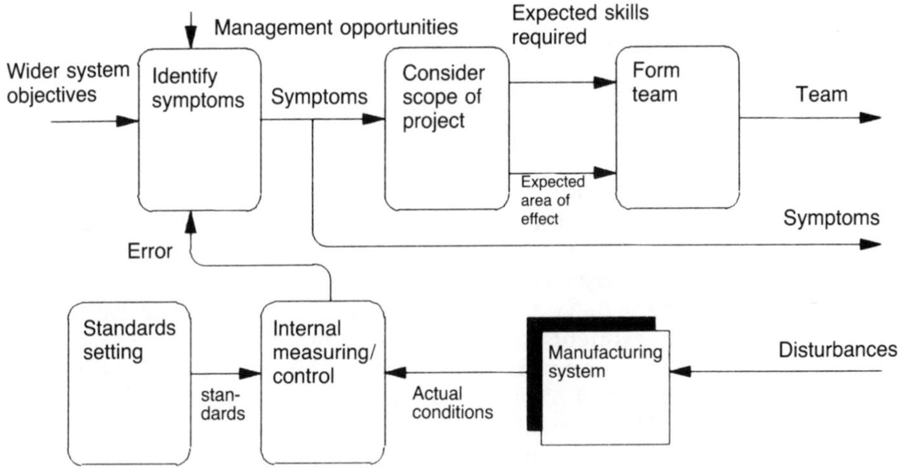

Fig. 7.12 Problem identification and initiation of investigation.

The team and the outputs of initiation

The purpose of individual managers within a system is clearly to optimize the functioning of the department concerned in line with certain objectives; this is normally a short-term and localized process. Compared to this, the manufacturing operation has been shown to be a system containing many departments, and these should have coherent goals. Furthermore, the full system operation must be understood and measures and levels of achievement defined if long-term success is to be achieved. Therefore, individual managers cannot undertake redesign for individual departments. Instead, a team approach is required which may draw on the managers' skills, but must deal with the whole system and not just parts of it. To achieve this, a team must first be established which should include representation from departments which will be affected, and the technical expertise expected to be required. It should also be remembered that any solution will have repercussions later in the system, and as such the possible areas which are to be influenced should also be represented. The level of representation must be balanced with the need for a team of a manageable size. A high-level management team must also be established to act as a steering committee, which will decide on the continuation and eventual approval of projects.

To establish exactly what the problems are, a preliminary investigation should be carried out. The DFD given in Fig. 7.13 illustrates this initiating process. The output from this phase should be a symptoms list containing all those items which are causing concern. Also a formal job specification should be given to each team member, setting out what must be achieved.

7.4.2 Inputs and objectives

The inputs to this phase are simple. A team of analysts is provided with the symptoms list which has been compiled. The team must then identify the root problems which occur. The approach discussed in Chapter 5 provides a good model of the processes which must be followed. That is, it can be thought of as a journey. The starting point and end point of the journey

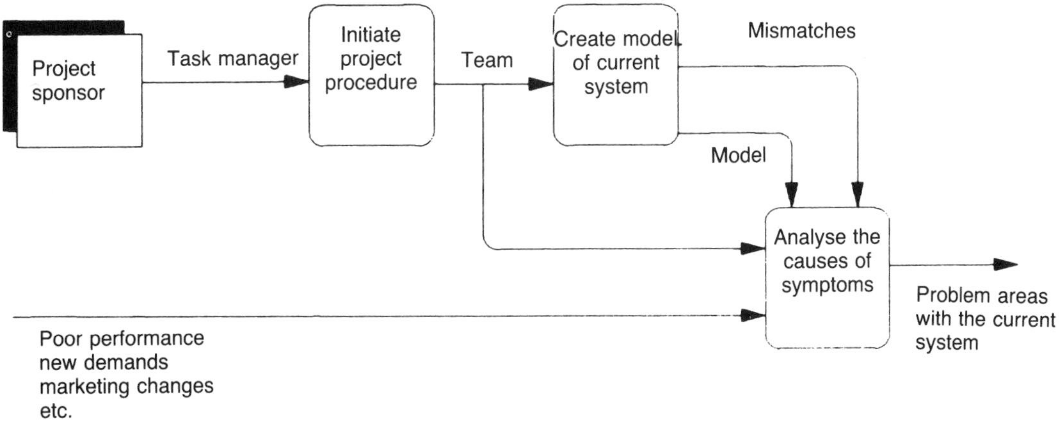

Fig. 7.13 Data flow diagram illustrating the functions of the analysis phase.

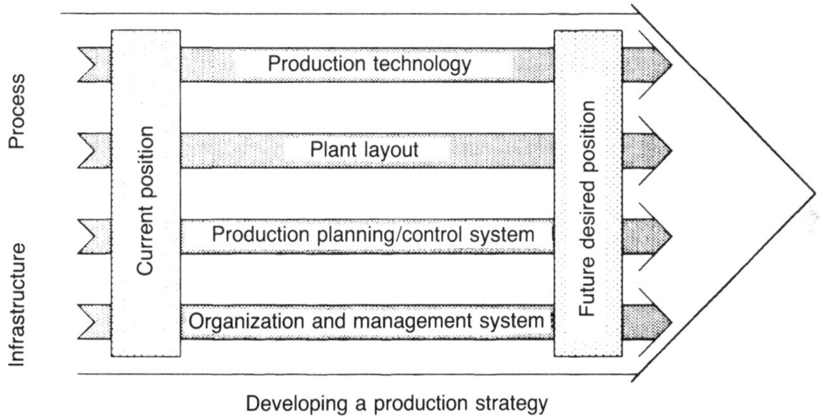

Fig. 7.14 Gudnasson and Riis strategy model (Hill, 1985).

must be known if the best route is to be selected. Similarly, to accomplish the best system changes, both the starting point and the desired end should be known. This requires an understanding of the current system, its components and their interaction. It is then necessary to understand how the current system fails to achieve the current or future requirements, that is, to identify the reasons for the problems. Figure 7.14 illustrates this concept by showing the current position, the end position and the different phases of development which will be required to achieve this.

The three areas in which analysis of the current position must occur are market analysis, product analysis and manufacturing system analysis. The first two are beyond the scope of this design frame, although their inputs are required and must be gained from the necessary sources. Figure 5.1 in Chapter 5 implies that they are regarded here as exogenous inputs to the decision process. However, full consideration will be given here to the analysis of the manufacturing system. Initially a model of the system should be created, together with a list of the current demands which are placed upon it. This model will be compared to the symptoms list so as to identify the constraining sub-systems. The information is then passed to the subsequent objective-setting function.

7.4.3 Analysis procedures

The system must be modelled to aid understanding and decision-making. The analysis team may have pre-formed ideas based upon their viewpoint, but these should not be allowed to restrict the model they attempt to create. In order to reduce the possibility of problem areas being hidden rather than highlighted due to bias, a structured procedure for system identification should be adopted – such as the 5W approach discussed in Chapter 2. This ensures that other perspectives of the system are considered and thus that the true problem areas are highlighted. This tool also helps to bring the analysts together as a team with a common understanding of the situation. The way in which the 5W method can be applied in this specific case is described below.

What

Detailed analysis of production technology

The first aspect to be considered is the production technology. All of the currently used production technology should be recorded in a plant register. The minimum information this is required to contain will be a name for each piece of machinery and a unique number to prevent ambiguity. If a group of items always operate together, they should be considered as a single item of plant. If, for example, a CNC machine has a set of tools and a CNC controller which are always used together within the manufacturing function, then these can be grouped together as a single item (Burbidge, 1971). This gives a more understandable picture of the resources available, and prevents the later separation of items which have essential mutual requirements for operational purposes. A map of the current layout of all equipment should be included. The functional range of each item of plant should be recorded in terms of its ability and capacity – for example, type of operation, size of components which can be processed, number of different tools which can be loaded, limits on accuracy which can be achieved, and availability of the machine in standard hours. Once completed, the plant register will provide a complete picture of the static capabilities of the system within the company.

In addition, a sub-contract function register must also be created in which the manufacturing functions which are available to the system, but are outside the organization, are formally recorded. This should be a record of actual functions, and not of suppliers. The sub-contractors which are used should be recorded in a separate sub-contractor register. However, these two registers should be cross-referenced to establish the relations between the sub-contractors and the functions they provide. Other relevant information about the sub-contractors should also be recorded in the sub-contractor register, such as speed and reliability of service for the particular function under consideration (Chapter 3).

Personnel infrastructure

A list of personnel, similar to the plant register, should be created. This should list all the employees of the firm, and should list all of their skills.

They can then be mapped together in an organizational structure, which will subsequently allow them to be located.

Engineering analysis of products

A current product catalogue should be created. This should identify the quantities of finished parts to be dispatched, showing cyclical variations if necessary. This information can be obtained from manufacturing records. However, it is more desirable to use information provided by the marketing department, which may currently turn away orders or smooth demand levels placed on manufacturing. Gathering such marketing data can best be handled by a functional department. Full use should be made of their expertise.

For each product listed in the current product catalogue, a part list should be produced. This allows a current part catalogue to be generated which specifies every part which must be manufactured or procured, together with demand levels. If cyclical variations are experienced, a graph can be used to illustrate this – the same applies if, for example, major growth is expected in the market.

Process analysis

Having carried out the above actions, it is then necessary to identify the different types and capacities of functions required in the manufacturing system. Identification of process plans for each part in the current part catalogue allows the functions to be identified. These functions should be recorded as a manufacturing function list, and cross-referenced to the current part catalogue. It is then possible, using estimated operation durations, to calculate the total capacity demands for each function, and this should be recorded in the manufacturing function list. It will also be possible to cross reference the manufacturing functions to the plant register, and thus give a reflection of the ability of the system to provide the currently required processes.

All of this information can be obtained by accessing a formal management information system such as MRP if there is one currently in operation in the company.

Why and whether

The next 5Ws question to arise is 'Why should I be interested?', and, assuming there is a reason, we must consider 'whether' this is suitable for a systems approach. In the case we are discussing, having reached this stage, these two questions should already have been answered. That is, the symptoms list is proof enough that a systems investigation is needed to improve the manufacturing situation.

Which and what

It is now necessary to identify 'which' areas are of interest, and to isolate and conceptually arrange these as components and/or sub-systems in order to aid the subsequent investigation. The insight gained should then be recorded in systems terms. The analysis can be carried out in the following steps:

Physical systems description

The manufacturing functions at this stage may be modelled and described using input–output models (or the $IDEF_0$ technique). The team should use

To / From	1	2	3	4	5	6	7	8	9	10	Total
1		15				12	8	5			40
2			10	5							15
3				10							10
4					8	7		5			15
5						5					5
6							12				12
7								12	8		20
8									12	8	20
9										20	20
10											
Total		15	10	15	5	24	20	20	20	26	

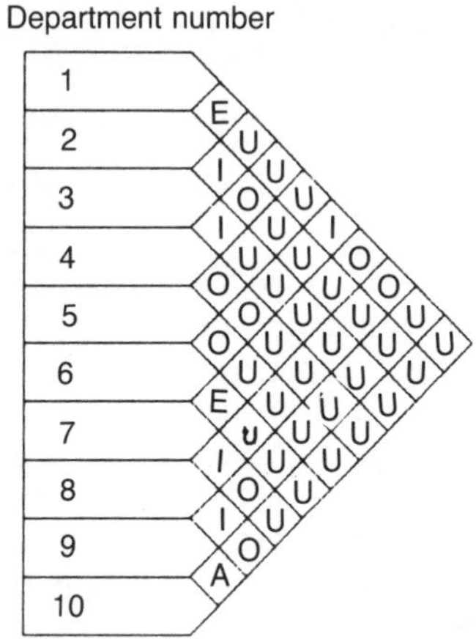

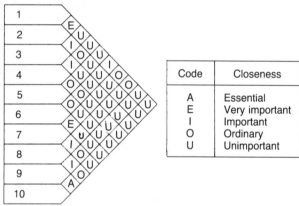

Code	Closeness
A	Essential
E	Very important
I	Important
O	Ordinary
U	Unimportant

Fig. 7.15 Tools of analysis of physical systems (Wild, 1989). (a) Cross chart and (b) Relationship chart.

the information contained in the process plans to represent flows accurately from one department or function to another. Other tools such as cross or relationship charts (Fig. 7.15) may also be used to supplement and clarify the diagrams if flows are complicated. As each level is completed, the team should assess the list of problems and where they are occurring in the system model. The departments identified will need further decomposition to allow full assessment of the problem. The lowest level of decomposition required for any set of functions will show the functional requirements for manufacturing the individual parts. At this stage, the flows on the model should be the same as those for the manufactured parts within the system.

Control systems description

The current control systems can then be added to the input–output diagrams, along with information flows. This enhancement should be in the form of a DFD or IDEF functional diagram. The company's operating procedures will provide the starting point for this kind of analysis. The formal control procedures can be incorporated into functional diagrams. However, the procedures can be open to different interpretations (for example, an $IDEF_0$ analysis of a particular firm revealed that five divisions operating under the same set of procedures had five different interpretations of these, and all conformed to the procedures). Furthermore, the actual operations are likely to be 'greased' by informal systems. These systems often provide methods of overcoming short-term problems, and occasionally allow an unworkable operating procedure to appear to work, and so it is essential that they are considered in the analysis. Failing to do this will cause vital information to be missed, and may cause some of the symptoms to be hidden. Also, these systems may not be coherent with overall system objectives.

To identify and model the informal system, the team should rely upon extracting information through interview with the employees in the system. These interviews must be aimed at finding out how jobs are carried out, and not to repeat what the procedures say. Interviewees should be selected carefully so that the functions concerned can be closely examined. Thus, if a department has not identified problems and is not closely linked to other problem areas, a high-level manager will probably be able to provide the necessary information. However, if a problem area involves lots of problems, it may be necessary to interview every employee in that particular department.

The system property of perspective may causes problems during this stage, as different viewpoints can cause conflict to arise. Areas where interfacing functions each have a different understanding of methods and reasons for operations must be highlighted. These should be recorded as system mismatches, that is, possible causes of problems. These will then require correction. To ensure these are identified, it is essential that the interviewing takes place across a broad cross-section of the workforce, and at all levels.

Finally, all of the items in the plant register, the personnel register and the sub-contractor register can be cross-referenced against the functions which have been identified. The result of the analysis will be a hard record of the functions of the system, in terms of both physical processing and control. This model will contain great detail in areas which are expected to require investigation of problems. Also highlighted will be the areas which have been shown by the analysis to be inconsistent, and therefore likely sources of problems. The functions will also have resources allocated to them, which in turn defines their capacities.

Static analysis

To identify the 'which' and 'what', it is next necessary to consider the resulting system model in relation to the symptoms list. Some items which may be on the symptoms list may be very specific. For example, if a corporate objective has been set to reduce WIP inventory by 10%, this can be directly transferred to a problem/requirement list. For other problems – for example, when new products are to be produced – the concern will be in terms of new functions required as well as increased capacity required from existing functions. All of these items should be identified and added to the problem/requirement list. Further analysis is required to establish where the system does not fulfil the requirements, and why. For example, increasing levels of WIP may have been the identified problem, but only by referring to the DFD and IDEF model is it possible to identify the control systems which might be the true source of the problem.

Dynamic analysis

The analysis so far carried out has been entirely static. It identifies whether the required functions or data flows are available. However, there may also be problems associated with the dynamic characteristics of a system; even if overall demand can be met by the overall capacity, it may be that the actual variation in demand over time cannot be met by the system.

The symptoms list has already been used to give direction to the functional model being developed, and a number of problems associated with symptoms will have been identified through the static analysis. The remaining symptoms may therefore be associated with dynamic factors, and as such must be investigated dynamically. There are a number of useful tools for dynamic investigation, and one of the favourites used for analysis of physical processing is discrete computer simulation, as discussed in detail in Chapter 6. The control system may also be dynamically modelled in a similar way, say, but using Petri net theory. Mathematical analysis of this type of model will identify certain characteristics of the system which can be problematic. Such dynamic analysis, when completed, should reveal the underlying problems causing the remaining symptoms. These should again be added to the problem/requirement list.

Output from situation analysis

Having carried out the above activities, a full picture of the current system becomes clear. The problems which give rise to the symptoms identified in the symptoms list will also have been identified. Further, the problem areas will be bounded by the analysis team during investigation, and the inputs, outputs and currently specified objectives identified. The results of the static and the dynamic analysis are combined together in the problem/requirement list. The resulting document will be used as the basis for the setting of realistic objectives. The information will also be used as an essential input to the system design process, if investment required is to be justified.

7.5 Setting objectives

Having identified the problem areas through the previous analysis phase, objectives should next be set to identify the variable/measure/level combinations which are relevant to the problem (section 5.2.3), and against which alternative solutions can be compared. These must adhere to the company's overall manufacturing strategy.

7.5.1 Manufacturing strategy

This stage of the design process is closely related to manufacturing strategy. In most cases, the system which is being designed or redesigned will be part of a wider system, such that there will be a hierarchy of strategies which will ultimately lead back to the decisions and strategies adopted by the corporate board. Therefore, the system's objectives must integrate with the aims of other parts of the enterprise.

The manufacturing strategy of a firm is the major input to the task of setting objectives. It is nowadays a widely accepted view that successful firms must concentrate on one or two aspects of their performance in order to achieve a competitive advantage. To achieve this, there should be a knowledge throughout a firm of what must be achieved and the priorities attached to these achievements. For example, the five competitive stances identified by Slack (1990) which a firm must balance are: quality advantage, cost advantage, response advantage, dependability advantage, and flexibility advantage. Defining the relative importance of these should be done at a high level of company management so as to provide the necessary business background for the design process.

In most cases investment in AMT is fragmented. Some items are approved because high tangible benefits are expected, while other supporting functions are rejected on the grounds of poor potential returns. Generally speaking, however, a 'CIM strategy' should be adopted to reduce this piecemeal approach, and hopefully make CIM operation a more feasible proposition. While the analysis team is carrying out investigations to define the objectives to be met, it is desirable that the steering committee negotiates at the same time with other levels of the hierarchy to ensure the long-term potential for further development, and that the need for a total system solution, rather than a fragmentary approach, is widely recognized. It is important, therefore, to set objectives which will balance the needs of individual projects with the long-term goals specified by the corporate strategy, thus providing a coherent direction of development for the whole.

The objectives selected will define the desired future situation, and thus the long-term infrastructure of the plant. An example is given in Fig. 7.16 of manufacturing strategy development by input–output analysis showing the possible inputs and outputs of the process.

7.5.2 Identification of variables

To deal with the problems identified in the problem/requirement list, it is first necessary to identify a method of measuring the problems, that is, to identify measurands which are relevant to the successful solution of the problems, and consistent with improved system performance when judged against the manufacturing strategy objectives. When selecting objectives, it should be remembered that the identified problems could affect various functions within the system. Also, the redesigned system must interface with components which may currently be considered to be outside the system. The objectives should be set in such a way so that the whole sphere of influence of the problem can be analysed. For instance, for a problem with inventories, it may be desirable not only that the shop-floor control system is included, but also that the whole organization of the shop floor is scrutinized.

Some of the general principles presented in Chapter 5 can be applied. For instance, the variables to be used should be as visible as possible, because a direct measure of efficiency is much easier to deal with than an indirect one. An example is the factory in which a target was set for the proportion of useful machine time to increase. After a poor start, the figures showed excellent improvements, although actual company performance remained the same. The improvement figures were later shown to be the result of reduced machining speed, causing the proportion of non-machining time to fall.

Also, it is usually necessary to use multi-measures, rather than a single one. In order to allow the system designers the freedom to develop the best possible system, they must know the importance of different factors to the overall success of the system. It is again necessary for the analysis team to carry out wide-ranging interviews to assess which of the measures are most important to the achievement of the manufacturing strategy. The different measurands can then be ranked according to their influence on the overall performance in a manner similar to that presented in Chapter 5.

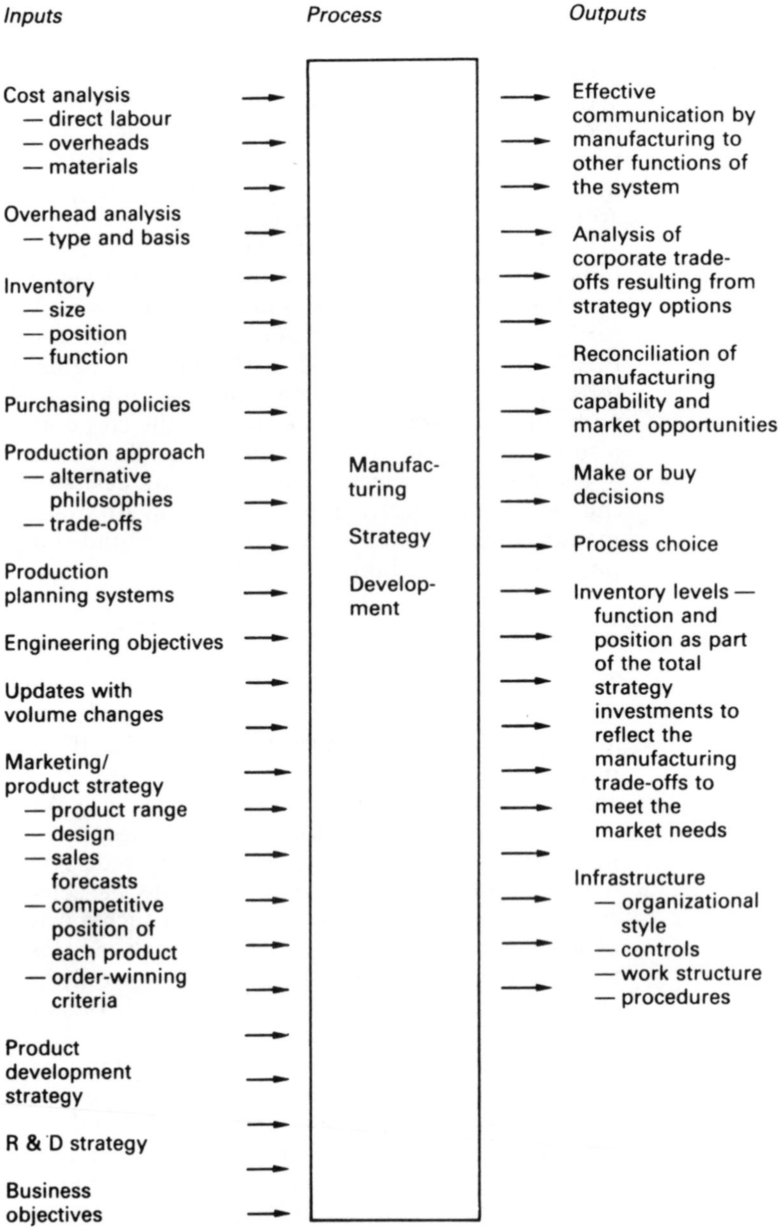

Fig. 7.16 Manufacturing strategy development.

Taking into consideration the fact that many of the benefits from AMT are often difficult to identify quantitatively, the prioritizing of the objectives should be consistent with: aiding in successful design by providing framework for trade-offs; the identification of contributing factors to success in the guiding manufacturing strategy; and the evaluation of possible solutions using the AHP (section 7.3.2).

7.5.3 Setting levels of achievement

As far as manufacturing operations are concerned, three types of variable are often used:

- *Nominal values*. Some variables will give best results at a single value. Threshold bands may be set on either side of this value so as to establish if performance is of an acceptable level. Alternatively, options may be considered by their distance from the nominal level. An example is a balanced line which runs efficiently at a certain output rate, yet individual machines work-stations on the line can work at different rates, but all must be within an acceptable allowance.
- *Threshold levels*. Some variables will have to achieve a threshold level before an option can be considered at all. An example of this is the maximum size of component which a particular machine can process.
- *Continuous improvement*. In this case, as the variable changes, the performance will continuously improve.

Having identified the key variables for the evaluation of the system, it is then necessary to define how the variables should be treated. It is difficult to generalize the process involved here, though it is possible to offer some examples. For instance, nominal values and threshold levels must be established by modelling the system in some way, and then measuring how the overall output varies with the identified measurand. There are many methods of modelling available for an analyst: these include the use of computer simulation to establish the required outputs from a particular cell, so that the overall factory objectives can be satisfied. A computer model could also be used to analyse the effect of quality upon sales, for instance, and help to define the minimum levels to be achieved. If the system under consideration is a sub-system of another system, it may also be possible to use a simulation model of the wider system, and allow the variables of interest to be altered so as to carry out sensitivity analysis. This would establish what nominal values are required in order to achieve the best performance of the wider system, and allow the identification of threshold levels at which the system performance would become unacceptable.

7.5.4 Statement of objectives

Finally, all the objectives should be collected into a formal statement of objectives. This should indicate the priority of each objective. Where appropriate, it should also include the nominal levels and the threshold levels which must be achieved if an option is to be an acceptable solution to the problem, along with the descriptions of the improvements to be made and of the scope of the probject (that is, the system boundaries). Approval of this statement of objectives must be obtained from the steering committee before the system design project can continue.

The output from this phase of objective-setting will indicate a gap between the current and the desired future situation. The following design phases will work towards developing a solution to bridge this gap.

7.6 Conceptual modelling

Following the analysis phases, the design phases of the methodology can be considered as providing the mechanism for transforming the firm from the current state into the desired state. These may be divided into two steps. The first develops the basic principles by which the system will work, while the second provides detailed accounts of what is required.

The first step – conceptual modelling of the desired manufacturing system – is necessary because although there are similarities between many firms, these tend to be most marked at higher organizational levels, while at lower levels they can disappear rapidly. A conceptual model is therefore needed to provide a framework of further decomposition for the particular system concerned.

The required outputs from this conceptual modelling phase are identified in Fig. 7.17. As can be seen, this phase must identify the building blocks of the system concerned. These blocks will be in terms of a combination of manufacturing functions, together with the necessary controlling functions. The relationships between these, which will allow them to fulfil the objectives previously set, must also be specified. Higher-level specification must be developed for the control functions and for the data design, which are to be used later to guide the detailed design phase.

7.6.1 Identification of functional requirements

Parts to be made

As the firm's markets and the product technology progress, the manufacturing system must develop to support them. Market requirements will have been defined through the objectives relating to the product range and the competitive stance. As the technology in the product will usually dictate the types of manufacturing technology choice, it is clear that the product range will have a major influence upon the system to be designed. This, therefore, will be a good starting point for the design process.

The desired product range will include new products, enhanced products, and different quantities of current products. The new or improved products should be identified by interviewing product development and marketing staff. The information gathered about these should include the products' expected parts lists. Each component part should be identified and recorded in a desired part catalogue. The current products which are also in the desired product range can be copied from the current part catalogue generated during the analysis phase. Estimates of demands for all products should also be obtained from the marketing department, and these figures should be recorded in the desired part catalogue.

Functions required

By considering the manufacturing process for each part identified in the desired part catalogue, it is possible to identify the manufacturing functions which must be included in the system (Fig. 7.17). For each of the parts which is also a current part, the required operations recorded on the process plans will reveal the required manufacturing functions. This information is recorded, along with the duration of the operation, in the desired part catalogue. The necessary functions are also recorded in a desired manufacturing function library. These two documents are cross-referenced. Information is also required for any products which are under development. This

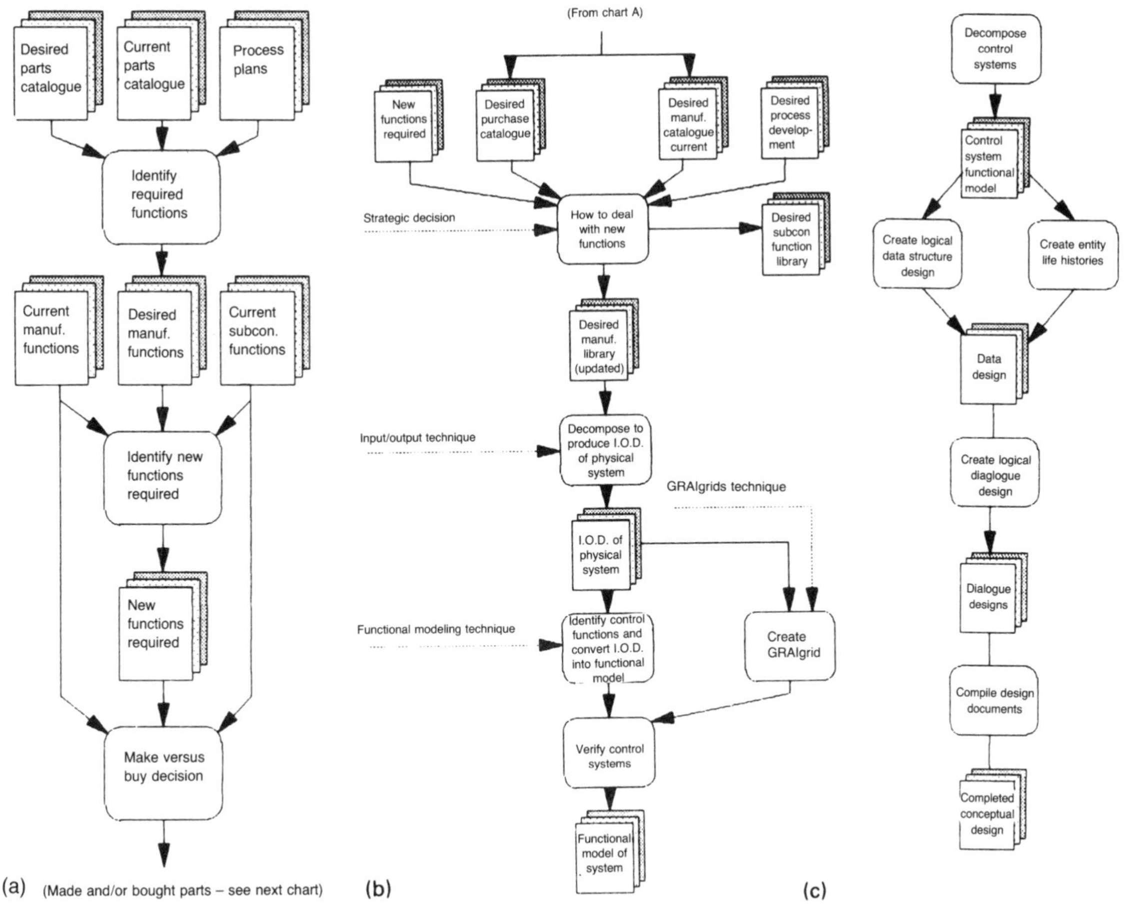

Fig. 7.17 Conceptual modelling flow chart.

should be gathered by interviewing the product development staff and then treated in a similar manner.

The capacities of functions required can now be roughly estimated by summing the products of demand values and operation durations, which have been cross-referenced to each manufacturing function. Set-up times must also be considered, but because of the 'fuzzy' nature of the data at this stage, it is suggested that a set percentage be added to all functions. This data should be recorded in the desired manufacturing function library.

New functions

By comparing the current sub-contract and manufacturing libraries with the desired manufacturing library, the currently 'unavailable' manufacturing functions can be identified. Before further action can be taken it is necessary to decide how the expertise to support these functions should be provided. There will be three options available: the expertise can be bought in from an outside company (for example, by taking over a specialist company or employing specialists who already have the expertise); the expertise can be developed in-house, by establishing a project team, using training courses, or through in-house experimentation; or expertise can simply be used from outside the company without attempting to create the necessary expertise within the company, that is, the work can be sub-contracted. The decisions made will define whether some of the parts in the desired part catalogue require in-house manufacture or sub-contracting. These parts should be recorded in the appropriate catalogue, and should be cross-referenced to the desired manufacturing function library (Fig. 7.17).

Make or buy

The make-or-buy decision is a major strategic decision involved in manufacturing systems design. It is not, therefore, possible to do it justice within the context of this book – for more detailed treatment, see, for example, Burt, 1984. However, an input–output outline of the make-or-buy decision process is shown in Fig. 7.18.

It should first be pointed out that sub-contracting has several advantages. It is a major method of increasing the flexibility of capacity. It can be used to provide extra capacity during peak periods or even provide the entire requirements of a particular function, thus allowing the company to develop and fully utilize its own expertise. It also allows the provision of manufacturing expertise outside the range of the current manufacturing system. A wider range of technologies can therefore be used in the products. The additional advantages include reduced inventory and reduced short-term risks, etc.

However, sub-contracting also has disadvantages. First, a certain level of control is lost. Second, costs can be higher than if the processing were carried out in-house. Third, there is the risk of expertise being made available to competitors.

Some of the problems which should be taken into consideration when making make-or-buy decisions are presented below:

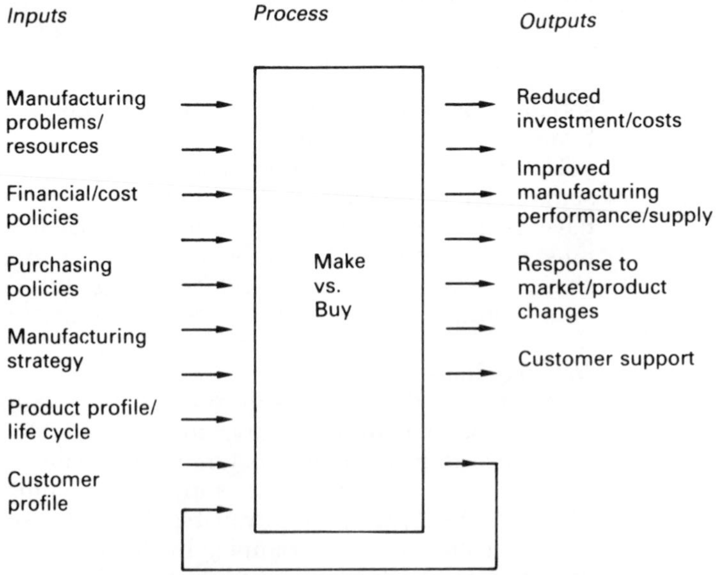

Fig. 7.18 The make-or-buy decision.

1. *The hidden costs of sub-contracting*. When sub-contracting single or a small number of operations, the costs of managing the infrastructure required for the order and delivery process, together with accounts and payments procedures, will usually make the practice relatively expensive.
2. *The hidden dangers of sub-contracting*. Few accounting systems provide facilities to deal with the overhead costs of sub-contracting. There is, therefore, a danger of contributing this to the in-house processes, making them appear more costly, so that sub-contracting appears more attractive than it should be.
3. *The management problems of sub-contracting*. Among these are the technical support required; the identification of actual quality; the cost of administering delivery differences; communications; geographical location; the cost of tooling; the control of lead times; security stocks; the risk of a sub-contractor becoming a sole source; the risk of 'bypass' by the customer; and pricing agreement of subsequent call-off.

Thus make-or-buy decisions are strategically important and require an understanding of policy issues and functional information such as costs, quality specifications and process tolerances. They should, therefore, involve a taskforce approach by the specialists concerned. A critical feature of a product, for example, may have such a great influence upon its quality that it may be considered necessary to process it in-house, although a cost analysis might suggest otherwise. If production technologies are rapidly developing, often a firm may be unable to pay back investment before the technology becomes obsolete, while a specialist would be able to use the equipment more efficiently. On the other hand, if the technology stabilizes it may be worth reconsidering previous 'buy' decisions. To summarize, a more detailed input–output analysis of the make-or-buy decision process is presented in Fig. 7.19.

Discrepancy between actual and desired functions

The make-or-buy analysis of the desired part list will have identified some parts which should be made in-house, and others which should be manufactured by sub-contractors. In these cases the capacity of functions required for processing must be 100% and 0%, respectively.

The actual in-house capacities of manufacturing functions presently available have been identified in the current manufacturing function library. The maximum required capacity for any function is the sum of the demands for the in-house parts already identified and those for all the undecided parts. Normally there is unlikely to be a good fit between the actual and the required capacities. If maximum demand is greater than capacity, two alternative solutions can be sought. That is, the excess demand can be fulfilled either by sub-contracting the work or by increasing the capacity of the factory.

The make-or-buy decision should be made for each undecided part. If the main criterion is to balance the factory demand to current capacity, then it may be possible to use average demand levels for the parts, and establish a range of parts whose total average demand is close to the available capacity. If, however, the demands have a significant but predictable amount of variation, then the dynamic case should be considered, by, for example, plotting demands and available capacities against time on a graph, and identifying combinations of make-or-buy decisions which offer the closest balance. Financial analysis and/or other considerations may indicate that

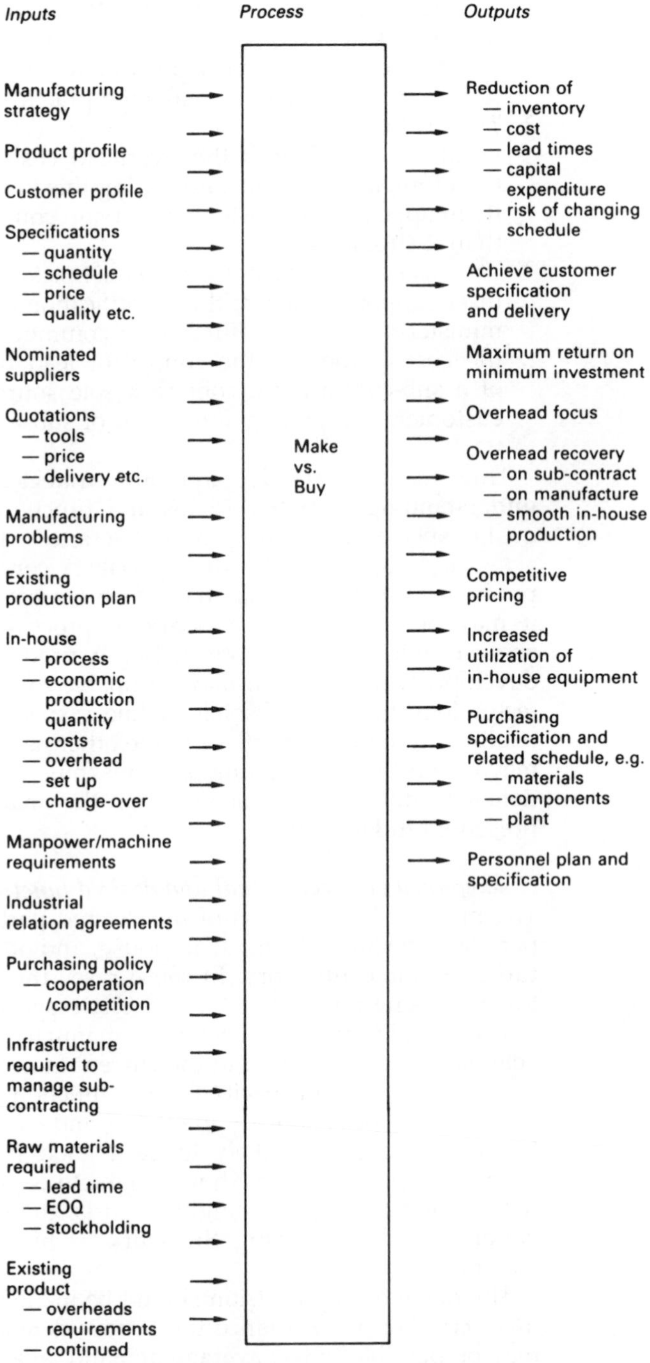

Fig. 7.19 Detailed input–output analysis of the make-or-buy decision.

certain parts should be manufactured in-house, even though the capacity is not yet available. This is of course an acceptable situation, which will be dealt with later in the detailed design. When the decisions are made, all of them, together with other necessary information, should be recorded in the relevant catalogue.

7.6.2 Organization of functions

By now catalogues will have been created to define how the parts are to be made in terms of the required functions and capacity allocation (in-house or externally). The identified functions must next be organized in such a way that the objectives laid down can be effectively fulfilled. This task can be achieved using a decomposition tool which will allow a function of the system to be examined in detail, but at the same time still keep an overall perspective and allow emergent properties to be recognized. The tools available for this task include the previously discussed graphical methods such as DFD, $IDEF_0$, HOOMA, SADT activity diagrams, GRAInets and input–output diagrams (IODs).

Creation and selection of manufacturing business system options
Since initially only the primary functions, their relationships to one another and the physical flows between them need to be considered, input–output analysis is one of the most frequently used tools at this stage. The process involved in describing the physical system will be similar to that of the manufacturing system model presented in Chapter 3. That is, the top layer of an input–output model will present the system operation as a single function with identified inputs and outputs. A description of the function will identify the purpose of the system and the competitive stance to be taken. The inputs identified on the model will be all the materials and parts which are bought in for the manufacturing process. The outputs will include a summary of the information as given in the desired part catalogue.

Next, the basic components identified for the system under consideration must be organized to produce the outputs of this IOD model. This is achieved by the decomposition of the IOD until the level concerning the component-manufacturing function is reached. It should be pointed out that a number of different groupings can result, dependent on the criteria applied to the decomposing process. The effect is that many different manufacturing system models can be created for the same situation, each with a different chance of success.

To deal with this problem, business system options (BSOs) should be created which are systems with different organizational structure and technologies, but all aiming to fulfil the same set of outputs. These will be used as the skeleton for the hierarchy of departments, cells, work-stations, and so on, within the system. They provide the basis not only for the collection and organization of physical resources, but later also for the control systems. It is suggested that in practice three or four such BSOs are generated.

A considerable body of knowledge can be drawn on from the manufacturing strategy literature in this regard. For example, AMT is most likely to be implemented successfully if systems are simplified. This is supported in such theories as the 'focused factory' or the 'plant within a plant'. A number of possible criteria are listed below which can be used in the decomposition of the top-level IOD.

1. *Decomposition by product/process matrix.* Manufacturing technology has been pictured as a continuum ranging from process industry through production lines, large and small batch and down to job shops (Chapter 1). Flexible manufacturing systems and group technology (GT), for example, are two ideas which have been applied to the mid-volume/mid-variety part of this continuum, and have in fact illustrated that a number of different technologies and approaches may be applied at the same volume/variety combination. For example, GT aims to reduce the variety by separating the parts into families, while FMS allow a higher level of automation to be applied to a more varied range of products.

   A framework for the selection of a structure has been established in the form of a product/process matrix (Hayes and Wheelwright, 1979). Although the exact position of a particular firm or product range on the product/process matrix cannot always be defined, this kind of approach does allow a range of alternatives to be considered, and so offers a certain amount of guidance as to the types of feasible manufacturing system structure. This method decomposes a product range at the top level according to product volume and the desired competitive stance. More detailed decomposition will have to utilize rules developed for the actual structure selected – for example, the coding techniques for GT (for example, Gallager and Knight, 1986).
2. *Decomposition by competitive characteristics.* Certain parts of the product range may fall into different categories because of differences in the ways in which the products compete in their marketplace. Thus some products require high quality and others require low cost, although the manufacturing functions are essentially the same. These competitive characteristics may be used as the criteria for the decomposition of a system, so that necessary facilities can be offered according to particular requirements. In this case, the outputs at the higher level of the system must first be divided into groups which consist of products competing in a similar manner. These groups will then be classified as the required outputs from the lower sub-systems in the hierarchy.
3. *Decomposition by process.* The manufacturing process required to make a component can be a good basis for decomposition. Group technology theories can be applied to a set of parts to break it into groups with similar manufacturing characteristics. If there is low repeatability in the product range, then functional specialization may have to be utilized (thus there may be a need for a milling section or a turning section). Also, some processes, by their very nature, are difficult to distribute. For instance, processes such as heat treatment, plating and casting involve capital-intensive equipment and a system can only have a limited number of such facilities. Under this circumstance a hybrid system could be used which organizes most of its facilities into GT groups, while treating those central functions as services to these groups.

To illustrate the use of these different types of decomposition criterion, let us assume that a manufacturing firm produces a range of products known as product 1, product 2 and product 3. The processes required for these involve three possible manufacturing functions, known as operation 1, operation 2 and operation 3. Each of the products may be produced in two possible ways: product 1 by operation 1 or operation 2; product 2 by operation 1 or operation 3; and product 3 by operation 2 or operation 3. Figure 7.20 illustrates two of the possible decomposition options according

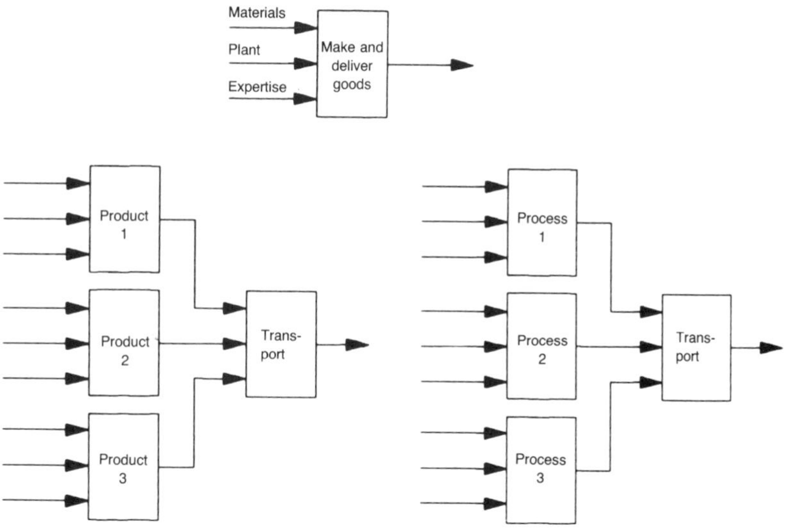

Fig. 7.20 Different decomposition approaches.

to different decomposition criteria (product or process). By following these two decompositions, it is obvious that different manufacturing facilities and also their configurations will later be required. A third operation might be to develop a flexible transfer system, which will move the products from any processing equipment to any other equipment. This, however, will obviously require a more complicated control system.

After this decomposition process a number of BSOs will have been recorded. Depending on the scale of the project, the final selection of a BSO must be taken by either the steering committee or the board of directors of the company, on a strategic basis, for further development.

Other considerations

There are a number of other considerations when modelling the physical system. The problem of transportation is one such consideration. It is necessary explicitly to draw attention to the logistic requirements of the system. Where there is a need for a transportation or transfer function, it should be included in the physical BSO model. This is to ensure that the control and communication implications are considered, and that hardware requirements are developed accordingly in the detailed design. Another area which will require attention is that of sub-contracting. The sub-contracting situations must be included at the highest level of decomposition. This will ensure that consideration is given to the implications which this function will have on future system performance. It will also allow the identification of control-system requirements later in the design process.

7.6.3 Analysis of control systems

Each sub-system defined in the IOD will require its own internal control system to be sorted out. The control systems are not an end in themselves, but are required to improve the effectiveness of the manufacturing operations concerned. As discussed in Chapter 2, different areas of the business will have different requirements as to the type of control needed. Some areas will change in a stochastic manner either due to variation in the inputs or due to disturbance from the wider system. For this type of operation it would be desirable to have a continuous online control function which responds immediately to changes. On the other hand, there are areas where things can be done in an actually or nearly deterministic manner. In this situation there may not be the need for outputs to be monitored and demands altered continuously, and decisions can be made on an offline, once-per-day or even once-per-week basis.

Important considerations here include the levels of synchronization required, the amount of information which the systems have to process and the time the processing takes. Wherever appropriate, smaller, distributed control systems should be used so that each level in the hierarchy need only deal with a small set of information, which can be rapidly processed. To achieve satisfactory operation, however, it is essential that the different control systems are effectively coordinated. It should be noted that the control systems need not always be computer-based. The kanban card method, for example, is a common non-computerized control approach.

Modelling the control system

First it must be remembered that normally the design of the control system, like the selection of a manufacturing structure, is not a clear-cut decision. For example, should a firm select MRP II or optimized production technology (section A.3.3 in Appendix A)? Therefore, again it is useful to have a number of options evaluated. However, this needs to be balanced against the budget and the time available.

The control functions can be incorporated into the model by converting the IOD into a model which will contain information-processing descriptions such as DFDs or $IDEF_0$. DFDs and $IDEF_0$ have specific methods of identifying control signals, and the conversion can be achieved quite easily. If a DFD is the selected tool, it may be desirable to use different types of arrow to distinguish between information and material flows. In all cases it may also be desirable to use slightly different 'function' symbols to allow easier identification of control and manufacturing functions. The GRAInet tool explicitly identifies decisional and processing functions, so that no modification is required.

The GRAIgrid has the ability to map the decisional time-scales as well as the control functions which are under consideration. The prototype grid shown in Fig. 7.1 may provide a start for discussion, although the design team should attempt to identify other functions which are essential to the environment in which the particular firm will operate. Having established their framework, the necessary functions with respect to different time-scales must next be analysed to identify their applicability to the system. Once a decision centre is positively identified, it should be indicated on the GRAIgrid with a cross. They may be duplicated in several sub-systems. For example, a short-term production control is likely to be required in all cells and workshops within the system.

Modelling approach

The representation of the required control functions on the physical system IOD can be achieved in a top-down manner, with a control function superimposed on each level of the decomposition, and the span of control being

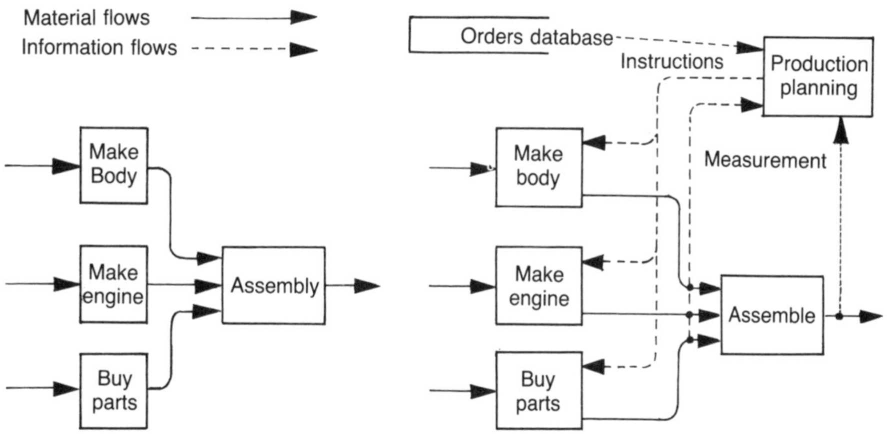

Fig. 7.21 Conversion of input–output diagram into data flow diagram.

the manufacturing functions at this level. The following steps may be followed:

1. *Identify variables.* The variables which affect the functional output must be identified, together with a process designed to utilize the information. It is important that the selected variable should have a relationship to a key objective and also be controllable by the system.
2. *Identify data flows.* The sources and destinations of the required data can then be identified from the functions on the IOD. These are then added to the model, converting it into a DFD or activity diagram (Fig. 7.21). A functional description should be prepared which includes the decision horizons and decisions period, i.e. time for a decision to affect the output from the system.
3. *Identify processes.* Having selected the variables to be monitored, it is now necessary to verify that they are the correct entities, and to understand how these variables can be used to control the system. If a direct relationship cannot be found, it may be necessary first to build up knowledge of the systems prior to system construction – for example, by using computer simulation modelling. Alternatively, the control system may be such that in-service monitoring and knowledge accumulation may be achieved – for example, by such techniques as statistical process control (SPC) or the use of expert systems. Regression analysis is another mathematical technique which can be applied to improve the selection of variables for monitoring. This analysis will provide an assessment of the applicability of the selected variable. If the variable is found to be unsuitable, the investigation must be repeated so that a different one can be found. An understanding of the relationships will be generated, and this should be added to the function description.

These control functions can be further decomposed to provide a more detailed description of the information processing involved. The resulting system can be either computer or manually based, both of which can be dealt with using the SSADM methodology.

Application of SSADM

As a control system should always be based on an information system, the current methodologies developed for information systems design and analysis, such as SSADM, should be able to deal with some of the design demands satisfactorily (see section 7.7.4).

SSADM can be used, for example, to aid the data-design process of a control system. The variables identified for control purposes will be the key entities involved here. SSADM will help clarify the information structure of the control functions and specify the data storage and management facilities. Logical dialogue design is also needed in this phase, to identify the necessary system interface and to structure these. The detailed techniques for this purpose are covered in detail in SSADM step 250 (Downs *et al.*, 1988).

The completion of the steps outlined in this section concludes the conceptual modelling of the system. All documentation should be gathered and sorted for use in the detailed design phase.

7.7 Detailed design

The conceptual design of the system has created and selected a model in terms of related functions from the desired manufacturing function library. It contains manufacturing and logistic functions each of which has a related catalogue of products, together with a hierarchy of control systems which process information. In addition, the conceptual modelling phase has specified the long-term production capacity to be achieved in terms of average or static capacity levels and levels of variation to reflect the dynamic characteristics. In accordance with the general frame outline, it is now necessary to prepare a detailed specification of the plant.

Again two interrelated areas of consideration are involved – physical and control. While the selected conceptual model dictates the type of functions and the relationships between them within the system, detailed design must decide on the detailed layout of the plant by selecting and allocating the required manufacturing equipment, as well as the physical infrastructure including transportation and storage facilities. At the same time, the controlling functions must also be developed by selecting specific algorithms to generate control signals together with all other facilities which are necessary to bring about the desired outputs from the system. Thus the detailed design of control systems will require the selection of hardware for the monitoring of selected variables, software for data processing and also the media through which personnel can interface with the decision processes.

In summary, the detailed design phase takes the conceptual model of the selected manufacturing system option and transforms this into detailed specifications which can be used later for implementation. The general structure of this phase is shown in Fig. 7.22.

7.7.1 Selection of production technology

Usually more than one machine will be available to fulfil a particular required manufacturing function. It is, therefore, necessary to provide detailed specifications for the selection of most suitable items of plant. The approach to this selection of production technology is illustrated in Fig. 7.22a.

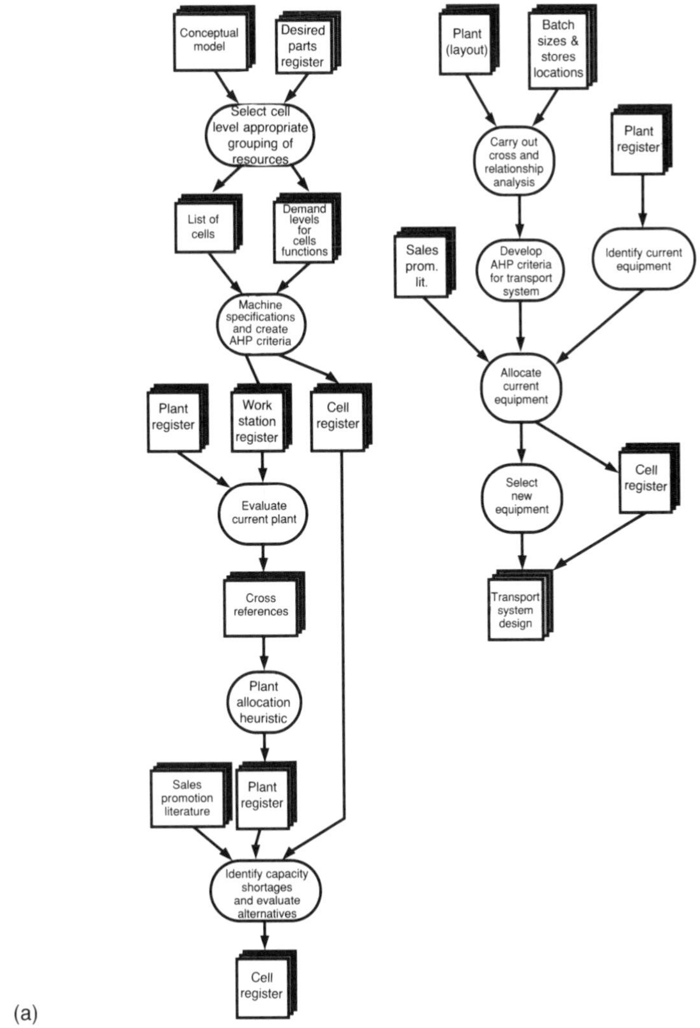

Fig. 7.22 Detailed design (a) physical systems, (b) organization of components and selection of batch sizes and buffers and (c) information structure.

Identification of cell requirements

Among other things, the conceptual design of the manufacturing system defines the manufacturing functions which must be available within its subsystems. It also defines the levels of manufacturing output which must be achieved, and the variation in output which must be possible. These requirements are brought together to provide constraints on the production to be selected. Several of the identified functions may be fulfilled by a single machine. Therefore, rather than considering single functions, we consider a higher 'cell' level so as to identify machines which will satisfy the cell requirements. In some cases the cell level will already have been defined in the conceptual modelling phase (for instance, when group technology is being applied). In general the cell level will have to be defined first. The

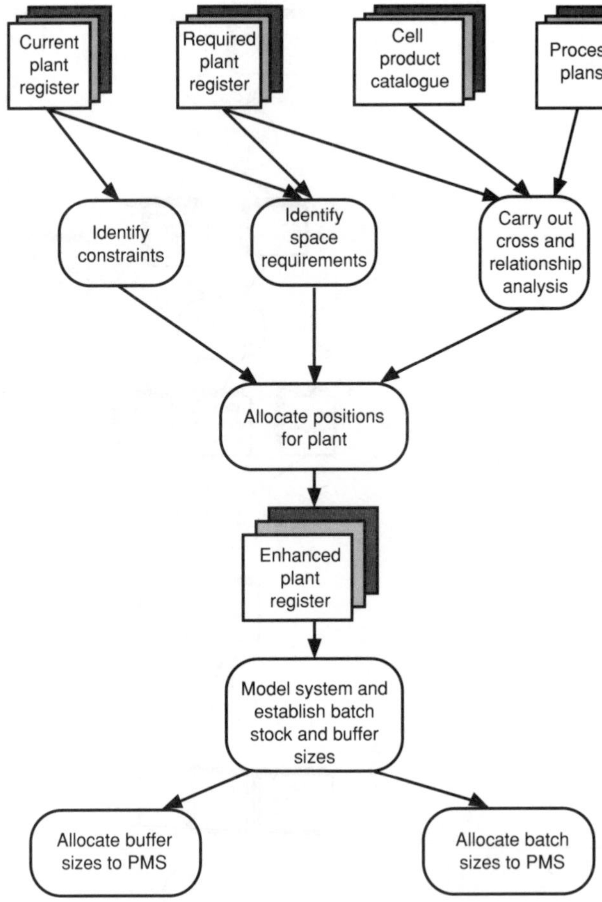

(b)

Fig. 7.22 *continued*

demand levels for the basic functions should be summed to give total static capacity requirements for each cell.

As the equipment is not being purchased at present, a full specification and list of decision criteria should first be created and recorded in a workstation register to provide accurate and detailed information, should implementation go ahead. The items selected should be recorded in a cell level plant register (Fig. 7.22a).

The total function demands, divided by the time available within a working period, will identify roughly how much machine capacity should be made available for the function. This will give some guidance, as if a single machine is required. In reality, if more than one machine is required to fulfil the capacity, it will then be necessary for the analysis to move lower down the hierarchy, and try to identify more specialized requirements. This process is completed when a range of machines has been identified which nearly match the capacity requirements.

The next step is to specify more rigidly the equipment needed. The most important factors to consider are those which allow the cell to fulfil the targets defined in the previous phase. They can be identified by utilizing the

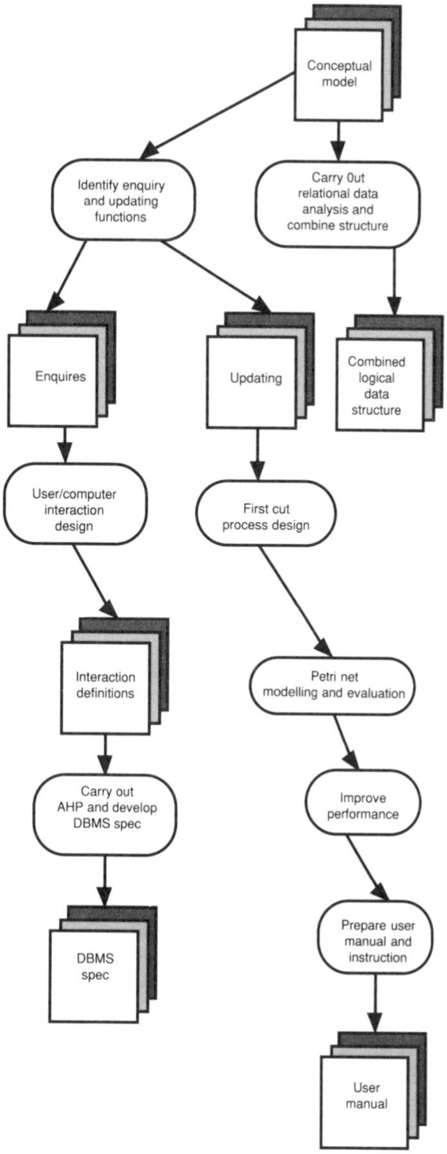

(c)

Fig. 7.22 *continued*

experience of the design team, by obtaining external expertise and by discussion with future users. The basic functions identified in the conceptual model will provide the basis for this task, although other less tangible benefits must also be included. It should always be borne in mind that non-quantifiable aspects should be taken into consideration in the selection of machines.

The identified objectives should be decomposed to give a hierarchy of requirements, an example of which is given in Fig. 7.23. The AHP technique can then be used to assess the production technology which is best suited to

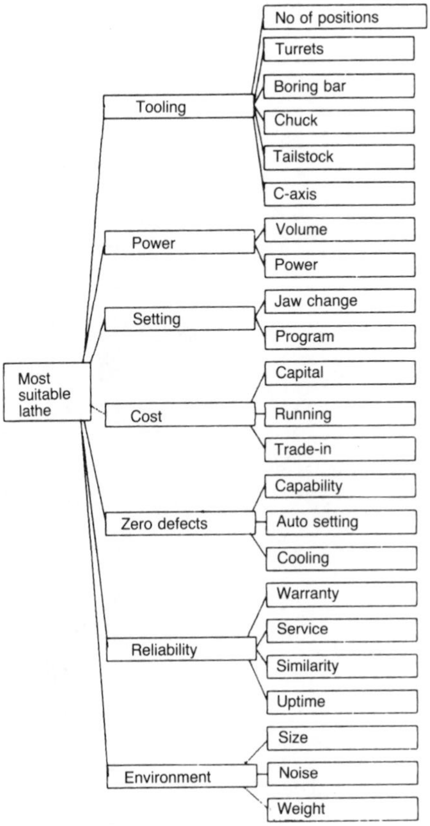

Fig. 7.23 Hierarchy of factors relating to the choice of a CNC lathe (source: Ajderian, 1989).

fulfilling the requirements of the cell. In each case, therefore, the design team must establish a priority rating for each of the factors identified. Although the use of these constraints may be subjective in many cases, it will help to provide a general framework for establishing the best distribution of current technology between the sub-systems.

Selection of machines

Current production technology

Since usually the resources available are limited, the design team must use current plant to minimize the costs involved whenever possible. The current production technology has been recorded in the plant register (from the analysis phase). It is necessary to allocate the usable, current production technology to specific functions. Care must be taken to ensure that equipment is not double-booked, otherwise there will be a shortage of machines and capacity later when the design is implemented.

Each item of plant should be considered for its possible applications by assessing it against the hierarchy of selection criteria which have been identified for the AHP. These factors are of two types. Qualifying factors are those criteria which must be fulfilled if the technology is to be acceptable. If

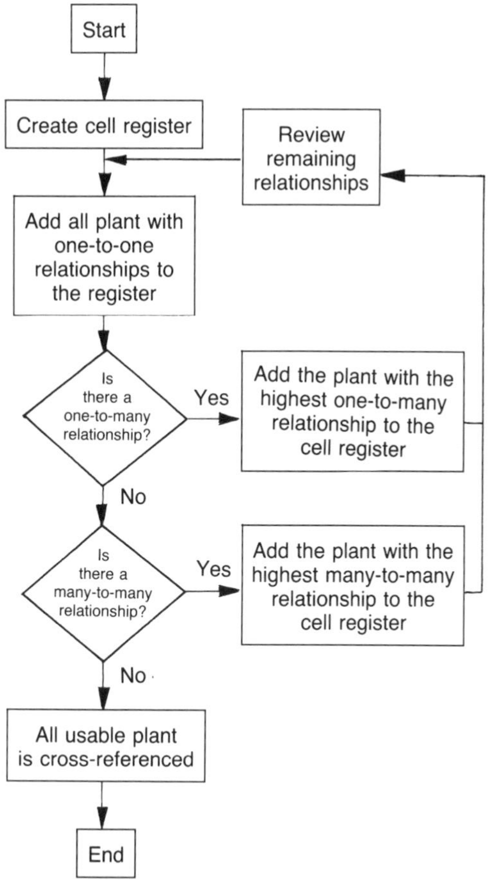

Fig. 7.24 Plant allocation heuristic.

the item is given a pass on each of the qualifying factors, then it is suitable for the process. The remaining factors are those which exceed a minimum level. The level of satisfaction will vary, according to how far this hurdle rate is exceeded. A level of satisfaction should be established for each item of plant against each selection criterion, so that it will be possible to quantify the best contender for any application.

Upon the completion of this analysis, all the machines which are capable of fulfilling a required function will have been cross-referenced to that function. In some cases this will be a simple one-to-one relationship. However, when there is a one-to-many or many-to-many relationship, it will be necessary to identify what allocation of equipment will provide the best operation for the system in terms of the given objectives.

A heuristic approach may be used for this problem, as shown by the example in Fig. 7.24. The allocation of one-to-one plant–cell relationship is relatively straightforward. Where relationships are one-to-many or many-to-many, the best combinations may be established by calculating the utility vector for each plant–cell relationship. One-to-many relationships should be given the highest priority, so as to ensure maximum utilization of current plant. Their associated facilities should therefore be identified and allocated first. As a piece of plant is allocated, the number of permutations available for other relationships is reduced. When all one-to-many relationships have been used, the same process is used for allocation for the remaining many-to-many relationships.

A spreadsheet model may provide one of the most suitable tools to allow the designer to keep track of information in this situation by allowing the same design and therefore the same layout to be used to assess various options being considered for a cell; by carrying out the calculation of utility-vectors; and by standardizing the display and quality of information being presented about different cells.

New production technology

Having allocated the current production technology, certain capacity requirements will remain unfulfilled or partially fulfilled. Therefore, it is next necessary to establish what new production technology should be used to fulfil the remaining requirements. The decision process concerned here is made slightly easier, for the hierarchy of decision factors has already been created.

First, it is necessary to assess what equipment is currently available on the market. This is one of the best opportunities to look for innovative options, since there are fewer constraints attached. Significant gains may be achieved by replacing some of the current technology already allocated, even though it is still usable. For example, some production technologies may have been developed which are capable of carrying out more than one function (a CNC machining centre, for example), although at extra expense. In general, the technology selected should be that which has lowest functionality while still fulfilling the operational and strategic objectives. The reason for this is obvious: machines with higher functionality can be unnecessarily expensive and the equipment would, therefore, be producing a lower return on investment. In this case, the lower cost of a less sophisticated but practical machine would allow more rapid payback, allowing the designer more freedom in the future.

Again for each of the requirements which must be fulfilled, it is necessary to prepare a list of different options. These should be assessed again in terms of the factors identified for the AHP.

7.7.2 Organization and layout of production technology

Figure 7.22b illustrates the processes involved in the tasks of equipment organization and layout, and selection of buffer and storage facilities. The objectives of these processes should be in agreement with the overall objectives, and will often fall into three categories (Wild, 1989): to minimize the cost of materials handling and movement; to minimize congestion and delay; and to maximize utilization of space, facilities and labour.

Simplicity

The key word to remember here is again *simplicity*. It is particularly important to simplify material flows where advanced manufacturing systems are concerned. This must be achieved within as well as between cells. The main reason for this is that there will be an inherent limit to the flexibility of the hardware involved.

An AMS is usually intended to increase flexibility. However, this does not mean that it must be possible to carry out any and every function, but instead tends to mean that it can achieve a range of functions efficiently. When related to materials flow, this implies that provision cannot usually be made for flows from any one node to any other node. Careful selection of cell products should, therefore, attempt to identify a relatively wide range of products, but which have a limited amount of variability in their flow patterns. The techniques and algorithms designed to help the analysis process in this regard are reviewed in the next section.

Requirements for the development and maintenance of software for control purposes may, in reality, also limit the degree of flexibility. For example, as the demands made on the system increase, there may be a need to increase the level of hardware required. For each transfer between two nodes, it is necessary to develop a piece of software to carry out the controlling tasks. This must be stored and transferred to the correct location when required. As the complexity of the required flow increases, the amount, complexity and cost of the associated software also increases. This tends to reduce the visibility of the system to the operators and potential problems will become increasingly more difficult to detect. As a result, the skill levels required to maintain the system will have to increase. Therefore, while the individual objectives for different projects vary, there is a general tendency to require system simplicity, and this should be emphasized throughout the design process.

Layout of components

To achieve the best layout of work-centres within cells, and the location of cells and departments in relation to one another, the space requirements of the previously identified functional groupings should first be established. These should be estimated on the basis of expected floor space for each of the items of plant listed in the desired cell plant register.

Cross and relationship charts can be used to give guidance in the location of departments (as well as cells or work-stations within cells). The cross chart is a two-dimensional matrix listing each department as a source (rows) and as a destination (columns), as shown in Fig. 7.15a. The significance of the transportation between each source–destination pair is quantified by the number of transfers between them. A figure for this can be achieved by counting the number of transfers which occur between current departments, for current products. However, a better approach, which will treat both current and future products by the same procedure, and will account for seasonal variations, is to identify the number of transfers from the product range's process plans and bills of materials. This is then multiplied by the expected annual sales quantity, as given in the sales forecasts. The figures produced can then be kept in absolute terms, or can be changed to relative values. Similar information is provided by the relationship chart, as shown in Fig. 7.15b. This codes the importance of relationships between individual departments. Either of these tools could be enhanced by considering a measure of transport costs – for instance, by the weight or volume of the parts. This can be multiplied by the number of transfers required, and will thus take account of the significance of the transfers. For instance, it would probably be easier to transfer ten electronic components than ten castings.

Following the above, the individual departments must now be located. Decisions can be made using both the quantitative information generated and the constraints identified earlier. These decisions must be recorded as a map of the locations, along with a report listing their justifications. The map, the decision chart, and a report for the selected layout should be added to the plant register for the sub-system concerned.

The process is repeated for more detailed layouts – for example, for work-stations within each cell. At these lower levels, it may also be necessary to identify sources and destinations outside the cell, for this will allow for the location of components in relation to the environment. Due to the specialized nature of work-station layouts it is not possible to define specific methods to be taken. There is, however, an extensive literature available in this particular subject area. For example, books on work study and ergonomics can be consulted should the work concerned have a high manual content (Wild, 1989).

Batch sizes and WIP stocks

Storage space, batch sizes and buffer stocks can all have a major effect on the ability of the firm to meet its overall objectives. These must be constrained to achieve long-term optimization because they are some of the most important parameters utilized by a production management system (PMS). For example, a firm's main strategy for competitive advantage may not be price, but the offer of a prompt response for customized products or a higher dependability on delivery dates. Therefore, in some cases it may be desirable to have a rapid throughput time which will require smaller batch sizes, while in the others the requirement may be for very precise delivery dates and so a certain amount of safe stock may be necessary. Consequently, the desirable characteristics have to be established. A balance must then be achieved between the desired levels, and the costs which will be incurred.

It is often not feasible to produce a lot size of one. For example, a machine-tool may require setting before a job, and the economic batch size will depend upon the cost of setting up the machine relative to the total cost of the component and the stock-holding cost. These are conflicting desires: to run large batches in order to achieve economies of scale, and smaller batches to reduce inventories.

To help establish the implications (in operational terms) of different trade-offs between batch size and in-process storage quantities, a discrete-event simulation model can provide a useful analysis tool, as illustrated by the examples given in Chapter 6. The level of aggregation of data is important here to reduce the effort and expense involved in the creation of a model and the analysis of results. The input data can be either a set of known 'typical' inputs or a randomly generated set of data known to possess the statistical characteristics of the true trends, as discussed in Chapter 6.

The investigation of the system using a simulation model allows the consideration of various alternatives without the requirement to commit resources to a full-scale model. There is also the benefit that the actual performance of the system can be tuned to provide the best performance, in terms of the system objectives. An additional benefit is that a similar model is likely to be used in the investigation of the control systems and practices required. As a result, the investment in the creation of the model will be better utilized, and experience will be gained which will help in the design of the control systems. (In fact, it is very likely that the two investigations will have to be undertaken simultaneously, as the control systems will not only affect the system performance, but will alter the way in which it will operate.) In addition, Petri nets are another possible tool for establishing the necessary buffer sizes for the initial investigation (Barad and Sipper, 1988; Ramamoorthy and Ho, 1980).

Materials handling

While, strictly speaking, materials handling should be considered as part of the selection and organization of production technology, it is given special consideration here because of the previous need for the establishment of batch quantities and in-process buffer stocks. Once these are specified, it is then possible to consider what are the best forms of transportation. In fact, by now a significant amount of the work concerned with developing material-handling systems will have been carried out in the analysis of the detailed layout of the system (such as the analysis concerned with simplifying flow patterns between work-stations). All that is needed at this stage, therefore, is to choose the particular forms of transportation. Publications are available which will provide an overview of the traditional mechanisms for the transportation and storage of components (for example, Department of Industry, 1978). In many cases, the decision will be constrained by the size and weight of products. However, robot controllers are available for different sizes of devices. Thus it is possible that an automatically controlled crane could be used for large parts; an AGV system used for smaller components; and a track with automated buggies used for the smallest of items. Other facilities may also be required – for instance, bar-loading machines to allow the automated running of CNC machines, or sequencing machines to order components for insertion into a printed circuit board. Finally, it is also necessary to consider what will happen with waste materials – swarf and trimmings, for example.

If transportation equipment is currently available, this should be recorded in the plant register along with any limitations. It should then be allocated to specific work-stations, or should be allocated to a general register of transportation equipment used between different cells, along with a map of the nodes (that is, work-centres or cells) which it will serve.

7.7.3 Algorithms for cell formation

A wide range of algorithms has been developed for design tasks such as process planning, equipment selection, cellular configuration and facility layout. In the following sections, those algorithms associated with cellular formation will be discussed with a view to highlighting their common characteristics when they are put into practice.

As the basis of FMS cell development and because of the need to simplify the control functions in an advanced manufacturing environment, cellular manufacturing is receiving increasing attention. In a conventional manufacturing layout the components pass through various functional sectors which specialize in specific operations such as milling, drilling and deburring. Although the physical layout of this kind of organization is relatively simple, the logical layout (i.e. production flow) is usually messy. Thus a complex control system is required to track and schedule each of the components around various sections in order to maintain their appropriate priorities. In contrast, a cellular-based manufacturing system is arranged into a series of small manufacturing units known as cells. These are a small collection of machines and operators which produce a defined group of components. Due to the similarities between the components produced by such a cell, the production facilities within each cell are operated like a flow line, hence greatly reducing the amount of material handling required and the complexity of the control system. In addition, such a layout creates a human-centred manufacturing environment. The main benefits that can be expected from cellular-based manufacturing include simplified planning and control, reduced lead times, reduced WIP level, improved quality and increased job satisfaction.

However, there are a number of problems with this approach. One of these is that the proper functioning of the cell depends to a large extent on their initial definition. Therefore, at the detailed design stage cell configuration is frequently one of the most important tasks to be performed. The ideal cell should be completely autonomous; that is, material flow between cells (known as the inter-cell movement) is entirely eliminated. In real life this is not always achievable and the aims of cellular formation are usually to obtain maximum cell autonomy and at the same time minimize the number of bottleneck components (those which need to visit more than one cell) and bottleneck machines (those which are required in more than one cell – these can be eliminated through duplication, sub-contracting or the use of a central service station). Cell formation algorithms can be classified into two broad categories: the product based methods and the product/process based methods. The former arrange components into sets according to their shape and size; the latter utilize production as well as product information to carry out cellular formation. Some of the product/process-based algorithms are discussed in more detail in the following text. The review given here is by no means exhaustive – many other methods are available. Comparative studies of the relative performance of these different approaches can be found in the literature (for example, Shafer and Meredith, 1989).

Production flow analysis

The production flow analysis (Burgidge, 1963, 1979) was the first to utilize product/process information for cellular formation. It is an integrated approach which considers the group technology problem at a number of levels. At the factory and the group levels, information of component flow and process is used to generate cell configuration. Material flow analysis utilizes operation and set-up times to balance production, and finally, tooling analysis attempts to rationalize the tooling requirements.

Similarity coefficient method

The similar coefficient-based methods exemplify a cell formation approach that has received significant attention recently. This is an approach which mainly operates on a component/machine matrix, as given by equation (7.1):

$$[a_{ij}] = \begin{bmatrix} a_{11} & a_{12} & \cdots & \cdots & a_{1n} \\ a_{21} & \cdots & \cdots & \cdots & a_{2n} \\ \vdots & & \ddots & & \vdots \\ a_{m1} & \cdots & \cdots & \cdots & a_{mn} \end{bmatrix} \tag{7.1}$$

where:

m = the number of machines
n = the number of components

$$a_{ij} = \begin{cases} 1 & \text{if component } i \text{ needs machine } j \\ 0 & \text{otherwise} \end{cases}$$

This method was first proposed by McAuley (1972) with the so-called single linkage similarity coefficient (SLINK) as defined by equation 7.2:

$$s_{ij} = \frac{\sum_{k=1}^{n} \delta_1(a_{ik}, a_{jk})}{\sum_{k=1}^{n} \delta_2(a_{ik}, a_{jk})} \tag{7.2}$$

where:

s_{ij} = similarity coefficient between machines i, j

$$\delta_1(a_{ik}, a_{jk}) = \begin{bmatrix} 1 & \text{if } a_{ik} = a_{jk} = 1 \\ 0 & \text{otherwise} \end{bmatrix}$$

$$\delta_2(a_{ik}, a_{jk}) = \begin{bmatrix} 0 & \text{if } a_{ik} = a_{jk} = 1 \\ 1 & \text{otherwise} \end{bmatrix}$$

The similarity coefficients are to be calculated for each of the machines (or machine groups), which can then be used by a cluster identification algorithm for cell formation in a number of ways. For instance, a threshold value of similarity coefficient can be used to identify the machines which should belong to a particular cell. However, one of the deficiencies of the SLINK approach is that it fails to recognize the 'chaining' problem caused by bottleneck machines, because with this analysis approach each machine can exist only in one group and hence there can be no duplication of machines (King and Nakornchar, 1982; Seiforddini, 1989). A number of attempts have been made to overcome the chaining problems including, for example, DeWitte's method (1980), the average linkage similarity coefficient (ALINK) (Seiforddini and Wolfe, 1986), and machine component cell formation (MACE) (Waghodekar and Sahu, 1984). MACE, for example, is a simple extension which uses three similarity coefficients to overcome some of the problems of single linkage cluster analysis. The method takes the component/machine matrix and produces a machine incidence matrix, which illustrates how many parts visit a pair of the machines. This is called the NCC (number of common components) matrix, and is given by A^*A^T, as shown in Fig. 7.25.

From this matrix three coefficients are used to compute the similarities.

Component/machine matrix (A)

Cmp't No.	M/C No. 1	2	3	4	5	6	7
1	0	1	0	0	1	0	0
2	0	0	1	1	0	0	0
3	1	0	0	0	0	1	1
4	1	0	0	0	0	1	1
5	0	0	0	0	0	0	1
6	0	0	0	0	1	0	0

A^*A^T matrix of machines

M/C No	M/C No. 1	2	3	4	5	6	7
1	2	0	0	0	0	2	2
2		1	0	0	1	0	0
3			1	1	0	0	0
4				1	0	0	0
5					2	0	0
6						2	2
7							3

Fig. 7.25 An NCC matrix.

These are: the additive similarity coefficient (equation 7.3) based on the total number of common components processed by a pair of machines, the product type coefficient (equation 7.4) based on the total number of components processed by the pair of machines and the total flow coefficient (equation 7.5) based on the total flow of the components. The first two coefficients are used initially to place the machines with the highest similarities alongside each other. This allows similar machines to be formed into groups. This data arrangement is then used and the inter-cell flows are calculated using the third coefficient, which determines the flow between the previously defined cells. From this the machines are arranged to minimize the inter-cell movement. The parts are then allocated to the cells by using a sorting routine.

$$ms_{i,j} = \frac{c_{ij}}{\min\left[\sum_{k=1}^{n} a_{ik}, \sum_{k=1}^{n} a_{jk}\right]} \tag{7.3}$$

where

c_{ij} = number of common components using both machines i and j

$$ms_{ij} = \begin{bmatrix} 0 & \text{if machines } i \text{ and } j \text{ are independent} \\ 1 & \text{if one is a sub-set of the other} \end{bmatrix}$$

$$ps_{i,j} = \frac{c_{ij} \cdot c_{ij}}{\sum_{k=1}^{n} a_{ik} \cdot \sum_{k=1}^{n} a_{jk}} \tag{7.4}$$

$$fs_{i,j} = \frac{c_{ij} \cdot c_{ij}}{TF_i \cdot TF_j} \tag{7.5}$$

where TF_i = total flow of components processed by machine i.

Some of the later developments have used a much wider range of manufacturing information including, for example, the sequence of operations and the expected work of the machines. In many cases, a_{ij} is no longer restricted to the binary values of 0 and 1, but represents the duration of the operations in question. One example of these is provided by the production data based similarity coefficient proposed by Gupta and Seiforddini (1990), as given by equation 7.6.

As can be seen, this particular coefficient uses the manufacturing information such as partwise production volume, pairwise production sequence and the operation time to define the similarity between a pair of machines. With this operational information taken into consideration, a more complete assessment of the situation can be expected. For instance, the inclusion of production sequence consideration in the formula recognises the fact that the material handling required to transfer a part between machines within a cell costs less than that needed to transfer the part between machines from different cells.

$$S_{i,j} = \frac{\sum_{k=1}^{n}\left[\delta_{1,k} t_k^{ij} + \sum_{o=1}^{n_k} \delta_{2,o}\right] m_k}{\sum_{k=1}^{n}\left[\delta_{1,k} t_k^{ij} + \sum_{o=1}^{n_k} \delta_{2,o} + \delta_{3,k}\right] m_k} \tag{7.6}$$

$$t_k^{ij} = \frac{\min\left(\sum_{o=1}^{n_{ki}} t_{k,i}^o, \sum_{o=1}^{n_{k,j}} t_{k,j}^o\right)}{\max\left(\sum_{o=1}^{n_{ki}} t_{k,i}^o, \sum_{o=1}^{n_{k,j}} t_{k,j}^o\right)}$$

where:

n = number of components
m_k = planned prod. volume of part k during a period
n^k = number of times k visits both machines in row
n_{ki} = number of times k visits machine i
$t_{k,i}^o$ = optn time for k on machine i during both visits

$$\delta_{1,k} = \begin{cases} 1 & \text{if part } k \text{ visits both machines } i, j \\ 0 & \text{otherwise} \end{cases}$$

$$\delta_{2,k} = \begin{cases} 1 & \text{if part } k \text{ visits both machines in row} \\ 0 & \text{otherwise} \end{cases}$$

$$\delta_{3,k} = \begin{cases} 1 & \text{if part } k \text{ visits either of machines } i, j \\ 0 & \text{otherwise} \end{cases}$$

The ROC and its extensions

The rank order clustering (ROC) method (King, 1980) sorts the rows and columns of a component/machine matrix according to binary weighting. The weight of a row or a column are given by equation 7.7:

$$\begin{aligned} &\text{Row } i: \quad \sum_{k=1}^{n} a_{ik} 2^{n-k} \\ &\text{Column } j: \sum_{k=1}^{m} a_{kj} 2^{m-k} \end{aligned} \tag{7.7}$$

The matrix is first sorted by the rows, and secondly by the columns. After the second step of this sorting cycle the initial order of the rows may be disturbed. The process is then repeated until there are no changes made to the orders of the rows and columns. An example of this iterative scheme is given in Fig. 7.26.

In comparison to other methods, ROC is relatively simple and easy to implement. However, it has a number of limitations and does not guarantee to produce the optimal solution. ROC has been subsequently enhanced. The later versions include that of the modified rank order clustering algorithm (MODROC) (Chardrasekharan and Rajagopalan, 1986), which initially sorts the component/machine matrix using ROC, and then segments the matrix in the top left corner of the resultant matrix, as shown in Fig. 7.27. The algorithm scans along the diagonal line to determine where the block which has been formed at this corner ends, and removes the identified block temporarily from the matrix. ROC is again applied on the reduced matrix and another block is sliced off. This process is repeated until no more slicing is possible. The pairwise similarities among the resultant blocks are evaluated by using similarity coefficients, so that some of these blocks may be rejoined. A similarity value of 1, for example, indicates that one of the blocks is a subset of the other (equation 7.3), and hence they can be joined together.

The bond-energy algorithm

The bond-energy algorithm was first described by McCormick *et al.* (1972). With this method the component/machine matrix is sorted using a measure of effectiveness, as defined by equation 7.8:

	Component number 1 2 3 4 5	Row	Resultant
Binary value->	(16)(8)(4)(2)(1)	value	row order
1	0 0 0 1 1	3	
2	1 1 1 0 0	28	2–4–1–3
3	0 0 0 1 1	3	
4	0 1 1 0 0	12	

(a)

	Component number 1 2 3 4 5	Binary value	Resultant column order
2	1 1 1 0 0	(8)	2–3–1–4–5
4	0 1 1 0 0	(4)	
1	0 0 0 1 1	(2)	
3	0 0 0 1 1	(1)	
Column value	8 12 12 3 3		

(b)

	Component number 2 3 1 4 5	Resultant configuration
2	1 1 1 0 0	Cell 1: components 1,2,3 machines 2,4
4	1 1 0 0 0	
1	0 0 0 1 1	Cell 2: components 4,5 machines 1,3
3	0 0 0 1 1	

(c)

Fig. 7.26 Rank order clustering. (a) Row ordering, (b) Column ordering and (c) The resultant cells.

$$E = \frac{1}{2} \sum_{i=1}^{m} \sum_{j=1}^{n} a_{ij}(a_{i,j-1} + a_{i,j+1} + a_{i-1,j} + a_{i+1,j}) \tag{7.8}$$

The objective is to arrange the columns and rows of the matrix so that the maximum value of E is obtained, and in doing so the formation of a block diagonal (or the nearest thing to a block diagonal) is achieved. One weakness of this method is the long computation time that is often required. Improvements made to the original algorithm include the work done by Bhat and Haupt (1976), who used a matching matrix A^*A^T (Fig. 7.25) as the basis for permutation rather than the original matrix A. Other attempts to use matrix A^*A^T include the Q analysis approach reported by Robinson and Duckstein (1986).

Component number

Machine number	1	2	3	4	5	6	7
1	1	1	1	1	1	0	0
2	1	1	1	1	0	1	0
3	1	1	1	0	0	1	1
4	0	0	0	0	1	1	1
5	0	0	0	0	1	1	1
6	0	0	0	0	1	1	1

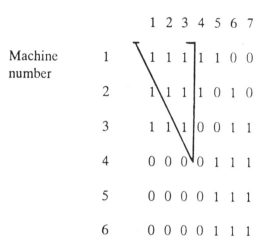

Fig. 7.27 The slicing process of MODROC.

Cluster identification algorithm

This was developed by Kusiak and Chow (1987) and operates on the component/machine matrix as follows:

1. Set $k = 1$ and generate the original matrix A_k.
2. Select any row i of A_k and draw a horizontal line.
3. For each entry of 1 on h_i draw a vertical line.
4. For each entry of 1 on v_j draw a horizontal line.
5. Repeat steps 3 and 4 until all crossing choices are exhausted.
6. All entries of 1 on a horizontal/vertical cross belong to machine cell k. Remove these entries from A_k to form A_{k+1}.
7. Set $k = k + 1$ and repeat steps 2–6 until $A_{k+1} = 0$.

An example of this iterative method is given in Fig. 7.28. This algorithm is simple and efficient in terms of program coding and computational speed. To overcome the problem of bottleneck parts, an extended version of this method was later developed and reported by Kusiak (1990).

Graph theoretic approach

This type of approach attempts to identify the weak relationships among the machines in terms of their similarity, and then rationalize the production routings to allow better groupings. First, seeds (i.e. small sub-sets of the matrix) are defined, which will act as the nucleus of cells to be defined. These seeds are then expanded by allocating their most similar rows and columns to them (Rajagopolan and Batra, 1975). Like the similarity coefficient based approach, this method depends on a threshold value selected. The size of the cells must be defined by the user. One such method was developed by Vannelli and Kumar (1986), which attempts to minimize the number of bottleneck components and bottleneck machines in the resultant configuration. With this method the components and machines are represented by a set of nodes, with their relationships shown by the appropriate

1) *k=1, i=1*, producing: h_1, and then v_4 and v_5, and then h_3; thus identifying entries of cell 1: $a_{1,4}$, $a_{1,5}$, $a_{3,4}$, and $a_{3,5}$.

Machine number		Component number 1	2	3	4	5	
	1	0	0	0	1	1	h_1
	2	1	1	1	0	0	
	3	0	0	0	1	1	h_3
	4	0	1	1	0	0	
					v_4	v_5	

2) *k=2, i=2*, producing: h_2, and then v_1, v_2 and v_3, and then h_4; no more iteration possible and thus identifying entries of cell 2: $a_{2,1}$, $a_{2,2}$, $a_{2,3}$, $a_{4,1}$, $a_{4,2}$, and $a_{4,3}$.

Machine number		Component number 1	2	3	
	2	1	1	1	h_2
	4	0	1	1	h_4
		v_1	v_2	v_3	

Fig. 7.28 Cluster identification algorithm.

linkages as defined by the component/machine matrix. The algorithm requires the user to define the maximum number of machines to be included in the cells, and then uses a simple formula to calculate the number of cells to be formed. The required number of seeds are then chosen on the condition that they must be disjointed. This can be achieved by using the A^*A^T matrix to find the entries which have no components in common. Once the independent seeds are identified, the rest of the entries are clustered around them using measures of effectiveness to analyse the quality of the resultant blocks. In addition, any set of data which has connection with more than one seed is considered a bottleneck item and is removed from the set and put into a bottleneck set, so that appropriate action can be taken to produce completely disconnected cells.

ZODIC (Zero One Data: Ideal seed method for Clustering) is another example of algorithms using this approach (Chandrasekharan and Rajagopalan, 1986, 1987). ZODIC is based on the observation that for a manufacturing configuration with n components, m machines and v visits, there can be no more than k independent product families:

$$k \leqslant 1 + \tfrac{1}{2}[(m + n - 1) - \sqrt{(m + n - 1)^2 - 4(mn - v)}] \tag{7.9}$$

The value of k, therefore, allows the initial number of cells to be determined. A number of k seeds can then be used to allow the entries of the matrix to be clustered around them. Three types of initial seeds can be used, i.e. the artificial seeds which represent the ideal configuration, the representative seeds which are sub-sets of the original matrix but identified according to an ideal configuration, and the natural seeds which are simply mutually dissimilar sub-sets selected from the matrix. Once chosen, the rows and columns of the matrix are assigned to these seeds by evaluating the distance of the rows (columns) from them (equation 7.10). The resultant blocks may vary in size because some of the seeds attract fewer vectors than the others. Finally the optimal solution is obtained by diagonalizing the resultant blocks.

$$Dr_{ij} = \sum_{k=1}^{n} |a_{ik} - a_{jk}|$$
$$Dc_{ij} = \sum_{k=1}^{m} |a_{ki} - a_{kj}| \quad (7.10)$$

where:

Dr_{ij} = distance between row i and row j
Dc_{ij} = distance between column i and column j

Linear and integer programming approach
These methods attempt to construct multi-variable, linear or integer equations which define certain measures of effectiveness of a cellular-based manufacturing system. Once a problem has been modelled by a set of such equations, the optimal solutions can be sought by using a computer based optimizing program, such as LINDO (Linear INteractive and Discrete Optimizer) (Schrage, 1980). For example, an integer programming model can use the information given in a component/machine matrix to form p cells (Kusiak, 1990). In this model, d_{ij} is a measure of the distance between any two components. It can take a number of forms, including the one given in equation 7.11. The objective of this model is to minimize the total sum of such distances (also known as dissimilarities), under the constraints that each component can belong to one cell only (equation 7.12); exactly p cells are to be formed (equation 7.13); and component i can belong to cell j only when this cell is formed (equation 7.14).

$$\min \sum_{i=1}^{n} \sum_{j=1}^{n} d_{ij}x_{ij} \quad (7.11)$$

subject to

$$\sum_{j=1}^{n} x_{ij} = 1 \qquad i = 1, 2, \ldots, n \quad (7.12)$$

$$\sum_{j=1}^{n} x_{jj} = p \qquad j = 1, 2, \ldots, n \quad (7.13)$$

$$x_{ij} \leqslant x_{jj} \qquad i = 1, \ldots, n, j = 1, \ldots, n \quad (7.14)$$

$$x_{ij} \leqslant 0,1 \qquad i = 1, \ldots, n, j = 1, \ldots, n \quad (7.15)$$

where

m = number of machines
n = number of components
p = number of cells to be formed
d_{ij} = distance between components i and j

$$x_{ij} = \begin{cases} 1 & \text{if component } i \text{ belongs to cell } j \\ 0 & \text{otherwise} \end{cases}$$

The end result from this particular model is rather like those of the seeding methods, in that a number, *p* of nuclei components are identified to which the rest of the components are attached to form the resultant cells. For example, suppose 3 cells are to be generated from a set of 6 components. If components number 1, 3 and 4 are identified as the key components, then three cells C1, C3 and C4 will be formed accordingly, as shown below:

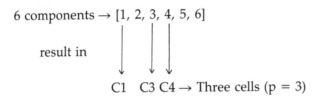

6 components → [1, 2, 3, 4, 5, 6]

result in ↓ ↓ ↓

C1 C3 C4 → Three cells (p = 3)

One advantage of this type of approach is that the objective function can be easily modified, and the constraints extended or relaxed so that the appropriate manufacturing information can be adopted to model a variety of situations. For instance, it is possible to construct a model in such a way that the cost is minimized or utilization maximized (Choobineh, 1988; Co and Arrar, 1988), or a component is allowed to visit a machine more than once. However, the attractiveness of this type of method decreases rapidly as the size of the problem involved increases.

The neural network approach

The fact that many of the manufacturing problems are not well structured has provided potential ground for neural network (NN) applications. So far, attempts have been made to use this approach in diversified areas to enhance the capabilities of manufacturing systems including, for example, diagnostics, tool condition monitoring, design optimization, collision detection, process control and group technology/cellular formation (Wu, 1992).

The structure and function of NNs are based on our current understanding of the biological nervous systems. NNs are built on a large number of simple and adaptable processing units (PU) which are interconnected in such a way that they can store experiential knowledge through learning from examples and, like biological systems, have the ability to take in hazy information from the outside world and process it without an explicit set of rules. This approach (parallel and distributed) is in contrast to the traditional computing approach which processes information sequentially according to a set of exact rules.

The processing units (PUs), as the building bricks of NNs, are basically logic processing devices with the fundamental function of producing an output signal as a function of the sum of their weighted inputs, and a certain threshold value, as shown in Fig. 7.29. The input to a PU_i can be either the output from other PUs, or an input directly from outside the NN (i.e. input to the NN). The output from PU_i can be used either as input to the subsequent PUs or as an output from the NN. The value of w_{ij} determines how strongly the output of PU_j influences the activity of PU_i. The total weight matrix **W** of an NN encompasses and reflects the NN's knowledge and skills that it has learnt through previous training, and is therefore referred to as its memory. The individual PUs can be arranged and connected in a number of ways to build different NNs, as shown in Fig. 7.30.

NNs can also be categorized according to the way in which they are trained to solve particular problems (Kosko, 1992). NN learning is mainly

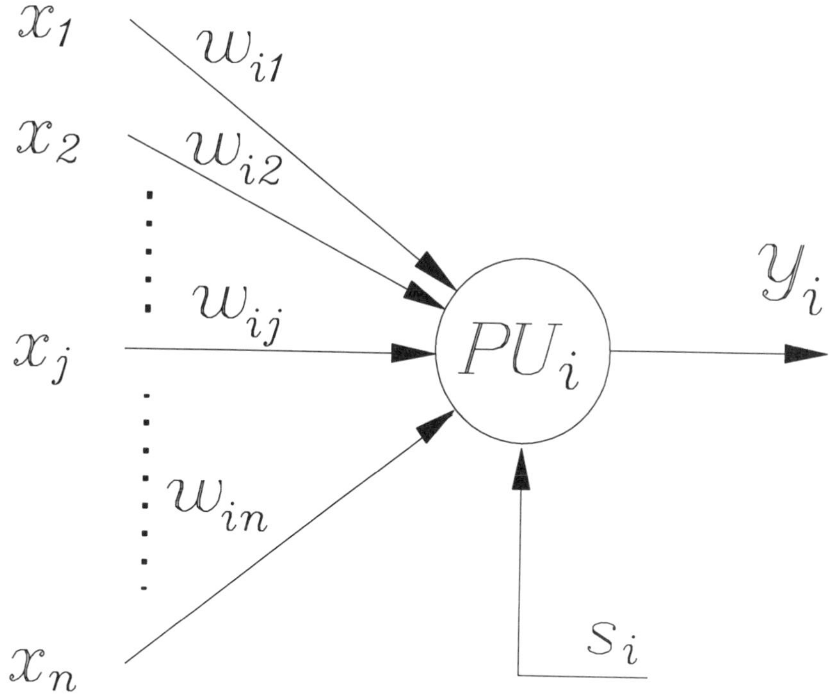

Fig. 7.29 The processing unit.

through the readjustment of the weights through the application of certain learning algorithms. According to how this is carried out, NNs can be classified as either *supervised* or *unsupervised*. A supervised NN will require necessary information to be provided from both within the NN itself (the local information) and from the environment (the external information) to guide its learning task. The tasks of this type of NN are often concerned with pattern association; that is, in response to a pattern presented to the input layer PUs, the NN attempts to produce an output pattern (by its output layer PUs) which is somehow associated with it. So far as the training of this kind of NN is concerned, the algorithm most widely used is known as error back-propagation (also known as the generalized delta rule):

1. Specify a set of desired input/output patterns.
2. Select an input/output pair from the set.
3. Present the input pattern to the NN to allow it to propagate to the output.
4. The change for weight w_{ij} is to be proportional to the error E_i produced by PU_i. First calculate the error values of the output PUs and adjust the relevant weights accordingly, then calculate the error values of the PUs on the next inner layer and adjust the weights accordingly. Repeat this until the layers are exhausted.
5. Repeat steps 2–4 until the training set is exhausted.
6. Repeat cycle 2–5 until the weight matrix W converges (i.e. stabilizes at a certain set of values).

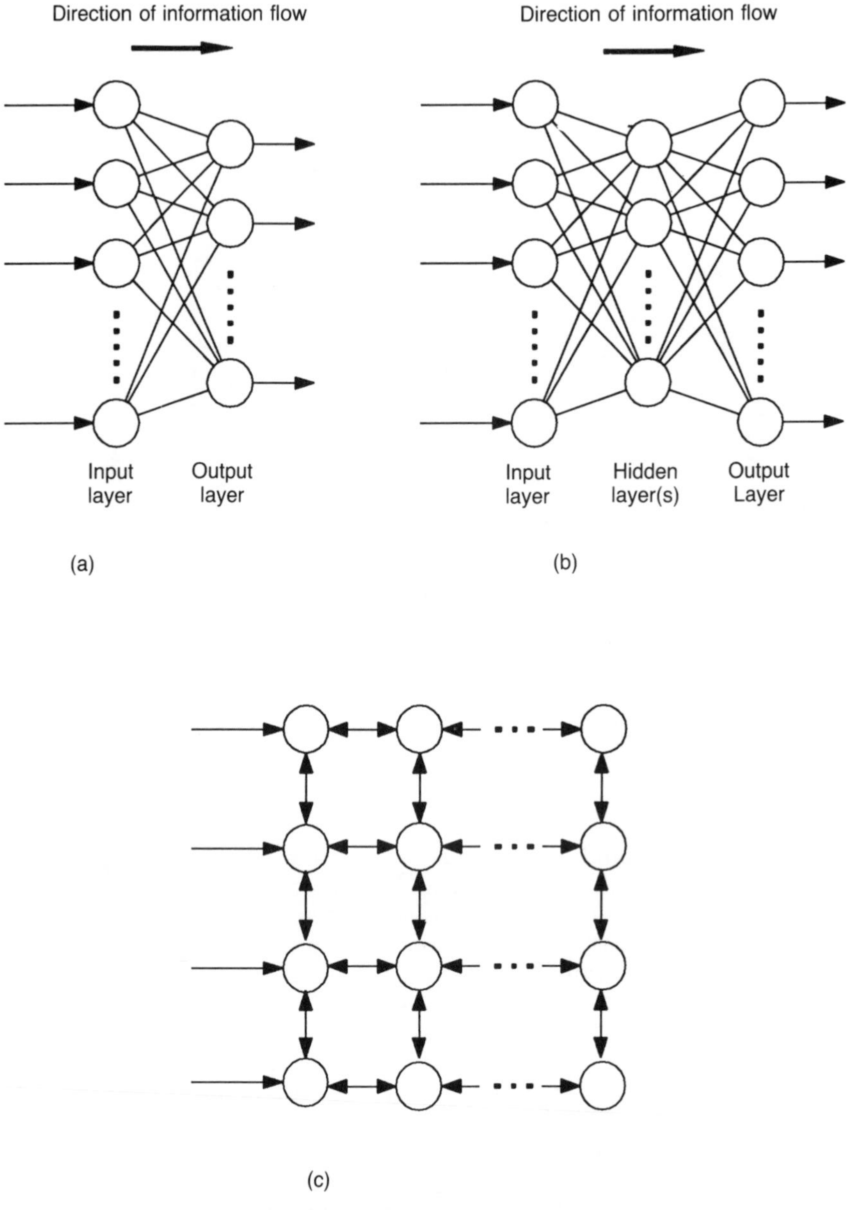

Fig. 7.30 Example of NNs. (a) Single-layer feedforward, (b) Multi-layer feedforward and (c) With feedback connection.

In contrast, with an unsupervised NN the source of learning information is to be entirely from within the NN itself, implying that the NNs involved must be self-adaptive without the help provided by external guidance. A competitive learning scheme, with a typically NN topological structure as shown in Fig. 7.31, provides a typical example of unsupervised NNs. Its training set consists of input patterns, as with those for supervised NNs, but its output is an indication of membership of the input pattern in a group

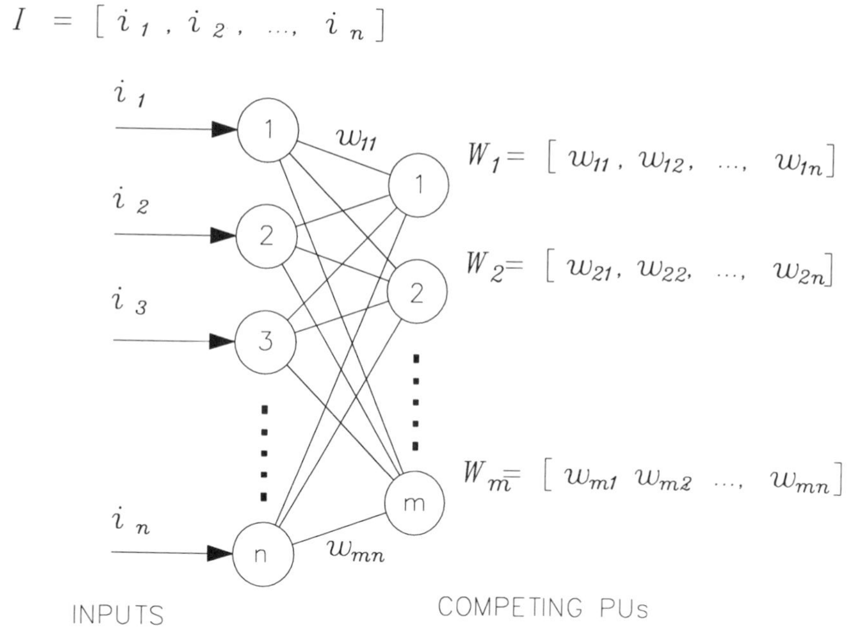

Fig. 7.31 A competitive NN.

with similar characteristics. For example, an output of [0,0,1] would indicate that it is likely that the current input pattern belongs to cluster no. 1. Similarly, [0,1,0] would indicate a membership in cluster no. 2, and [1,0,0] in no. 3. The NN learns to clusterize the input patterns by making the output PUs compete with each other for the right to respond to a stimulating input. The output PU which has the weight vector that is most similar to the input vector, determined by the 'distance' between them, claims this input pattern by producing an output of 1, and in the same time inhibits other output PUs by forcing them to produce 0s. The learning mechanism is primarily according to the competitive rule of 'learn only if you win'. That is, only the winning PU is permitted to alter its weight vector, and this weight vector is modified in such a way that it is brought even nearer (or more similar) to the representative input pattern in the cluster concerned.

It is easy to see that this type of NN provides the potential for an ideal tool for cellular formation. An attempt in this regard was reported by Malave and Ramachandran (1991). The NN used operates on the classical machine/components matrix, from which either the machine vectors or the component vectors can be used as the input patterns (representing respectively the machine characteristics and component characteristics). The authors tested this approach on a number of such matrices, and it was claimed that the results produced by this approach compare well with the classical clustering techniques.

7.7.4 Development of manufacturing information system

The major tasks involved in the detailed design of the control system include design of the processes, the data bases, the selection and location of hardware and the specification of managerial roles which will be responsible for certain decision centres. The procedure is illustrated in Fig. 7.22c.

As discussed in Chapter 2, one of the main elements here is *communication*. A manufacturing operation can only be truly integrated and effectively controlled if the machines, operators and managers have the means to communicate to each other effectively. This in turn depends on two important elements: *communication media* and the *same language*. Therefore, the following text will concentrate on the analysis and specification of the language, i.e. the *organised data structure* within a computer-integrated manufacturing environment, aiming to examine the general approach in the design and implementation of manufacturing information systems and to illustrate the typical structure, essential linkages and flows of data within a manufacturing organisation.

Data base development

According to the definition of ANSI/SPARC, an efficient data base system should be divided into three levels:

- The *internal* level. This is concerned with how data are physically stored within a computer system. For example, data can be packed with flags, pointers or indexes to facilitate data manipulation. The data format defined at this level can also be used for mapping the data structure to the next (conceptual) level.
- The *conceptual* level. This specifies the way in which data is presented within the data base system, and defines the *static* integrity constraints of data. A schema, as it is often called, of a data base is developed and presented at this level in a graphical format – such as the Entity/Relational diagrams of a CIM data model to be given in the following text. In practice, a schema is implemented to the internal level through an interpreting command provided by a data base management system (DBMS), which creates appropriate computer files according to the layout specified by the conceptual schema.
- The *external* level. This is the level closest to the end user, where specific application programs are defined. The correspondence between the conceptual level and external level is established through a mapping process which defines *transitional* integrity constraints of data to avoid violation of data integrity during a transaction. An example of such constraints is

  STOCK (PRODUCT_NUMBER) >= DISPATCH (PRODUCT_NUMBER)

  The user at this level accesses the data sets through various search facilities provided by the application programs without having to worry about the data implementation method at the internal level. Again in practice a DBMS is often used to create user interfaces according to an external schema, as well as to provide support for data integrity.

The design of the schema, therefore, plays an important role in the successful implementation of an application data base. It is particularly crucial to CIM design and implementation, which in general requires the development of a global conceptual schema in order to integrate the data effectively among a number of participating data bases into a single cohesive system. Dependent on the type of the data bases involved, a global schema can be created either through schema integration (if the participating data bases have the same forms of data type) so that the data sets from a local schema are integrated within the global schema, or schema conversion (for distributed data base systems in which the participating units have different forms of data type) which translates the data sets from a local schema to the global schema.

Data modelling can in general be carried out through two approaches. With the traditional *transaction approach* a set of standard tools, such as a procedure flowchart, is used explicitly to specify the processes and their operating sequences within a system. A data base is then designed to fulfil the requirements of individual processes. The transaction-oriented approach is relatively easy to adopt in practice, since the data models used are simply an immediate extension from their associated data flow diagrams. However, since the structure of these data base systems is very much dictated by the application programs, they tend to contain redundant data and lack the ability to cope effectively with future changes. When applied to a CIM environment, therefore, one of the major problems with this approach is that of data integrity – or the lack of it. For example, the product information used by the purchasing department of a company could be different from that used by, say, the production department, because the files designed for material ordering are not exactly the same as those designed for production planning. In contrast, with the *data approach* the abstract properties of the data themselves are used as the basis of system design. Data are grouped and used as an 'information pool', with individual processes accessing them in any way required. As a result, since they are independent of the application tasks, they can be changed without violating the data base integrity. This is a very desirable feature for a manufacturing information system to have, since the provision of data independence is an important requirement for achieving flexibility and the ability to accommodate future organisational changes. However, data-oriented systems are more difficult to construct, demanding a thorough understanding of the whole system from the analyst.

SSADM, for example, is one of the data-based approaches for the analysis and design of information systems. This method uses a complete analysis and design framework which consists of a set of modelling tools, including:

- Data flow diagram;
- Entity model;
- Relational model;
- Entity life history;
- Data dictionary.

These tools are used for the main tasks which are normally involved in a development cycle for an information system including, for example:

- Define functions involved in various operational areas (business function diagram, BFD).
- Specify data flows between these functions, as well as data links within them (data flow diagram, DFD).
- Identify relationship between data (entity relationship diagram) and develop relational data model (relational model).
- Define ownerships of data within the system and user interfaces (views).

Data flow of manufacturing information

The first step in the analysis of an information system specifies the overall structure of the system in terms of BFDs. Having established the functional hierarchy, the next step is to examine the data necessary for their operation by using DFD analysis to find:

- what data are needed to perform the functions,
- how data enter and leave the functions,
- where the data are stored,
- which functions generate changes of the data, and
- who provides, uses and modifies the data.

A complete picture of a data base specified in such a manner can be used to ensure that all data flows, data sources and data stores are accounted for (together with a relational model, it is also helpful for the purpose of ensuring data integrity). In order to reveal the typical features of a manufacturing information system, we will examine in more detail the data flow pattern within a few key functional areas of a manufacturing operation.

Since DFDs are usually developed to map their associated BFDs, as shown in Fig. 7.32, their development also follows a top-down process. Thus at the

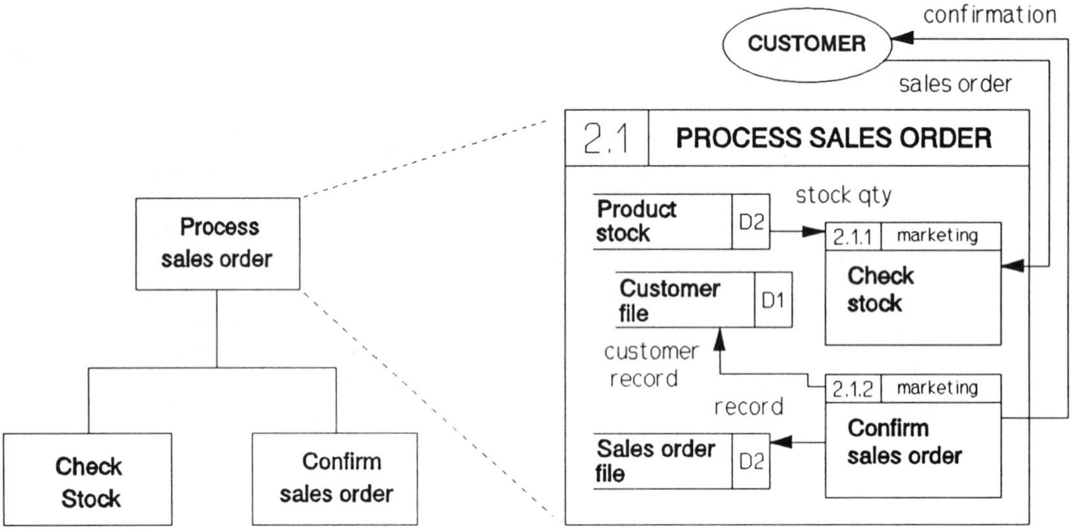

Fig. 7.32 BFD/DFD mapping.

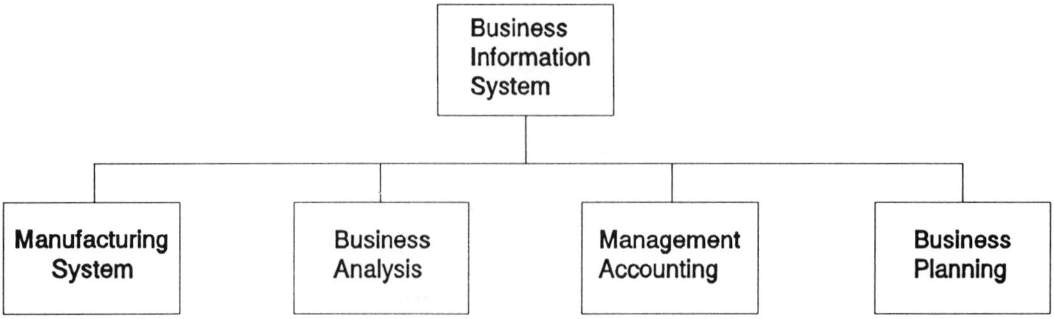

Fig. 7.33 BFD of a business information system.

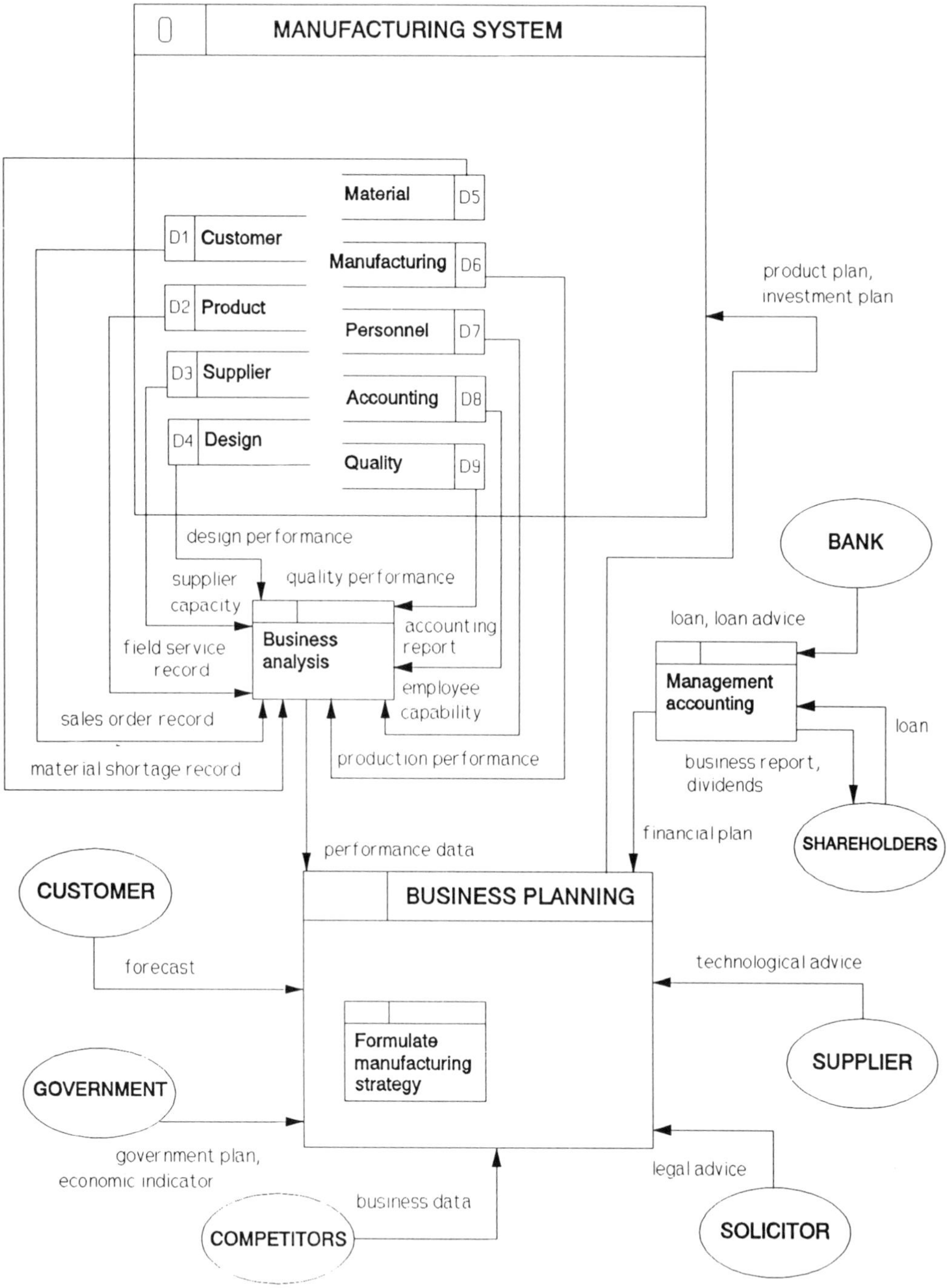

Fig. 7.34 DFD of the business information system.

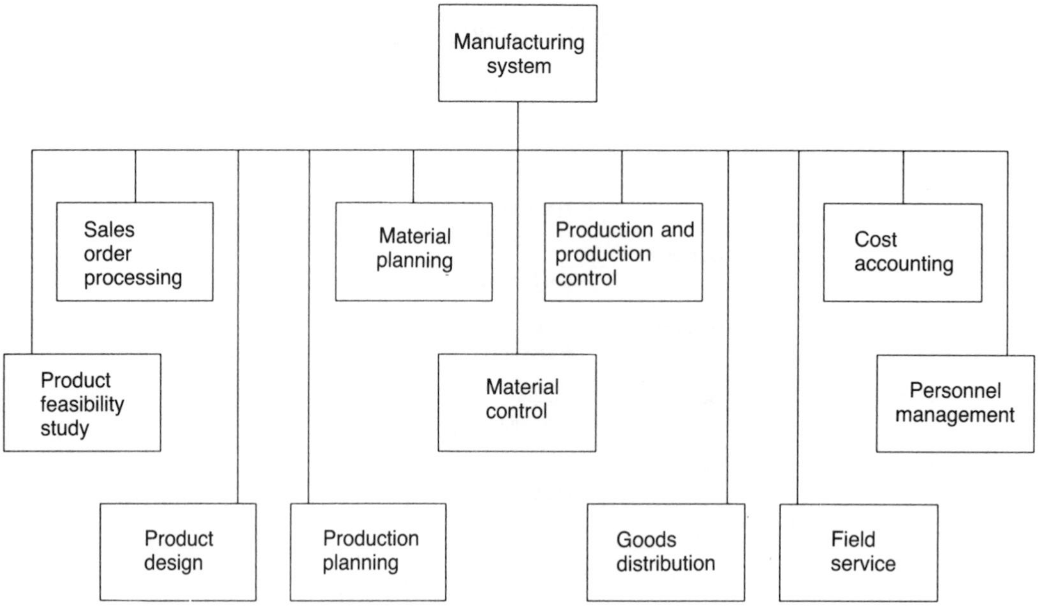

Fig. 7.35 BFD of a manufacturing system.

highest level of a company's information structure is the *business information system*, as shown in Figs 7.33 and 7.34. One of the most important activities at this level is to analyse what the market needs and how to satisfy those needs. Therefore, the main function areas at this level are concerned with the development of a manufacturing strategy and a financial strategy which are to be coherent with the business strategy of the organisation. This overview is then developed into its lower levels of decomposition. According to our discussion on the general structure of a manufacturing operation, for example, Figs 7.35 and 7.36 provide a more detailed view of the typical functions and data flow involved in it.

This decomposition of the manufacturing information structure can continue until the desired level of detail is reached. The information structure for a product feasibility study, as shown in Fig. 7.37, can be used as an example to illustrate this process. This function can be initiated by the receipt of a customer enquiry and product specification, and based on both external and internal information available, it ensures that:

- It is possible to manufacture the product so that the customer's particular requirements are satisfied.
- The company is in a position to gather all necessary manufacturing resources for that product.
- The product is profitable to make.

The information structure of this application is illustrated in Fig. 7.38, while Fig. 7.39 provides a more detailed view of the data flow within one of its subfunctions, as specified by Fig. 7.37. One of the most important issues that deserves careful consideration here is the design of product and part numbering systems. Each product manufactured must have its own, unique

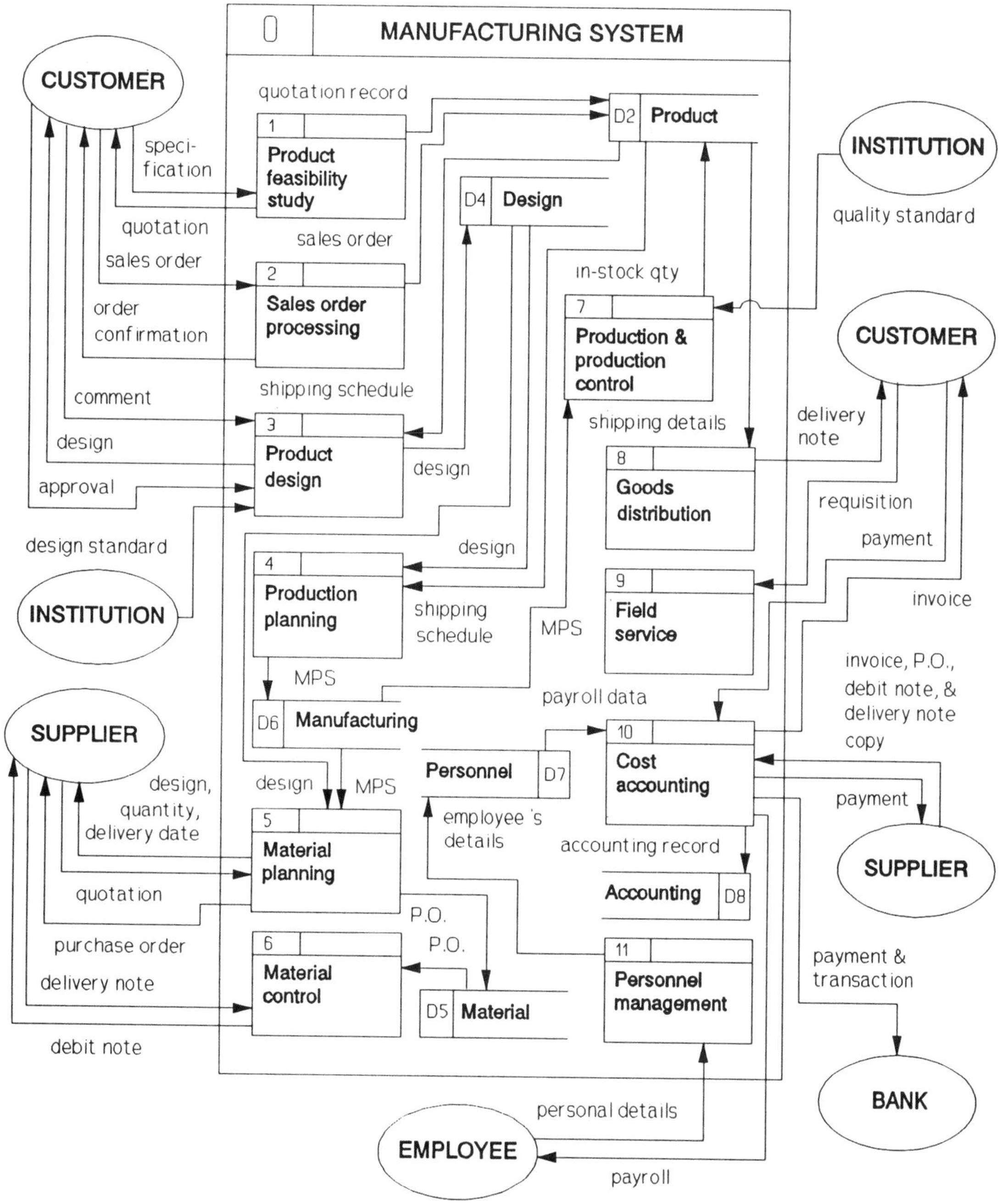

Fig. 7.36 DFD of a manufacturing system.

product number as an identifier so that the correct data can be referred to when needed. The importance of this is due to the fact that many accesses to the database will have to use such a number as the identifier. Care must therefore be taken so that the numbering system is capable of uniquely

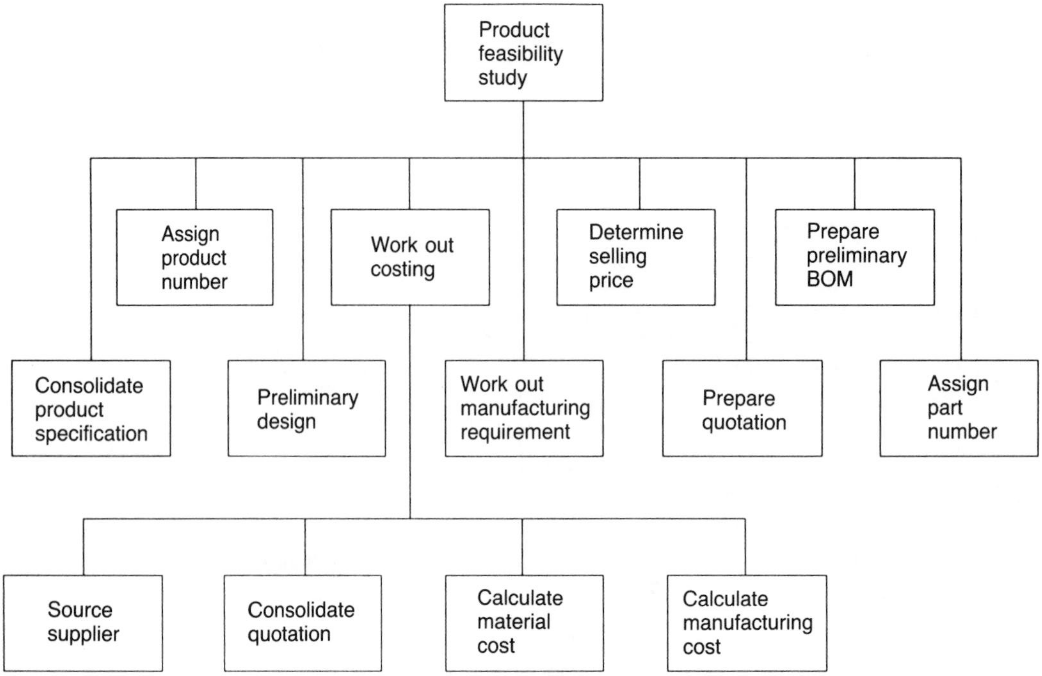

Fig. 7.37 BFD of the product feasibility study function.

identifying a particular product even if it is just another product model which is different from other models in the same product group only in its colour. In addition, the long-term perspective of the manufacturing system concerned must be taken into consideration when designing the numbering system. The company's manufacturing strategy can be used as a good reference in this regard. Generally speaking, a well designed numbering system should be clear and precise, and also expandable so as to accommodate future business growth.

In contrast to the product number, a part number is used to identify a single item of material within the data base system. This is another of the most frequently quoted data items in a manufacturing information system. Several coding schemes can be used in practice, such as GLOCODE (Ingram, 1984), OPTIZ (Optiz, 1970), DCLASS (Allen, 1979), PNC, CODE, MICLASS and TEKLA (Eckert, 1984).Whatever scheme is used, the format of the part number is always one of the key elements in a manufacturing data base system. A carefully designed part coding system can be used to facilitate, for example, the procedures of process planning, production and material planning, and cellular formation. A part number system is therefore similar to a product number system but with additional requirements. At the present stage, part coding is still typically a manual process, although attempts are being made to automate the process. For instance, this could be achieved through the use of an expert system embedded in a CAD system. Since digitized information about a part in terms of its size and form are readily available from the design, this can be used for automatic code generation. All the designer has to do is fill in the remaining unknown parameters through the system's user interface.

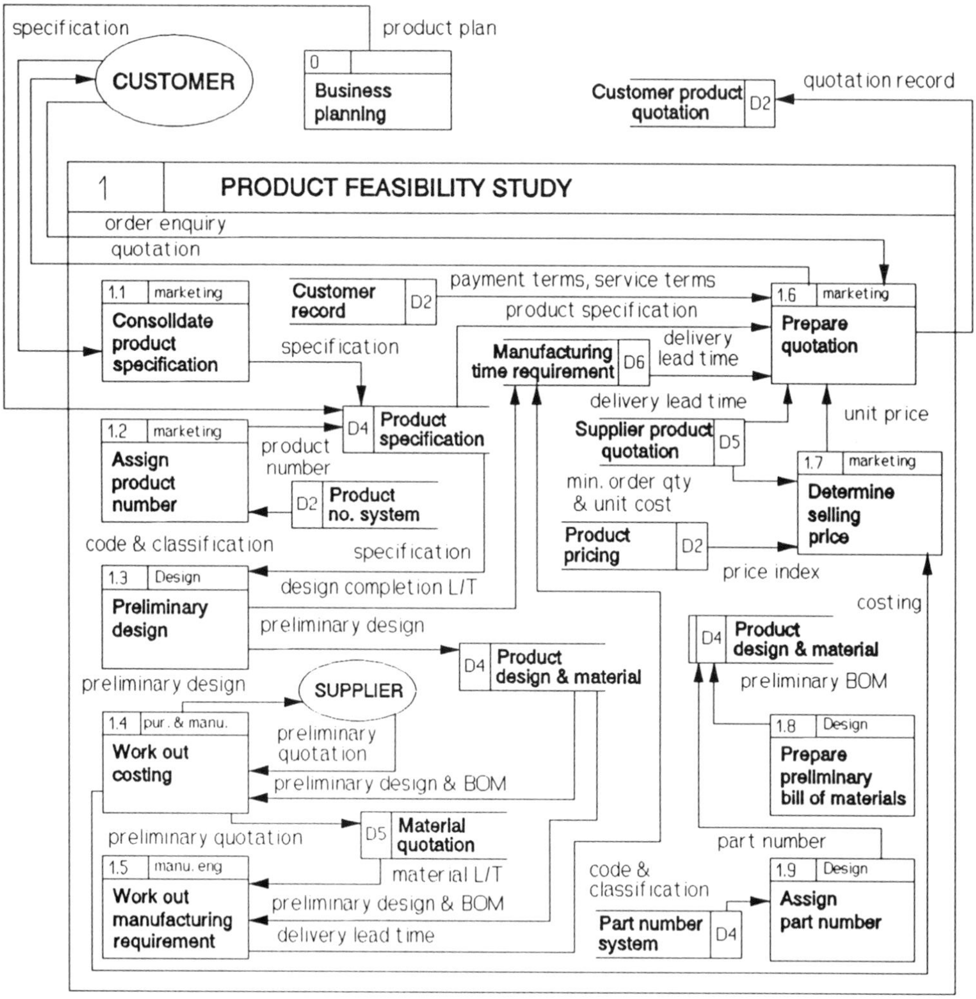

Fig. 7.38 DFD of the product feasibility study function.

The information structure of the production planning function (Fig. 7.40) has to support some of the most dynamic and uncertain activities within a manufacturing system, due to the fact that production plans have to be continuously reviewed and updated. Therefore a feedback loop in the data base system is necessary, which goes through the various correlated functions to reflect what is really happening so as to plan what needs to be done next, as shown in Fig. 7.41. The outputs from this function include the master production plan, the capacity plan and the process plan. Generally speaking, the two main categories of production (make-to-stock and make-to-order) involve different planning practices. To some extent, the data structure shown in Figs 7.42 and 7.43 reflect the concept of *optimised production technology* (see Appendix A), which can be seen as a compromise between these two categories. The main feature of OPT is the use of the critical capacity of an operation as the basis for determining a production schedule. For non-critical items of the operation, schedules are defined subject to the

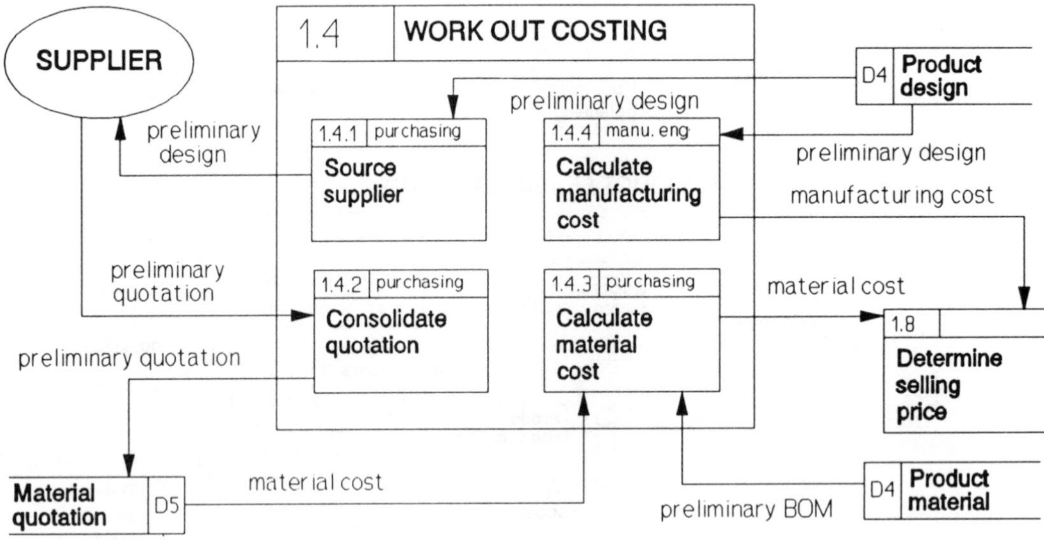

Fig. 7.39 DFD of function level, 1.4 – work out cost.

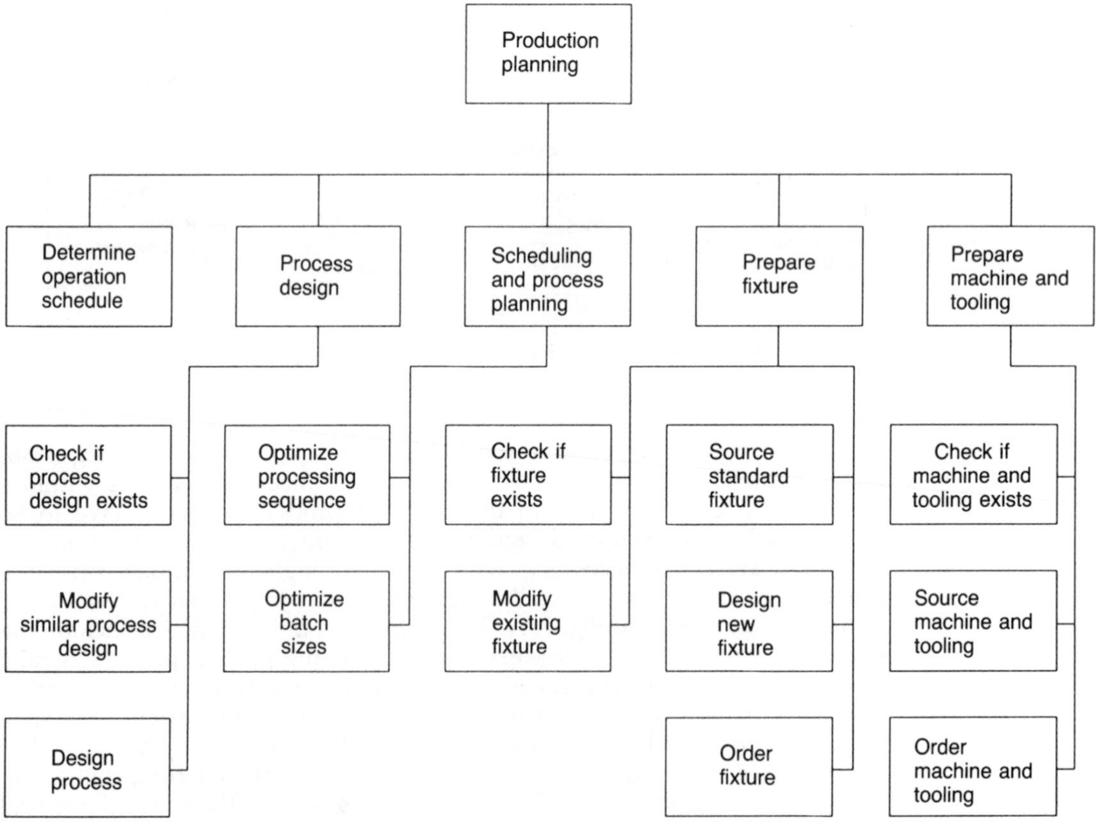

Fig. 7.40 BFD of production planning function.

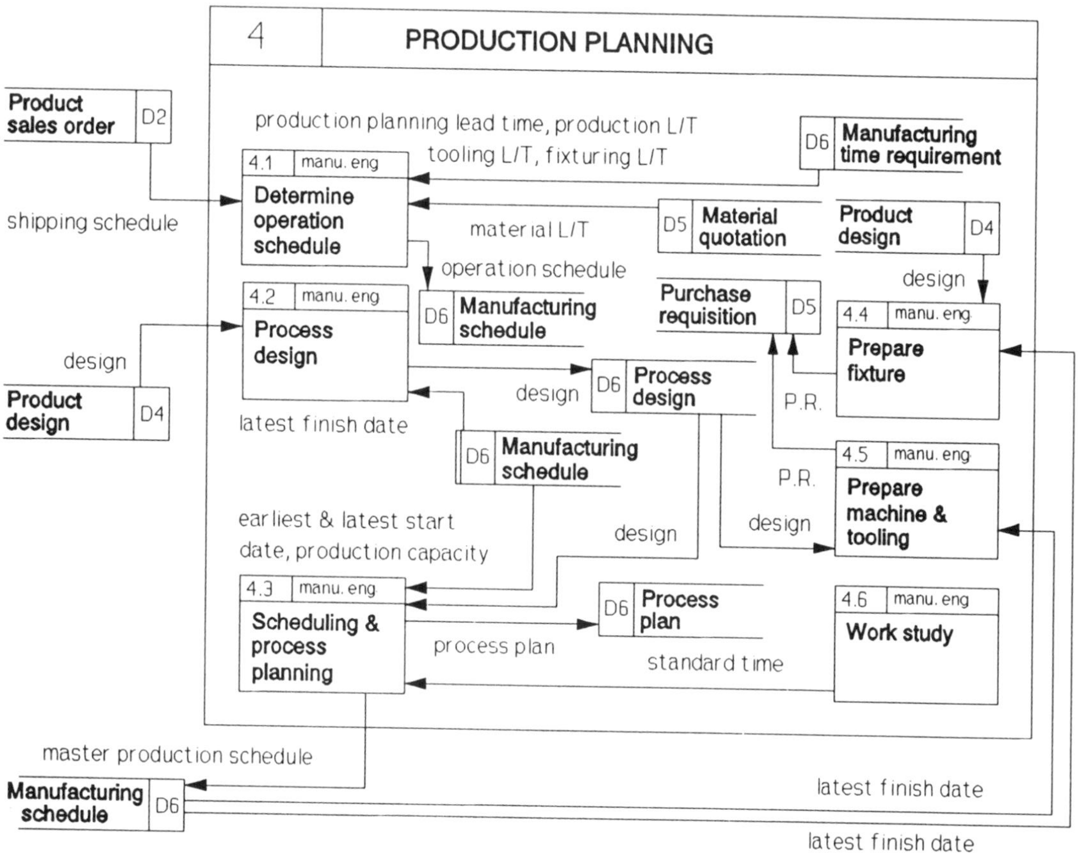

Fig. 7.41 DFD of production planning function.

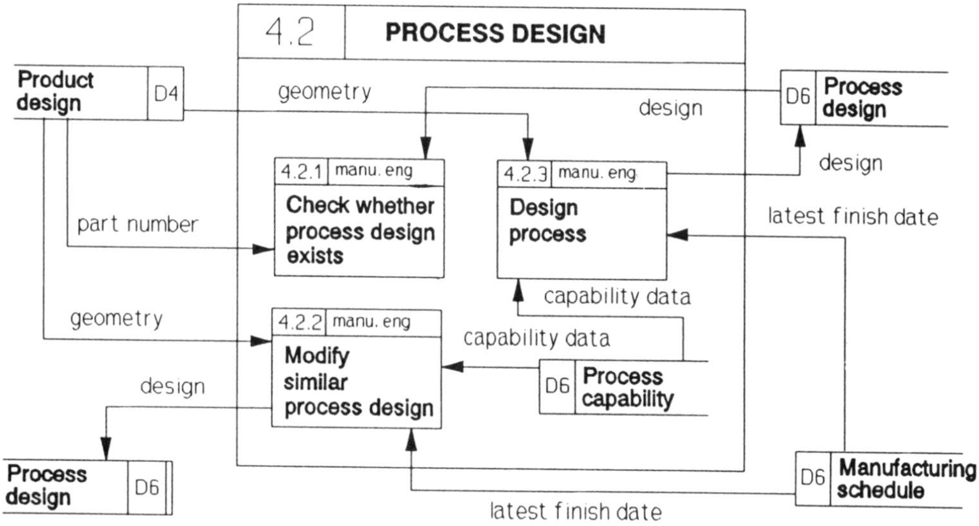

Fig. 7.42 DFD of function level 4.2 – process design.

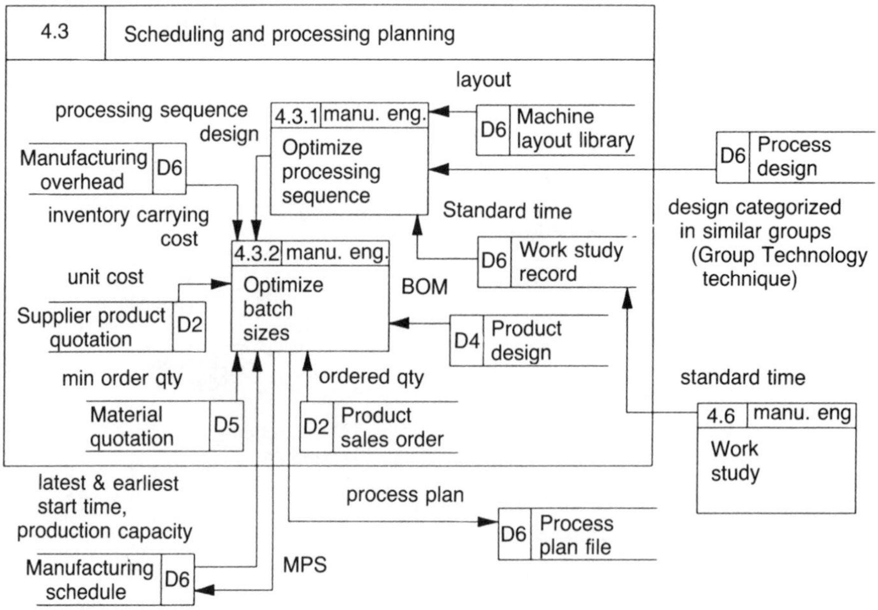

Fig. 7.43 DFD of function level 4.3 – scheduling and process planning.

balance of the cost of keeping stocks and late deliveries. Figure 7.41 also implies that process planning and production scheduling must be treated as an interrelated pair. The production schedules are determined from the production order/forecast information, according to capacity utilization, material delivery dates and estimated operation lead times. Process planning is involved when information is required about the estimated production durations. Ideally process planning should play a more active and dynamic part in the decision-making process, so that capacity planning, material planning and process planning can be carried out simultaneously. In this regard, the importance of the format of the part number can now be restated. If the part number format is designed to be capable of presenting an overall picture of the items concerned (e.g. the form, the major details and outside dimensions), the information can be readily utilized for the preparation of the necessary tooling, as shown in Figs 7.44 and 7.45.

There is always the danger that communication may break down during the production stage. The consequence of this is that the data within the system do not really reflect what is happening in the production process, which may eventually cause the final collapse of the total system. To reduce such risks, the efficient implementation of feedback loops is one of the major concerns of a production control system. These feedback loops are necessary to ensure that:

- The master production schedule is timely and accurate.
- The product is properly designed for manufacture.
- The product is properly designed to achieve the right quality.
- The process plan is properly designed for manufacture.
- The production operation is up-to-date and satisfactory.

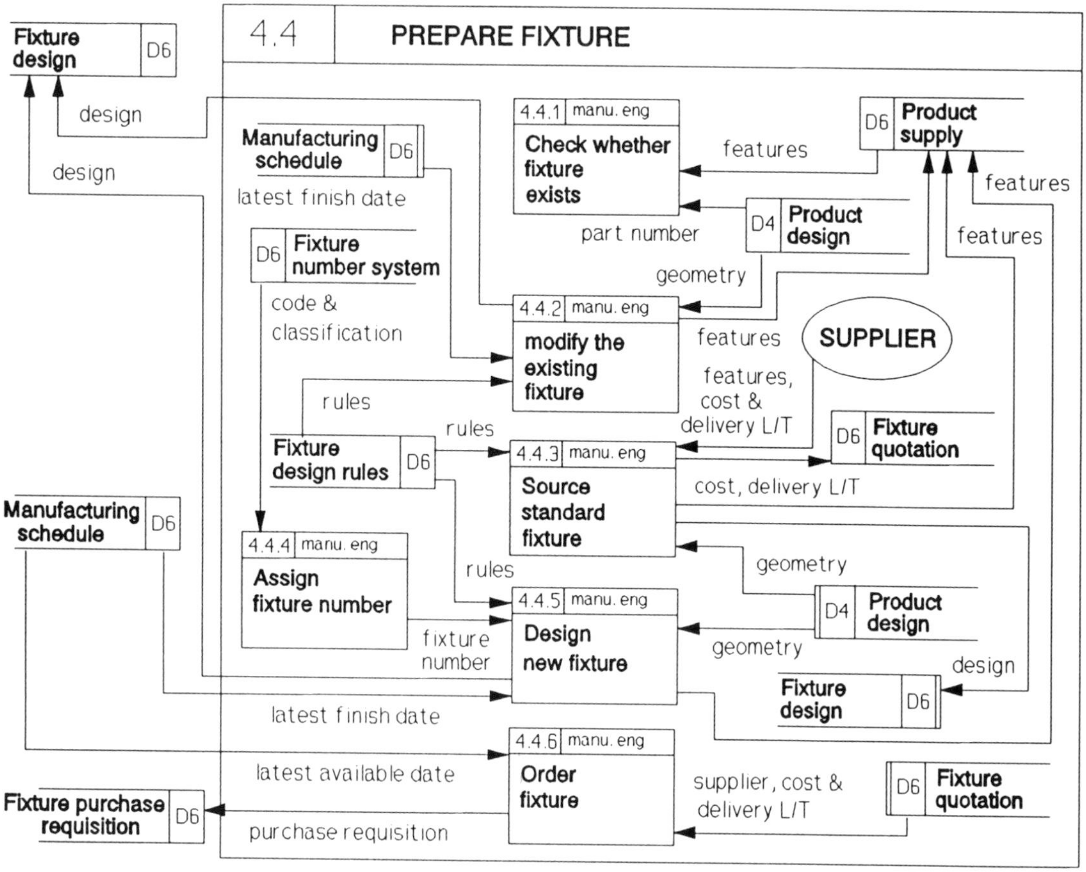

Fig. 7.44 DFD of function level 4.4 – prepare fixture.

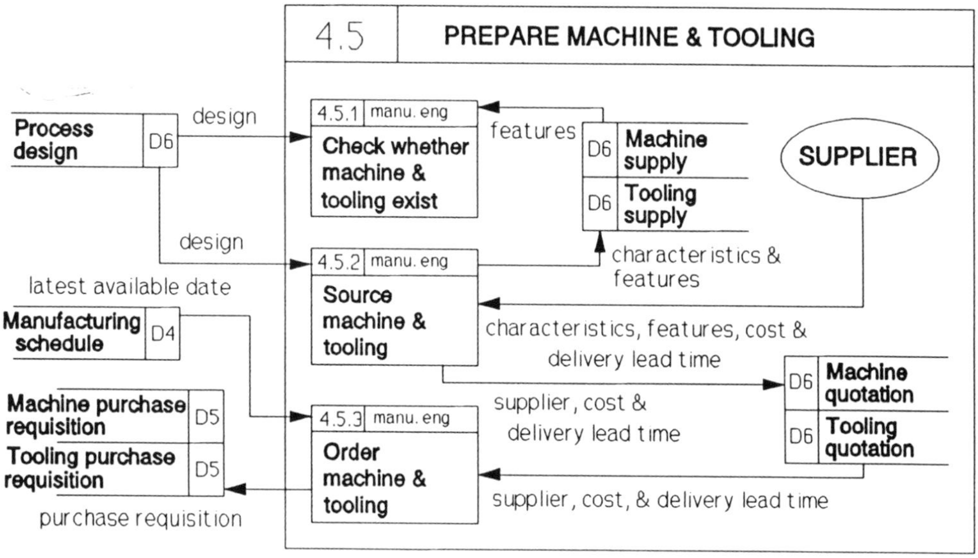

Fig. 7.45 DFD of function level 4.5 – prepare tooling and machine.

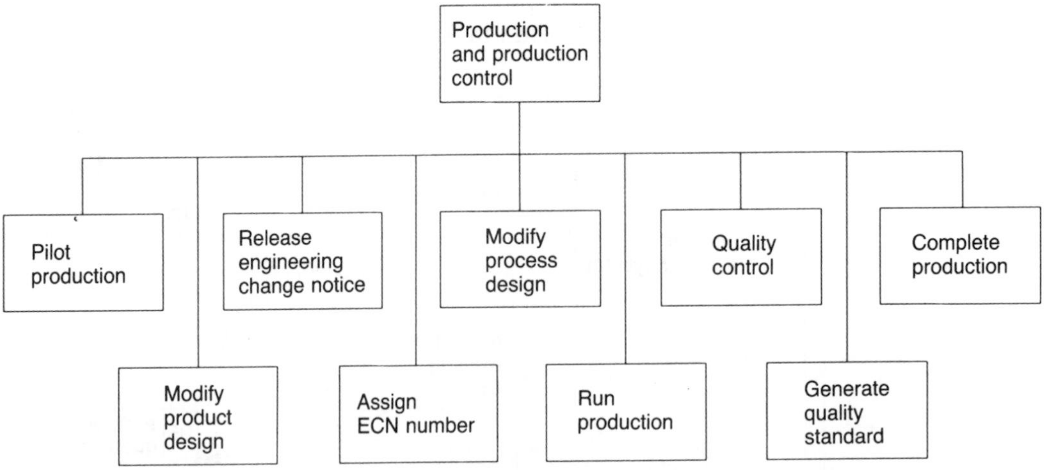

Fig. 7.46 BFD of production and production control function.

The above requirements should be reflected in the information structure of the data base, as shown in the sample diagrams of Figs 7.46 and 7.47.

Relational data structure of manufacturing information
Codd (1982) has defined a data model as consisting of at least three components: a collection of data structure types, a collection of data manipulation rules and a collection of data integrity rules. Having illustrated the data flows within an integrated environment, the typical data-oriented structure of an integrated manufacturing information model will be discussed with the *relational model* as the main theme (the issues related to data manipulation, user interfaces and the implicit integrity of data will be dealt with later). *Entity/relationship* diagrams will be developed to present the model, which can be used to cross-check with the previous data flow diagrams in order to increase the reliability of the data base.

It is perhaps safe to say that the relational data model is so far the most widely adopted in data base applications. In the simplest form, a relational structure presents its data in tables which list their relevant attributes. Relationships between these attributes are then used to link mutually related entities. In a relational structure, for example, an entity 'customer' would have these attributes:

*customer (**customer number #**, name, address, payment term)*

with customer number as the unique identifier. *Entity/relationship* (E/R) diagrams are usually used for presenting a relational data structure. An example E/R diagram is given in Fig. 7.48, which describes a sales order.

An important issue associated with relational data modelling is normalization, which is a process to examine and convert the list of attributes to a specific format with an aim to minimize duplication, and avoid data redundancy and violation of data integrity. Five normal forms have been developed for normalization, and generally a data structure will fulfil these requirements after being normalized to its third norm (the fourth and fifth normal forms have been introduced by researchers for advanced applications).

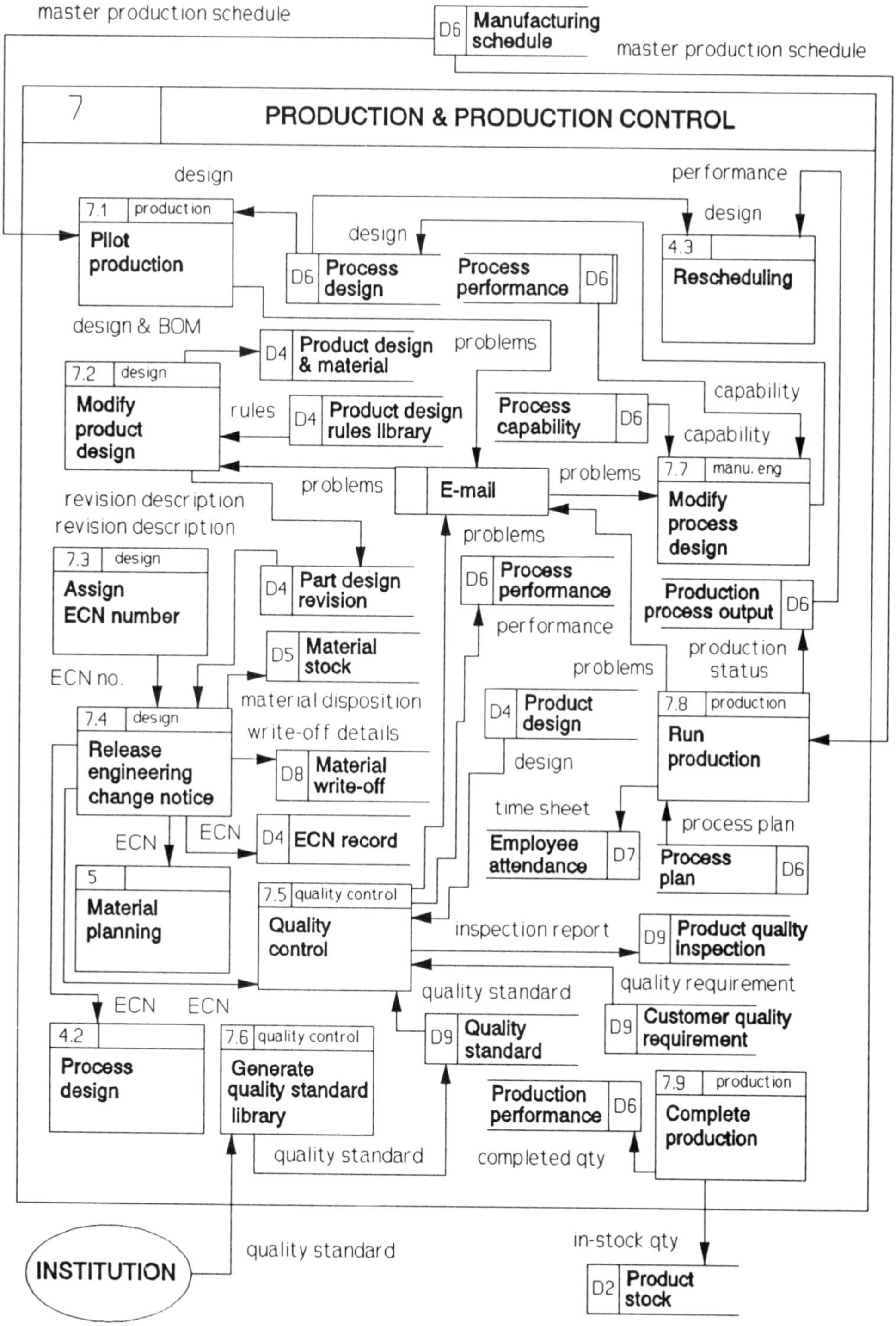

Fig. 7.47 DFD of production and production control function.

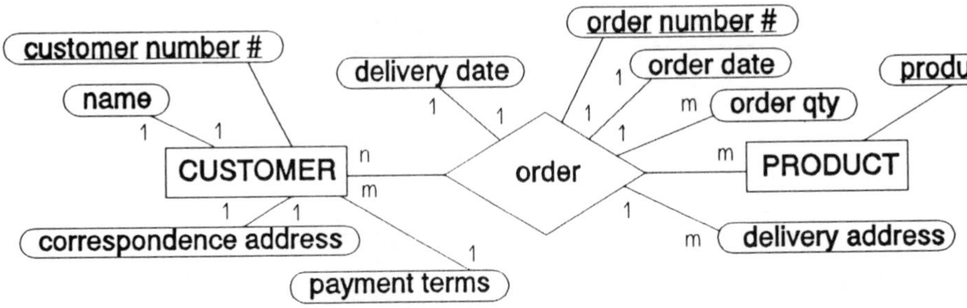

Fig. 7.48 An example of entity/relationship diagram.

In these text, therefore, examples of third form relational structures are developed by following these normalizing rules:

1. If the relationship between an entity and one of its attributes is one-to-one or many-to-one, then this relationship is in the third normal form. The same applies to the relationships between and entity and other entities.
2. If the relationship between an entity and an attribute is one-to-many, then an additional line is added.
3. If the relationship between an entity and another entity is many-to-many, an additional line is formed with two identifiers. The associated attribute(s) which has one-to-many relationship with the entity is included in this line.

Sometimes, a relationship may have its own attributes. In that case, the rules are applied by treating the relationship itself as an entity.

To see how these rules are applied in practice, suppose a customer can only have:

- one customer number (#);
- one name;
- one corresponding address;
- one payment term.

Then the relational data structure in the third normal form can be defined as (rule 1):

*Customer (**customer no. #**, name, corresponding address, payment term)*

However, if now the customer has:

- one customer number (#);
- one name;
- one corresponding address;
- one payment term;
- many delivery addresses;

then by applying rules 1 and 2 this customer's record in third normal form will be:

Customer (***customer no. #***, *name, corresponding address, payment term)*
Customer Line (***customer no. #***, *delivery address)*

In practice, of course, there is no need for a delivery address unless there is a customer order (Fig. 7.48). If a customer order can only have:

- one order number (#);
- one order date;
- one customer number (#);

then its relational data structure in third normal form is given by (rule 1):

Order (***order no. #***, *customer no. #, order date)*

On the other hand, if an order can specify:

- many product numbers (#);
- many product quantities;
- many delivery addresses;

then an order line must be added and the relational data structure in the third normal form is as follows (rule 3):

Order Line (***order no. #, prod. no. #***, *quantity, delivery address)*

with the product number as the secondary identifier. That is, in order to specify uniquely the particular quantity or the particular delivery address, the user must key in both the order and product number with the data base constructed in such a way.

In relation to the information model developed earlier in this section, a complete set of E/R diagrams to represent the relational data model of the product of a manufacturing operation, as indicated in Fig. 7.36 (D2: Product), is provided in Appendix D. This aims to present the reader with an view of the typical data attributes, entities, and their mutual relationships within an integrated manufacturing information environment. A relational data base system can be implemented in practice by using commercially available relational data base management systems (DBMS) which provide tools for relational data manipulation such as data table definition, projection, selection, division and joining. For example, the following is an example of a SQL (structured query language) statement which will create the appropriate computer file for CUSTOMER:

```
CUSTOMER
  (CUSTOMER_NUMBER   SMALLINT               NOT NULL,
   NAME              CHAR (15)              NOT NULL,
   PAYMENT_TERM      CHAR (2)               NOT NULL,
   STREET            CHAR (15)              NOT NULL
   HOUSE_NO          CHAR (4)               NOT NULL,
   POST_CODE         CHAR (6)               NOT NULL,
   TOWN              CHAR (10)              NOT NULL,
     UNIQUE          (CUSTOMER_NUMBER))
```

With the conventional relational structure outlined above, a data base is treated only as a flat structure of entities related by their identifiers and described by their static attributes. Consequently, although relational data base systems offer the possibility of managing large amounts of data due to their logical and physical independence, they are limited as far as managing and structuring information are concerned. The object-oriented approach is now being seen as a way of overcoming this shortcoming. The idea behind the so-called object-oriented data base management systems (OODBMS) is to combine the relational data structure with an object-oriented programming approach, so that the flexibility and power of the relational data structure can be further extended (Alagic, 1989; Hughes, 1991). OODBMSs overcome the traditional dichotomy between data and operations. Since a large number of operations can be centralized an object-oriented data base, the application programming is simplified.

Since there is no single, agreed object-oriented data model as for the other models, three approaches have been attempted for the construction of OODBMSs:

1. With a programming language neutral object model, which extends the relational or semantic data models with object-oriented features.
2. With a data base programming language, and a new programming language that is designed from scratch to support data base capabilities.
3. With a programming language specific approach.

Since the exact make up and features of a truly object-oriented data base system are still not clear, formalization work is still being carried out. However, a number of OODBMSs are currently available on the market.

Definition of user interfaces

Having identified the data required, together with the relationships among them, to construct a manufacturing data base system, the next step is concerned with how the data are to be presented to, and used by, the user. Two types of interaction will take place with a manufacturing data base: enquiries, which simply view the data; and updating processes, which modify the data.

At the external level of the ANSI/SPARC architecture are the external 'views', which define the way in which data are presented to the user. According to the definition of the SQL standard (Van der Lans, 1989), for example, a view is a table of data whose contents are derived from one or more other tables and views, created by the following command:

```
CREATE VIEW <table name>
  [ <column list> ]
  AS <query specification>
  [ WITH CHECK OPTION ];
where,
<query specification> ::=
  SELECT [ ALL | DISTINCT ] <select list>
  <from clause>
  [ <where clause> ]
  [ <group by clause> ]
  [ <having clause> ]
```

For example, the following is a view created to get a list of customers with payment term equal to '30':

```
    CREATE VIEW Customer_Payment_term
(Customer_name, Payment_term)
```

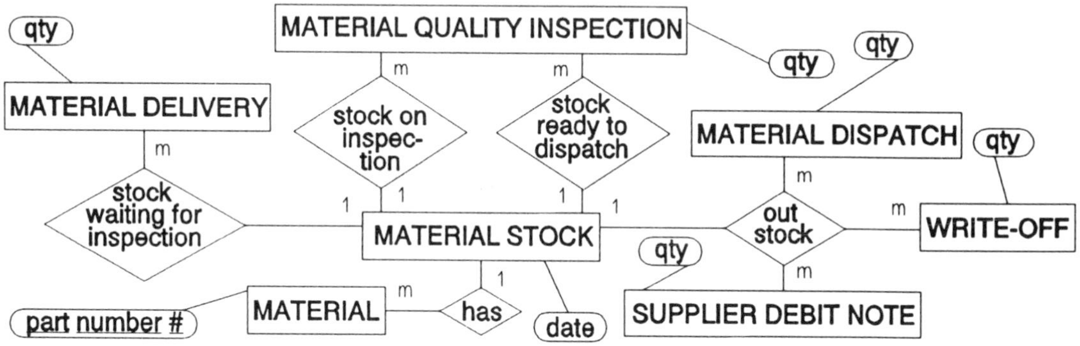

Fig. 7.49 E/R diagram of material stock.

```
    AS SELECT Customer_name, Payment_term
  FROM Customer_File
  WHERE Payment_term = '30';
```

In the following two examples, views are given which are developed in relation to their associated E/R diagram.

Material stock

This view can be used to monitor the inventory level of the material as part of an MRP system. The associated E/R diagram is shown in Fig. 7.49, and the SQL statement to create this view (with MODULE linked to the Pascal language) is as follows:

```
    CREATE VIEW Material_Stock (part_number_#, quantity_waiting_for_
inspection,
          quantity_on_inspection, quantity_ready_to_dispatch)
    AS SELECT part_number_#, SUM (qty)
       FROM Material_delivery,
     Material_released_for_inspection,
     Material_quality_inspection,
     Material_dispatch,
     Material_write-off
       AND  Material_quality_inspection . disposition
              = 'accept'
    MODULE Material_Stock
    LANGUAGE Pascal
    AUTHORIZATION Mike
    PROCEDURE Calculate;
  VAR Material_delivery_qty,
    Released_for_inspection-qty,
    Material_quality_inspection_qty,
    Material_quality_inspection_approved_qty,
    Material_dispatch_qty,
    Material_write-off_qty,: INTEGER;
    A, B, C, D, E, F: INTEGER;
    Quantity_waiting_for_inspection,
```

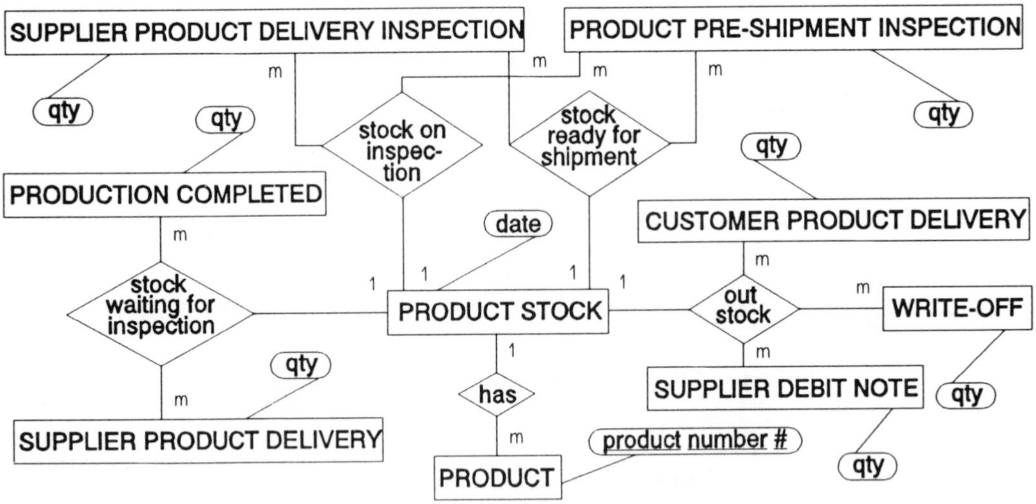

Fig. 7.50 E/R diagram of product stock.

```
    Quantity_on_inspection,
    Quantity_ready_to_dispatch: INTEGER;
  BEGIN
    Material_delivery_qty := A;
    Released_for_inspection_qty := B;
      Quantity_waiting_for_inspection := A − B;
    Material_quality_inspection_qty := C;
      Quantity_on_inspection := B − C;
    Material_quality_inspection_approved_qty := D;
    Material_dispatch_qty := E;
    Material_write-off_qty := F;
      Quantity_ready_to_dispatch := (D − E) − F;
  END;
```

Product stock

This is a view that may be created to monitor the inventory level of a product. The associated E/R diagram is shown in Fig. 7.50. The SQL statement used is similar to that for the material stock:

CREATE VIEW Product Stock (*Part_number_#*, date, qty_waiting_for_inspection, qty_on_inspection, qty_ready_for_shipment)

The design of both the enquiries and the updating processes are dealt with adequately by SSADM. The final phases of SSADM are specifically concerned with the design tasks in this regard. By following its relevant procedures, the draft physical and process designs can be created. Their performances are then predicted to ensure that the processing fulfils the requirements. Upon completion of the performance appraisal, the basis of the detailed specification for the data base management system is created, together with file specifications. Finally, it is necessary to ensure that the correct data processing devices are made available. This will require the selection of transducers, PCs and networking equipment, etc. For user interactions, the number of terminals or networked PCs and their locations must also be established. The consideration of location of terminals should include identifying the location of manufacturing functions being controlled. The physical locations and layout should be recorded in the *plant register*.

7.7.5 Summary

There are three main areas to be considered in the detailed design stage of AMS. These are the selection of production technology together with the selection of transportation and storage facilities, the organization and layout of the technology, and the detailed design of the control system, including both the hardware and software required. These processes are unlikely to occur sequentially, for decisions for each will have implications for the others. The output from this stage will be a design which is accurate to a high level, and detailed enough for actual implementation. However, these designs must still be assessed against the objectives which were originally set, so as to ensure that they are fulfilled and that the investment involved is viable. The methods available to achieve this will be considered next.

7.8 Evaluation and decision

In Chapter 5, the general approach and techniques of the evaluation and decision process were discussed in detail. Therefore, only a brief review is needed here to complete the design framework.

While it is recognized that design is fundamentally the process of creating, evaluating and selecting from alternatives, the actual decisions to be made usually differ significantly throughout and between projects. However, there are two phases in the design of manufacturing systems which are consistently required and can be dealt with in a similar way each time. These evaluation and decision phases follow the end of the conceptual modelling and the detailed design phases, respectively, as shown in Fig. 7.11. They are slightly different in their requirements, and are therefore dealt with separately below.

7.8.1 Selection of business system option

The inputs to this section are various BSOs from the conceptual modelling phase. The decision to be made is to select the best option available from these. Since the information currently available is not very detailed, it would be unwise to rely entirely on the use of the quantitative, financial appraisal methods discussed in section 7.3. Instead, the AHP method is more suited to such situations where judgemental or intuitive assessment is required in addition to the limited quantitative analysis that may be applied.

It is necessary to assess each of the BSOs with respect to each of the criteria which have been laid out in the objective-setting phase. Most of the information will have to be generated through discussion within the design team, or by the use of external experts who will be able to rank the alternatives available, and offer advice on the levels of differentiation between these. In some cases it is possible to carry out more rigorous assessment. For example, although detailed work-station designs have not been made available at this stage, it can still be feasible to utilize an aggregated simulation model to establish the acceptability of different groupings within the system, as well as their ability to meet objectives.

If a clear front-runner can be identified, this will be chosen automatically. However, frequently the steering committee or board of directors will have to select between several close options. Occasionally there may seem to be no acceptable solution at all. When this happens it is necessary to establish how easily a design can be modified to make it acceptable. This will form the basis for a decision to return to the conceptual modelling phase, to modify the objectives, or to terminate the project altogether.

7.8.2. Acceptance of detailed system design

The second evaluation and decision point is after the detailed design process. The decision is now concerned with whether or not the necessary investment should actually be made. Initially the project is considered against its initial objectives. It is then necessary to consider the financial justification, as this is a measure of the project against all other investment opportunities open to the firm.

Measurement of the system's fulfilment of objectives can be achieved by evaluation using computer simulation, queuing theory, or Petri net theory (other techniques are also available). Much of the system will already have been modelled during the detailed design process, or at higher levels during the earlier phases. However, at this stage a thorough evaluation of the system is essential, and so the remaining sub-system must be included in the evaluation frame at a disaggregated level.

Both tangible and intangible factors should be included in the evaluation. The list of possible intangible benefits presented in section 7.3.2 will provide some guidance for this. Computer support tools may also be used for this purpose (for example, Primrose and Leonard, 1985).

Conditions for evaluation

Before any information is collected and the 'hurdle rate' set, the conditions for evaluation must be fixed. This will include decisions on how to deal with the factors such as risk and inflation. For example, if an estimate is to be made with inflation included, the design team should take responsibility for reducing the values by the best current estimates for inflation. Also, it may be decided that the risk factors would be best dealt with using the probability approach. Again, in this case the design team should take responsibility for estimating the necessary values.

Having fixed the evaluating conditions, a realistic hurdle should be calculated for the project. This should be in line with the real returns required by shareholders on a risk-free investment, plus an allowance for the rate of risk attached to the project. The proposed hurdle rate should be approved by the steering committee or the company board, to ensure that no problems arise if the rate is below the company norm.

Assessing the project

To assess the proposed system, opinion must be invited from a diverse range of sources. For example, discussions with marketing personnel may identify a possible increase in sales which could be expected from an improved level of quality. Should some items be considered as of low significance, or too difficult to quantify, they can be recorded in a list of further benefits for later use.

The total costs of the project should be estimated, along with a cash-flow profile. Very accurate hardware costs will be available through the data gathered during the detailed design phase, whereas some other costs may have to be estimated. The timing of the spending is an important factor to consider. Expenditures should not all be considered to occur at the beginning of the project, as distant prices will have a lower net present value, and thus the total costs will fall. The actual calculating process may be carried out using either a special-purpose package or a spreadsheet-based model.

Should the proposal fail to achieve the hurdle rate, the value of the shortfall should be estimated. This is then presented to the board, along with the list of further benefits. The board can then assess whether it feels that the extra benefits provide acceptable returns for the shortfall in revenue, and thus will either accept or reject the project in its current condition.

Dealing with failures

If the project fails on any of the criteria, it must be decided whether to continue with the development of the design, to modify the objectives, or to terminate the project. It is impossible to define specifically how to make a good decision in this regard, but the logic is straightforward. The additional cost of modifying the design, and the expected change in cost of implementation of the system, should be added to the cash-flow estimate. If the targeted rate of return can still be achieved, then it is obviously desirable to continue with the project. If the rate of return cannot be achieved, but the project will still make positive in-flows, the project should be compared to the 'do nothing' case. If, however, the project will clearly cost the firm more than it will earn, it must be terminated before further costs are incurred. The results of this analysis should be presented to the board for a final decision.

7.8.3 Conclusion

This completes the picture of the design and analysis of manufacturing systems. The approach identified in this particular chapter, together with the relevant concepts, techniques and methods previously presented in Parts One, Two and Three, are aimed at improving the quality of the design produced, and thus reducing the problems and costs associated with poor systems specifications.

In summary, if it is properly designed and conforms to the appropriate standards, the implementation and operation of the actual system should be relatively trouble-free and provide a good return on the capital invested.

7.9 Manufacturing systems redesign: a case study

(Based on the Fordhouses case study in *Manufacturing Systems Redesign*, Hobsons Scientific, 1992. Reproduced by kind permission of Lucas Engineering & Systems Ltd and Hobsons Publishing PLC.)

We shall now look at the application of the principles explained above to a specific business, the Actuation Division of Lucas Industries at Fordhouses in Wolverhampton, UK. The current business is the design and manufacture of advanced hydraulic actuation systems for military and civil aircraft.

The division supplies systems and components to the major manufacturers of aircraft and engines for both civil and military markets. It also supports over 200 customers, such as airlines, in the aftermarket.

The Actuation Division's programme involves sharing work on each project with up to three international partners. The vast majority of equipment (90%) is exported.

The aim of the business is expressed as:

> To be recognised by our customers as the best supplier of actuation systems through the excellence of our people, products and services.

The business objectives are stated as:

- To respond to our customer requirements.
- To change the business from 70:30 military:civil, to at least 50:50.
- To make the business grow profitably.
- To be a world leader in the secondary flight control and the engine actuation market sectors.

7.9.1 The need for change

In 1987, it was forecast that military demand would decrease, and that, to maintain sales, the Division must change to the US-dollar-based civil business, which was highly competitive. New products were introduced in this market, for example hydraulic thrust reversers for civil engines: these replaced competing pneumatic systems which caused problems. There were a large number of competitors fighting for a share of this market and therefore to be a winner it was necessary to improve performance and reduce costs.

Civil airlines were also aiming to reduce their costs by reducing their spares stock, and relying on shorter lead times for the supply of spares and repair of units. Engine and aircraft constructors expected shorter lead times for new product design, in addition to excellent product performance and a rapid solution to any problems they encountered.

The business had evolved in an attempt to respond to these demands, but there was a need for further improvement:

- Lead times needed to be shortened.
- Excessive re-scheduling, modification, re-work and administration needed to be reduced.
- Disruption caused by the introduction of new products needed to be minimized.

Current programmes needed to be accelerated to accommodate changes in customer's requirements.

The executive recognized the need for a major business redesign, with these aims:

- Better response to customer problems.
- Reduction of customer arrears.
- Reduction of repair turn-around times.
- Cost reduction.
- Reduction of design and manufacture lead times.
- Change of culture.

- Reviewing organisation structure.
- The streamlining of internal process flows.
- The harmonizing of terms and conditions of employment.
- The replacement of work activities previously performed in series with parallel activities.

7.9.2 Diagnostic study

An eight-person business team of members of the Divisional Executive, senior managers and Lucas Engineering & Systems manufacturing systems engineers was established. Members were allocated to work on the team full-time for two months to:

- Define the essentials of the business.
- Specify the most effective method of operation.
- Develop an action plan.
- Develop a mechanism for managing change.

The team concluded that:

- The existing structure contributed to complexity, fragmentation, long lead times and unnecessary cost.
- The existing processes were over-complex, costly and unreliable, with excessive non-value-added activities.
- Work was performed in series, resulting in lack of ownership and accountability, with excessive re-work.

7.9.3 Change programme management

The resistance of people to change makes it necessary to coach, present, train and vanguard simple projects first, to win over hearts and minds. It is important to remember that most people have worked at the same job for 20 years whilst managers have generally been in their positions for 2–5 years. The workforce has a genuine fear that managers come and go and this can create great damage. The business executive therefore has to demonstrate truth and commitment and show that he or she gets things right.

The team's proposals for change were discussed at a management meeting, and communicated to all employees. Taskforces were established to review and implement the necessary changes and were given specific achievement objectives.

Taskforces were co-ordinated through the 'problem owner' and the leader of each team together with a steering committee which reported to the Division's Executive. The Personnel Manager acted as the 'facilitator' or 'change manager' in order to respond to the human resource requirements arising from change.

7.9.4 Matrix organisation

The business development team recommended matrix organisation principles:

- All line functions to be part of a business-wide matrix organisation.
- Stable project teams as an intermediate step from previous 'no change' culture.

- Teams developed for the future.
- Project managers of project teams report to the General Manager and negotiate with functional managers for resources.

The matrix organisation recommended is shown in Fig. 7.51.

The ownership of responsibilities has improved the visibility of producing within the agreed lead time and cost.

7.9.5 A solution to the problem

The team's deliberations led them to look at the following areas as they formulated a solution to their business problem.

Product Introduction Process – PIP

The Product Introduction Process or PIP is defined as: the process of concept design, detailed design, design for manufacture, design to target cost, design of the manufacturing system and introduction to manufacture.

To help the technical specialists make a full contribution, the design and detail engineering sections were moved from two separate floors to work together in the re-organised engineering floor. On a larger scale, production engineering and development engineering were previously located in separate buildings, disjointing communications and process flows, with major rework loops increasing costs, extending lead times and causing customer dissatisfaction. They were moved into the engineering floor.

A technical taskforce was assembled representing:

- Manufacturing engineering.
- Production engineering.
- Customer support engineering.
- Contracts engineering.
- Quality engineering.
- Purchasing.
- Computer engineering services.
- Manufacturing systems engineering.

The taskforce used the matrix management concept to establish a technical process flow, including input-output analysis and deliverables, from 'bid proposal' through to 'aftermarket' stages. The flow diagram shown in Fig. 7.52 incorporates the principles of simultaneous engineering, and of involvement from the start of any discipline that can effect change at any stage.

The Division's Executive had earlier agreed a computer integrated business strategy which was also introduced at the same time, linking the technical and manufacturing organisations through a common data base. Communication between technical and manufacturing organisations now takes place through the Product Engineering Support data base – PES. The former separate administration teams for product engineering and production engineering were combined to service the team.

Benefits of the project team approach

The change from the traditional functional, series ways of working, in which production staff only saw drawings and specifications at a late stage of design, to simultaneous engineering under a project team, has shown considerable benefits.

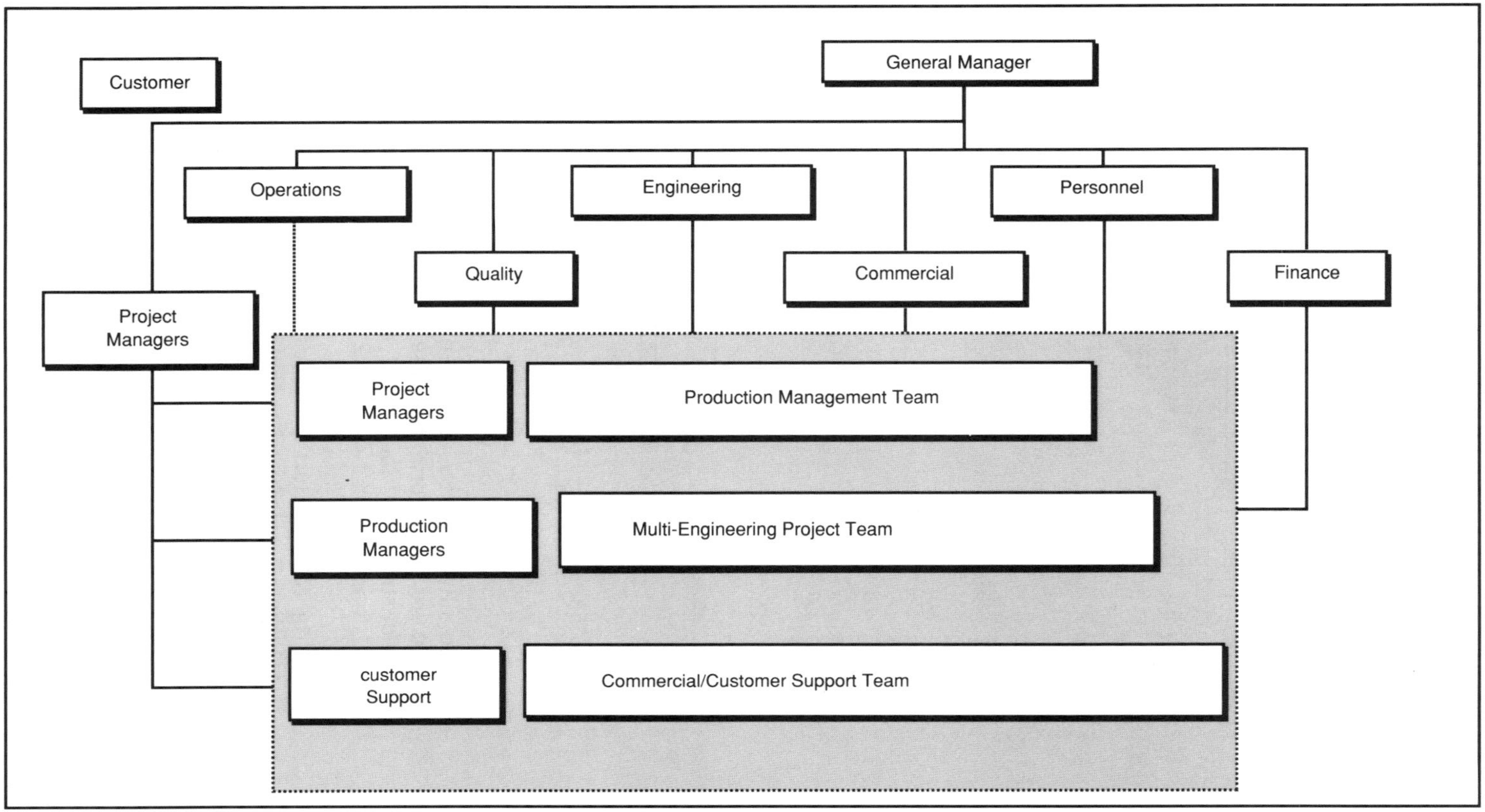

Fig. 7.51 Matrix organization.

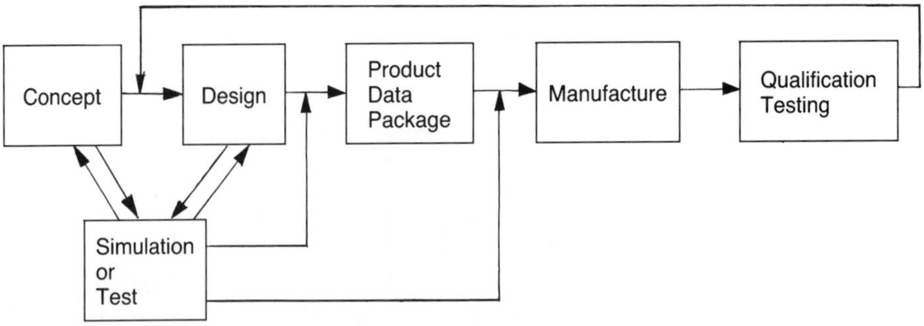

Fig. 7.52 Technical process flow.

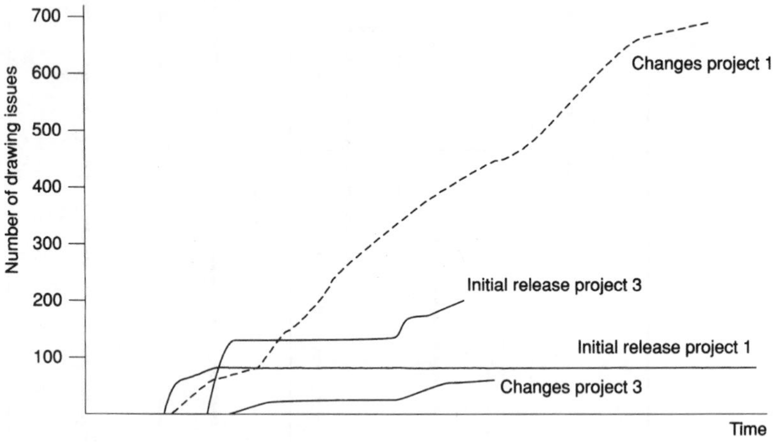

Fig. 7.53 Numbers of drawing changes.

To illustrate the effectiveness of the redesign we can consider a key measure of performance: the number of modifications made. This can be demonstrated by looking at three thrust reverser projects which we shall refer to as Project 1, Project 2 and Project 3:

- Project 1, produced under the old system.
- Project 2, produced at the time the new system was being introduced.
- Project 3, produced when the new system was fully effective.

Key measures of performance are the numbers of modifications and drawing changes.

Figure 7.53 shows that, for similar points in the programme, Project 1 has 90 drawing releases and 370 changes, compared with 175 drawing releases and 45 changes for Project 3, which is however more complex.

Even though the products have become more complex due to enhanced technological requirements, lead times have been reduced by the 'right first time' approach as shown in Figs 7.54 and 7.55.

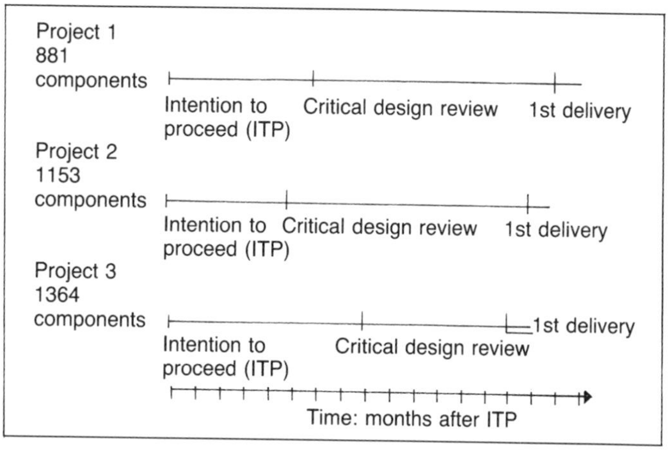

Fig. 7.54 The different lead times of project 1, project 2 and project 3.

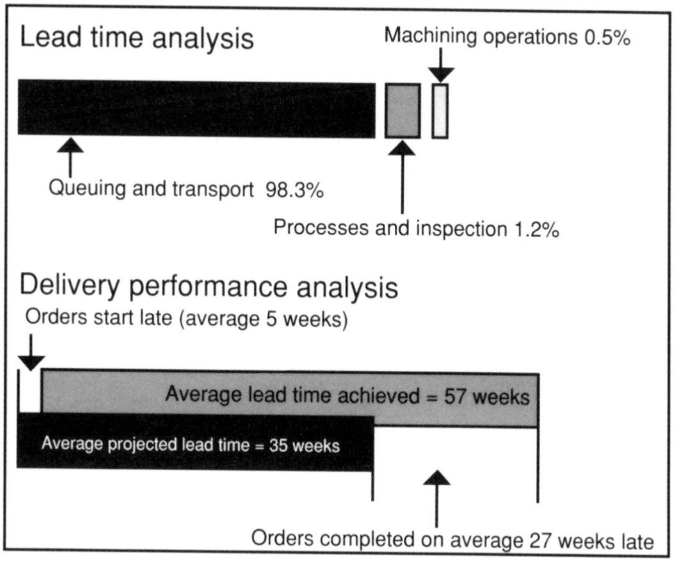

Fig. 7.55 Analysis of lead times and delivery performance in 1987.

Manufacturing cells

The existing machining operations were carried out in the traditional pattern, in which each batch progressed through the functionally specialized machine groups in turn: turning, milling, grinding etc. Components moved large distances, and were prone to loss or damage. Problems were only discovered some time after the event which caused them, making it difficult to find out what had gone wrong and how to avoid a recurrence. The piece-work system caused customer schedule and quality to be regarded as secondary to 'standard hours produced': as a pre-requisite to change, this was replaced by flexible working practices. There were too many changes of ownership and overcomplex material flows.

The operations taskforce set out to:

- Re-organize the machine shop into product or process units.
- Set unit operating targets.
- Define the natural group cell organization.
- Establish and monitor a unit introduction programme.
- Co-ordinate the taskforces for each unit.

When the taskforces, which consisted of multi-functional personnel including machine operators and trade union representatives, had been designed and trained, they set up the following cells:

- Transmission.
- Offset gearbox.
- End fitting.
- Thrust reverser.
- Valve and guide.
- Rotary actuator.
- Computer numerically controlled machining.
- Repair.

The benefits gained from reorganisation were

- Adherence to master project schedule.
- Focused manufacturing and investment.
- Reduced non-conformance.
- Product ownership.
- Reduced complexity.
- Increased control.
- Improved lead times.
- Reduced levels of work in progress.

The overall organisation of a cell now has a flat structure, new flexible job structures and the mix of required skills. Examples of the targets that were set for two of these cells are shown in Figs 7.56 and 7.57.

CNC machining centres, installed under a parallel capital investment programme running at 3% of sales, carry out a number of operations previously requiring a variety of machines. All existing machines were renovated and recalibrated. Machine to machine movement has been reduced by about 80%, as shown by Fig. 7.58.

Customer repair and overhaul service business

One of the taskforces reported the following conclusions from flowcharting:

> The structure of the business needed to be less fragmented, less complex, have fewer than 12 departments. There needed to be better accountability and control and changes in ownership needed to be kept to a minimum.
>
> In order to avoid major delays more ownership of the process had to be established, the production priorities needed to be reviewed and parts shortages had to be avoided.
>
> Twenty-five routine documents needed to be reviewed to sort out common errors, to reduce their complexity and to add value to the

Measure of performance	Units	Prior	Mid-1991	Future target
Output	sets/month	5	15	19
Manpower	people	14	30	30
Work in progress	shafts	4500	300	300
Leadtime	weeks	30	1	1
Annual quality cost	£1,000s	120	20 (35% rework)	<20
Schedule adherence	% delivered on time to customer	<50%	96%	100%
Annual stock turn over ratio (physical) times	number of times	0.85	28	32

Fig. 7.56 Transmission cell targets.

Measure of performance	Units	Prior	Mid-1991	Future target
Output	sets/month	7	14	19
Manpower	people	29	37	48
Work in progress	£1,000s	517	628*	408
Leadtime	weeks	29	18	9
Annual quality cost	£1,000s	89	40	<30
Schedule adherence	% delivered on time to customer	<50%	100%	100%
Annual stock turn over ratio (financial) times	£turnover/ £average stock	2.7	4.2	8.6

Fig. 7.57 Rotary activator cell targets.

activities. Problems needed to be taken care of as they arose, rather than being passed on.

These improvements were designed to cut back on:

- Excessive leadtimes.
- Poor visibility of problems.
- Lost hardware.
- Lost documentation.
- Labour-intensive control and monitoring.
- Major customer dissatisfaction.

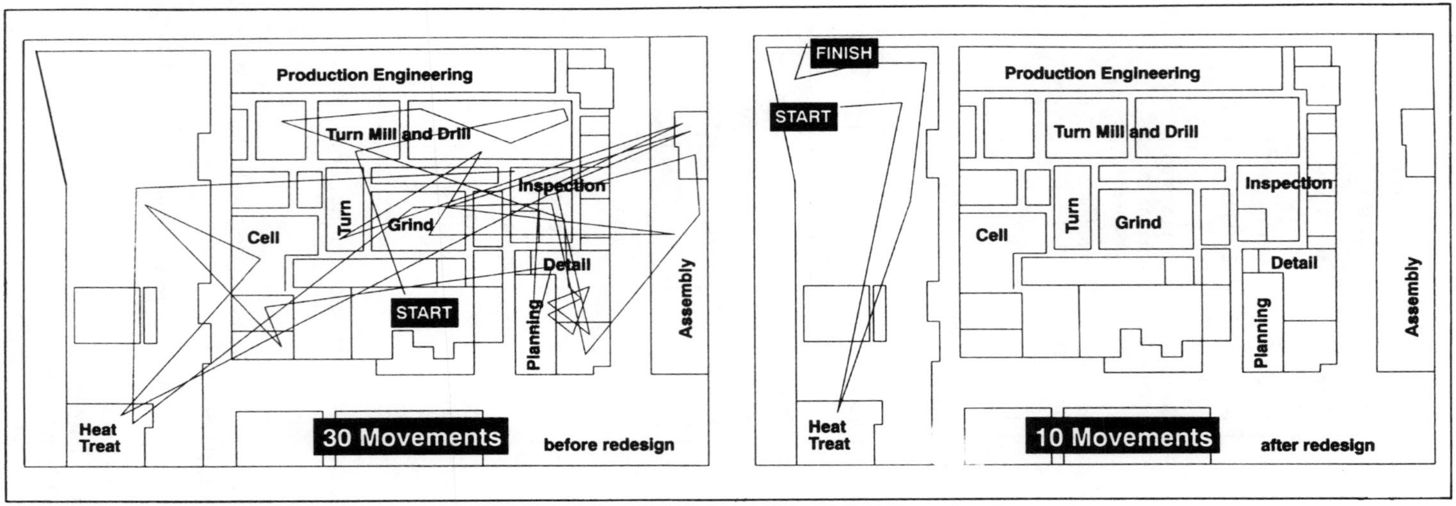

Fig. 7.58 Comparison of process route.

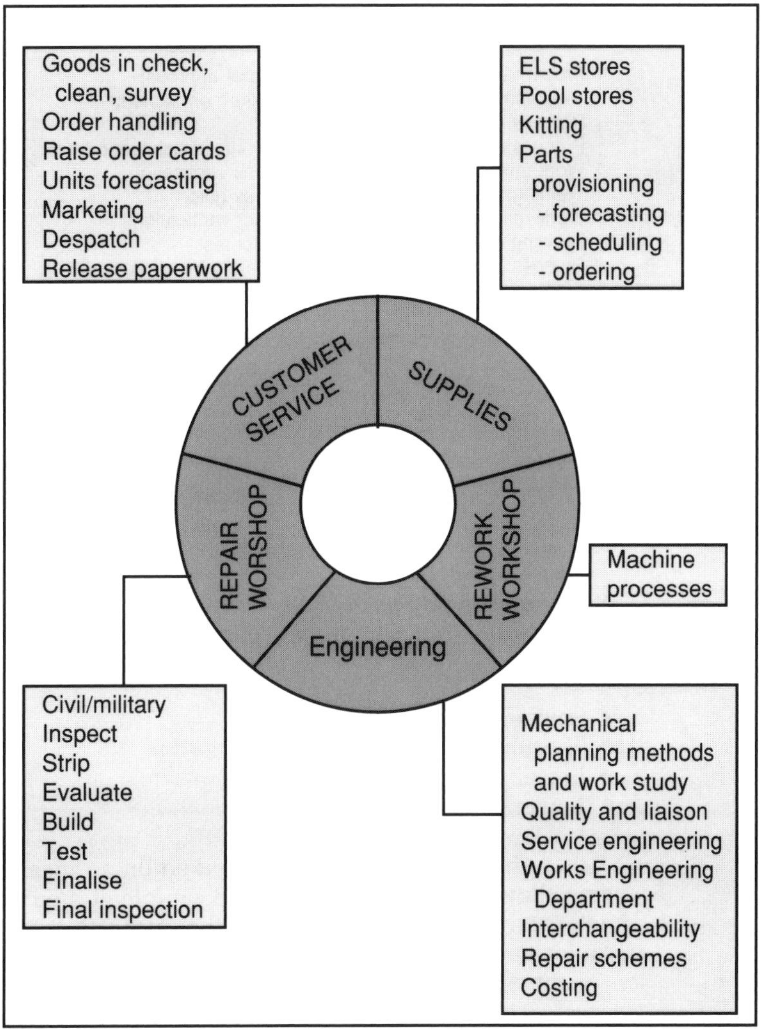

Fig. 7.59 Business definition.

The business was redesigned to achieve:

- Maximum autonomy of the product unit.
- Simplified structure.
- Structure based on work flow and tasks.
- Minimum turn-round time.

A Repair Business Manager runs an integrated unit achieving the specified performance (Fig. 7.59). For example, in the repair manufacturing unit, 10 multi-skilled manufacturers operating 25 machines/pieces of equipment are responsible for their own quality. Repair turn-round times have dramatically decreased. This makes possible Repair Station Approval from the US Federal Aviation Authority and other business opportunities.

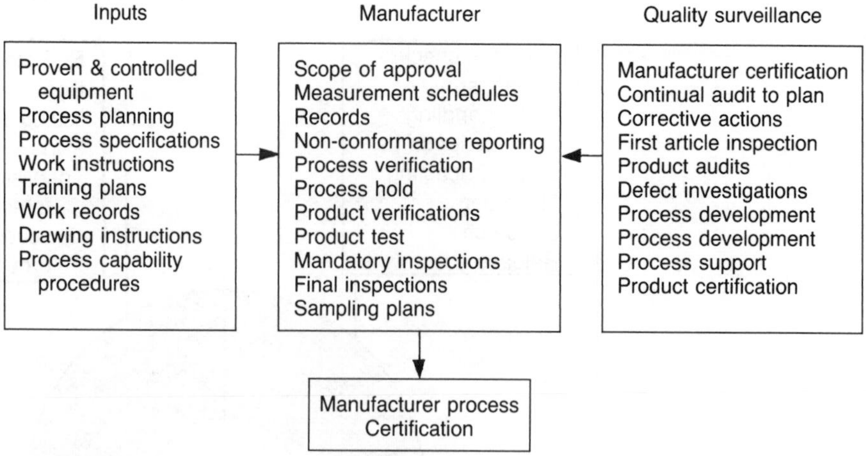

Fig. 7.60 Manufacturer self-inspection.

Manufacturer self-inspection
The previous practice of inspection after the event needed to be changed to ensure:

- Smaller queues for inspection.
- Shorter lead times.
- Lower levels of rectification/scrap.
- Lower costs.

Although the skilled operators worked on an incentive scheme containing a time allowance for them to check a sample of their own production, they did not carry this out, relying on the inspection 'backstop' to detect non-conformances.

General overall industry figures indicate that:

- 85% of non-conformances are due to inadequate repeatability in the production process.
- The remaining 15% can be ascribed to errors in manufacturer setting and adjustments.

A detailed examination of the relationships between external process inputs, manufacturer influence, monitoring and quality assurance led to a model for further development, as shown in Fig. 7.60. The most consistent CNC machining processes were selected for a pilot programme, which included major training and review of all process inputs: for example, calibration of machine tools to ensure slide, bearing and control system accuracy. Manufacturers are 'licensed' by Quality Assurance for specific tasks; when required, they record all non-conformances for further action.

The manufacturers were enthusiastic about the new system, accepting their authority with a great sense of responsibility. They no longer had the frustration of work flow being interrupted by waiting for an inspector; but assistance was immediately available if needed. The introduction was reviewed weekly by a group with technical, supervisory, trade union and

	1987	1991
Staff	116 inspectors	21 surveyors
Queues	2000 batches	Nil
Rectification costs	£1,800,000	£339,000
Scrap costs	£943,000	£500,000
Customer returns (% of units delivered)	1.6	0.32
Cost of quality as % of sales	7	4.6

Fig. 7.61 Improvements achieved in quality standards.

manufacturer members chaired by the Personnel Manager. The programme of manufacturer self-inspection was then extended to cover the whole business. The independent team of inspectors was replaced by a Quality Assurance department (QA). In this, a small team of surveyors conduct continuing audits of the inputs (including machine tool calibration) and manufacturer tasks, advise where necessary, and prepare summary reports for operations management. Any extension to a manufacturer's task requires further approval by QA. The manufacturer has the authority to stop the process, and either the manufacturer or the surveyor can impose a hold notice. As the manufacturer can identify problems at the outset, and take action, the level of non-conformance has been dramatically reduced already, and continues to fall (Fig. 7.61).

Supplier development

36% of revenue is spent on supplies. A supplier development programme was started in 1989 with a target of reducing the number of suppliers from 500 to 250. This would allow:

- A 'two-way commitment' to share risks.
- Involvement at the design stage.
- Potentially larger volumes, bringing benefits from being further along the 'learning curve.'
- Reduction of cost to both parties.

200 suppliers were invited to a conference, the purposes of which were:

- To present forecasts of the future of the business, including markets, risks and current and future spending.
- To present intended changes to the procurement strategy.
- To explain the intended introduction of 'product assured suppliers'.
- To discuss delivery and non-conformance.

Representatives from the National Economic Development Office (NEDO) were present at the conference to assist smaller suppliers. Since the conference, the number of suppliers has been reduced from 500 to 260. The

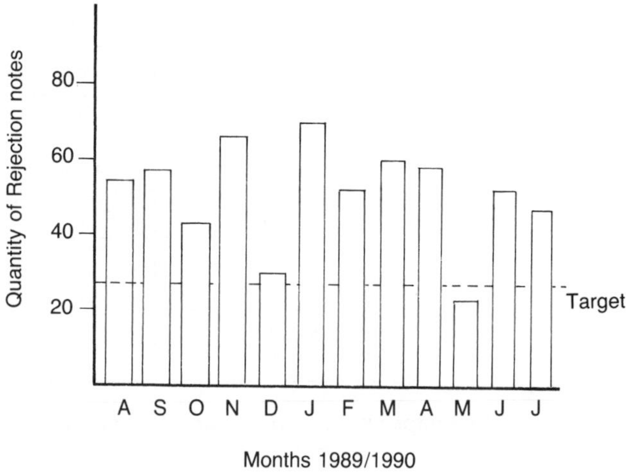

Fig. 7.62 Number of rejection notes received 1989–90.

target for rejects was a reduction from the previous average of 50 per month to 28, which is only occasionally met as yet – see Fig. 7.62.

Following the conference, an operating procedure for appointing product assured suppliers was developed and tested on a pilot basis with 5 suppliers; it was subsequently revised in the light of this experience.

A company may be considered for the product assured supplier scheme if it meets the following criteria:

- Quality performance rating in excess of 96%.
- Proven record of price and delivery.
- Satisfactory site audit.

It will be required to demonstrate:

- A resource and investment commitment extending over the life cycle of the product.
- A contribution to product design, where appropriate.
- Willingness to accept responsibility and liability for non-conformance with Lucas customer requirements.

The quality, price and delivery from the selected suppliers is continually monitored. A further benefit is that goods from product assured suppliers can be sent direct to stores or assembly point, without the need for incoming goods inspection. Further product approved suppliers are being appointed under a phased programme.

The benefits of this programme should become apparent in the medium to long term when current purchase orders are re-negotiated and new long-term contracts agreed.

Continuous improvement

A continuous improvement programme, outlined in Fig. 7.63, has been set up to continue the change for the better.

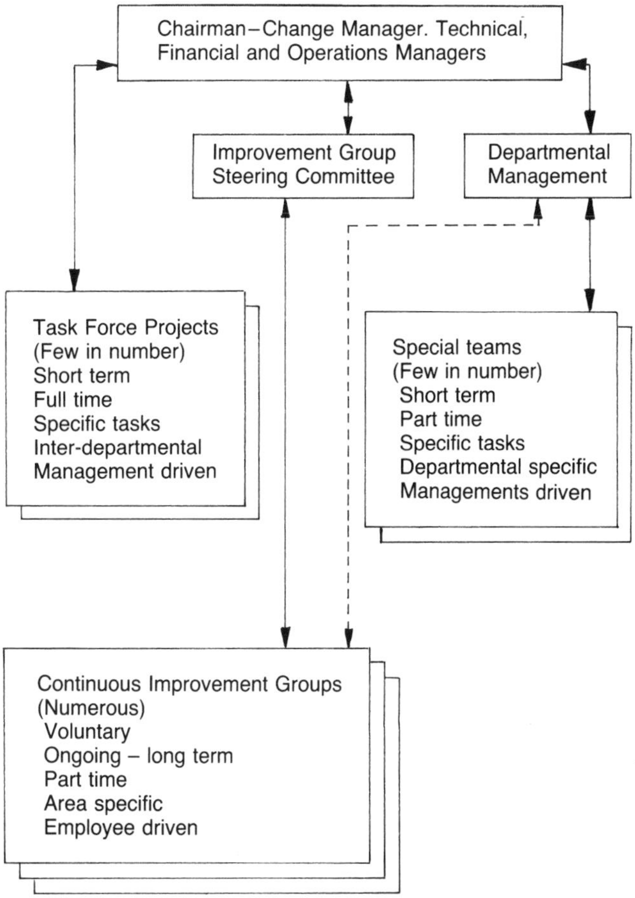

Fig. 7.63 Continuous improvement programme.

A suggestion scheme, forming part of a Lucas-wide programme, rewards employees' ideas for improving business activities; in 1990, two Chairman's awards and one Director's award were given to a total of £10 000 in response to a saving of £70 000.

Flexible job roles
The major changes described in the previous sections have affected all areas by merging tasks and skills, resulting in increased job flexibility. The effect of this re-organisation on personnel was assessed to maximise the benefits of these changes, and to allow effective training.

The previous system with numerous skill/grade levels and rates of pay, a division between 'staff' and 'hourly paid' terms and conditions, and a piece-work scheme in the manufacturing facility, made it extremely difficult to introduce task merging and job flexibility. Any minor change to an established task was used as a negotiating ploy to move pay up to the next level, so much time and effort was expended in negotiations to implement any changes. New packages were negotiated with the trade unions to harmonize terms and conditions for 'hourly paid' and 'staff' personnel. Flexible job profiles were negotiated to reduce the numbers of skill levels:

Manufacturing cells:	12 to 4
Technical and supervisory:	8 to 3
Clerical:	10 to 5

The individual incentives of the piece-work scheme were made obsolete by the introduction of manufacturing cells with performance monitored through:

- Lead time.
- Work in progress.
- Cost of quality.
- Schedule adherence.
- Annual stock turnover.

It was replaced by a flexible 'manufacturer' job profile. With people now working in flexible job roles, they would appreciate the need for training programmes, and obtain maximum benefit from them.

All executives now have a higher profile within the business, attending and leading group sessions. The role of a manager has evolved from that of a traditional supervisor to one where he or she leads a team or business unit. Shop floor personnel have taken on more flexible roles, with the introduction of manufacturing cells and manufacturer self-inspection. Supplies personnel now give more attention to the reduction of lead times, and to supplier development.

The above changes demanded a training programme which addressed both the 'technical' and 'people' aspects. The Division therefore increased its training resources to meet these demands, enabling the changes to be effectively introduced.

Communications

Good communications and employee participation was a major factor in the successful implementation of the business redesign process to achieve the stated objectives, and continues in order to maintain the improvements.

Internal strategies and other developments are presented in a quarterly video, and feedback is sought via a questionnaire. After a showing of the video to a cross-functional group, the Division's General Manager reviews the current business performance, followed by a question time with other members of the executive team. Management meetings enable the 120 management and professional specialists to discuss specific issues affecting the management strategy of the Division. Briefing meetings at the local team, departmental, sectional and manufacturing unit level are held every one or two weeks. Departmental and team managers have been trained in team briefing. Strategically located notice boards communicate information relating to the whole business.

Summary

The case study illustrates how the Actuation Division of Lucas Industries at Fordhouses needed to make significant changes to its structure and operating procedures to respond to the demands of its customers.

The objectives of the programme of change were:

- Improved customer service.
- Reduction of customer arrears.
- Reduction of repair turn-around times.
- Cost reduction.
- Reduction of design and manufacture lead times.
- Change of culture.
- Reviewing organisation structure.
- Streamlining of internal process flows.
- Harmonizing of terms and conditions of employment.
- Replacement of work activities previously performed in series with parallel activities.

The programme of change was designed by an eight person business team who looked at the Product Introduction Process (PIP), the Manufacturing Cells, the Repair Business, Manufacturer Self-inspection and Supplier Development. Taskforces were appointed to implement the necessary changes which resulted in the following improvements:

- Flexible working practices were introduced, new manufacturing cells were established and performance was improved considerably.
- New product development and after-support was improved by co-locating the development and manufacturing engineering departments so that they could work together. The traditional series way of working was replaced by concurrent engineering under a project team. Both changes resulted in shorter lead times and better response to customer demands.
- The structure of the repair business was simplified and improved, dramatically reducing repair turn-around times, ensuring Repair Station Approval from the US Federal Aviation Authority and other business opportunities.
- The practice of manufacturer self-inspection was introduced, together with a Quality Assurance department, replacing the old system of independent inspectors. Manufacturers now have more responsibility and authority, allowing problems to be identified at the outset, rather than later in the process and the level of non-conformance has been dramatically reduced.
- The number of suppliers to the process was reduced from 500 to 260 and companies were only considered as suppliers if they could demonstrate certain imposed criteria. Such suppliers could deliver directly to the assembly point, thus reducing the time spent in inspecting supplies.
- A continuous improvement programme and a scheme which rewards employees' suggestions has been put into operation, ensuring a continued move towards better business activities.
- Flexible job roles and better training programmes were introduced to reward employees fairly whilst avoiding lengthy negotiations.
- Communication throughout the organisation was improved, leading to increased participation by personnel at all levels.

Further reading

Ajderian, N. (1989) Do You Really Want That New Machine?, *Professional Engineering*, November.

Ashworth, C.M. (1988) Structured Systems Analysis and Design Method (SSADM), *Information and Software Technology*, **30** (3).
Askin, R.G. and Standridge, C.R. (1993) *Modeling and Analysis of Manufacturing Systems*, John Wiley.
Bennett, D.J., *et al.* (1989) A Model for Analysing the Design and Implementation of New Production Systems, paper presented at the Xth ICPR, Nottingham, England.
Department of Industry, Committee for Materials Handling (1978) *Materials Handling – An Introduction*, HMSO.
Doumeingts, G., *et al.* (1987) Design Methodology for Advanced Manufacturing Systems, *Computers in Industry*, **9** (4) December.
Downs, E., *et al.* (1988) *Structured Systems Analysis and Design Method – Application and Context*, Prentice-Hall.
Gallager, C.C. and Knight, W.A. (1986) *Group Technology Production Methods in Manufacturing*, Ellis Horwood.
Gudnason, C. and Riis, J. (1984) Manufacturing Strategy, *International Journal of Management Science*, **12** (6).
Hall, R.W. (1987) *Attaining Manufacturing Excellence*, Dow Jones-Irwin.
Hendricks, J.A. (1989) Accounting for Automation, *Mechanical Engineering*, February.
Hill, T. (1985) *Manufacturing Strategy*, Macmillan Education.
Kusiak, A. (1990) *Intelligent Manufacturing Systems*, Prentice-Hall.
Marca, D.A. and McGowan, C.L. (1988) *SADT – Structured Analysis and Design Technique*, Prentice-Hall.
Peterson, J.L. (1981) *Petri Net Theory and the Modelling of Systems*, Prentice-Hall.
Slack, N.D.C. (1990) *Achieving a Manufacturing Advantage*, W.H. Allen.
Swann, K. (1988) Investment in AMT – a Wider Perspective, *Production Engineer*, September.

Exercise: Go Fast Cars Ltd – a mini project

1 Introduction

This project is designed to help bring the topics presented so far in this book together by applying a manufacturing system design framework in a practical environment. On the completion of the project, a report should be produced, presenting the final design and including detailed discussion of each of the major tasks completed.

Project objective
It is required to carry out a complete analysis to redesign the manufacturing system of Go Fast Cars Ltd, whose present company profile is provided in the next section. The new manufacturing system should be designed to:

- Satisfy the increased demand level.
- Be more flexible and efficient – by adopting advanced manufacturing facilities/organisation/operations.
- Be capable of producing a more diversified range of product models.

Project approach

It is advised that the structured design framework as presented in Chapter 7 should be followed, and the suggested tools such as $IDEF_0$, HOOMA, simulation and various algorithms be used to help the analyst carry out the tasks involved. In addition to this project briefing, relevant sections of Chapter 7 will need to be referred to frequently as the project progresses, to provide detailed discussion on the manufacturing design tasks within the design framework and to give technical information about the design tools. Reading materials from this chapter are specified for each of the tasks outlined below. In addition to the parts of text specifically highlighted, other chapters in the book, together with their lists of references, can be used as useful sources of information.

In addition, attempts should be made to find and utilise as much information as possible through for example, searching the literature and contacting equipment/system suppliers.

2 Profile of Go Fast Cars Ltd

The importance of effective approaches and techniques for the design and evaluation of modern manufacturing systems is reflected by some of the common problems faced by manufacturing industries, as discussed in Chapter 7: either pulled by market demands or pushed by new technology development, firms are having to restructure more frequently than ever before. Such is the new manufacturing environment in which Go Fast Cars Ltd has to operate. Although this manufacturing company's market is a growing one (as reflected by its production forecast for the next few years), it will have to introduce new products, and improve its operational efficiency through introduction of AMTs as well as by re-organisation. It is therefore necessary for the company to adopt a new manufacturing system in the near future.

Go Fast Cars Ltd is at present a small sized engineering company, which manufactures and markets a range of small toy cars. These are collectable items manufactured to a high engineering standard. A traditional, rather old-fashioned, functional layout exists in the factory. For example, all the lathes are located in one area, and all the mills in another. Batches of work are moved from one work centre to the next when the whole batch is complete. General purpose machines are available in each area and it can be assumed that any of the machines within a specific area can do a required job. Details of the factory, store locations, the present product range, components and assemblies and sales production forecasts can be found at the end of this briefing. (Assumptions may be made where sufficient information is not provided. These, however, must be spelt out clearly.)

3 Project procedure

By following the combined top-down/bottom-up approach, the current and the desired future positions of Go Fast Cars should first be analysed. The current position of the system should always be given first consideration. Analysis of the desired future position leads to the identification of a set of new objectives, and then a new system model can be developed which will fulfil these objectives. The structure of the design process is as summarized in Fig. 7.11. The following indicate the tasks which must be completed for this mini-project:

1. **Preparatory work.** Begin by studying sections 7.1 to 7.3 to develop a general understanding of the purposes of design methodologies and what is currently available. Other relevant material can be found in Chapters 1, 2 and 5.
2. **Analysis of situation.** In order to understand how the current system fails to achieve the current or future requirements, a preliminary investigation should be carried out to produce a *Symptoms List* containing all those items which are causing concern, and to identify the roots of the problems. For this purpose, refer to section 7.4 and apply the procedures/tools to analyse the manufacturing environment in which Go Fast Cars Ltd is operating. The material presented in Chapters 2 and 5 should also be useful.
3. **Setting objectives.** Having identified the problem areas through the previous analysis phase, and in relation to the general project objectives outlined above, realistic objectives with nominal and threshold levels should be set for Go Fast Cars' future position. The resultant statement of objectives will indicate the direction of development for the next design phases. Refer to section 7.5 for further guidance.
4. **Conceptual modelling.** Next, a conceptual model is needed to identify the building blocks of Go Fast Cars' future plant. A detailed discussion on this important design task is provided in section 7.6. Follow the steps outlined to complete the conceptual modelling of the system. All documentation should be gathered and sorted for future use. The material presented in Chapters 2, 3 and 4 should also be useful.
5. **Detailed design.** The completion of conceptual design should create a system model for the future Go Fast Cars in terms of interrelated manufacturing functions, together with a hierarchy of control systems which will process key information associated with the effective performance of these functions. This should next be transformed into detailed specifications which could be used later for implementation. The following are some of the major areas which are likely to require consideration:

   - Product design and process planning.
   - Equipment selection and justification.
   - Facility configuration (e.g. cellular formation).
   - Facility layout.
   - Material handling.
   - Production planning and control system.

   A detailed discussion on this design phase can be found in section 7.7. The result of this should be a design for the Go Fast Cars' new manufacturing system, which is accurate to a high level and detailed enough for actual implementation. However, this design must be assessed against the objectives which were originally set, to ensure that they are fulfilled and the investment involved is viable (Fig. 7.11). For this purpose the procedures and tools suggested in section 7.8 can be applied. The materials presented in Chapter 5 (structured decision-making) and Chapter 6 (computer modelling and simulation) should also be of use.

4 Information on Go Fast Cars Ltd

Factory details

Work centre number	*Description*	*Capacity (hrs per day)*	*Variable overhead rate (£ per hr)*
WC[R]100	Turning Cell	16	15.00
WC[R]200	Drilling Cell	8	10.00
WC[R]300	Grinding Cell	8	18.00
WC[R]400	Painting Booth	8	10.00
WC[R]500	Milling Cell	16	17.00
WC[R]600	Wheel/Axle Assy	4	8.00
WC[R]700	Plating Shop	24	10.00
WC[R]800	Final Assy Area	4	8.00
WC[R]900	Final Test Area	4	8.00

Normal working week: 5 shop days
Normal shop day: 8 hrs
Efficiency factor: 100%
Direct labour cost: £5.00 per hr

Store locations

Because Go Fast Cars Ltd try to keep their organisation simple, and do not at the moment intend to track WIP, they have adopted a policy of not storing stock in different bins within their stores. This makes monitoring stock simpler because it can only be stored in three locations. These are:

Store/Bin [STK][BIN]		*Description*	*Typical part no.*
F	1	Finished Goods Stores	5000–5999
R	1	Raw Material Stores	1000–1999
S	1	Sub-Assembly Stores	2000–2999

The present product range

Go Fast Cars make a range of three model cars. These are all based on the same chassis (Base Plate) but have different bodies. The components fit together as shown on these bills of materials.

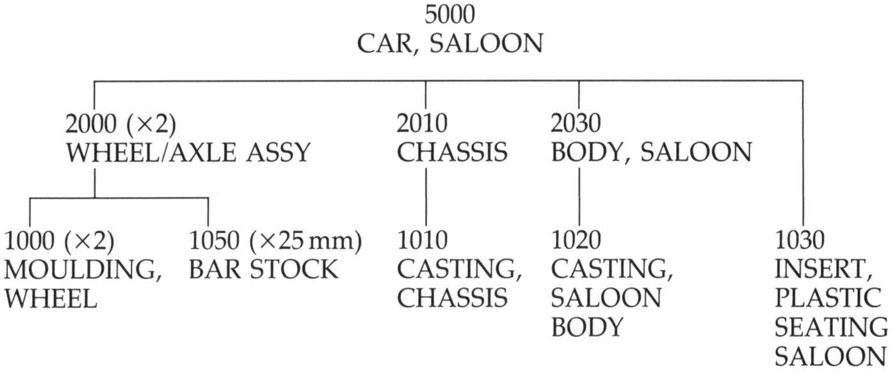

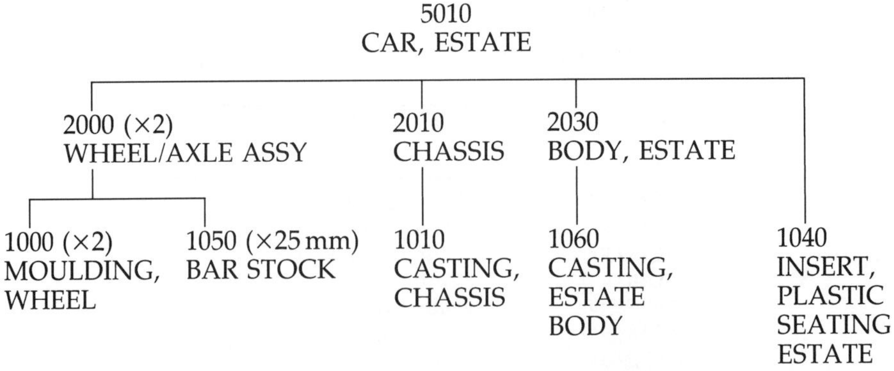

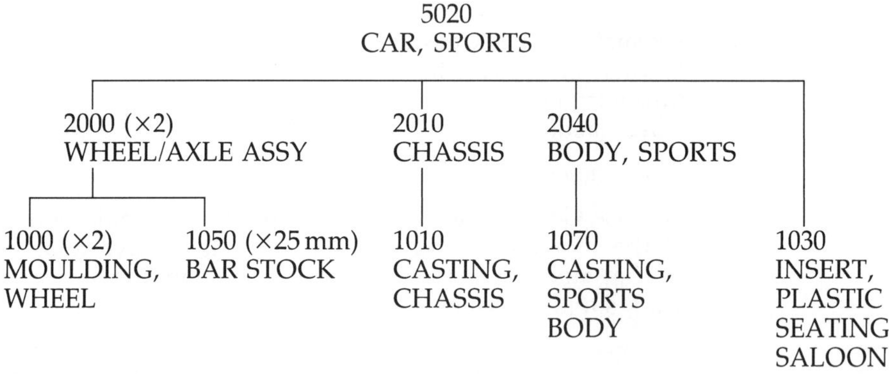

The components

1000 Moulding, Wheel

Purchased from Injection Moulders (I.M.) [Vendor Cd 002], at a price of £0.01 each including delivery. Produced by an injection moulding process, the minimum purchase quantity is 200, with increments above that of 50.

1010 Casting, Chassis

Purchased from Die Cast Injection Specialists (D.C.I.S.) [Vendor Cd 003], at a price of £0.05 each including delivery. Produced by a pressure die cast process, the minimum purchase quantity is 250, with increments above that of 50.

1020 Casting, Saloon Body

Purchased from Die Cast Injection Specialists (D.C.I.S.) [Vendor Cd 003], at a price of £0.05 each including delivery. Produced by a pressure die cast process, the minimum purchase quantity is 250, with increments above that of 50.

1030 Insert, Plastic Seating Saloon

Purchased from Injection Moulders (I.M.) [Vendor Cd 002], at a price of £0.02 each including delivery. Produced by an injection moulding process, the minimum purchase quantity is 200, with increments above that of 50.

1040 Insert, Plastic Seating Estate

Purchased from Injection Moulders (I.M.) [Vendor Cd 002], at a price of £0.02 each including delivery. Produced by an injection moulding process, the minimum purchase quantity is 200, with increments above that of 50.

1050 Bar Stock

Purchased from B.S.C. [Vendor Cd 001], at a price of £1.00 per metre including delivery. The minimum purchase quantity is 2 m, with increments above that of 1 m.

1060 Casting, Estate Body

Purchased from Die Cast Injection Specialists (D.C.I.S.) [Vendor Cd 003], at a price of £0.05 each including delivery. Produced by a pressure die cast process, the minimum purchase quantity is 250, with increments above that of 50.

1070 Casting, Sports Body

Purchased from Die Cast Injection Specialists (D.C.I.S.) [Vendor Cd 003], at a price of £0.05 each including delivery. Produced by a pressure die cast process, the minimum purchase quantity is 250, with increments above that of 50.

Purchase lead time is 10 days for all bought-in components.

2000 Wheel/Axle Assy

Manufactured by Go Fast Cars Ltd on the following routing (at present no set-up times have been included in the machine times provided. These are assumed to be included in the run times):

Cell name	*Op no.*	*W/C no.*	*Description*	*No. off hrs (qty)*
M/C	010	WC[R]100	Turning Cell	0.01
M/C	020	WC[R]200	Drilling Cell	0.01
M/C	030	WC[R]400	Painting Booth	0.05
M/C	040	WC[R]500	Milling Cell	0.005
M/C	050	WC[R]100	Turning Cell	0.02
M/C	060	WC[R]600	Wheel/Axle Assy	0.02

Average batch manufacturing lead time is 10 days.

2010 Chassis

Manufactured by Go Fast Cars Ltd on the following routing:

Cell name	*Op. no.*	*W/C no.*	*Description*	*No. off hrs (qty)*
M/C	010	WC[R]500	Milling Cell	0.02
M/C	020	WC[R]200	Drilling Cell	0.01
M/C	030	WC[R]300	Grinding Cell	0.01
M/C	040	WC[R]700	Plating Shop	0.25

Average batch manufacturing lead time is 10 days.

2020 Body, Saloon

Manufactured by Go Fast Cars Ltd on the following routing:

Cell name	*Op. no.*	*W/C no.*	*Description*	*No. off hrs (qty)*
M/C	010	WC[R]500	Milling Cell	0.01
M/C	020	WC[R]300	Grinding Cell	0.005
M/C	030	WC[R]400	Painting Booth	0.15

Average batch manufacturing lead time is 5 days.

2030 Body, Estate

Manufactured by Go Fast Cars Ltd on the following routing:

Cell name	*Op. no.*	*W/C no.*	*Description*	*No. off hrs (qty)*
M/C	010	WC[R]500	Milling Cell	0.01
M/C	020	WC[R]300	Grinding Cell	0.005
M/C	030	WC[R]400	Painting Booth	0.15

Average batch manufacturing lead time is 5 days.

2040 Body, Sports

Manufactured by Go Fast Cars Ltd on the following routing:

Cell name	*Op. no.*	*W/C no.*	*Description*	*No. off hrs (qty)*
M/C	010	WC[R]500	Milling Cell	0.01
M/C	020	WC[R]300	Grinding Cell	0.005
M/C	030	WC[R]400	Painting Booth	0.15

Average batch manufacturing lead time is 5 days.

5000 Car, Saloon

Assembled by Go Fast Cars Ltd on the following routing:

Cell name	*Op. no.*	*W/C no.*	*Description*	*No. off hrs (qty)*
ASSY	010	WC[R]800	Final Assy Area	0.05
ASSY	020	WC[R]900	Final Test Area	0.01

Average batch manufacturing lead time is 5 days.
Selling price: £20.00

5010 Car, Estate

Assembled by Go Fast Cars Ltd on the following routing:

Cell name	*Op. no.*	*W/C No.*	*Description*	*No. off hrs (qty)*
ASSY	010	WC[R]800	Final Assy Area	0.05
ASSY	020	WC[R]900	Final Test Area	0.01

Average batch manufacturing lead time is 5 days.
Selling price: £20.00

5020 Car, Sports

Assembled by Go Fast Cars Ltd on the following routing:

Cell name	*Op. no.*	*W/C no.*	*Description*	*No. off hrs (qty)*
ASSY	010	WC[R]800	Final Assy Area	0.05
ASSY	020	WC[R]900	Final Test Area	0.01

Average batch manufacturing lead time is 5 days.
Selling price: £20.00

Sales Forecast & Production Schedule

Month No.	*Date*	*Forecast no. off required*
1	1-5-94	700
2	1-6-94	750
3	1-7-94	800
4	1-8-94	800
5	1-9-94	800
6	1-10-94	900
7	1-11-94	950
8	1-12-94	1200

Forecast for 1995–97
Available at approx. 12 000 per annum.

Appendices

APPENDIX A Overview of traditional and current methods used for production planning and control of manufacturing systems

A.1 Overview of production planning and control

Production planning and control constitutes one of the main subject areas of the science of production management. The management of production involves many decision-making processes, whose outcomes determine the present and future operations of the manufacturing system. These decisions are made within a diverse range related to the operational activities of production. For OHMS production in particular, these are dealt with by function blocks A_{12}, A_{21}, A_{31} and A_{41}. The main concerns of this overview are the traditional and current methodologies used to aid the decision-making associated with the planning and control of production. These illustrate some of the most important operational aspects of manufacturing systems.

It has been emphasized several times in this book that because of inherent systems properties, the control functions within a system should be hierarchical in nature. Production planning and control functions are no exception. Traditionally, they have been hierarchically divided into three levels – the long, medium and short term – with each of the different management levels participating in these planning and controlling processes with different emphasis and scope (Fig. 3.27). The medium- and short-term levels of planning and control are of the greatest interest to the production management of a manufacturing organization, as they deal with the more immediate production activities of the system with a planning horizon usually between a few months and one year.

A.2 Traditional methodologies

A.2.1 Aggregate production planning

In our OHMS model, aggregate planning is the concern of function block A_{12}. The process of planning the quantity and timing of output from a manufacturing system over the medium term is usually achieved by adjusting the production rate, inventory level, and other controllable variables. Aggregate planning is a preliminary step to the more detailed planning of capacity and material requirements. The information inputs to this planning process include product demand forecasts, customer orders, inventory levels, and factory capacity levels. Since it is usually impossible to consider every detail associated with production over a long planning horizon, the information is often aggregated at this level of the planning process. This aggregation may take place by arranging similar items into product groups, different machines into machine centres, and different labour forces into labour pools. The planning process can be aided by the use of suitable analytical or computer simulation models. A wide variety of such planning models exist – linear decision rules, linear programming and integer programming models, fixed cost models, feedback-control models, queuing theory models, heuristic decision rules, search decision rules, and simulation models.

Linear programming (LP), for example, is a mathematical modelling approach which has found widespread use to aid decision-makers in resolving complex operational alternatives. It is applicable to the general category of problems that require the optimization of a linear effectiveness function subject to linear constraints. The theory and applications of LP are discussed in many technical papers and textbooks.

When applied to the problems of aggregate production planning, an LP model treats the problem as one of allocating plant capacity to meet the requirements of customer demand. A typical and simple LP model for this purpose would have a structure such as that shown in equation A.1, where it is desired to minimize

$$E = \sum_{i=1}^{N} \sum_{t=1}^{T} (\alpha_{i,t} P_{i,t} + \beta_{i,t} Q_{i,t}) + \sum_{t=1}^{T} (c_t L_t + d_t O_t) \tag{A.1}$$

subject to:

$$P_{i,t} + Q_{i,t-1} - Q_{i,t} = e_{i,t} \tag{A.2}$$

$$\sum_{i=1}^{N} p_i P_{i,t} - L_t - O_t = 0 \tag{A.3}$$

$$0 \leqslant L_t \leqslant (lt)_t \tag{A.4}$$

$$0 \leqslant O_t \leqslant (ot)_t \tag{A.5}$$

$$P_{i,t}, Q_{i,t} \geqslant 0 \tag{A.6}$$

$$t = 1, \ldots, T$$

$$i = 1, \ldots, N$$

where

- E is the total cost of production (£)
- $P_{i,t}$ is the quantity of product i to be produced in period t (units)
- $Q_{i,t}$ is the quantity of product i to be left over as inventory in period t (units)
- L_t is the labour used in period t (person-hours)
- O_t is the overtime labour used during period t (person-hours)
- $\alpha_{i,t}$ is the unit production cost (excluding labour) for product i in period t (£/unit)
- $\beta_{i,t}$ is the inventory carrying cost per unit of product i held in stock from period t to $t + 1$ (£/unit)
- c_t is the cost of labour in period t (£/person-hr)
- d_t is the cost of overtime labour in period t (£/person-hr)
- $e_{i,t}$ is the expected demand for product i in period t (units)
- p_i is the labour required to produce one unit of product i (person-hrs/unit)
- $(lt)_t$ is the total labour available in period t (person-hrs)
- $(ot)_t$ is the total overtime labour available in t (person-hrs)

T is the planning time horizon
N is the total number of products.

The objective function (equation A.1) expresses the minimization of the total cost function including the costs of production, inventory, and regular and overtime labour. Constraint (A.5) represents the typical production–inventory balance equation; (A.6) defines the total effort to be used in every period; (A.7) and (A.8) impose the lower and upper bounds on the use of regular and overtime effort in each time period; and (A.9) implies that no back-ordering is allowed. This simplified model is represented here only as an illustration of the application of LP in the area of production planning. Addition or relaxation of constraints may occur in real applications according to the particular system in question.

The main advantages of LP are that the models thus constructed are relatively simple to understand. Furthermore, computer software packages are readily available so that solutions of the models can be easily obtained. However, these methodologies have two serious disadvantages – their linear approximation and the deterministic assumption.

In the above example, the set-up costs of production are not taken into account in the planning objective functions. However, frequent set-ups are a special feature of batch and make-to-order types of production, therefore it may be necessary to include this cost element in the total cost functions of these types of production system. Since batch and make-to-order production account for more than 70% of total manufacturing output (O'Grady, 1986), the set-up problem deserves particular attention.

The inclusion of a set-up cost element in the planning process tends to disaggregate the problem, that is to say, more detailed information about the production process is needed, and the aggregation of items may take place only to a small extent. Furthermore, a more realistic planning model will also incorporate the workforce as a decision variable whose value may change when persons are hired or discharged during the planning horizon, as a way of dealing with fluctuations in demand. Therefore, a typical lot-size model with more comprehensive fixed-cost elements, variable workforce, and capacity constraints would typically take the form of equation A.7, where (using previous notation) the problem is to minimize

$$\begin{aligned} E = \sum_{i=1}^{N} \sum_{t=1}^{T} [s_{i,t}\delta(P_{i,t}) + \alpha_{i,t}P_{i,t} + \beta_{i,t}Q_{i,t}] \\ + \sum_{t=1}^{T} (c_t L_t + d_t O_t + h_t H_t + f_t F_t) \end{aligned} \tag{A.7}$$

subject to:

$$P_{i,t} + Q_{i,t-1} - Q_{i,t} = e_{i,t} \tag{A.8}$$

$$\sum_{i=1}^{N} [a_i\delta(P_{i,t}) + p_i P_{i,t}] - L_t - O_t \leqslant 0 \tag{A.9}$$

$$L_t - L_{t-1} - H_t + F_t = 0 \tag{A.10}$$

$$-qL_t + O_t \leqslant 0 \tag{A.11}$$

$$P_{i,t}, Q_{i,t} \geqslant 0 \tag{A.12}$$

$$L_t, O_t, H_t, F_t \geqslant 0 \tag{A.13}$$

where:

$$\delta(P_{i,t}) = \begin{cases} 0, \text{ if } P_{i,t} = 0 \\ 1, \text{ if } P_{i,t} > 0 \end{cases} \tag{A.14}$$

Besides the notation used in the above discussion of LP, the following additional quantities are used in this model:

H_t is the labour capacity hired in period t (person-hrs)
F_t is the labour capacity fired in period t (person-hrs)
h_t is the cost of hiring labour in period t (£/person-hr)
f_t is the cost of firing labour in period t (£/person-hr)
a_i is the labour needed to carry out a set-up for product i (person-hrs)
q is a constant
$s_{i,t}$ is the cost of a set-up for product i in period t (£).

The main differences between this model and the one previously presented are clear. Because of the set-up operation, a non-linear element $\delta(\ )$ is introduced into the model, and therefore fixed-cost components are now present in the objective function. Also, the workforce level is included here as a decision variable.

Since a lot-size model can more closely represent the real production processes than an LP model, in theory it should produce more accurate and more detailed solutions. However, a great deal of detailed information throughout the planning horizon is needed, making the process very costly. Also, exact-solution methods are computationally limited to relatively small-sized problems and thus have little practical value at present. Consequently, both the theory and applications of this type of aggregate planning model are still being pursued by researchers.

In a continuous feedback model, the information about the workforce, inventories and production within a manufacturing system are aggregated to form continuous information flows. Information and materials flow through different sectors of the model, with flow rates determined by managerial decision. Several approaches have been proposed to deal with aggregate production planning and to study the characteristics of various production systems in a continuous time horizon. Due to the nature of this analytical methodology, a high level of aggregation is usually required to model a production system. Cost-objective functions of production policies are not explicitly included in this approach and the models thus constructed do not lead in general to practical computational procedures. The approach is either analytical or simulational.

The most representative of these is the so-called system dynamics (SD) (see, for example, Legasto *et al.*, 1980; Richardson and Pugh, 1981; Coyle, 1982). Unlike models that produce optimal solutions under various assumptions, SD attempts to model the managerial behaviours of a particular system and produce models responsive to the complexities introduced by the specific decision environment. The models developed by this approach should be more representative of the particular system under study, but they do not guarantee to produce an optimum solution. However, the computer simulation approach enables the decision-maker to test various alternatives in an iterative fashion. The methodology involves the following steps: objective and boundary definition; analysis of present system; development of mathe-

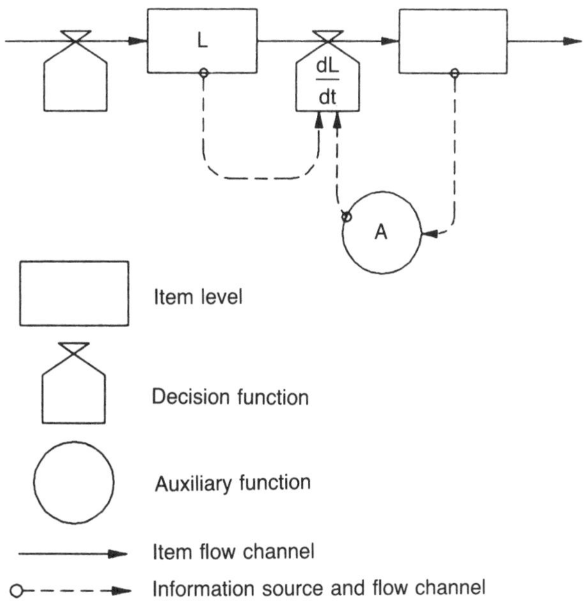

Fig. A.1 Basic elements of the SD model.

matical model; computer simulation and interpretation of simulation results; and system revision and repeated experimentation.

The basic elements of an SD model are shown in Fig. A.1. Their meanings and functions are as follows:

1. *Level*. The levels are the integration of flow within a system – for example, the parts waiting to be processed in a workshop or the number of workers hired. Levels could be the integration of any of the six types of flow in a production system – material, information, orders, money, personnel and equipment.
2. *Flow rate*. This defines the present flows between the levels in the system. The rates of flow are determined by the levels of the system according to rules of the decision functions. The rates in turn determine the magnitudes of the levels.
3. *Decision functions*. The decision functions are the statements of policy that determine how the available information about levels (the current states) leads to the decisions (the current rates). In real life this would reflect a manager's decision rules. That is, a manager would convert the available information into decisions according to these rules.

A simple example will illustrate the interrelationship of these basic SD components. To determine the number of workers required on a production line, W, a shop manager would need the following information: the number of units waiting to be processed, L (units); the productivity of workers, P (units/worker-week); and the required lead time, T (weeks). The manager's decision rule would be to adjust the output rate of the production line such that a desired production lead time is maintained. Therefore, the number of workers to be assigned to the line should be proportional to the workload L:

$$W = L/TP \tag{A.15}$$

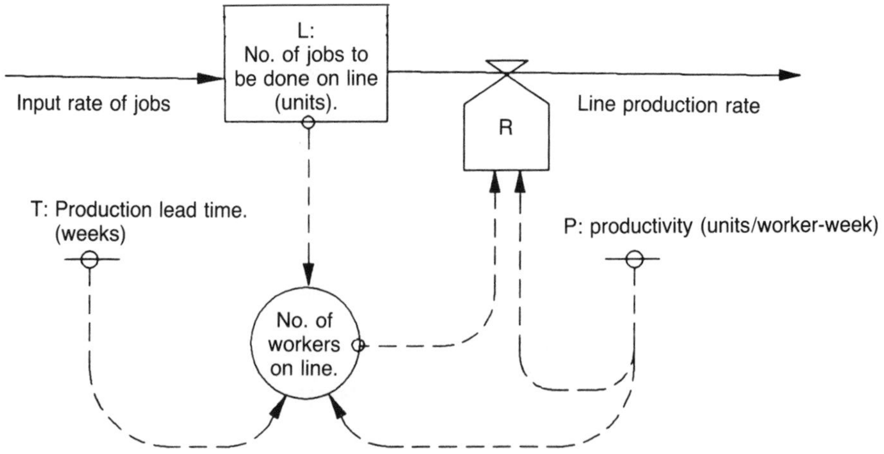

Fig. A.2 SD representation of the decision rule of a production line manager.

As the number of workers (a level) is adjusted according to this decision rule, the output rate of the production line (a flow rate) changes also:

$$\begin{aligned} \text{Output rate} &= \text{No. of workers on line} \times \text{Productivity} \\ &= [L/(TP)]P \qquad\qquad (A.16) \\ &= L/T \end{aligned}$$

This output rate in turn determines the levels of the waiting jobs and the finished jobs at the end of the evaluation period. An SD representation of this decision-making process is shown in Fig. A.2. An SD model is made up by many of these basic blocks consisting of these basic system components (level, rate, and decision function). One of the 'classical' examples of an industrial SD model has been already described in Chapter 2.

A.2.2 Short-term production planning

Job scheduling occupies the lowest level in the hierarchy of production planning and control. The output from the aggregate planning process – the master production schedule – is used here as the input to produce detailed workload schedules for each work centre or process. In our OHMS model, this task is shown in function block A_{41}. The task of production scheduling is to determine the sequence in which a set of jobs should be processed on one or more facilities under a set of capacity or other constraints.

A production scheduling problem can in general be described as $N/M/T/S$, where N is the number of jobs to be scheduled; M is the number of processors involved; T describes the type of the production shop; and S shows the type of measure to be used. Thus a problem in which N jobs are to be scheduled to minimize their mean flow time through a two-machine flowshop is denoted $N/2/F/F_M$.

The following notation is usually used in the study of scheduling problems:

- N is the number of jobs to be processed
- J is the number of processors in the shop
- T is the length of the planning horizon
- t is the present time
- J_i is job i, where $i = 1, 2, \ldots, N$

r_i	is the ready time of J_i
$p_{i,j}$	is the processing time of J_i on processor j, where $j = 1, 2, \ldots, J$
d_i	is the due date of J_i
a_i	is the allowance for J_i ($a_i = d_i - r_i$)
C_i	is the completion time of J_i
F_i	is the flow time of J_i ($F_i = C_i - r_i$)
$W_{i,j}$	is the waiting time preceding the jth operation of J_i
W_i	is the total waiting time of J_i ($W_i = \Sigma W_{i,j}$)
L_i	is the lateness of J_i ($L_i = C_i - d_i$)
T_i	is the tardiness of J_i ($T_i = \max(L_i, 0)$)
$N_w(t)$	is the number of jobs waiting at time t
$N_p(t)$	is the number of jobs being processed at time t
$N_c(t)$	is the number of jobs completed by time t
$N_u(t)$	is the number of jobs to be completed by time t

The measurement criteria used for studying scheduling problems are usually defined as follows:

Measurements related to completion times:
- maximum flow time : $F_M = \max \{F_i\}$
- maximum complete time: $C_M = \max \{C_i\}$
- mean flow time : $F_m = \Sigma F_i/N$
- mean complete time : $C_m = \Sigma C_i/N$

Measurements related to due dates:
- maximum lateness : $L_M = \max \{L_i\}$
- maximum tardiness : $T_M = \max \{T_i\}$
- mean lateness : $L_m = \Sigma L_i/N$
- mean tardiness : $T_m = \Sigma T_i/N$
- number of tardy jobs: $N_t = \Sigma\delta(T_i)$, where

$$\delta(T_i) = \begin{cases} 1, & \text{if } T_i > 0 \\ 0, & \text{if } T_i = 0 \end{cases}$$

Measurements related to utilization:
- mean number of jobs being processed: $N_{m,p} = [\Sigma N_p(t)]/T$
- maximum machine idle time : I_M
- mean machine idle time : I_m

Measurements related to WIP:
- mean number of jobs waiting : $N_{m,w} = [\Sigma N_w(t)]/T$
- mean number of jobs uncompleted: $N_{m,u} = [\Sigma N_u(t)]/T$
- mean number of jobs completed : $N_{m,c} = [\Sigma N_c(t)]/T$
- mean job waiting time : $W_m = [\Sigma W_i]/N$

Scheduling problems have been studied extensively for the $N/1//B$ model, that is, the single-processor environment. Extensive discussions on this type of problems can be found in several sources (for example, Conway *et al.*, 1967; Baker, 1974; and French, 1982). Some of the more important conclusions obtained for the static case may be summarized as follows:

1. Mean completion time C_m, mean flow time F_m, mean lateness L_m, mean number of unfinished jobs $N_{m,u}$, and mean job waiting time W_m are minimized by the *shortest imminent* processing-time rule (SI).
2. The maximum job lateness L_M and the maximum job tardiness T_M are minimized by the *earliest due-date* dispatching rule (EDD).
3. There are as yet no simple rules that minimize the mean tardiness T_m. However, T_m can be minimized by using *dynamic programming* and *branch and bound* procedures.

In a flow shop model, more than one processor is involved in the scheduling process and the problems become more complicated than the single-machine system. Although the flow-shop problem is the simplest scheduling problem in the general $N/M/A/B$ case, it has been shown to be a difficult problem to solve. At present, only $N/2/F/F_M$ problems may be solved by constructive algorithms.

Johnson's (1954) algorithm for scheduling the static $N/2/F/F_M$ is the earliest and best known in the research of flow-shop problems. The evaluation process of this algorithm may be expressed simply as follows:

Job J_i should precede Job J_{i+1} if:
$\min \{p_{i,1}, p_{i+1,2}\} < \min \{p_{i+1,1}, p_{i,2}\}$

It has been shown that this algorithm also generates optimal schedules for a special case of the $N/3/F/F_M$ problem under the condition that the second processor in the system never acts as a bottleneck. To put it mathematically, that is:

$$\min \{p_{i,1}\} \geqslant \max \{p_{i,2}\}$$
$$\min \{p_{i,3}\} \geqslant \max \{p_{i,2}\}$$

The $N/M/G/B$ problem is the most difficult scheduling problem. No exact specialized solution procedures are yet known. Although some models for deriving optimal plans for a general workshop do exist, the exceedingly demanding requirements on computational effort limit their usefulness in any practical situation. A model for this purpose is given below.

Since each processor in a workshop can process only one job at a time, the general situation may be symbolized by a series of sequence constraints:

$$\sum_{j=1}^{M} Q_{i,j,k}(S_{i,j} + p_{i,j}) \leqslant \sum_{j=1}^{M} Q_{i,j,k+1}S_{i,j} \tag{A.17}$$

$$S_{z,j} - S_{i,j} \geqslant p_{i,j} - (1 - Y_{i,j,z})(H + p_{i,j}) \tag{A.18}$$

$$S_{i,j} - S_{z,j} \geqslant p_{z,j} - (H + p_{z,j})Y_{i,j,z} \tag{A.19}$$

$$i, z = 1, 2, \ldots, N$$
$$j = 1, 2, \ldots, M$$
$$k = 1, 2, \ldots, M - 1$$

where $S_{i,j}$ is the starting time of J_i on processor j; the operation indicator, $Q_{i,j,k}$ is defined as:

$$Q_{i,j,k} = \begin{cases} 1 \text{ if processor } j \text{ is required for the } k\text{th operation of } J_i \\ 0 \text{ otherwise} \end{cases}$$

the sequence indicator, $Y_{i,j,z}$ is defined as:

$$Y_{i,j,z} = \begin{cases} 1 \text{ if } J_i \text{ precedes } J_z \text{ on processor } j \\ 0 \text{ otherwise} \end{cases}$$

and H is a constant.

Inequality (A.17) implies that the kth operation of a job must be completed before the $(k + 1)$th operation can start, while expressions (A.18) and (A.19) set the capacity (or timing) constraints for the system. With this set of constraints, different objective functions may be applied to study various problems. For example, for the $N/M/G/F_m$ model, the objective function would be given by:

$$\text{Minimize } E = \sum_{i=1}^{N}\sum_{j=1}^{M} (S_{i,j} + p_{i,j})Q_{i,j,M}$$

while the objective function for the $N/M/G/C_M$ problem may be written as:

$$\text{Minimize } C_M \geqslant \sum_{j=1}^{M} (S_{i,j} + p_{i,j})Q_{i,j,M}$$

$$i = 1, 2, \ldots, N$$

Unlike the constructive scheduling methods presented before, dynamic programming is an enumeration process, that is, it lists all possible combinations and then eliminates the non-optimal ones. However, the most widely used enumeration procedure in the area of production scheduling seems to be the branch and bound approach. The methodologies of the branch and bound and dynamic programming techniques are closely related – both of them rely on listing and eliminating processes. Like the dynamic programming approach, the application of branch and bound techniques to scheduling is also based upon a simple observation about the building of a permutation schedule – it has the structure similar to that of a tree. The structure starts from an initial node at which no job has yet been scheduled. Then additional jobs are attached to the sequence at each node until the whole set of jobs is scheduled. This approach has been used to solve many types of scheduling problem.

In addition to the above, heuristic dispatching rules select the next job to be processed from a set of jobs awaiting service on a facility in a production shop. Like the optimal solution algorithm, these rules are normally used in the attempt to minimize certain objective criteria. Dispatching rules can be very simple or extremely complex.

The operation-time-based rules involve the job operation times in one way or another and use them as the system parameter to determine the priorities for the associated jobs. The best-known rules in this category are listed below, where $A_{i,j}$ indicates the priority of job J_i at processor j, such that the job with the smallest A-value will possess the highest priority.

SI Shortest imminent operation – select the job with the shortest operation time, that is,

$$A_{i,j} = p_{i,j}$$

LI Longest imminent operation – select the job with the longest operation time, that is,

$$A_{i,j} = -p_{i,j}$$

FRO Fewest remaining operations, that is,

$$A_{i,j} = \sum_{i,k} Q_{i,j,k}$$

where $Q_{i,j,k}$ is an operation indicator, $j \in$ {all remaining operations of J_i}, $k \in$ {all processors in the system}.

MRO Most remaining operations, the converse of FRO, that is,

$$A_{i,j} = -\sum_{i,k} Q_{i,j,k}$$

SR Shortest remaining operation time – select first the job with the shortest remaining processing time, that is

$$A_{i,j} = \sum_j p_{i,j} \qquad j \in \{\text{all remaining operations of } J_i\}$$

LR Longest remaining operation time – the converse of above – select first the job with the longest remaining processing time, that is,

$$A_{i,j} = -\sum_j p_{i,j} \qquad j \in \{\text{all remaining operations of } J_i\}$$

STW Smallest total work – select the job that has the smallest total processing time, that is,

$$A_{i,j} = \sum_j p_{i,j} \qquad j \in \{\text{all operations of } J_i\}$$

LTW Largest total work – the converse of the above, select the job that has the largest total processing time, that is,

$$A_{i,j} = -\sum_j p_{i,j} \qquad j \in \{\text{all operations of } J_i\}$$

The SI has been considered the most promising rule in this group. The advantages of the SI rule were outlined at the beginning of this sub-section. It minimizes the make span and the mean lateness in the single-processor environment, exhibits the same advantages when used in the multiple-machine environment and is less sensitive to uncertainties in processing times.

The due-date-based rules use the due date as the primary variable to determine the priorities for the associated jobs. Examples of this type are given below:

EDD Earliest due date – select the job with the earliest due date, that is,

$$A_{i,j} = d_i$$

SL Least slack – select the job with the least free time available before the job due date, that is,

$$A_{i,j} = d_i - t - \sum_j p_{i,j}$$

where $j \in$ {all remaining operations of J_i}.

S/OP Job slack per operation – select the job with the least free time per operation before the job due date, that is,

$$A_{i,j} = \left(d_i - t - \sum_j p_{i,j}\right) \Big/ \sum_{i,k} Q_{i,j,k}$$

where $k \in$ {all processors in system}.

S/T Job slack ratio – select the job with the minimum ratio of remaining free time to total remaining time, that is,

$$A_{i,j} = \left(d_i - t - \sum_j p_{i,j}\right) \Big/ (d_i - t)$$

An extensive survey of static scheduling research involving due-date-based rules has been carried out by Sen and Gupta (1984). In general, these due-date-based rules can be expected to produce a smaller number of tardy jobs than operation-time-based rules. Furthermore, a smaller variance of job lateness is often reported. Among these rules, the S/OP rule has been found consistently to perform better than the others. However, unlike the widely accepted advantages of the SI rule, not all researchers have considered the S/OP to be the best of the due-date-based rules.

A.3 Integrated production planning and control

As can be seen from section A.2, considerable research effort has been devoted to the problems of aggregate production planning and shop production scheduling. However, it has now been recognized that the development of optimal aggregate production plans or shop schedules alone is not sufficient for solving real problems. Such plans must be integrated so that the aggregate plans may be disaggregated into specific schedules for specific products. Therefore, attempts have been made to relate production planning decisions at the aggregate and disaggregate levels (for example, Bergstrom and Smith, 1970; Zoller, 1971; Hax and Candea, 1984). At the same time, the development of computer-aided production management (CAPM) has permitted production managers to plan and control shop activities heuristically to strike a reasonable balance between capacity and demand at both levels. While the problem of disaggregation of optimally generated aggregate production plans remains a challenging research task, more and more firms are now introducing CAPM systems into their management practice. An example of each of the optimal and the heuristic approaches will be given in this section.

A.3.1 Hierarchical production planning system

The hierarchical production planning system (HPP) was first proposed by Hax and Meal (1975). This is an optimal approach which attempts to coordinate the overall decision-making process of production management. As the name suggests, the methodology adopts a hierarchical approach. Decisions of different hierarchical levels are made in sequence, with the aggregate decisions imposing constraints within which more detailed decisions are made. The overall production planning system is divided into sub-systems by defining the following three product levels: items, the final products to be delivered; families, groups of items with common set-up cost; and types, groups of families with similar production cost per unit and similar demand patterns.

Aggregated production planning models are first used to produce production plans at the highest level – the product-type level. Family disaggregation models are then used to yield production plans at the product-family level, with the condition that the sum of the productions of the families in a product type equals the amount of production determined by the plan at the product-type level. Similarly, specially designed algorithms are used to disaggregate the production plan once again to the product-item level.

Hax and Candea (1984) have described this methodology in detail. They carried out extensive computational experiments to assess the performance of a proposed two-stage hierarchical planning system based on this approach. The performance was compared with that yielded by an MRP model in terms of costs and back-orders. It is claimed by the authors that during the experiments their method in general appeared superior to the MRP approach of planning in all but the most unusual cases. However, different views exist. For instance, Lee *et al.* (1983) conducted a simulation experiment to compare the effectiveness of such an 'aggregate–disaggregate' planning approach and the MRP approach (the LDR model was used as the optimal aggregate technique in this study). It was reported that the results appeared to favour the MRP approach. Even the authors of the HPP system warned against definite conclusions on the relative merits of these two different approaches before further studies. They themselves do not consider that HPP should be viewed as an alternative to MRP, but suggested that some elements of the hierarchical framework could be constructively used to enhance the MRP system.

A.3.2 MRP and CRP

The output from the aggregate planning function is usually an optimal or feasible workload of the product which can be further broken down into its dependent demand items. Material requirements planning (MRP) is a systematic method for determining the quantity and timing of dependent-demand items (Orlicky, 1975; Corke, 1985). Since the introduction of the concept some 20 years ago, the development and implementation of MRP have both made substantial progress. However, since it has become evident that MRP alone is not sufficient to cope with balancing plant capacity with demand adequately, further development has produced the so-called manufacturing resources planning system (MRP II), which consists of both the material requirements planning and the capacity requirements planning functions. Capacity requirements planning (CRP) is the function of determining what capacity resources are needed to carry out the production plan. The MRP and CRP together are used to determine which, when and what quantity of materials and capacities are needed over a finite planning horizon. The functions of MRP and CRP and the relationship between them are shown in Fig. 3.27. MRP is essentially a computer-based information system and a simulation tool. Unlike the optimal-searching 'aggregate–disaggregate' approach, it does not deal directly with the optimization criteria associated with various production issues. The major features of MRP systems are as follows: generation of lower-level requirements for dependent items; time-phasing of the items; planning order releases for the items; scheduling/rescheduling production. The logic used in these MRP systems is quite straightforward. The inputs to MRP systems are the information of end-product (independent items) requirements from the master production schedule and the inventory information from the status files. The net quantities of the dependent items (materials, components and sub-assemblies) required are calculated and 'time-phased' in a level-by-level manner, so that orders are released for these items to ensure that they arrive when needed. The time-phasing process is done using the following logic:

$$RN_{i,t} = RP_{i,t} - (IH_{i,t} + IP_{i,t})$$

where $RN_{i,t}$ is the net requirement of item i in period t; $RP_{i,t}$ is the projected requirement of item i in period t; $IH_{i,t}$ is the on-hand inventory of item i in period t; and $IP_{i,t}$ is the planned receipts of item i in period t.

MRP systems on their own assume that capacity is available when needed unless told otherwise. CRP, on the other hand, takes material requirements from the MRP system (the horizontal 'yes' path from the MRP block in Fig. 3.27) and converts them into standard hours of load of labour and machine on the various work-stations. By utilizing the management information system, CRP attempts to develop loading plans that are in good balance with plant capacities. Thus CRP is an iterative process that first simulates loads on the work-stations and then feeds back the necessary information to suggest changes if the plans are not feasible. Production orders are issued to workshops only when the material and capacity requirements are both satisfied, as illustrated by the vertical 'yes' path from the MRP block in Fig. 3.27.

It is interesting to compare the different thinking behind JIT and MRP. MRP is a formal, expensive, complicated but elegant solution which is focused on solving a particular set of problems (though this may require a company-wide involvement) while JIT is a less formal, simple, yet flexible and efficient approach which involves not only company-wide involvement, but also company-wide considerations. In the 1970s the MRP approach to manufacturing planning and control was considered a universal answer to all manufacturing planning and control problems in the United States and the Western European countries. Unfortunately, reality has proved otherwise – as many stories of failure have been reported as of success, and people have gradually come to realize just how difficult it is to implement an MRP system successfully. This may be explained through a systems perspective. For example, in a complex MRP system, too many channels are sometimes required to ensure efficient communications between the central control and the operational units. These channels are inevitably subjected to failure in practice. When some of them do fail, efficient control becomes impossible. For example, the author of this book recently heard a story of how a shop foreman had thrown a works schedule generated by the central control computer in the bin, because the machines listed on the schedule no longer existed in the shop! Rather than relying on only a handful of planners to maintain data and respond to action messages, the kanban system motivates the line people who will participate in the process of achieving the overall corporate goal by determining themselves whatever corrective action is needed.

Does this mean that the so-far seemingly more successful JIT approach is a better solution to manufacturing planning and control problems than MRP? The answer is almost certainly 'not necessarily'. It has been concluded by researchers that the significant efficiency improvements made by JIT were not due to the kanban method itself, but rather to the complete manufacturing environment – that is, its systems approach. If used properly, an MRP system is equally capable of producing spectacular results. The important thing is not the tool itself, but the way it is used. Finally, it should be pointed out that the kanban approach is more suited to the situation where the production involves high-volume and low-variety products, or the system structure has been simplified through 'cellularization'.

A.3.3 Optimized production technology (OPT)

OPT is a production control system which was originally developed by Dr E. Goldratt. It may be viewed as a mixture of MRP II and JIT, with the potential combination of the benefits of both systems. Like MRP, OPT uses a sophisticated, computer-integrated formal scheduling system to control production. On the other hand, the OPT method has several similarities to the kanban approach – for example, like JIT it focuses on some of the important organizational (systems) aspects of production with a coherent corporate goal, that is, to improve profits both now and in the future. To achieve its goal, the OPT approach attempts to reduce throughput time, inventory level, and operating expenses.

OPT is based on the idea that the bottlenecks which occur in a production environment are the 'gating factors' which limit the efficiency of the line. Therefore, maufacturing resources should be directed to them to eliminate their effects. OPT is based on nine rules, each dealing with a specific aspect of production at a bottleneck:

1. Balance flow, not capacity.
2. The level of use of a non-bottleneck is determined not by its own potential but by another constraint within the system.
3. Utilization and activation of a resource are not synonymous.
4. The time lost at a bottleneck is time lost for the whole system.
5. The time saved at a non-bottleneck is just a mirage.
6. Bottlenecks govern throughput and inventories.
7. The transfer batch can be less than the process batch.
8. The process batch should be variable, not fixed.
9. Schedules should be established by considering all system constraints simultaneously. Lead times are the result of a schedule and cannot be predetermined.

Therefore, OPT attempts to overcome the effect of bottlenecks by any means available. These include forward and backward scheduling around the bottlenecks, using increased capacity, and obtaining better efficiency at the gating factor. Based on a network model of production which defines how a product is manufactured, including the flow paths of materials and resource constraints, OPT utilizes the production requirements to perform the following tasks: to identify the critical or bottleneck resources; to use a specially developed finite scheduling algorithm to produce an optimized schedule for these resources; and to use this optimized, critical resource plan as the input to schedule non-critical resources, in such a way that a reserve of excess capacity exists at all times.

The tools provided by OPT are also flexible in the sense that the specific objectives used for optimizing the scheduling may be tailored by the manufacturing management, so that the schedules generated by the system support a coherent cooperative strategy. For example, if maintaining good delivery performance is regarded as the primary corporate target, then OPT can be set up to favour meeting product due dates (perhaps at some cost to other performance levels). If the company strategy is to market the cheapest product, adjustments can be made in a similar way by setting the relevant management parameters in OPT so as to generate a schedule which minimizes production costs.

As well as for manufacturing planning and control, OPT can also be used as an analytical technique for simulating and optimizing production operations. This ability of the system can be utilized to solve system design problems by asking 'what-if' questions, much as in a computer simulation investigation.

There are several difficulties associated with its implementation. Like any other formal manufacturing planning and control approach, OPT specifies a set of preconditions which must be met to ensure its successful operation, including an understanding of basic finite scheduling concepts, sound basic systems, adequate education, management commitment, and the willingness and ability to unlearn some ingrained habits.

A particular issue that makes OPT difficult to implement is its proprietary and hence unpublished scheduling algorithm. As a result, the system is not 'transparent' to its user. Since the basis for the schedules produced is not clear to the people who are responsible for their execution, difficulties in implementation are bound to arise.

APPENDIX B

Sample company document defining the system requirements for the control of supplies and services

B.1 Introduction

This document sets out to inform and establish the working procedure necessary to achieve the company's objective of supplier Quality Assurance.

Quality Assurance has been defined by the British Standards Institute (BSI) as 'all activities and functions concerned with quality'. An important aspect of Quality Assurance is that quality is best controlled by those responsible for manufacture. This company, therefore, is seeking to apply this fundamental principle in order to make certain that the quality of the product delivered to us is being maintained by vendors.

B.2 Objective

The objective of this document is to help our vendors establish operational procedures of Quality Assurance, so as to maintain a production quality which is acceptable to this company.

Although the control of bought-out items and services to the specified requirements is the responsibility of this company, the vendors are to be encouraged to fulfil their commitment to produce items, or provide services, in line with the technical specification. The appropriate department of this company will assist the vendors in understanding these specifications, and providing assistance where possible.

B.3 Requirement of quality system

The vendors should establish, document and maintain an effective system to ensure and prove that product quality conforms to the specified requirements. The following areas should be included and documented in such a quality system:

System Structure,
System Monitoring,
Documentation Maintenance,
Measuring and Test Equipment Maintenance,
Production Control,
Inspection Control,
Supplier Quality Assurance,
Product Quality Preservation,

Corrective Action, and
Training.

B.3.1 System structure

A management representative should be appointed who will be responsible for the efficient operation of the quality system. The representative should possess both the necessary authority and relevant technical knowledge to carry our his/her duties, and have access to top management. In addition, all personnel whose activities affect quality should be identified and clearly documented.

B.3.2 System monitoring

The operation of this quality system should be continuously monitored so as to ensure its effectiveness. Reviews for this purpose should be documented. The documentation should be made available for this company if required.

B.3.3 Documentation maintenance

All documentation that relates to the vendor's quality system, together with all documents issued to the vendor by this company regarding specification, instruction and modification, etc., should be properly maintained and controlled to ensure that:

- all changes to quality documents are recorded; and
- all quality documents are promptly updated.

B.3.4 Measuring and test equipment maintenance

The vendor should ensure that its measuring equipments are accurately calibrated and measurement standards properly controlled. These should be accomplished in accordance with British National Standards. Documents proving that this has been achieved should be maintained and made available if required.

B.3.5 Production control

The vendor should ensure that its production operations are conducted under controlled conditions, that is, that:

- production method and sequence are properly defined and documented;
- batches are clearly identified and recorded; and
- inspection control is properly executed and documented.

B.3.6 Inspection control

3.6.1 The vendor should carry out and document all inspection and tests on both materials, parts or sub-assemblies, and the finished product batch or service, necessary to provide complete evidence of full conformance to specified requirements.

3.6.2 Sampling procedures used by the vendor are to be in line with BS 6001, General Inspection Level 2, single-sample plan, but with no allowable rejects permitted within the sample. Should a reject be found during sampling, revert to 100% sorting for defective features.

3.6.3 A system for identifying the inspection status of work in progress should be established and maintained, which at all times makes known whether a work:
- has not been inspected;
- has been inspected and passed; or
- has been inspected and rejected.

B.3.7 Supplier quality assurance

The vendor should establish and maintain procedures for controlling supplies and services from its suppliers.

3.7.1 Suppliers should be provided with a clear and specific definition of quality requirements.

3.7.2 No procured supplies and services are to be used or processed until they have been inspected or tested, and approved as conforming to specified requirements.

B.3.8 Product quality preservation

The vendor should establish and maintain a system to control the preservation of product quality. For this purpose, the vendor should establish and maintain:

3.8.1 an effective material-handling system which prevents damage and deterioration of work through all phases of production and transportation until accepted by this company;

3.8.2 secure and adequately protected storage areas where necessary, which are effectively controlled by established operation procedures; and

3.8.3 packing standards to ensure the safe arrival and ready identification of the delivery.

B.3.9 Corrective action

Documented formal procedures should be established to guarantee prompt and effective corrective action when:

- unacceptable supplies are received;
- defective work operations occur during production; and
- items delivered to this company are rejected.

3.9.1 Procedures for controlling non-conforming supplies should be established and maintained. All reject supplies should be promptly removed from production.

3.9.2 For items which do not conform to quality requirements, a concession can only be obtained with the formal approval of this company.

B.3.10 Training

The vendor should ensure that all personnel involved with work for this company have the appropriate experience and training.

APPENDIX C The PCModel instruction set

PCModel uses routing instructions to define the flow of objects through the simulation. Routings can be defined through the use of the routes or links as follows:

BR and ER respectively mark the beginning and the end of a route. All other instructions in this route must be included between this pair of commands. A route usually describes the entire experience of an object in the system, that is, it directs the object concerned through the system from the time it enters the simulation until the time it exits.

BL and EL mark the beginning and the end of a link, which is PCModel's equivalent to the sub-routines of a general-purpose programming language such as BASIC or FORTRAN.

All PCModel instructions that are used to build routes and links can be classified into five categories according to their usage, i.e.:

object movement,
routing control,
arithmetic operation,
data input/output, and
execution control.

Each of these will be briefly described below.

C.1 Object movement

This group of instructions, when included in a route or a link, drive objects moving around in a defined manner on the computer screen. The following are examples of these instructions:

MR(#no,time)	Move to right #no steps with a delay of time between each move
ML(#no,time)	Move to left #no steps with a delay of time between each move
MU(#no,time)	Move to up #no steps with a delay of time between each move
MD(#no,time)	Move to down #no steps with a delay of time between each move
MA(*xy,time)	Move to location *xy with a delay of time

C.2 Routing control

This group includes the instructions used for the logical control of routing sequences. Many of the instructions here are similar to the control statements used in general programming languages. For example, IF and JUMPs in PCModel are similar to IF and GOTO in BASIC, and are used in the same manner. The instructions in this group are as follows:

IF IF (true jump 1 else jump 2)
JP Jump to a routing label
JB Jump if a position is blocked
JC Jump if a position is cleared
PO Post a location (as busy)
CL Clear a location
TP Test a position
SE Simulate an event for other objectives
SC Set background/foreground colour

C.3 Arithmetic operation

This group of instructions is used for numerical manipulation. They set or calculate the values of variables. Examples are:

SV Set value of a variable
AO Arithmetic operation (+, −, *, /)
DV Decrement value of a variable
IV Increment value of a variable
RV Generate a random value

C.4 Data input/output

This group includes instructions used for data input/output between the simulation model and screen or disk files. Again, many of the instructions here are similar to the data input/output statements used in general-purpose languages. For example, the OF and CF instructions in PCModel are very similar to those of BASIC and FORTRAN and are again used in the same manner.

In addition, a group of instructions is provided to control the execution of the simulation.

APPENDIX

D Sample E/R diagrams of manufacturing data model

In relation to the information model developed in section 7.7.4, a relational data model to represent the product of a manufacturing operation, as specified in Fig. 7.36 (D2: Product), is given here using the graphical presentation of E/R diagrams (one-to-one relationships are specified without indicator in these diagrams in order to make them tidier).

D.1 Product file

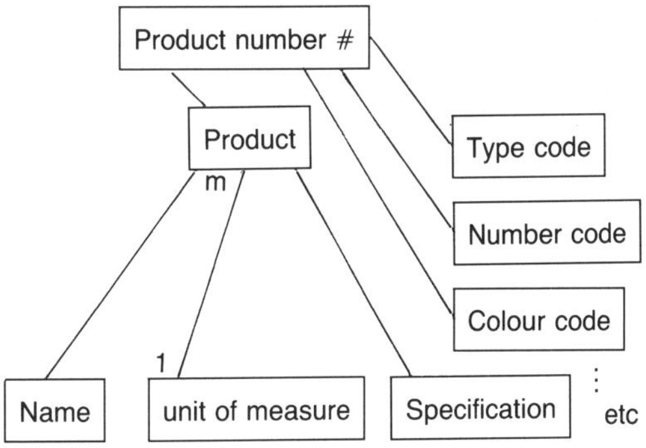

Fig. D.1 E/R diagram of product.

*Prod (**prod. no.#**, prod. name, unit of measure, specification)*

D.2 Product supply

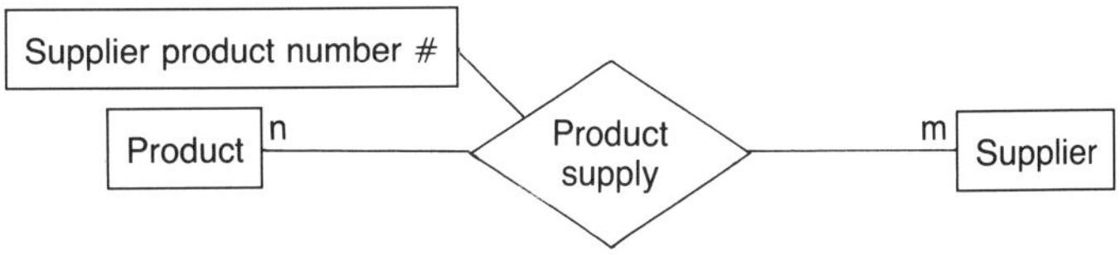

Fig. D.2 E/R diagram of product supply.

*Prod Supply (**prod. no.#**, **supplier no.#**, supplier prod. no.#)*

D.3 Product purchase order

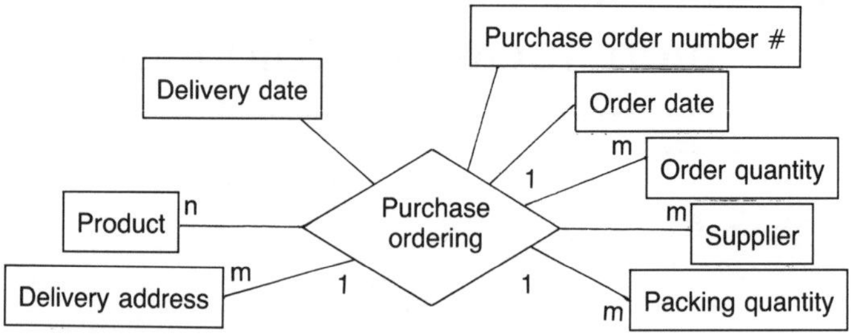

Fig. D.3 E/R diagram of product purchase order.

Prod Purchase Order	*(**purchase order no.#**, supplier no.#, order date, delivery date)*
Prod Purchase Order Line	*(**purchase order number#**, **prod. no.#**, order quantity, packing quantity, delivery address)*

D.4 Product sales order

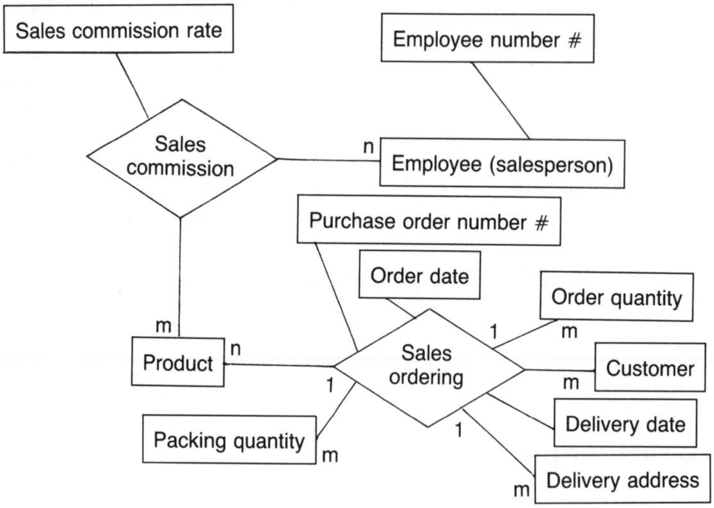

Fig. D.4 E/R diagram of product sales order.

Prod Sales Order	*(**sales order no.#**, customer no.#, order date, delivery date, sales employee no.#)*
Prod Sales Order Line	*(**sales order no.#**, **prod. no.#**, order quantity, packing quantity, delivery address)*
Prod Sales Employee Commission	*(**prod. no.#**, **sales employee no.#**, commission rate)*

D.5 Product pricing

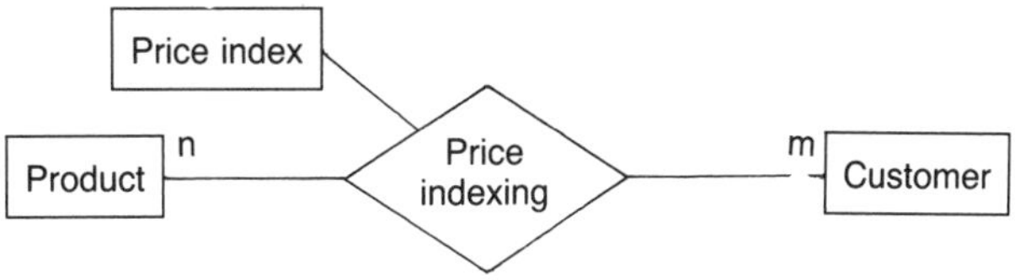

Fig. D.5 E/R diagram of product pricing.

Prod Unit Price ***(prod. no.#, customer no.#****, price index)*

D.6 Product quotation

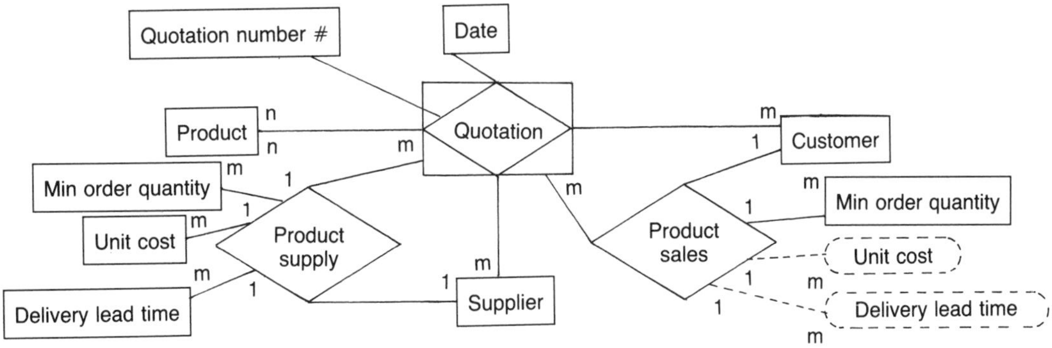

Fig. D.6 E/R diagram of product quotation.

Supplier Prod Quotation	***(quotation no.#, supplier no.#****, date)*
Supplier Prod Quotation Line	***(quotation no.#, supplier no.#, prod. no.#****, minimum order quantity, unit cost, delivery lead time)*
Customer Prod Quotation	***(quotation no.#****, customer no.#, date)*
Customer Prod Quotation Line	***(quotation no.#, prod. no.#****, minimum order quantity, unit price, delivery lead time)*

Again, due to the risk of duplicating supplier numbers, the supplier number is used here as a secondary identifier.

D.7 Product delivery

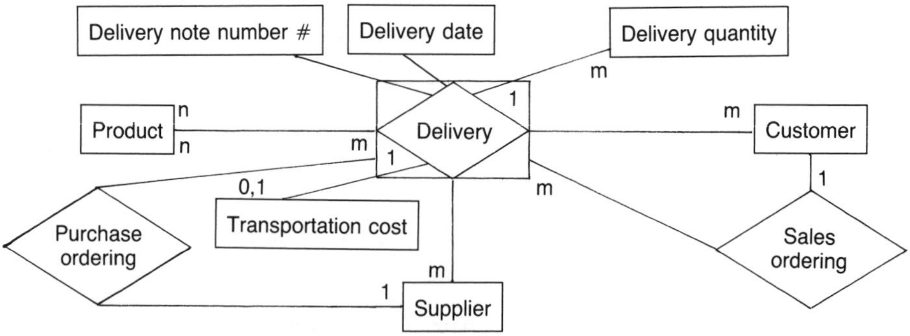

Fig. D.7 E/R diagram of product delivery.

Supplier Prod Delivery	*(**delivery note no.#**, **supplier no.#**, delivery date, transportation cost)*
Supplier Prod Delivery Line	*(**delivery note no.#**, **supplier no.#**, **supplier no.#**, delivery quantity)*
Prod Purchase Delivery	*(**delivery note no.#**, **supplier no.#**, purchase order no.#)*
Customer Prod Delivery	*(**delivery note no.#**, customer no.#, delivery date, transportation cost)*
Customer Prod Delivery Line	*(**delivery note no.#**, **prod. no.#**, delivery quantity)*
Prod Sales Delivery	*(**delivery note no.#**, sales order no.#)*

D.8 Product quality inspection

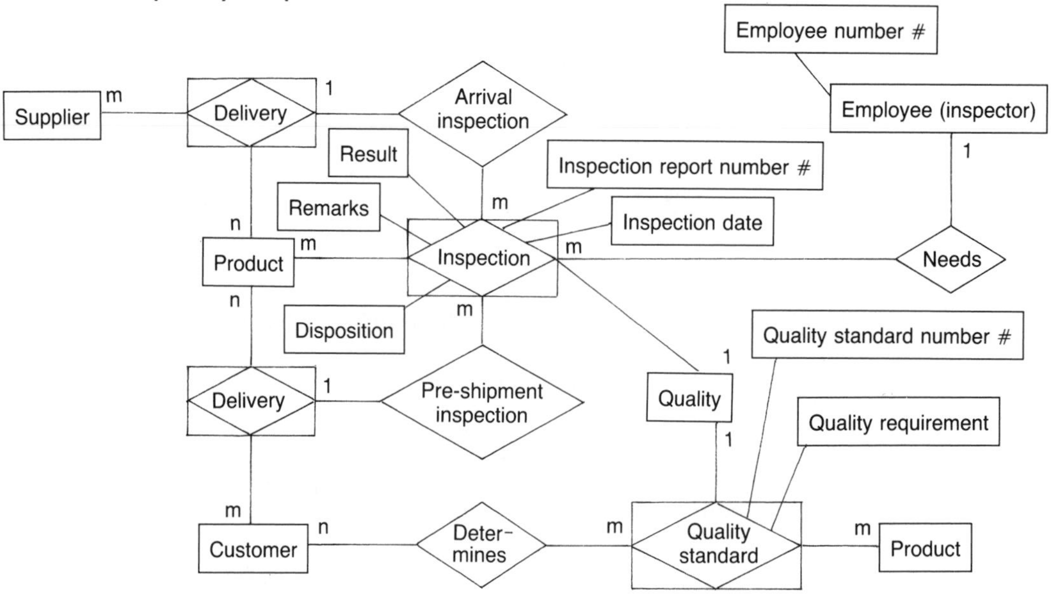

Fig. D.8 E/R diagram of product quality inspection.

Prod Quality Standard	*(**quality standard no.#**, quality requirement)*
Customer Prod Quality Requirement	*(**prod. no.#**, **customer no.#**, quality standard no.#)*
Prod Pre-shipment Inspection	*(**inspection report no.#**, delivery note no.#)*
Prod Pre-shipment Inspection Line	*(**inspection report no.#**, prod. no.#, inspection date, result, disposition, remarks, inspector's employee no.#)*
Supplier Prod Delivery Inspection	*(**inspection report number**, supplier no.#, delivery note no.#)*
Supplier Prod Delivery Inspection Line	*(**inspection report no.#**, prod. no.#, inspection date, result, disposition, remarks, inspector's employee no.#)*

D.9 Product field service

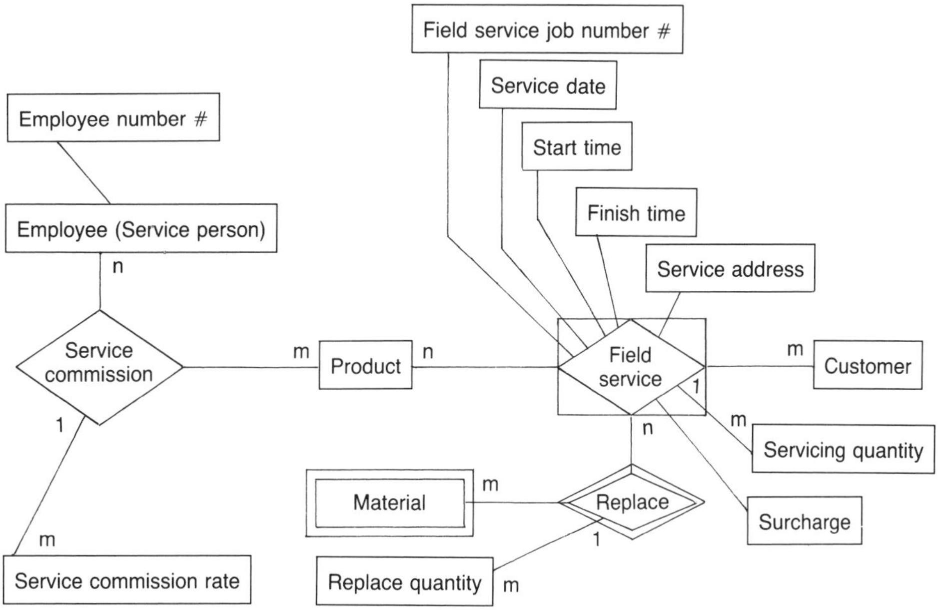

Fig. D.9 E/R diagram of product field service.

Prod Field Service	*(**field service job no.#**, customer no.#, service date, start time, finish time, service address, sub-charge, service employee no.#)*
Prod Field Service Line	*(**field service job no.#**, **prod. no.#**, servicing quantity)*
Field Service Material Replacement	*(**field service job no.#**, **part. no.#**, replace quantity)*
Prod Service Employee Commission	*(**prod. no.#**, **service employee no.#**, commission rate)*

D.10 Product design

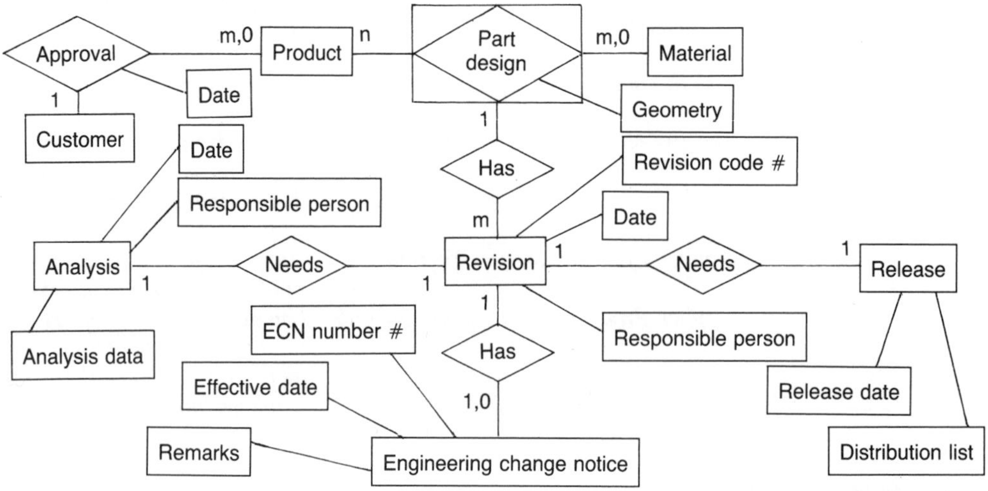

Fig. D.10 E/R diagram of product design.

Prod Design	*(**part. no.#**, revision code #, geometry)*
Prod Customer Approval	*(**customer no.#**, **prod. no.#**, approval date)*
Prod Design Revision	*(**part. no.#**, revision code #, revision date, revision description, responsible person employee no.#)*
Prod Design Analysis	*(**part. no.#**, revision code #, analysis date, analysis data, responsible person employee no.#)*

D.11 Product material

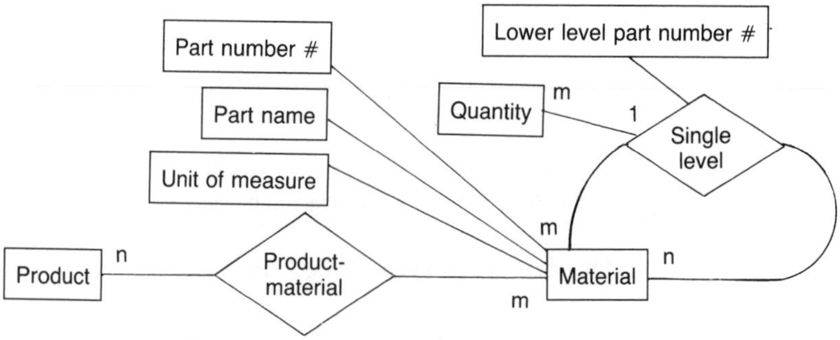

Fig. D.11 E/R diagram of product material.

Material	*(**part. no.#**, part name, unit of measure)*
Prod-Material	*(**part. no.#**, part. no.#)*
Single Level Material Where Use	*(**product no.#** or **part. no.#**, lower level part. no.#, quantity)*

References

Chapter 1

Bedworth, D.D. and Bailey, J.E. (1987) *Integrated Production Control Systems* (2nd edn), John Wiley.

Don Swann, P.E. (1984) Execution is the Key to Success of any System for Manufacturing Material Flow Control, *Industrial Engineer*, October.

Ettlie, J.E. (1988) *Taking Charge of Manufacturing*, Jossey-Bass.

Goddard, W. (1982) Kanban versus MRP II – Which is Best for You? *Modern Materials Handling*, November.

Hall, R.W. (1987) *Attaining Manufacturing Excellence*, Dow Jones-Irwin.

Harrington, J. Jr (1984) *Understanding the Manufacturing Process*, Marcel Dekker.

Hitomi, K. (1979) *Manufacturing Systems Engineering*, Taylor & Francis, London.

Huang, P.Y., Rees, L.P. and Taylor, B.W. (1985) Integrating the MRP-Based Control Level and the Multi-Stage Shop Level of a Manufacturing System via Network Simulation, *International Journal of Production Research*, **23** (6), 1217–31.

Kaldor, N. (1966) Inaugural lecture, in *Manufacturing Systems* (ed. V. Bignell *et al.*) p. 51, Basil Blackwell.

Lupton, T. (1986) The management of change to advanced manufacturing systems, in *Human Factors*.

Miller, J.G. (1981) Fit Production Systems to the Task, *Harvard Business Review*, **59** (1), 145–54.

Miller, J.G., *et al.* (1981) Production/Operations Management Agenda for the 80s, *Decision Sciences*, **12**, 547.

Morecroft, J.D.W. (1983) A Systems Respective on Material Requirements Planning, *Decision Sciences*, **14**, 1.

Parnaby, J. (1979) Concept of a manufacturing system, *International Journal of Production Research*, **17**, 123–35.

Parnaby, J. (1981) Concept of a Manufacturing System, in *Systems Behaviour* (3rd edn), The Open University Press.

Parnaby, J. and Donnovan, J.R. (1987) Education and Training in Manufacturing Systems Engineering, *IEE Proceedings*, **134** (10).

Seitz, D. and Volkholz, V. (1984) Perspective and Social Implications of Assembly Automation in West Germany – Some Results of Representative Inquiry, *Proceedings of the 5th ICAA*, 333–42.

Shaw, R.J. and Regentz, M.O. (1980) How to Prepare Users for a New System, *Management Focus*, **27** (33).

Sirny, R.F. (1977) The Job of the P and IC Manager – 1975, *Production and Inventory Management*, **18**, 100.

Skinner, W. (1978) *Manufacturing in the Corporate Strategy*, John Wiley.

Smith, B.M.D. and Rogers, N.M. (1985) The Demise of the British Motorcycle Industry, in *Manufacturing Systems* (ed. V. Bignel *et al.*) Basil Blackwell.

Smith, C.A., Halliday, J.S. and Hall, D.J. (1985) Analysis of Order-Handling Systems as a Means of Determining Influences on Throughput Times and Productivity, *International Journal of Production Research*, **23** (5), 867–78.

Towill, D.R. (1982) Dynamic Analysis of an Inventory and Order Based Production Control System, *International Journal of Production Research*, **20** (6), 671–87.

Towill, D.R. (1984) Information Technology in Engineering Production and Production Management, in *The Management Implications of New Information Technology* (ed. N. Piercy) pp. 15–47, Croom Helm.
Voss, C.A. (ed.) (1983) *Current Research in Production/Operations Management*, London Business School.
Voss, C.A. (1984) Production/Operations Management – a Key Discipline and Area for Research, *OMEGA*, **12** (3), 309–19.
Wu, B. and Price, D.H.R. (1989) Advanced Manufacturing Technology and Production Planning and Control: A Case Study, *Proceedings of the 3rd European Simulation Congress*, Edinburgh, pp. 537–9.

Chapter 2

Checkland, P.B. (1981) Science and the Systems Movement, in *Systems Behaviour* (3rd edn), Harper & Row.
Forrester, J.W. (1961) *Industrial Dynamics*, The MIT Press.
Magee, J.F. (1958) *Production Planning and Inventory Control*, McGraw-Hill.
O'Grady, P. (1986) *Controlling Automated Manufacturing Systems*, Kogan Page.
Richards, R.J. (1979) *An Introduction to Dynamics and Control*, Longman.
Richardson, G.P. and Pugh, A.L. III (1981) *Introduction to System Dynamics Modelling with DYNAMO*, The MIT Press.
Sandquist, G.M. (1985) *Introduction to System Science*, Prentice-Hall.
Williams, D.J. (1988) *Manufacturing Systems – An Introduction to the Technologies* Chapman & Hall.

Chapter 3

Barber, K.D. and Hollier, R.H. (1986) The Effects of Computer-Aided Production-Control Systems on Defined Company Types, *International Journal of Production Research*, **24** (2), 311–27.
Choobineh, F. (1988) A Framework for the Design of Cellular Manufacturing Systems, *International Journal of Production Research*, **26** (7), 1161–72.
Electrical Review (1990) Putting a human face on CIM, February.
Forrester, J.W. (1975) *Collected Papers of Jay W. Forrester*, The MIT Press.
Groover, M.P. (1987) *Automation, Production Systems, and Computer Integrated Manufacturing*, Prentice Hall.
Henry, C. Co. and Araar, A. (1988) Condifuring Cellular Manufacturing Systems, *International Journal of Production Research*, **26** (9), 1511–22.
Hitomi, K. (1979) *Manufacturing Systems Engineering*, Taylor & Francis, London.
Langmoen, R. and Ramsli, E. (1984) Analysing Products with Respect to Flexible Assembly Automation, *Proceedings of the 5th ICAA*, pp. 277–86.
Lotter, B. (1985) Flexible Assembly Lines for Precision Engineering, *Assembly Automation*, **5** (2), 103–9.
Machulak, G.T. (1985) An Examination of the $IDEF_0$ Approach Used as a Potential Industry Standard for Production Control System Design, in *Automated Manufacturing* (ed. L.B. Gardner) pp. 136–49, ASTM STP 862.
Parnaby, J., Lucas lecture, Presented in 1989, Locus Manufacturing Systems Engineering Group.
Richardson, G.P. and Pugh, A.L. III (1981) *Introduction to System Dynamics Modelling with DYNAMO*, The MIT Press.
Wild, R. (1989) *Production and Operations Management – Principles and Techniques* (3rd edn) Holt, Rinehart and Winston.

Chapter 4

Abbott, R. (1983) Programming Design by Informal English Description, *Communication of the ACM*, **26** (11).

Booch, G. (1986) Object Oriented Development, *IEEE Transactions on Software Engineering*, **12** (2).

Booch, G. (1989) What Is and What Isn't Object-Oriented Design? *American Programmer*, **2** (7–8), pp. 14–21.

Booch, G. (1991) *Object Oriented Design*, Benjamin Cummings.

Coad, P. (1991) Why Use Object Oriented Development?, *Journal of Object Oriented Programming*, **4** (3).

Coad, P. and Yourdon, E. (1991) *Object-Oriented Analysis* (2nd edn) Prentice-Hall.

Graham, I. (1991) Object Oriented Methods, Addison-Wesley.

Hill, T. (1989) *Manufacturing Strategy*, Macmillan Education.

Jackson, M. (1983) *System Development*, Prentice-Hall.

Jacobson, I. (1987) Object Oriented Development in an Industrial Environment, *Addendum to the Proceedings of the Conference on Object Oriented Programming: System, Languages and Application – OOPSLA'87*, Orlando, Florida, October.

Mayer, R.J., *et al.* (1992) Information Integration for Concurrent Engineering (IICE): $IDEF_4$ Object-Oriented Design Method Report, AL-TR-1992-0056, Wright-Patterson Air Force Base, Ohio, USA.

Marca, D.A. and McGowan, C.L. (1988) *SADT – Structured Analysis and Design Technique*, Prentice-Hall.

Manfredi, F., Orlando, G. and Tortorici, P. (1989) An Object-Orientated Approach to the System Analysis, *ESEC'89, Lecture Notes in Computer Science*, **387**.

Rumbaugh, J., Blaha, M., Premerlani, W., Eddy, F. and Lorensen, W. (1991) *Object Oriented Modelling and Design*, Prentice-Hall.

Scientific American (1991) Special Issue on Communications, Computers and Networks, September.

Seidewitz, E. (1989) General Object Oriented Software Development: Background and Experience, *The Journal of Systems and Software*, **9** (2).

Shlaer, S. and Mellor, S.J. (1988) *Object Oriented Systems Analysis: Modelling the World in Data*, Prentice-Hall.

Wasserman, A., Pircher, P. and Muller, R. (1990) The Object Oriented Structured Design Notation for Software Design Representation, *Computer*, **23** (3).

Wu, B. (1993) Information Integration – an Essential Element of Integrated CAMSD/CIM Environment, *Control and Dynamic Systems*, **48**, Academic Press, Inc.

Wu, B. and Fung, P. (1993) HOOMA – An Object-Oriented Methodology for Manufacturing Systems Analysis and Definition, *Proceedings of The Second Conference on Methods in Manufacturing Systems Engineering*, Southport, England.

Yourdon, E. (1989) Object-Oriented Observations, *American Programmer*, **2** (7–8), pp. 3–7.

Chapter 5

Blanchard, B.S. and Fabrychy, W.J. (1981) *Systems Engineering and Analysis*, Prentice-Hall.

Box, G.E.P. and Draper, N.R. (1969) *Evolutionary Operation*, Wiley.

Checkland, P. (1981) *Systems Thinking, Systems Practice*, John Wiley.

Cleland, D.I. and King, W.R. (1983) *Systems Analysis and Project Management*, McGraw-Hill.

Dandy, G.C. and Warner, R.F. (1989) *Planning and Design of Engineering Systems*, Unwin Hyman.

Pidd, M. (1984) *Computer Simulation in Management Science*, John Wiley.

Sandquist, G.M. (1985) *Introduction to System Science*, Prentice-Hall.

Wilson, B. (1984) *Systems: Concepts, Methodologies, and Applications*, John Wiley.

Wu, B. (1990) WIP Cost-related Effectiveness Measure for the Application of an IBE to Simulation Analysis, *Computer-Integrated Manufacturing Systems*, **3** (3).

Chapter 6

Adshead, N.S. and Price, D.H.R. (1986) Experiments on Stock Control Policies and Leadtime Setting Rules, Using an Aggregate Planning Evaluation Model of a Make-For-Stock Shop, *International Journal of Production Research*, **24** (5), 1139–57.

Browne, J. and Davies, B.J. (1984) The Design and Validation of a Digital Simulation Model for Job Shop Control Decision Making, *International Journal of Production Research*, **22** (2), 335–57.

Carrie, A. (1988) *Simulation of Manufacturing Systems*, John Wiley.

Conway, R.W., Maxwell, W.L. and Miller, L.W. (1967) *Theory of Scheduling*, Addison-Wesley.

Ellison, D. and Wilson, J.C.T. (1984) *How to Write Simulations Using Microcomputers*, McGraw-Hill.

Ford, D. and Schroer, B. (1987) An Expert Manufacturing Simulation System, *Simulation*, May.

Forrester, J.W. (1975) *Collected Papers of Jay W. Forrester*, The MIT Press.

French, S. (1982) *Sequencing and Scheduling: An Introduction to the Mathematics of the Job-Shop*, John Wiley.

Garey, M.R. and Johnson, D.S. (1979) *Computers and Intractability: a Guide to the Theory of NP-Completeness*, Freeman, San Francisco.

Gass, S.I. (1977) Evaluation of Complex Models, *Computer & Op. Res.*, **4**, 25.

Gottfried, B.S. (1984) *Elements of Stochastic Process Simulation*, Prentice-Hall, New Jersey.

Graham, A.K., *et al.* (1979) Optimization and Approximation in Deterministic Sequencing and Scheduling: a Survey, *Ann. Dis. Math.*, **5**, 287–326.

Hax, A.C. and Meal, H.C. (1975) Hierarchical Integration of Production Planning and Scheduling, in *TIMS Studies in Management Science, Vol. 1, Logistics* (ed. M. Geisler), North Holland, New York.

Hill, T. and Roberts, S. (1987) A Prototype Knowledge-Based Simulation Support System, *Simulation*, May.

Ketcham, M., *et al.* (1989) Information Structures for Simulation Modelling of Manufacturing Systems, *Simulation*, February.

Kiran, A., Schloffer, A. and Hawkins, D. (1989) An Integrated Simulation Approach to Design of Flexible Manufacturing Systems, *Simulation*, February.

Law, A.M. and Kelton, W.D. (1982) *Simulation Modelling and Analysis*, McGraw-Hill.

Mellichamp, J. and Wahab, A. (1987) Process Planning Simulation: An FMS Modelling Tool for Engineers, *Simulation*, May.

Murray, K. and Sheppard, S. (1988) Knowledge-Based Simulation Model Specification, *Simulation*, March.

Pidd, M. (1984) *Computer Simulation in Management Science*, John Wiley.

Poole, T. and Szymankiewicz, J. (1977) *Using Simulation to Solve Problems*, McGraw-Hill.

Pritsker, A.A.B. (1979) *Modelling and Analysis Using Q-GERT Networks*, John Wiley.

Pritsker, A.A.B. and Pegden, C.D. (1979) *An Introduction to Simulation and SLAM*, Halstead Press.

Shannon, R.E. (1975) *Systems Simulation – the Art and Science*, Prentice-Hall.

Smith, R. and Platt, L. (1988) Benefits of Animation in the Simulation of an Assembly Line, *Simulation*.

Ullman, J.D. (1976) Complexity of Sequencing Problems, in *Computer and Job-Shop Scheduling Theory* (ed. E.G. Coffman jr) pp. 134–64, John Wiley, New York.

Van Horn, R.L. (1971) Validation of Simulation Results, *Management Science*, **17** (5), 247–58.

Wu, B. (1990) Large-scale Computer Simulation and Systems Design of Computer-Integrated Manufacturing, *Computer-Integrated Manufacturing Systems*, **3** (2), 100–10.

Wu, B. and Price, D.H.R. (1989) Advanced Manufacturing Technology and Production Planning and Control: A Case Study, *Proceedings of the 3rd European Simulation Congress*, Edinburgh, pp. 537–9.

Zeigler, B.P. (1976) *Theory of Modelling and Simulation*, Wiley-Interscience, New York.

Chapter 7

Ajderian, N. (1989) Do You Really Want That New Machine? *Professional Engineering*, November.

Alagic, S. (1989) *Object-Oriented Database Programming*, Springer-Verlag, New York.

Allen, K. (1979) *Generative Process Planning System Using DCLASS Information System*, Monograph 4, Utah: Computer Aided Manufacturing Laboratory, Brigham Young University.

Ashworth, C.M. (1988) Structured Systems Analysis and Design Method (SSADM), *Information and Software Technology*, **30** (3).

Barad, M. and Sipper, D. (1988) Flexibility in manufacturing systems: definitions and Petri-net modelling, *International Journal of Production*, **26** (2).

Barekat, M. (1990) DMRP – CIM without Tears, *Computerised Manufacturing*, May.

Bennett, D.J., *et al.* (1989) *A Model for Analysing the Design and Implementation of New Production Systems*, Presented in the Xth ICPR, Nottingham, UK.

Bhat, M.V. and Haupt, A. (1976) An Efficient Clustering Algorithm, *IEEE Transactions on Systems, Man, and Cybernetics*, **SMC-6**.

Burbidge, J.L. (1971) *The Principles of Production Control*, MacDonald Evans.

Burgidge, J. (1963) Production Flow Analysis, *Production Engineer*, December.

Burgidge, J. (1979) A Manual Method for Production Flow Analysis, *Production Engineer*, December.

Burt, D.N. (1984) *Proactive Procurement*, Prentice-Hall, New Jersey.

Chardrasekharan, M.P. and Rajagopalan, R. (1986) An Ideal Seed Non-Hierarchical Clustering Algorithm for Cellular Manufacturing, *International Journal of Production Research*, **24**.

Chardrasekharan, M.P. and Rajagopalan, R. (1986) MODROC: an Extension of Rank Order Clustering for Group Technology, *International Journal of Production Research*, **24**.

Chardrasekharan, M.P. and Rajagopalan, R. (1987) ZODIAC – An Algorithm for Concurrent Formation of Part-Families and Machine Cells, *International Journal of Production Research*, **25**.

Checkland, P.B. (1981) Science and the System, Movement, in *Systems Behaviour* (3rd edn), Harper & Row.

Choobineh, F. (1988) A Framework for the Design of Cellular Manufacturing Systems, *International Journal of Production Research*, **25**.

Co, H.C. and Arrar, A. (1988) Configuring Cellular Manufacturing Systems, *International Journal of Production Research*, **26**.

Codd, E.F. (1982) Relational Database: A Practical Foundation for Productivity, *ACM CACM*, **25** (2).

Department of Industry, Committee for Materials Handling (1978) *Materials Handling – An Introduction*, HMSO.

De Wittee, J. (1980) The Use of Similarity Coefficients in Production Flow Analysis, *International Journal of Production Research*, **18**.

Doumeingts, G., *et al.* (1987) Design Methodology for Advanced Manufacturing Systems, *Computers in Industry*, **9** (4).

Downs, E., *et al.* (1988) *Structured Systems Analysis and Design Method – Application and Context*, Prentice-Hall.

Eckert, R.L. (1984) Codes and Classification Systems, in *Group Technology At Work* (ed. Nancy L. Hyer) Michigan: Society of Manufacturing Engineers, pp. 43–61.

Favrel, J. and Kwang, H.L. (1985) *Modelling, Analysing, Scheduling and Control of FMS by Petri Nets*, Proceedings of the IFIP wg 5.7 Working Conference on Modelling Production Management Systems, Elsevier.

Gaine, C.P. (1979) *Structured Systems Analysis*, Improved Systems Technology.

Gallager, C.C. and Knight, W.A. (1986) *Group Technology Production Methods in Manufacturing*, Ellis Horwood, Chichester.

Gupta, T. and Seiforddini, H. (1990) Production Data Based Similarity Coefficient for Machine-Component Grouping Decisions in the Design of a Cellular Manufacturing System, *International Journal of Production Research*, **27**.

Hayes, R.H. and Wheelwright, S.C. (1979) Link Manufacturing Process and Product Life Cycles, *Harvard Business Review*, January/February.

Hendricks, J.A. (1989) Accounting for Automation, *Mechanical Engineering*, February.

Hill, T. (1985) *Manufacturing Strategy*, Macmillan Education.

Hughes, J.G. (1991) *Object-Oriented Databases*, Prentice-Hall.

Ingersoll Engineers (1982) *The FMS report*, IFS.

Ingram, B.F. (1984) Group Technology, in *Group Technology At Work* (ed. Nancy L. Hyer) Michigan: Society of Manufacturing Engineers, pp. 31–42.

King, J.R. (1980) Machine-Component Group Formation in Production Flow Analysis: An Approach Using a Rank Order Clustering Algorithm, *International Journal of Production Research*, **18**.

King, J.R. and Nakornchar, V. (1982) Machine Component Group Formation in Group Technology – Review and Extension, *International Journal of Production Research*, **20**.

Kosko, B. (1992) *Neural Networks and Fuzzy Systems*, Prentice-Hall.

Kusiak, A. (1990) *Intelligent Manufacturing Systems*, Prentice-Hall.

Kusiak, A. and Chow, W.S. (1987) An Algorithm for Cluster Identification, *IEEE Transactions on Systems, Man, and Cybernetics*, **SMC-17**.

McAuley, J. (1972) Machine Grouping for Efficient Production, *Production Engineer*, February.

McCormick, W.T., Schweitzer, P.J. and White, T.W. (1972) Problem Decomposition and Data Reorganisation by a Clustering Technique, *Operations Research*, **20**.

Marca, D.A. and McGowan, C.L. (1988) *SADT – Structured Analysis and Design Technique*, Prentice-Hall.

Malave, C. and Ramachandran, S. (1991) Neural Network-based Design of Cellular Manufacturing Systems, *Journal of Intelligent Manufacturing*, **2** (5), pp. 305–14, October.

Optiz, H. (1970) *A Classification System to Describe Work Pieces*, Pergamon Press, New York.

Peterson, J.L. (1981) *Petri Net Theory and the Modelling of Systems*, Prentice-Hall.

Primrose, P.L. and Leonard, R. (1985) Evaluating the Intangible Benefits of Flexible Manufacturing System by Use of Discounted Cash Flow Algorithms within a Comprehensive Computer Program, *Proceedings of the Institution of Mechanical Engineers*, **199** (B1), 23–8.

Rajagopolan, R. and Batra, J.L. (1975) Design of Cellular Production Systems – A Graph Theoretical Approach, *International Journal of Production Research*, **13**.

Ramamoorthy, C.V. and Ho, G.S. (1980) Performance and Evaluation of Asynchronous Concurrent Systems Using Petri-nets, *IEEE Transactions on Software Engineering*, **6**.

Robinson, D. and Duckstein, L. (1986) Polyhedral Dynamics as a Tool for Machine Part Formation, *International Journal of Production Research*, **24**.

Schrage, L. (1980) *Linear Programming Models with LINDO*, The Scientific Press, California.

Seiforddini, H. (1989) A Note on the Similarity Coefficient Method and the Problem of Improper Machine Assignment in Group Technology Applications, *International Journal of Production Research*, **26**.

Seiforddini, H. and Wolfe, P.M. (1986) Application of the Similarity Coefficient Method in Group Technology, *IIE Transactions*, **18**.

Shafer, S.M. and Meredith, J.R. (1989) A Comparison of Selected Manufacturing Cell Formation Techniques, *International Journal of Production Research*, **28**.

Slack, N.D.C. (1990) *Achieving a Manufacturing Advantage*, W.H. Allen.

Swann, K. (1988) Investment in AMT – a Wider Perspective, *Production Engineer*, September.

Van der Lans, R.F. (1989) *The SQL Standard: A Complete Reference*, Academic Service and Prentice-Hall International, UK.

Vanelli, A. and Kumar, K.R. (1986) A Method for Finding Minimal Bottleneck Cells for Grouping Part Machine Families, *International Journal of Production Research*, **24**.
Waghodekar, P. and Sahu, S. (1984) Machine Component Cell Formation in Group Technology: MACE, *International Journal of Production Research*, **22**.
Wild, R. (1989) *Production and Operations Management* (4th edn), Cassell Education.
Wu, B. (1991) Computer-Aided Manufacturing Systems Design: Framework and Tools, in *Advances in Control and Dynamic Systems*, **45/48**, International Academic Press.
Wu, B. (1992) An Introduction to Neural Networks and their Applications in Manufacturing, *Journal of Intelligent Manufacturing*, **3** (6), pp. 391–403.

Appendix A

Arrow, K.J., Karlin, S. and Scarf, H. (1958) *Studies in the Mathematical Theory of Inventory and Production*, Stanford University Press, Stanford, California.
Bahl, H.C. and Ritzman, L.P. (1984) An Integrated Model for Master Scheduling, Lot Sizing and Capacity Requirements Planning, *J. Opl. Res. Soc.*, **35** (5), 389–99.
Baker, K.R. (1974) *Introduction to Sequencing and Scheduling*, John Wiley, New York.
Bedini, R. and Toni, P. (1980) The Planning of a Manufacturing System: a Dynamic Model, *International Journal of Production Research*, **18** (2). 189–99.
Beged-Dov, A.G. (1983) Analysis of Productions as Closed Self-Regulating System, *International Journal of Production Research*, **21** (6), 859–68.
Bergstrom, G.L. and Smith, B.E. (1970) Multi-Item Production Planning – an Extension of the HMMS rules, *Management Science*, **16** (10), 614–29.
Burton, R.M., Damon, W.W. and Obel, B. (1979) An Organizational Model of Integrated Budgeting for Short-Run Operations and Long-Run Investments, *J. Opl. Res. Soc.*, **30**, 575–85.
Conway, R.W., Maxwell, W.L. and Miller, L.W. (1967) *Theory of Scheduling*, Addison-Wesley.
Corke, D.K. (1985) *A Guide to CAPM*, The Institution of Production Engineers, UK.
Coyle, R.G. (1982) *Management System Dynamics*, John Wiley.
Eilon, S. and Pace, A.J. (1970) Job Shop Scheduling with Regular Batch Arrivals, *Proceedings of the Institution of Mechanical Engineers*, **184** (17), 301–10.
Fisher, M.L. (1973) Optimal Solution of Scheduling Problems Using Lagrange Multipliers, Part I, *Ops. Res.*, **21** (5), 1114–27.
Forrester, J.W. (1980) System Dynamics – Future Opportunities, in *System Dynamics*, pp. 7–21, The MIT Press.
Garey, M.R., Johnson, D.S. and Sethi, R. (1976) Complexity of Flow Shop and Job Shop Scheduling, *Math. Ops. Res.*, **1** (12), 117–29.
Gere, W.S. (1966) Heuristics in Job Shop Scheduling, *Management Science*, **13** (3), 167–90.
Gonzalez, T. and Sahni, S. (1978) Flowshop and Jobshop Schedules: Complexity and Approximation, *Ops. Res.*, **26** (1), 36–52.
Graham, A.K. (1980) Parameter Estimation in System Dynamics Modelling, in *System Dynamics*, pp. 125–60, The MIT Press.
Gunther, H.O. (1981) A Comparison of Two Classes of Aggregate Production Planning Models under Stochastic Demand, *2nd International Working Seminar on Production Economics*, Innsbruck.
Hax, A.C. and Candea, D. (1984) *Production and Inventory Management*, Prentice-Hall.
Hax, A.C. and Meal, H.C. (1975) Hierarchical Integration of Production Planning and Scheduling, in *TIMS Studies in Management Science, Vol. 1, Logistics* (ed. M. Geisler) North Holland, New York.
Khoshnevis, B., Wolfe, P.M. and Terrell, M.P. (1982) Aggregate Planning Models Incorporating Productivity – an Overview, *International Journal of Production Research*, **20** (5), 555–64.
Krajewski, L.J., Mabert, V.A. and Thompson, H.E. (1973) Quadratic Inventory Cost

Approximations and the Aggregation of Individual Products, *Management Science*, **19** (11), 1229–40.

Kriebel, C.H. (1967) Coefficient Estimation in Quadratic Programming Models, *Management Science*, **13** (8), 473–86.

Lee, W.B., Steinberg, E. and Khumawala, B.M. (1983) Aggregate versus Disaggregate Production Planning: a Simulated Experiment Using LDR and MRP, *International Journal of Production Research*, **21** (6), 797–811.

Legasto, A.A., Forrester, J.W. and Lyneis, J.W. (Eds) (1980) *System Dynamics*, MIT's Studies in the Management Sciences, Vol. 14, North-Holland.

Magee, J.F. (1958) *Production Planning and Inventory Control*, McGraw-Hill, New York.

O'Grady, P. (1986) *Controlling Automated Manufacturing Systems*, Kogan Page.

Orlicky, J.A. (1975) *Material Requirements Planning: The New Way of Life in Production and Inventory Management*, McGraw-Hill, New York.

Sen, T. and Gupta, S.K. (1984) A State-of-Art Survey of Static Scheduling Research Involving Due Dates, *OMEGA*, **12** (1), 63–76.

Simon, H.A. (1978) On the Application of Servomechanism Theory in the Study of Production Control, in *Managerial Applications of Systems Dynamics* (ed. E.B. Roberts) pp. 95–115, The MIT Press.

Ullman, J.D. (1976) Complexity of Sequencing Problems, in *Computer and Job-Shop Scheduling Theory* (ed. E.G. Coffman jr) pp. 134–64, John Wiley, New York.

Welam, U.P. (1975) Multi-Item Production Smoothing Models with Almost Closed Form Solutions, *Management Science*, **21** (9), 1021–8.

Zoller, K. (1971) Optimal Disaggregation of Aggregate Production Plans, *Management Science*, **17** (8), 533.

Index

Page numbers appearing in **bold** refer to figures and page numbers appearing in *italics* refer to tables.